D0202813

Nature: An Economic History

Nature: An Economic History

Geerat J. Vermeij

PRINCETON UNIVERSITY PRESS

PRINCETON AND OXFORD

Published by Princeton University Press, 41 William Street,
Princeton, New Jersey 08540
In the United Kingdom: Princeton University Press,
3 Market Place, Woodstock, Oxfordshire OX20 1SY

Library of Congress Cataloging-in-Publication Data

Vermeij, Geerat J., 1946–
Nature : an economic history / Geerat J. Vermeij
p. cm.
Includes bibliographical references and index.
ISBN 0-691-11527-3 (cl : alk. paper)
1. Natural history—Economic aspects. I. Title.
QH45.2.V47 2004
576.8—dc22 2004044521

British Library Cataloging-in-Publication Data is available.

This book has been composed in Electra

Printed on acid-free paper. ∞

pup.princeton.edu

Printed in the United States of America

1 3 5 7 9 10 8 6 4 2

CONTENTS

PREFACE

It is a lucky accident of personal history that I came to Panama early in my career. On this twisted isthmus, where the Atlantic Ocean almost meets the Pacific, and where lush forest grows alongside the Panama Canal—a geographic symbol of the enormous economic power of the human species—one can experience in a single day the works of nature in undiminished glory and the works of human enterprise on a grand scale. And there is history, too. Fossils tell of a time when there was no isthmus, when the economy of marine nature was different from the contrasting relationships we see on the two coasts of Central America today.

My first exposure to this extraordinary place came while I was a graduate student in 1969. There was talk of building a canal that would connect the two great oceans at sea level and that, unlike the present freshwater thoroughfare, would be filled with seawater. The politicians and engineers who built the original canal early in the twentieth century never gave the biological consequences of this waterway a thought, but in the late 1960s biologists were beginning to worry about how a marine connection would affect the rich biotas of the Caribbean and Pacific coasts of tropical America. Although this matter was not the focus of my first visit, it gnawed at me as I glimpsed the distinctive marine faunas firsthand. A year earlier, I had noticed that Caribbean snails from Curaçao have shells less well endowed with features of armor—a narrow impenetrable opening, thick ribs and tubercles, and a compact build with few protruding parts—than their counterparts on the wildly exotic shores of Hawaii and Guam in the tropical central and western Pacific. I now observed the same contrast, albeit a less dramatic one, between the snails of Caribbean Panama and the more armored versions on the Pacific coasts of Panama and Ecuador.

These shells posed an intriguing puzzle. Why were shells from the various tropical faunas of the world so differently endowed with what I came to understand to be resistance defenses against predators? Would the fortified Pacific snails have an advantage in survival over their Atlantic counterparts once they encountered each other if an interoceanic seaway were reestablished in Central America? Physical differences between the coasts—a greater tidal range in the Pacific, less seasonality on the Caribbean side—seemed inadequate to explain the architectural contrasts. Life itself—the complex interactions between predators and prey—must play a key role in setting the rules for what works and what doesn't work in nature, and for how living things choose among mates and repel their rivals.

That message was communicated even more forcefully by the forest. I can't think of a more dramatic place in which to witness the economic clout of life than Barro Colorado Island in the rainy season. With every breath of warm, maximally humid air, I smell the work of plant photosynthesis and fungal decay. Orchid bees engage in an elaborate, faintly threatening dance of low buzzing flight. Frogs, toads, ant shrikes, cicadas, crickets, and howler monkeys call, sing, rasp, chirp, and growl as they continually advertise to each other, communicating, seeking mates, choosing among mates. Leaves, with all but a few of the veins eaten away by insects, catch what little light filters through to ground level. Lines of hard-bodied leaf-cutter ants carry a burden of leaf fragments back to the nest, an underground refuge where domesticated fungi transform the greenery into edible food for the insects. An agouti noisily gnaws at a fallen nut. The strikingly smooth trunks of canopy trees, whose leaves vie with those of lianas and epiphytes for a place in the sun, are clothed in a community of climbers, mosses, fungi, and microfauna. Other trunks are untouchable by virtue of long spines that discourage would-be vines and herbivores. Even the dust—pollen, spores, and tiny insects—is alive. Here are millions of life forms, every single one the latest in an unbroken line of descent of successful beings stretching back to the impossibly distant beginning of life, struggling to transform resources into themselves and their young and defending those resources and themselves against a host of rivals. All around me, going concerns create and respond to an environment, much of it of their own making.

Had I been walking in a lowland rain forest during the coal age, some 300 million years ago, I would no doubt have been just as awed by the adaptations of the residents and by the apparent harmony of the ecosystem they collectively forge, but the forest would have been utterly different. The canopy trees did not bear flowers, nor were they festooned with woody climbers. Gigantic insects with wingspans of 75 cm or more flew, but there were no birds; and the ground, thickly carpeted with the very slowly rotting remains of the forest vegetation, was alive with huge scorpions and millipedes and other assorted unfamiliar arthropods that fed on decaying plant matter. There were insect-eating amphibians, but no ants, no mammals, no big mushrooms. By today's standards, the forest's creatures moved and grew slowly, many lacking the colors and the behavioral complexity that so delight naturalists like me lucky enough to look in on what remains of the great rain-forest ecosystem. Yet, these ancient creatures and the economies they created were successful. They were going concerns, fully capable of transforming raw materials into growth and progeny, defending themselves against enemies, competing with their neighbors for light and mates and food, and engaging in all the activities that constitute economic work.

My experiences in Panama and elsewhere, together with my exposure to the literature of evolution and geology, made me see living things as parts of a larger whole. In my boyhood, I could love the objects of natural history—shells, leaves, feathers, seeds, pinecones, and minerals—for their abstract beauty and for the sense of wonder and bliss they elicited; but I gradually came to realize that these works of architecture are parts of living things that must make their way in a challenging world. Shells—the objects I came to study most intensively—were no longer just things pleasing to the senses and hailing from lands and seas I dreamed of visiting someday, but in addition revealed a context in which organisms live and evolve. That context proved to vary greatly from place to place as well as through time. Informed by my observations on the workings of organisms and their communities in different parts of today's biosphere, I set about interpreting the evolution and the conditions of life of shell-bearing animals and other living things from the distant past. These forays into comparative history and comparative biology led me to propose that predation and other manifestations of competition had intensified over the course of Earth history, eliciting increasingly sophisticated adaptations by organisms everywhere.

All science, no matter how arcane or irrelevant it may appear to outsiders, has broader implications, which can lead the willing scholar into some quite unfamiliar territory. Early in my career, I was an unlikely candidate for stepping outside the confines of natural history, the study of whole organisms and their environments modern and ancient. Human history was fun to read, but the concatenation of wars, repressive regimes, and wanton exploitation of people and resources repelled me. It was all just so much human folly, which I found easy to dismiss as trivial and unappealing, certainly of less consequence than the greater truths the natural world and its history had to offer. I had avoided economics for similar reasons. Money, power, banks, stock markets, business, interest rates, and trade policy did not arouse my curiosity; and abstract mathematical models, with which I was already not terribly enamored in biology, attracted me even less in economics. The field of economics seemed utterly divorced from the affairs of nature. As I saw them, economists cared about growth and development in the human sphere, and about satisfying or even creating human wants, but about little else. The goals of their field seemed largely antagonistic to the preservation of the biosphere's pleasing variety and sensual beauty that to me gave life so much meaning.

In the end, however, the broad economic and historical implications of my esoteric work on fossil and living shell-bearers and their enemies became irresistible. Nobody living during the latter half of the twentieth century could ignore the military arms race between the superpowers, the United States and the Soviet Union. When President Ronald Reagan unilaterally ratcheted up the arms race by vastly increasing outlays for nuclear weapons and missile-defense technology in 1983, it became clear to me that an exploration of the obvious parallels between the arms races I had been studying and the weapons buildup in which the superpowers seemed to be inextricably engaged might be worthwhile. Ancient fossils and modern threatened ecosystems might, I hoped, tell us something about how arms races begin and end, and about long-term political and economic interests that drive rivals to spend vast sums on weapons of aggression and defense.

As I began to read outside my discipline, I quickly discovered that there were some awfully smart historians and economists out there who, even if their sensibilities differed from my own, had a great deal to offer the disciplines I liked. It became just as clear that the human-oriented disciplines, which had traditionally been isolated as the "soft sciences" and humanities, had much to gain from the evolutionary perspective and, most of all, from the patterns and the long historical perspective that an economically informed reading of the fossil record could offer. Whether our field is natural history, human history, evolutionary biology, or economics, we grapple with similar phenomena, with growth and decline, competition and cooperation, economic inequality, and the disproportionate influence of the rich and powerful. We deal with stabilizing trade-offs and destabilizing positive feedbacks, and the destructive effects of rare disturbances.

We ask similar questions. Adam Smith and David Landes inquired why some nations become rich while others stay poor. Biologists wonder why some predators exercise strong control over the ecosystems in which they work, and how such predators evolved. All of us wonder about how innovations arise and what their role is in the unit of our choice—cell, organism, ecosystem, the human economy. All of us want to know how disruptions of various kinds affect our systems and the processes and interactions that characterize those systems. But always the questions are considered in a narrow context, within the bounds of established academic disciplines. The generality of the answers is either denied or, more commonly, never acknowledged. My claim in this book is that common principles underlie all these diverse fields, and that these principles imply broadly predictable patterns of change through time, patterns with interesting implications for our future.

This book draws together ideas and observations from economics, evolution, and history. I argue that the universal processes of competition and cooperation create successful economic entities—living things capable of growth, proliferation, work, and adaptation—which together forge still larger economies. On all scales of space and time from the molecular to that of the biosphere and beyond, those economic units that gain more or lose fewer resources during interactions exercise disproportionate influence over the characteristics, activities, and distribution of other units, and in so doing impart an overall directionality to economic history. Despite periods of disruption and episodes of status quo, the long-term trend is toward greater reach, greater power, and greater autonomy among those economic entities that exercise the greatest control and that, as a by-product of actions on their own behalf, increase the rate and reliability of supply of essential raw materials for the larger economies in which they live and which they helped to create. As the most powerful economic agents yet to have evolved on Earth, we humans have in broad outlines and over a vastly shorter time recapitulated this same trend discernible in the development of the biosphere over the last three and a half billion years. We continue to obey the fundamental principles of economics set down at life's beginning by inanimate, unintelligent processes, even as we invent startlingly novel abilities and institutions. Nothing in the historical record or in the arsenal of economic principles suggests that humanity can alter the directionality inherent in history even as we overwhelm the biosphere. But given our unprecedented ability to anticipate, to learn, and to modify, we may collectively find ways to prevent internally generated disruptions from destroying us and the mechanisms of production that support us.

A synthesis of this kind may strike many readers as impossibly ambitious, even presumptuous. My response is that there is now sufficient theory and observation available to warrant such a synthesis, however imperfect it may be. Moreover, the power of evolutionary and economic thinking as applied to history and to present-day change is so great that an intellectual fusion among fields with separate traditions and distinct languages seems to me essential if the scientific world view of testing and improving hypotheses in the light of observation set in a framework of mutually consistent theory is to prevail over supernaturally inspired dogma and irrefutable doctrine as we meet the future.

Perhaps the greatest challenge in writing a book of this kind is to read and absorb enough of the right kinds of information. The world is awash in information, stuff worth knowing and pondering, insights and data that shape one's viewpoints. Although there probably never was a time when an individual could know it all, the accumulated knowledge and wisdom of today far surpass what any one individual can meaningfully absorb. This problem will only get worse in the future. Nearly every source I have read refers to critically important sources that I have not read. Why haven't I read them? It isn't just the lack of time, nor the fact that essentially everything must be read aloud to me by another person while I transcribe the parts I need into Braille. After all, I could have decided not to read the hundreds of monographs dealing with the classification of obscure fossil molluscs, or the long lists of species found in particular places, or so much other obscure and tedious scientific literature that is known and important to less than a handful of people. But the arcane detail that this literature documents, and to which I have made my own esoteric contributions, is utterly essential. The devil is, after all, in the details. Grand principles and generalizations emerge only after the hard work of discovering who did what, when, and how, and which factors ranging from the laws of physics and chemistry to the events of celestial space are at work. The expertise gained through careful empirical investigation is essential as the foundation of the more sweeping theory that unites and explains the disparate facts as we know them. My choices of reading might not be anyone else's, but all scholars must find a compromise in which they read some great works and many minor ones. By necessity, I must resort to the solution that all economic entities have arrived at: to do the best you can with limited information. None of this is meant to excuse major oversights or blunders; I am only trying to explain how my shortcomings of knowledge come about. Any flaws in thought are, of course, more directly personal, and I cannot so easily hive off responsibility for mere ignorance to an appeal to overload.

Three women—Janice Cooper, Elizabeth Dudley, and my wife Edith Zipser—have shouldered the bulk of the reading aloud on which my knowledge of scholarly literature rests. The volume of material, the tedium, the arcane jargon, the bizarre nomenclature, and the enormous diversity of subjects with which these wonderful readers have had to deal strain credulity. It takes great intelligence to read aloud, and these three women have it in spades.

Knowledge is cumulative. In writing this book, I have been keenly aware that my own contributions are massively dwarfed by those of others. Not only have I learned a great deal by reading the works of others, but I have also benefited enormously from discussions and critiques. My wife Edith Zipser has been a constant and invaluable intellectual companion. Richard Cowen, Greg Herbert, Michael Kirby, Lindsey Leighton, Egbert Leigh, David Lindberg, Joel Mokyr, Peter Roopnarine, and Simon Levin have immensely enriched me with thoughts, facts, criticisms, and insights. My brother, Arie Pieter Vermeij, has given me valuable perspectives from his vantage point in the world of banking. Parts or all of the manuscript received extremely thoughtful and insightful comments from Eric Chaisson, Richard Cowen, Egbert Leigh, Joel Mokyr, and Kenneth Pomeranz. Leigh and Pomeranz in particular forced me to think deeply about matters I had previously glossed over. Janice Cooper, who for years has read to me and prepared and edited my work for publication, is breathtakingly broad in her knowledge and insight, and has likewise given me a great deal to think about.

My views have been shaped not only by the works I have read and the discussions I have had, but to an even greater extent by my own experiences. I am grateful to the National Science Foundation for funding much of my research on the shapes, lives, and evolution of molluscs. I have also received research funds from the John Simon Guggenheim Foundation and the National Geographic Society. As a MacArthur Fellow, I have keenly felt the incentive and the support to explore fields beyond my immediate research interests.

Above all, I live in a time and place where economic conditions gave me opportunities that would have been unthinkable for someone in my place even fifty years earlier. Education for the blind did not become compulsory in the Netherlands until just after World War II, a few years before I began schooling in 1950. There was enough money, and enough social compassion, to provide Braille books and itinerant teachers when I entered public school in the United States in 1955. Past expectations that the blind could do at most menial work such as basketry were giving way to the possibility that, with proper training, a blind person could realize the same high aspirations as those who were sighted. Science and freedom of thought have flourished in a golden age of enlightened education, public health, high standards of living, rapid communication, and reliable transportation. Opportunity has loomed large in my life, as indeed it has done so often in the economic history of life on Earth.

Nature: An Economic History

Chapter 1

ECONOMY AND EVOLUTION: A ROAD MAP

WE LIVE IN A CHANGING WORLD, in an economic context of competition and resources to which we and our forebears have both adapted and contributed. It is impossible to separate us and other life forms from that context, for the context determines our character and we determine its character. If we wish to understand the past and the future, we must discover how the link between life and its context changed in history, to what extent economic change is predictable, and how the principles governing competition-driven demand and resource-driven supply intersect to yield laws of history.

At its core, this book is an economic analysis of history. In it, I argue that economic principles applicable to humans are the same as those that govern all other forms of life. The patterns of history that emerge from these principles are therefore universal, even if the details of timing and actors vary according to time and place.

It is easy to get lost in the fascinating phenomenology of economic history—in the particulars of places, participants, events, behavior, relationships, patterns of descent, and the like—and to lose sight of the structure of the arguments and of the principles on which these arguments rest. I therefore begin with an outline of the main points, with a road map to the sometimes tortuous path of the narrative that follows as it takes us through past and present ecosystems around the world.

Although we usually think of the word *economy* in such human terms as trade, profits, markets, and finance, it applies just as aptly to the systems of which living things other than humans are constituents and architects. In chapter 2 I lay out what I hold to be the fundamental characteristics and principles of economies. An economy is a collective whole, a system of metabolizing, interacting, smaller units or entities that are themselves economies. The constituent units adapt to, and bring about changes in, their environment as they compete locally for energy and material resources. Economies are built on living things, which complete cycles of work by coupling chemical transformations that alternately use and release energy in the context of an architecturally and organizationally constrained physical structure. Economies thus have knowable properties not possessed by any one of their individual members. The work of life—growth, replication, and activity—creates meaning and information and ultimately leads to a history in which self-interested parties who cooperate to fashion larger wholes give rise to and replace each other. Evolution—descent with modification—is thus an expected and universal historical process in economic systems. It occurs because economic units compete locally for resources, and because only those entities that acquire and retain the necessities of life in the face of such competition and of uncertainty persist. Cooperation among economic players reduces rivalry at one level, but creates more potent competitors on a larger scale. Trade and cooperation (or mutual exploitation) thus lead through self-organization, or co-construction, to regulation of resource supply and consumption, and to complex interdependencies that emerge as the common good for the larger economy and for many of its constituents, especially for those that wield disproportionate power.

Energy or its equivalent is what economic systems or their members exchange, but power—the amount of energy produced, consumed, or retained per unit time—is the best measure of absolute performance of living things. To the evolutionary biologist, absolute performance is usually expressed as fitness, the number of offspring in the next generation; to the economist, it usually translates into profits. Although we often think of words like economy or economical in terms of efficiency—the amount of energy gained per amount of energy invested—it is absolute power rather than efficiency that matters in the economies of nature and human affairs. Selection (or nonrandom elimination) operates whenever entities differ in performance-related characteristics that are heritable in some way. The universe works, and life works and persists, because we co-construct our universe through the combined processes of modification and selection. Adaptation, the process resulting in a better fit between entities and their environment, is universal among living things, which create and improve hypotheses about their surroundings much as scientists propose and test hypotheses explaining observations and regularities in the world.

For a variety of reasons, adapted entities are inevitably imperfect, and thus in principle are capable of improvement. Some challenging circumstances are simply too infrequent for an entity of limited lifespan to predict; others call for adaptive solutions that are in conflict with the demands imposed by different challenges, so that an imperfect compromise reflecting functional tradeoffs becomes the most likely option. Moreover, information about the environment is never either complete or accurate, a situation often exacerbated by the actions of competitors. Economic units and the coalitions they form create hypotheses about their environment that are as good as they have to be relative to the hypotheses of their competitors. Where competition is intense—where the consequences of losing are dire and the benefits of winning are great—the standards of performance and the predictiveness of adaptive hypotheses must be high. Where the stakes are lower, or where there is more tolerance of error, trade-offs are less constraining and persistence, or economic success, is achievable with lower levels of performance and with more generalized hypotheses. In other words, living things do only as well as they have to rather than optimize, pretty much as students do in classes.

One of the most pervasive and far-reaching realities in economic systems is inequality, the tendency for one party involved in an interaction over resources to gain more, or lose less, than its rival. Dominants exercise disproportionate influence and accumulate disproportionate power and wealth. Through top-down control, they affect not only the characteristics and distribution of other members of the economy, but they define the structure and workings of the economy as a whole. Inequality and dominance occur at all scales of economic life, from the cell to the biota, from the household and firm to international relations. They are expressed at all scales of time, from the life cycles of cells to the long-term replacements observed among empires and the great evolutionary branches of life. Inequality does not mean that the winners succeed while the losers fail. Winners can be, and often are, replaced, and although the losers are often restricted to the economic margins where resources are less plentiful and less reliable, the entities with subordinate economic positions may enjoy long periods of stability. In fact, the refuges in which they find themselves are often created by the winners, and as long as such refuges exist, the losing parties and their adapted descendants not only survive, but are often at the forefront of economic expansion into underexploited domains of the biosphere.

The fundamental processes operating in economic systems—competition, cooperation, selection, adaptation, and the feedback between living things and their environment—apply to all such systems, from those as small as a cell to human societies and to the biosphere as a whole. They work regardless of the particulars of how performance-related characteristics are introduced, inherited, modified, or eliminated. Importantly, they have characterized life since its beginning some 3.5 billion years ago, and they have been responsible for unbroken evolutionary lines of descent chronicling continuous success in the face of startling long-term change, devastating disruptions, and ever-present struggles over resources.

It is all well and good to characterize the domain of nonhuman life as a hierarchy of economies ruled by competition, cooperation, adaptation, and inequality, but does this perspective have anything useful to say about human economic life and civilization? There is a long history of claiming for humanity all sorts of unique attributes—language, intentionality, cultural transmission of knowledge, anticipation of future conditions, moral codes, the ability to use previously unexploited sources of energy, and more—to say nothing of the unique institutions that dot the landscape of civilization: schools, corporations, governments, stock markets, farms, factories, shops, prisons, churches, banks, amusement parks, and so on.

I argue in chapter 3, however, that the human species and the human economy do not differ fundamentally from units encountered in the rest of the biosphere. Humans are without question the most powerful economic entity that has yet evolved on Earth; we accomplish things on spatial scales vastly larger than any other species, and our unparalleled power enables us to do everything much faster by orders of magnitude than is possible in any other economy that arose on Earth. We anticipate and predict; we modify environments globally, and exercise control over every ecosystem on this planet. Yet, in spite of all these unique qualities and institutions, our species and the economic and social system we have created follow all the same fundamental rules that govern other life forms and their economic structures. Like other living things, we too are ruled by conflicts of interest, cooperative behavior, adaptation, unequal outcomes of trade, the disproportionate influence of the rich and powerful, and the vagaries of resource supply that dictate when and where opportunities are created and constraints are imposed. None of our attributes has thus far enabled humans to violate the principles that apply to all other life forms. Life and its history therefore have much to say about ourselves, including our future. Moreover, all of our unique characteristics are derived from precursors observable in nonhuman life forms. Cultural (or at any rate nongenetic) transmission of performance-related information, machinery, and behavior has existed since life's origin; and our ability to modify and to anticipate has obvious, if less spectacular, precedents in the adaptations and the mechanisms of adaptation in other organisms.

Having laid out the rules common to all economies, I turn to the two activities—consumption and production—that together describe what living things do. Predation—the consumption of living things by other living things—invests the abstract ideas of chapters 2 and 3 with real life-and-death significance in chapter 4. Here we meet potential victims who face daily confrontations with, and risks from, enemies. Their defenses range from the passive—cryptic coloration, toxicity, and heavy armor—to more active means such as fleeing and aggression. Those victims without effective defenses are restricted by their enemies to refuges, safe sites where predators are themselves physically constrained. Thus, we find life forms on the bodies of well-defended large organisms, or deeply buried in soil, or floating in the plankton, or cowering in the inhospitable border between sea and land, or populating temporary habitats.

We also meet the predators themselves, animals among which competition for food is intense and whose feeding methods and abilities are dictated largely by performance during encounters with rivals. Predator and prey respond to one another directly behaviorally and even architecturally, but in an evolutionary sense it is the predators in particular, and enemies in general, who hold the upper hand. In all ecosystems, at least some predators exceed all potential prey in energy-demanding performance. They have more acute senses, move or maneuver more rapidly, and exert greater forces than do their victims. It is only in passive performance—toxicity, armor, and large body size—that potential victims exceed their enemies in performance. Given that passivity is usually linked to inactivity, predators exercise far more control over the characteristics of living things—their behavior, form, and distribution—than do victims. Powerful consumers, with high per capita and collective rates of consumption, drive economic systems and impart structure and direction to them. They impose pervasive top-down evolutionary and economic control, because adaptation to enemies characterizes all members of all economies. In short, adaptive evolution is largely a process of escalation in which potential victims adapt to enemies, which in turn adapt or accommodate to their enemies.

Consumers may evolutionarily control the characteristics and distribution of life, but they cannot exist without producers. In chapter 5, I explore how primary producers—microbes, plants, and phytoplankton, which make organic matter from inorganic components—compete for energy and raw materials. Resources often appear to be globally superabundant—think of energy from the sun, nitrogen in the atmosphere, and silicon in the sand, for example—but they limit productivity because competition for them is local, and because access to them is constrained by the technology of extraction, storage, conversion, and distribution. Resources and the creatures that convert them to biomass exercise pervasive, diffuse, bottom-up controls over economic systems and their constituents. All living things are in some sense producers, and all provide potential resources to other life forms. A surplus of production—that is, the amount produced beyond the requirements of the producers themselves—determines how much consumption an economy can support sustainably. It, in turn, depends on the ability of victims to thwart consumption, often through passive means. Resource supply and anti-consumer adaptation thus set limits to the amount and nature of consumption through a series of complex positive and negative evolutionary and economic feedbacks. Abundance and access provide evolutionary and economic opportunity for adaptation and growth; scarcity and lack of access impose constraint.

Forms of consumption that stimulate production will in general be favored over those that interfere with it. By consuming living plant matter, for example, herbivores circumvent some decomposers and therefore recycle nutrients faster than if nutrient transfer from plants were exclusively in the hands of organisms that consume dead tissues. Organisms tend to regulate and often improve the supply of resources so that the supply becomes reliable and high enough to support populations of consumers. For example, predators that eat snails without destroying the shells of their victims create a reasonably steady supply of shells for hermit crabs, which in the absence of such consumers would be at the mercy of rare storms or other unpredictable events to make empty shells available. Functional links between producers and consumers thus create a more productive, more opportunity-rich economy in which conditions for the common good of both dominant and subordinate members emerge.

Energy is the currency of economic exchange and value, but power—the rate at which energy and raw materials are used—is the measure of economic performance. More power means more control. In part, power is derived from the environment. The ability to respond adaptively, to improve hypotheses about the environment, and to capitalize on opportunity depends on how rapidly and how reliably the environment delivers resources, but it also depends on the technology that economic entities possess. In chapter 6, I explore the architectural and organizational qualities that bestow power, wealth, and adaptability on living things. Powerful entities are large, metabolize rapidly, have a wide individual or collective reach, and possess a flexible, hierarchical organization characterized by semiautonomous interacting parts subject to diffuse central control. Cooperation, division of labor, effective communication among parts and among individuals, and an ability to respond rapidly and flexibly under a very wide variety of circumstances, including rare ones, characterize all economically powerful entities. Power provides access to resources, and this power stimulates production and all other processes that confer greater power in the first place. It is therefore a great amplifier of economic life, an attribute which depends on, and in turn promotes, intense high-stakes competition.

The technological and organizational qualities that bestow economic and physical power will increase only if the benefits of greater power exceed the high costs of investment in power-enhancing innovations. Such large outlays become feasible only when resources are abundant, accessible, and predictable in economies of large effective size in which competition is intense. It is only under such circumstances that entities are able to modify patterns of allocation and where large imperfections of innovations are able to survive long enough without lethal consequences for selection to improve them. Conditions favoring increases in the sizes and numbers of individuals and groups thus provide opportunities for improvement and growth by reducing the cost of imperfection and by relaxing the constraints of trade-offs that under economic stagnation or decline enforce an adaptive status quo.

Four interrelated factors and processes in the environment—temperature, the effective size of economies, heat-powered eruptions and tectonic movements of Earth's crust, and life's metabolism—together determine the supply of resources and therefore the distribution of opportunity and constraint in space and time on our planet. I consider this environmental component in chapter 7.

Temperature—the average kinetic energy of molecules moving in a fluid or gas—affects almost every chemical reaction and material property on which life depends. A rise in temperature from the freezing point of water to somewhere between 35 and 40°C, above which the molecules of life begin to suffer heat-related loss of function, enables organisms whose body temperatures approximately match those of the surroundings to accomplish the work of life faster and often more cheaply, and thus to draw and deliver more power. Animals can move and feed faster, growth rates are potentially higher, and microbes and plants extract limiting mineral nutrients more quickly as they chemically break down rock and soil. At higher temperatures, such viscosity-related activities as filter-feeding and swimming in water become cheaper because the surrounding medium becomes less sticky and therefore less resistant to flow. Perhaps most importantly, they make available a much wider range of adaptive possibilities. In the cold, everything is slow, and there is little difference between the fastest and slowest members of a community. Under warm conditions, adaptive solutions emphasizing rapid movement and growth coexist alongside solutions featuring inactivity or slow accumulation. Higher temperatures therefore permit more extreme patterns of energy allocation and functional specialization in more directions, and allow the realization of adaptations that are unachievable in cold-bodied creatures.

Although maintenance costs tend to rise with increasing temperature, the power-related benefits of operating under warm conditions are greater as long as sufficient resources exist to maintain the fast-running engine of life. For this reason, warm-blooded animals—birds, mammals, some insects, and even some fish—capable of producing capable internal body heat for extended periods hold prominent economic positions of power in most of the world's ecosystems, despite their great inefficiency. Moreover, by achieving a degree of thermal independence, they can operate effectively as active competitors under a very wide range of circumstances.

Small economies set severe upper limits to the per capita and collective power of their members. For large-bodied or rapidly metabolizing creatures, small economies imply small populations, which are more vulnerable to disruptions than large ones. The effective size of an economy—the area or volume of "suitable" habitat—depends both on the physical structure of the environment and on the locomotor and physiological responses of living things to it. For wide-ranging individuals or species, the economy is effectively larger than for a sedentary one. All else being equal, the largest land masses and water bodies tend to be the most permissive locales for entities of high performance, whereas isolated small islands support more slowly metabolizing types.

Materials cycle into and out of the biosphere on time scales dictated by both geological and metabolic processes. Volcanoes and undersea vents carry nutrients, including the greenhouse gas carbon dioxide, from the mantle and lower crust to the surface. Collisions between continental blocks, and sites where one block rides over and subducts another, are associated with mountain building. Wind, ice, and water physically break down the new surface rock and transport the eroded debris into rivers, lowlands, and the coastal zones of oceans, where organisms further break the fragments down to release soluble mineral nutrients. The formation of new crust beneath the seafloor raises sea level and indirectly creates large areas of high productivity in moist lowlands and shallow coastal seas. The release of carbon dioxide attending the formation of new crust warms the world and thus stimulates production and consumption still further. High temperatures and high production conspire to cause not only more consumption, but also the burial of a larger amount of organic matter, a process that in turn releases more free oxygen to the atmosphere by preventing oxidation of some of the biomass that is buried. Positive feedbacks among producers, consumers, and the geologically and celestially controlled environment of Earth magnify the stimulatory effects of external inputs and create productive ecosystems in which selection due to competitors and other enemies is intense.

The conditions of life—risks, rewards, opportunities, and constraints—vary greatly from place to place on many spatial scales, over millimeters for microbes to thousands of kilometers for ecosystems. In chapter 8 I examine this variation on the scale of geography. Although the physical structure of Earth's surface—the distribution of oceans, land masses, rivers, sediments, rainfall, and temperature, among other features and factors—in part determines where particular types of organisms and ecosystems exist, living things themselves profoundly influence the geography of life. Competition and specialization create and dissolve barriers that define the limits of distribution of species and the economies to which they belong. The geography of life therefore reflects a complex interplay between Earth's intrinsic heterogeneity and the superimposed, competition-related heterogeneity imposed by life itself, much as the geography of humanity reflects cultural and political patterns on the physical landscape.

Regions and ecosystems in which productivity is high, competition is intense, and adaptation is least constrained by energetic and material limitations, occur at low to middle latitudes—the tropics and subtropics, in other words—in moist lowlands and the shallow waters of oceans near large, topographically complex land masses. These environments and the species living in them economically and evolutionarily subsidize less productive parts of the biosphere with raw nutrients, food organisms, and evolutionary lineages; they thus act as donor regions and dominant species on a geographic scale. Species invade from the tropics to the temperate zone, not in the opposite direction. There are far more invasions from the sea to freshwater than in the opposite direction, except among vertebrates capable of functioning actively under a wide range of circumstances. Initially, the sea economically and evolutionarily subsidized the dry land, but beginning some 300 million years ago, after land plants had achieved the stature of forest trees, more lineages invaded from land to sea than from sea to land. Large regions—land masses such as Eurasia, Africa, and to a lesser extent North America—and marine regions such as the Indo-West Pacific tropics produced many competitively superior entities that have spread far outside their original ranges into smaller, less productive regions.

In line with this geographic pattern, the origin of hominids and of modern humanity in competitively rigorous Africa, and the rise of early agriculture in the extensive fertile lands of the Fertile Crescent, south Asia, and China is not surprising. Productivity, power, and innovation continue to characterize large, opportunity-rich economies and societies in the human domain.

There is also a historical dimension to the pattern of subsidy among the world's major biological regions. Areas whose biotas have suffered less extinction of species tend to support competitively more sophisticated species, which have spread to biota that have suffered significantly more disruption. The removal of well-adapted incumbents thus allows plants and animals to become established even if these species are not initially well adapted to the conditions of life in the newly available habitats. As we humans increasingly eliminate incumbent species while transporting many others around the world, opportunities for successful establishment are expanding greatly, sometimes with disastrous consequences.

These relationships underscore the central importance of disruption. I treat this universal phenomenon and its consequences in chapter 9. Because economic entities tend to be well adapted to their surroundings, any change in those surroundings is apt to be harmful. All economic systems and their members are vulnerable to disruption, that is, to instability imposed either from the outside or from within the system. Storms topple trees, fires devastate forests, hunters extinguish populations of large animals, droughts bring societies to their knees, and cataclysmic geological upheavals—massive volcanic eruptions, exhalations of methane gas from great reserves beneath the ocean floor, and especially impacts of asteroids and comets—buffet and sometimes cause the collapse of economies large and small.

I argue that bottom-up disruptions, those interfering with the delivery and regulation of resources and the means of production, are far more destructive than are the major instabilities caused by the elimination of only top consumers. Interruptions in primary production not only remove the necessary resources for those herbivores, suspension-feeders, and predators that depend directly on primary producers for food, but also destroy the three-dimensional structure in which individuals find refuge from consumers. If the machinery of production remains largely intact, and chiefly the top consumers—large herbivores, carnivores, and suspension-feeders—are removed, as has been the case up to this point in the human-dominated biosphere, the dramatic changes in ecosystem composition are not accompanied by widespread collateral extinctions. For example, the human-caused disappearances of most of the large mammals in North and South America and Australia following the arrival of humans on those continents led to great irreversible alterations in the vegetation, but most other species were able to persist, albeit in a markedly different biological environment. The same applies to the oceans, from which humans have eliminated or drastically reduced populations of most larger predators and herbivores.

All of the mass extinctions, and probably most of the so-called minor episodes that punctuate the history of life, are associated with the disappearance of much of the phytoplankton in the ocean and the land-plant producers on the continents. Similarly, the demise of ancient human civilizations is often linked to droughts and other circumstances that reduce the primary production on which human society ultimately depends. Interference with sunlight through the formation of giant dust clouds following the impact of celestial bodies seems to be the most plausible mechanism for precipitating global collapses in primary production during the most severe economic disruptions.

But there is also a bright side to disruption. Short-term disturbances enable fast-growing species with high metabolic rates and with inert life stages capable of withstanding adverse conditions to capitalize on briefly favorable circumstances, and it is these opportunists that are in the best position to spawn highly competitive dominants in postcrisis ecosystems from which well-adapted incumbents have been eliminated. The rare, extinction-causing disturbances thus give opportunities to new invaders for which there is a premium on rapid establishment and rapid exploitation. High-powered survivors therefore have a critical advantage during the recovery from crises.

This stimulatory role of disturbance in spreading rapidly metabolizing opportunists and their competitively superior descendants exemplifies the central thesis of this book, laid out in chapter 10: inequality, as manifested by the power and wealth of economic players with disproportionate control over others, imparts an overall directionality to the historical development of economic systems, including our own. Successive dominants at every scale of organization become more powerful, more productive, and internally more diversified, and they increase both their reach of control and their independence from their environment. Life itself increases opportunities for growth and adaptation while reducing constraints. Disturbances interrupt and occasionally reverse these trends, but in the long run they enhance rather than thwart these trends. Events that trigger or disrupt overall trends determine the time course of historical processes. We cannot know their timing in advance, nor can we predict the precise nature of the participants and the interactions in which they are engaged; but these accidents and contingencies of history, important as they are, do not undermine or invalidate the claim that economic history is inherently directional.

In the history of life, directionality and contingency can be seen over the time span of billions of years. We see plants and animals partially eclipse bacterial-grade microbes, warm-blooded vertebrates succeed cold-blooded ones as the dominant consumers of their age, fast-growing flowering plants push slower-growing ferns and conifers out of low latitudes and rich soils, and heavily armored or exceptionally fast molluscs increasingly replace older groups with less specialized shell defenses and slower life-styles. The parallel history of humanity plays out on the much shorter time scale of decades to thousands of years, but the patterns as chronicled by the rise and fall of empires, the invention of new means of extracting energy, and the increase in the magnitude and reach of trade mirror those seen on the long time scale of the history of the biosphere as a whole.

What do the conclusions about power and history portend for our future? I consider this question in the final chapter. On the one hand, nature has given us innumerable past and present examples of economic structures that endure and adapt sustainably in the face of predictable and unpredictable change. On the other hand, all economic activity brings about change, and it does so in broadly predictable ways—toward greater productivity, diversity, and power—all the while creating higher-order emergent entities. Human history has mainly fallen in line with these long-term trends, but our unprecedented power has enabled us to throw the economic system out of a state of orderly change and harmony into one of instability and chaos. During the past three centuries, there have been vast and rapid increases in human population size, per capita and collective energy use, individual lifespan, trade, material wealth, the rate of invention, the directed gathering of scientifically acquired knowledge, and opportunity. We engage in arms races of unprecedented magnitude, and we are changing the economic structure of the biosphere so rapidly that our and others' ability to adapt is increasingly in doubt. We have come to depend not just on growth, an attribute that has always enabled adaptation in economic systems, but on very rapid growth.

In the very long term, such rapid growth is unsustainable for those of us who remain on and depend on the Earth. Sooner or later, despite all the economic substitution of resources and all the efficiency we can still build into our economic system, we will reach a limit. When we reach it, this limit will impose a regime of rigid conservatism in which adaptability and opportunity, including the scientific acquisition of knowledge, the solution of problems, and funding for the finer things of life will be severely curtailed. This is a type of stability that few would find desirable.

The preservation of economic growth well below the threshold between order and chaos would seem most ideal, not least because this is the regime that has endured and enriched the biosphere from the very beginning. It stimulates individual opportunity, and permits investment in science, the only method of discovery and problem-solving that precisely parallels adaptive evolution. Based on 3.5 billion years of history, I find it inconceivable that humanity and any of its most powerful individuals or institutions will voluntarily choose to cede power and to forego economic expansion. Instead, we must endeavor to create a system in which no part of humanity, nor the human species as a whole, gains so much monopolistic, centralized power that regulation and competition disappear. Given that all sustainable natural systems, from developing embryos to physiologically functioning bodies, are characterized by diffuse regional regulation and local control, it would be wise for us to model our own institutions, from governments to corporations, on such adapted economies. We cannot eliminate power or inequality, but we can regulate and use them to create a common good at the largest possible scale—that of the Earth as a whole—by intentionally designing flexible, responsive, accountable structures in which all elements participate and in which opportunity and adaptation are allowed to lead to slow growth. We are unlikely to break the laws of economies or to alter the course of economic history. We should therefore use them to our best collective advantage by learning how nature provides opportunity, adaptation, and constraint.

Chapter 2

THE EVOLVING ECONOMY

IN THE WORLD WE INHABIT—a world of structure—everything consists of parts, and everything is part of something else. The parts interact and thus affect each other as well as the wholes they comprise and the parts they contain. Science endeavors to understand this structure and the interactions imparting that structure in two complementary traditions. The reductionist tradition emphasizes parts and how they work. The whole is dissected, atomized, and broken down into its elementary particles. General laws are expected to emerge because there are just a few kinds of elementary particles and just a few fundamental forces and interactions among these particles. Younger and less widely respected, the holistic tradition emphasizes the whole, whose properties and interactions arise from, but are not the same as, those of its component parts. The more inclusive the whole, and the further removed it is in its organization from its elementary components, the more it is perceived to be complex, to be invested with unique characteristics that resist generalization and systematic inquiry. Yet, as Edward O. Wilson points out in his book *Consilience*, science has matured to the point where the search for general principles governing wholes is both feasible and profitable.[1]

The whole with which this book is concerned is the economy. The word *economy* immediately brings to mind earning and spending money, when the next recession or boom will occur, what next month's unemployment figures will reveal, thriftiness, corporate fusions and takeovers, capital investments, and the like. For Gary Becker, the economy is "the allocation of scarce means to satisfy competing ends." These human-centered conceptions, however, are so narrowly drawn that they obscure a central truth: the human economic system is built on the same principles of competition for locally scarce resources as are all other economic systems composed of living things and their environments. Just as the human species evolved from nonhuman primate ancestors, our economic system is an outgrowth of systems of interacting nonhuman species that have persevered and changed for billions of years of tumultuous Earth history, and these systems in turn were built from yet simpler economies of interacting cells, interacting macromolecules, and ultimately nonliving components. By stripping away all the particulars and standing back from the day-to-day and year-to-year concerns about our economic well-being, we can discern how economies of all kinds work, how they arise and develop through time, and how the conflicts and common interests that characterize the members of all economies intersect to produce entities of increasing power and independence. Then, by putting the particulars back in, we can apply the fundamental principles to the past and to the future.[2]

I shall use the term *economy* to mean a system of living, self-interested, interacting beings or entities that compete for locally scarce resources. An economy may vary in size from a living cell, studied by people who in their lighter moments might call themselves sell-buy-ologists, to the biosphere. Developing embryos are growing economies of cells that move, divide, interact, self-organize, and specialize to form a complexly differentiated body. Communities and ecosystems are assemblages of producers and consumers, divisible into species, which in turn comprise individuals. Biotas are economies on a geographic scale comprising many different ecosystems. To the human economists of yesteryear, the economies of nature seemed irrelevant to the markets and the people who produce, trade, and consume in those markets; but all economies are part of a single global biosphere, in which all the academic specialists with their system-specific languages and cultural traditions must come together. To use David Sloan Wilson's phrase, the members of any functionally integrated economy, at all scales of inclusion, have a shared fate. The study of economies is therefore the study of self-interest of members and common interest of the whole.[3]

Economic systems, regardless of their size or sophistication, are founded on life. They consist of living things and their creations and interactions in an environment whose living and nonliving components vary from place to place and over time on many scales. To understand economies is to understand life and its context.

The Economy of Life

Although we all have an intuitive grasp of what it means to be alive or dead, an unambiguous definition of life is elusive. The chemical differences that set life apart from nonlife are subtle and gradual, and the transition from the inorganic to the organic is seamless and arbitrary. Not one attribute that we typically associate with life—activity, growth, replication, interaction, competition, cooperation—is unique to it. In fact, many of the principles that underpin the economic systems of life apply also to the establishment and development of nonliving systems, starting with the universe itself and working down to galaxies, stars, planets, and the many kinds of nonliving systems we encounter on Earth: fires, patterns of circulation in the atmosphere and oceans, movements of materials within the crust and mantle, the growth of crystals, and so on. Despite our often mystical characterizations of life, we need no vital forces or supernatural powers to distinguish or explain life.

I like Stewart Kauffman's definition of a living thing as an autonomous agent, "a self-reproducing system able to perform at least one thermodynamic work cycle." In other words, an autonomous agent reproduces itself as a collective enterprise of interacting molecules, and carries out work on its own behalf through metabolism in a cycle in which chemical reactions that require energy are coupled with reactions that produce energy. Thus, in Kauffman's words: "Life is an expected, emergent property of complex chemical reaction networks." Through metabolism, life modifies—in fact, it creates—its environment.[4]

Life and its environment on Earth collectively comprise the biosphere. The divisions of the biosphere into biotas, ecosystems, communities, and the many smaller units of life and nonlife are in part enforced by the physical structure of Earth and its celestial surroundings, but they are greatly modified, forged, and obliterated by life itself. The biological and economic division of the world is, in other words, partly due to processes beyond the control of autonomous agents, and partly a reflection of how interacting agents respond to and alter this existing nonliving foundation. I return to this point in chapter 8.

I use the somewhat abstract term *entity* to denote an economic unit, that is, any unit or collection of units of life. The word *phenotype* refers to the characteristics or properties of an entity.

The fundamental phenotype of life is metabolism. Through a series of linked chemical reactions, living things extract energy from some source—light, electrons flowing from one substance to another, and bonds within energy-rich molecules—and convert it to do the work of life: growth, activity, and replication. Metabolism powers the construction of nucleic acids, proteins, carbohydrates, and fats. These serve to encode and translate genetic instructions, to store energy, to fuel activity, to stimulate (or catalyze) chemical reactions so that they proceed rapidly at the relatively low temperatures of much of the biosphere, and to build the physical architecture—the tissues and organs—of the body. Energy is captured and put to work before it leaves the body in degraded, unusable form as heat.[5]

Because of metabolism, life exists in a state far from equilibrium with its surroundings. Metabolism imposes an irreversible trajectory from randomness to structural complexity, a trajectory maintained as long as chemical energy passes through the living entity and is converted to do work. The establishment of order through metabolism would seem locally and temporarily to violate the second law of thermodynamics, according to which all systems tend toward an equilibrium with their environment with the consequent loss of order. As Eric Chaisson points out, however, the complexity and order that characterize living things and their creations are acquired with the concomitant disposal of degraded energy and of waste matter or pollution. These degraded products are no longer usable to the living entity. The second law appears to be violated only if we unrealistically think of the living being in isolation, but the law is strictly obeyed once we realize that life exists in a larger context.[6]

Growth is the basic ability of life forms to increase in size. This process, powered by metabolism, is made possible by a loop of chemical reactions in which the initial substances (reactants) are combined and modified to form products which, through further reactions, yield more of the reactants. The chemical loop, or autocatalysis, is a prime example of a positive feedback, in which the effects of a given chain of causal events reinforce the cause. Positive feedbacks pervade all economic systems, and I shall have many occasions to return to them throughout this book.

The ability to replicate, or make copies, sets life apart from nonlife. Replication gives life continuity and the ability of entities to expand in numbers. Like other life functions, it requires energy and is therefore powered by metabolism. This all-important process requires two kinds of polymers, or long chains of simpler molecular components. Nucleic acids are composed of nucleotides, whereas proteins are chains of amino acids. With the help of proteins, whose amino-acid components would have been present without the intervention of life on the early Earth, short nucleic acids in the presence of positively charged ions were not only able to replicate, but also able to code for the proteins that aided replication. As Kauffman points out, however, replication is not the accomplishment of a molecule, but of a set of cooperating molecules interacting within the confines of a distinctly bounded simple economy, the cell.[7]

Competition, Cooperation, and Trade

From the beginning, life and the economies it launched have been characterized by interactions among components and among wholes. Interactions transfer energy, material resources, and information among economic players. We can think of the system as points—cells, individuals, species, communities, or ecosystems—connected by links to other points. The points exert influence and exchange commodities with each other by two-way communication across the links. Interactions can then be described by the magnitude and sign or direction of the effect, being positive if beneficial and negative if harmful. Almost always, the effect of unit A on unit B differs from the effect of unit B on unit A. An economy is thus a network of interacting entities, a whole whose properties—so-called emergent properties—differ from those possessed by the components individually. Small economies interact to forge larger economies. Some economies are highly connected, each unit being linked directly to every other point. In this case, with n points, there will be $n(n-1)$ interactions across $n(n-1)/2$ links. At the opposite extreme, the system may approach the linear or open-branched condition, in which there are $2(n-1)$ interactions across $n-1$ links. Regardless of the geometry of the economy, interactions dictate, and are dictated by, the members of the system and by the economy as a whole.

Interactions are, of course, not unique to life. Gravity ensures that celestial bodies interact at a distance and collide. Atoms interact within molecules, and molecules continually bump into each other in liquids and gases. Life's interactions are distinguished from those of nonliving particles in that meaningful information is exchanged. By *meaningful information* I denote a sequence of data, or observations, that impart to the participants a hypothesis about their environment, a hypothesis about economic resources and their acquisition and retention. The notion of *meaning*—information translated into hypothesis, function, or action—is an emergent property of life. This property is again powered by metabolism. In Daniel Dennett's forceful words: "An impersonal, unreflective, robotic, mindless little scrap of molecular machinery is the ultimate basis of all agency, and hence meaning, and hence consciousness, in the universe." Information and meaning can ultimately be expressed purely in terms of energy, but their central role in living systems warrants a distinction among free energy, material resources, and information as economically important commodities exchanged in interactions among living things.[8]

The fundamental interaction of economic systems is competition arising from conflict of interest. Whenever resources are potentially available to more than one economic entity, there can be competition for those resources; that is, resources can move between entities. Competition implies that when one entity is enriched with resources, other entities have failed to gain access to those same resources.

To most economists, the defining interaction of a growing economy and one that promotes the creation of wealth is trade, the exchange of goods or services among entities. Trade in my view is an important form of competition, in which two self-interested parties exchange goods that one party has in excess and the other party wants. It is a mechanism for relieving what we might call reciprocal inequality, a way of satisfying demands and of redistributing supply to the benefit of both participants. It requires that there be at least two resources over which interacting entities compete. Trade thus involves cooperation, or the suppression of conflict between partners, but it redistributes resources in such a way that both partners end up as better competitors than they would be without trade.

The variety of resources for which economic entities compete is vast. In the earliest ecosystems, competition would have been for energy-rich compounds to power metabolism and to build the components of life. Light from the sun became an important resource when bacteria evolved the ability to make organic compounds by trapping photons from the visible part of the sun's radiant energy. For still more sophisticated organisms, there was competition for safe places from enemies and for food organisms. Land plants compete for water and for the nutrients dissolved in it in the soil. Plants and animals that practice internal fertilization and reproduction compete fiercely for mates. Humans have enormously expanded the repertoire of resources to include fossil fuels, ores, labor, capital, opportunities to advertise, and know-how, among many others. As more and more resources are added, opportunities for and benefits from trade multiply.

It has become fashionable for some economists and biologists to deny the primacy of competition. Echoing the writings of economists from the early twentieth century, Geoffrey Hodgson claims: "Neither biology nor anthropology give support to the universal presupposition for competition and scarcity." This kind of misunderstanding arises for at least three reasons. First, competition (or conflict of interest) is often thought to operate only when resources are scarce, meaning that the supply of resources limits the number (population size) of economic agents. If resources are plentiful, or for all practice purposes infinite, there cannot be competition for them.[9]

This argument misses the important point that competition is local, not global. Individuals compete with each other, and it is this competition that determines the outcome of an interaction. If, for example, children quarrel over a toy, or corals fight over a particular space in which to expand their growing colonies, there is no indication that toys or space are limiting resources, but there is strong evidence for competition for locally scarce goods. The characteristics that enable one entity to gain more of a resource than another entity may in fact be crucial to the survival and reproduction of the interacting parties, even if other members of the population do not experience the shortage. Klaus Rohde gives another example from his studies of the parasites of fishes. The worms and copepod crustaceans parasitizing the gills do not compete for food, because many parts of the gill are left unoccupied. Each parasitic species is very specific about its position on the gills. This peculiar arrangement, Rohde suggests, reflects intense competition for mates. If individual parasites could wander everywhere at will, they might encounter plenty of mates; but they are sedentary, so that mates are effectively extremely scarce and rarely encountered. By occupying only a few sites, the parasites effectively increase the likelihood of encountering a member of their own species, whose site preferences will be similar. Here again, it is local competition among individuals that matters.[10]

The second objection to competition as a prime economic agency is that, over the long run, great disruptions such as mass extinctions and economic collapses are more important to survival than are local rivalries. As Stephen Jay Gould sees it, competition acts inconsistently and temporarily, whereas global calamities affect a great number of organisms everywhere. I would respond that every individual of every population is likely to compete for each of several resources during its lifetime. Competition may not regulate the number or density of individuals in a population, but as an agency of selection it can nonetheless be extremely powerful. Without denying the importance of catastrophes, I would argue that competition is a ubiquitous phenomenon that lies at the heart of economic life.[11]

A third argument against the universality of competition is that cooperation often appears to be more important, and to many observers certainly more desirable. The common good replaces conflict of interest in the process we call cooperation and the relationship we call mutualism or interdependence. In this view, competition and cooperation—antagonism and harmony—are alternative means of distributing resources. This argument, often tinged with a dash of hopeful sentimentality for a less violent and contentious world, misses two crucial points, first that cooperation often accompanies antagonism, and second that, although cooperation implies the suppression of selfishness among the parts of entities, it leads to the formation of more inclusive entities whose competitive abilities far surpass those of the cooperating members individually.

To illustrate the first point, consider some examples from the annals of bizarre natural history. In many interesting cases, species have become dependent on their exploiters for continued growth and reproduction. Rocks on wave-washed shores in the tropics and on the crests of coral reefs are covered with a thick crust of coralline algae, red seaweeds whose delicate reproductive organs are hidden beneath a resistant outer layer of mineralized cells. The corallines evidently cannot reproduce unless this protective layer is scraped away by such herbivores as fish, sea urchins, and limpets. The two parties are clearly antagonistic—one eats the other, one protects itself against being eaten—yet they also have become mutually dependent; there is enforced cooperation.[12]

Browsing by elk and deer increases the number of inflorescences of scarlet gillia (*Ipomopsis aggregata*) in Colorado almost two and a half times over the number in unbrowsed plants. Similarly, on the Serengeti plain of Tanzania in East Africa, grazing by a diverse array of hoofed mammals keeps grasses short, but it also increases the productivity—the rate at which biomass is produced—of these plants. Claire de Mazancourt and her colleagues have shown that such mutualisms can evolve whenever grazers or browsers are able to recycle the plant's nutrients faster than would occur if nutrients were made available to the next generation of plants only through the decomposition of dead plant parts by bacteria and fungi. This condition is likely to be common. Noteworthy in all these cases is that the plants have evolved mechanisms to slow herbivory. Grasses, for example, have a high silica content, which has the effect of extending the time necessary for herbivores to chew and digest these plants. Again, apparent cooperation is forced by underlying antagonisms.[13]

Cooperation as a means of forging formidable units of competition has unquestionably been a powerful agency in history. In fact, it is probably fundamental to the origin of life itself. Stewart Kauffman cogently argues that the development of living agents—systems capable of self-reproduction and of performing a work cycle—requires a set of molecules that together (but not separately) catalyze the reproduction, of the whole. A living thing can be likened to a tiny engine whose parts must cooperate to do all of the work of life. In organisms in which the genetic code resides in DNA (deoxyribonucleic acid), the code is translated by RNA (ribonucleic acid) into proteins. Cooperation between the macromolecules is essential to ensure that the components are in the proper positions to fulfill their roles at the appropriate times. Genes—segments of DNA that code for specific proteins or regulate the machinery of life—cooperate to form what Richard Dawkins calls "survival machines," individual organisms that transmit these genes according to how well individuals compete.[14]

Economically independent organisms have very frequently come together to form competitively powerful symbioses, or partnerships. The origin of Cyanobacteria ("blue-green algae"), which produce oxygen as they fix carbon in organic matter from carbon dioxide, may lie in the close association or union of two organisms, a purple or green bacterium and a green sulfur bacterium, which both photosynthesize without liberating oxygen. The eukaryotic cell—the basic unit of plants, animals, and fungi—is a highly integrated economy consisting of a controlling nucleus, whose origins lie in the kingdom Archaea, and various organelles with bacterial origins. These organelles include energy-producing mitochondria derived from free-living proteobacteria, photosynthesizing chloroplasts (in plants) derived from Cyanobacteria, and beating cilia (in many animals), perhaps derived from spirochaete-like organisms. When these parts came together at various times and in various combinations, they eventually created an integrated unit within which competition among parts was suppressed in favor of cooperation, to the benefit of the new larger entity. Eukaryotic cells were able to compete very effectively with free-living bacteria and Archaea in many environments, especially through their novel ability to engulf those smaller cells.[15]

Symbiosis has produced many other competitively superior composite organisms. An association between soil fungi (mycorrhizae), capable of taking up minerals from the soil, and photosynthesizing plants made possible the effective exploitation of dry land by plants. It allowed plants for the first time to tap resources from the soil, a capability that is absent in most algae, presumably including the charophytes that are thought to be ancestral to land plants. Nitrogen-fixing bacteria on the roots of legumes and some other flowering plants enrich the soil and promote the growth and competitive ability of their hosts. Symbiotic dinoflagellates—single-celled algae of the genus *Symbiodinium*—live and photosynthesize inside the light-exposed tissues of corals, giant clams, and various other marine animals. Not only do these guests transform their hosts into functional plants—net producers of oxygen—but the symbiotic partnership dominates other animals that lack symbionts, at least in clear, well-lit waters and where grazers and predators are common.[16]

Cooperation among genetically similar (though not necessarily identical) individuals produces societies or coalitions whose collective power to compete can be very great. This effect is observable even among bryozoans, colonial invertebrates that attach to and grow on hard surfaces like rocks and shells on the seafloor. In experiments carried out at Woods Hole, Massachusetts, Leo Buss showed that individual colonies of *Bugula* are competitively inferior to colonies of *Schizoporella*, another bryozoan. When *Bugula* colonies grow cooperatively, however, the competitive tables are turned, and *Bugula* can overgrow *Schizoporella*.[17]

Societies of cooperating individuals give ants, bees, termites, thrips, aphids, dogs, naked mole rats, and humans enormous competitive advantages over solitary animals. Cooperative hunting dogs, for example, succeed in more than 80 percent of attempts to catch fleeing prey mammals in Africa, a figure that is much higher than in solitary hunters. The power of cooperation in human society as a means of increasing influence and competitive clout can hardly be understated, and is manifested in a great profusion of organizations—governments, religions, corporations, trade unions, political parties, and consumer groups, among many others—which populate our political and social landscape. Many, perhaps most, cases of mutualism and altruism, in which one party helps another, arise as effective responses to combat a greater common enemy. In Simon Levin's words: "It would be nice to know that altruism could arise without the common enemy, but it is hard to think of examples where that is clearly true."[18]

Competition is, in Darwin's words, a "struggle for life," a struggle for continuity, which ensues inevitably from the existence of multiple metabolizing entities in an environment of locally limited resources. Success means perpetuation in the face of predictable and unpredictable change, achieved by entities that control the right resources for a time long enough to leave descendants.

Performance and the Currency of Economic Activity

The copying of information is rarely precise. Like all atoms and molecules at temperatures above absolute zero, the molecules involved in replication move relative to each other. The precise geometric configuration necessary for the faithful transmission of one polymer's sequence to a daughter sequence is therefore subject to slight variation, which can result in the introduction of errors in the sequence. Multiple editions of macromolecules, as well as of the more complex manifestations of life—cells, individuals, and species, for example—come to differ among themselves as errors accumulate at each replication. Moreover, larger molecules often change shape as their environments change. Protein molecules and single-stranded RNA chains are not simply linear sequences of components, but complex structures such as helices and folded edifices whose shape and stability depend on the physical environment (temperature, pressure) and the chemical environment of electrically charged ions. DNA is more stable, forming a double helix, but it interacts with RNA and with proteins and is therefore also subject to these environmental variations. Thus, like all forms of transmitting information, replication is inevitably accompanied by some degradation, or at least change, in content.[19]

The variation introduced by imprecision in copying and by environmental influences provides the foundation for additional processes—selection, evolution, and adaptation—that characterize organized systems in general and economic systems in particular. Variant entities differ not merely in composition—in the sequences of genetic material, or the pool of species in ecosystems and entire geographic regions—but also in performance. By performance I mean an absolute measure of production, consumption, and defense. In a given environment, metabolizing entities compete to convert energy and raw materials into economic work, an outcome expressed in continuity—survival and reproduction.

How should performance be measured? Economic entities succeed or fail through competition for energy (the product of force and distance, expressed in joules) and its material equivalents. It is energy that entities exchange during interactions. Energy is therefore a currency, or means of exchange, for valuing economic activity. In the monetized economy, money substitutes for energy. Money acts as a means of valuing resources; we buy and sell goods and services with it, but money does no economic work by itself. Economic activity, however, is not just about energy; it is also about time. Being first in the market with a product can be more important than being the best; the risks of injury or death can be minimized by quick action. The appropriate measure of performance is therefore power, expressed in watts, and defined as energy (or work) per unit time.[20]

Power as a measure of performance and money or energy as a measure of exchange are obviously linked. Price—the amount of money paid or received for goods and services—is determined by the rate of production or supply (productivity), the rate of consumption (demand), and the extent to which producers and consumers control information about these rates. Producers, for example, may know more about (or have greater influence over) the market than do consumers, and may often be in the position of creating demand by advertising. Price is thus determined by power, even if we pay it only in units of money or energy. Price integrates availability, use, risk, and information as these relate to buyer and seller.[21]

It is well known that rates are extremely sensitive to the amount of time over which they are measured. Just as phenomena vary in space, they vary through time. The longer the interval over which a rate is measured, the more this small-scale temporal variation is integrated and smoothed out. In other words, information is lost as the interval over which rates are calculated lengthens. Fluctuations in price depend in part on how often critical information about supply and demand is assessed. Power and price measured over the long run will not be the same as power and price calculated in the short term. The calculations used should match the time scale of lifespans of the entities whose performance is being measured. This issue of compatibility of time scale has rarely been raised in the economic works I have consulted.[22]

Measuring economic performance in units of power will strike many economists and biologists as peculiar and unfamiliar. The reasons for this unease arise from common word usage. The word power as employed in everyday speech refers to influence, control, the implied or actual use of force, or the imposition of the will of one party on others. Whether this power is used or only implied, it is based on power in the strict physical sense, as energy per unit time. I shall therefore use the word power throughout this book in both the strict physical meaning and, to my mind equivalently, in the more metaphorical sense familiar to politicians, historians, and sociologists. Furthermore, each discipline has its own jargon for power as a measure of performance. To economists, performance is usually expressed as profit. In the currency of money, profit can be employed to enrich the producer, to reinvest in the producing firm, to fund new ventures either by the firm itself or by its investors, or to be redistributed as tax by the state for the real or perceived public purpose. To ecologists, performance is measured as net productivity, the power beyond that needed for subsistence. The excess thus becomes available for growth, reproduction, or consumption. Evolutionary biologists are accustomed to fitness as a measure of performance. Fitness in this technical sense means the surplus of power devoted to offspring. This reproductive power can be allocated either to large numbers of poorly provisioned young or to a few offspring that receive extensive parental care or investment. All three terms—profit, net productivity, and fitness—are special cases of power. The details of measurement vary according to discipline and context, but energy and time are integral to all measures of performance.

Selection, Evolution, and Adaptation

Performance affects, and is affected by, interactions among entities. These interactions do not occur in an economic vacuum, but instead take place in a context of opportunities, limitations, risks, and rewards. By virtue of phenotypes that affect performance, some variants in a given environment last longer or leave more copies (offspring) than others. This advantage also accrues to the components of the entities in question. Entities whose characteristics best match or fit the challenges and opportunities of their environment are thus selected in favor of those in which the match is less good. Selection operates whenever variants differ in characteristics that are transmitted through replication or some other form of reproduction or retention, and whenever these characteristics are consistently related to performance. Variants may arise through genetic mutations or genetic rearrangements, or by recombination of existing components, or culturally through the creation of new ideas, products, or social groups. Reproduction may proceed genetically, as in organisms, or by cultural transmission. Ideas are replicated and modified through language; cultural traditions are passed on and modified often with the help of social institutions; and previous designs are imitated from fresh components to resemble previous designs. Regardless of how variants arise or how information is transmitted, selection occurs through differential culling of variants according to the performance of entities in which these variants are expressed. I reserve Darwin's term "natural selection" for the differential representation of genetically based variants from one generation of organisms to the next.[23]

The process of selection (or nonrandom elimination) applies not just to economic systems or to life forms, but to the working of all organized systems. We marvel at how Earth is just the right environment for life, the place with just the right temperatures and the ready availability of liquid water; and some see the hand of a supernatural Creator in crafting a universe where the fundamental constants of physics have just the right values allowing the order we see to emerge. A selectionist's view, however, implies that life and the universe work not because of some lucky accident or the labors of an omniscient deity, but because of an accommodation—or, in the case of life, a partial modification—to prevailing laws. With different physical constants, or a planet in which the force of gravity were greater or lesser and where conditions of temperature and water differed, an entirely different order would have emerged, which would have appeared in just the same complete harmony with the universe as the order we actually observe. What is of interest is not that selection takes place—after all, selection seems to be universal to all organized systems—but how it comes about and what its consequences are.

Stewart Kauffman makes the important point that the process of selection cannot exist in a system dominated by chaos. In a chaotic system, the rate of introduction of new states or new entities is too rapid for the weeding-out process of selection to operate. Put another way, the error rate, or mutation rate, overwhelms the culling phase, or selection based on performance. There must be a degree of order—organization, partition, and constraint—imposed by work in order for selection to prevail. Economic entities—living things and the larger units and environments they create—collectively construct their ordered environment, which therefore includes selection as the dominant process determining what will propagate and what will not.[24]

Selection, due ultimately to local competition among entities varying in their ability to acquire and defend resources, is probably the most important mechanism of evolution and adaptation, two related processes that change all economic systems. By evolution I mean descent with modification. In the continuity of life, as expressed in the transmission and conservation of traits from generation to generation during successive episodes of replication (or reproduction), some traits will evolve, often in ways that improve the competitive performance of the replicators. Adaptation is the process resulting in a better fit between the replicators and their environment. The word "adaptation" is also used for a trait that confers on its bearer some improvement in performance in a given environment, a benefit that is manifested as greater fitness (survival and reproduction) when compared to the performance of entities without the adaptation. Through competition, selection determines which traits work and which do not, which characteristics and which entities evolve and which ones remain unchanged or disappear. Selection, evolution, and adaptation are inextricably linked through competition, regardless of how variants are generated or transmitted, regardless of which resources are in locally short supply or which entities engage in it. It is competition that makes adaptation, selection, and evolution fundamentally economic concepts.[25]

Notions of evolution and adaptation apply to all systems with a history. The apparent harmony between physical laws and the systems in which these laws apply attests to the same phenomenon that biologists recognize when they speak of a good fit between an organism and its environment. The details of the mechanisms and agencies responsible for adaptation and evolution may differ among systems, but the results are broadly the same: systems change through time and do so in ways that are in harmony with the fundamental laws and constants of energy and matter. Because this book is concerned with economic systems, I shall emphasize the processes and circumstances that apply directly to living systems and their constructs, but at no time do I wish to imply that the central concepts of selection, adaptation, and evolution are unique to such systems.

In an adapted unit, most variation introduced by errors (mutations) in copying are harmful. For an adapted entity, therefore, increasing fidelity in copying, or mechanisms that concentrate error in parts of the code where they will be least harmful or most helpful, will be favored. If one or more errors occurs in a sequence of symbols during each replication event, the sequence will exceed the so-called error threshold, the point above which the replicated sequences consist of random fragments without meaning. RNA, which consists of a single strand of nucleotide bases, lacks mechanisms to correct errors. The error rate of RNA replication is 10^{-3} to 10^{-4}, meaning that RNA can form only very short genomes. The lower the mutation rate, the larger is the potential size of the genome. Without machinery to reduce the mutation rate, a gene-based coding system cannot transmit much information.

This is why two-stranded DNA has a decided advantage over single-stranded RNA. In DNA, one strand can be checked against the other by an enzyme and, if need be, repaired by other enzymes. Such mechanisms of control and repair require cooperation among molecules and help to separate the role of replication from that of transcribing and translating the code. One-stranded RNA has mutation rates of 10^{-3} to 10^{-4} per nucleotide base pair; with repair mechanisms in place, DNA has a far lower mutation rate of about 10^{-9}.[26]

All participants in an economic system are adapted to that system in the sense that, by virtue of possessing specific traits, they are able to persist and propagate in their surroundings. In the words of Wolfgang Sterrer, referring specifically to adaptation in the evolutionary context of organisms: "An organism represents a hypothesis of its environment, continually tested by selection for its predictive value and modified by adaptation for a better fit."[27]

Biologists traditionally view natural selection as the primary mechanism of adaptation, and some have gone so far as to define adaptation as being caused by natural selection. In this view, adaptation is achieved through the differential representation of genes. In other words, the information on which adapted entities base their hypotheses is transmitted and sorted through genetic means.

Adaptation through genetic change characterized the early stages in the evolution of life and remains important to all organisms today, but it has never been the only mechanism. In human societies, cultural transmission through education, advertisement, propaganda, and entertainment have overwhelmed genetic means of achieving change. In mammals, including us, response to novel disease agents is made possible by the nongenetic creation of an enormous diversity of proteins that recognize, attack, and disable pathogens. These components of the immune system are not directly coded, but arise through combination. Nerves and muscles in animals permit instantaneous responses to challenges and opportunities. The basic design of the immune system, the nervous system, and the system of muscles and bones resides in the genes, but the details are too numerous and too complex to be specified by linear arrays of nucleotides.[28]

Selection will operate and adaptations will arise regardless of the particulars of how information is transmitted and modified. Where the particulars matter is in the speed, extent, and flexibility of response. Nongenetic means, especially those involving the nervous system and culture, make extremely rapid response possible. Genetically based adaptation must perforce be quite general, to cover adequately a rather large class of circumstances; whereas nongenetic means allow entities to respond to highly specific situations and, in the case of learning, even to anticipate circumstances. It is not just the amount of information that is important in adaptation, but also the rate at which it is transmitted, modified, and used. It is power that matters.

Imperfection

The adaptive hypothesis need not, and in fact cannot, be completely accurate. Imperfection is mandated by at least four universal and overlapping circumstances: unpredictable or rare events, insufficient time, misleading or false information, and the high cost of improvement imposed by functional tradeoffs.

There is a great deal of predictive information in the environment, and this information is translatable into adaptation. For example, some events—predation, competitive bouts, opportunities to mate—occur once or several times during the lifetime of an average individual organism. Most individuals will therefore be exposed to, or tested by, these events, so that an adaptive response is possible. Other events and circumstances occur so infrequently that, although they are predictable, the intervals between them far exceed the lifespan of most economic entities. For such entities, therefore, these circumstances are effectively unpredictable, and indeed are often catastrophic precisely because adaptation to them is difficult to achieve. As lifespans lengthen, rarer conditions find their way into an adaptive hypothesis. Alternatively, entities can evolve a kind of memory, enabling them to respond by drawing on hypotheses retained from ancestors that survived conditions for which the same hypothesis was adequate.

Even at the appropriate time scale, however, there is almost always a yawning information gap between the environment and the adapted economic unit. Classical economic theory is founded on the assumption that households, firms, and all the other elements of human society make decisions based on complete and accurate information. Acquiring and evaluating all this information, however, takes time, a commodity often in short supply when sudden dangers or opportunities call for immediate action. Economic activities therefore proceed with incomplete information. Although this limitation can never be wholly overcome, there are ways in which isolated bits of information can be supplemented and to an extent replaced by an overall theory of how things work. This has been one of the principal contributions of scientific inquiry. The development of a unified and consistent body of knowledge has enabled us to acquire, organize, and interpret facts so that we can better anticipate and react to situations that without the theory would have seemed incomprehensible.

Deception and falsehood complicate the task of acquiring and evaluating reliable data. Many tasty butterflies have the colors and behavior characteristic of unpalatable species, and are therefore afforded a measure of protection from wary predators. Fish are tricked into attacking what they perceive to be live prey, only to be hooked by a fisherman. Advertisers exaggerate claims and occasionally mislead people into buying or selling items. The dissemination of misleading or false information occurs throughout nature as an effective means of competition among rivals. Finally, the costs of improvement—of gaining more and better predictive information, and of increasing performance—are often prohibitively high. Every economic activity and every investment comes with costs. With a fixed budget, greater expenditure on one item means cutting back on another. This is the principle of the trade-off, a manifestation of competition that leads to functional compromise, imperfection, specialization, and the division of labor. The pluses and minuses—benefits and costs, advantages and disadvantages, rewards and risks—vary from place to place as well as over time, and they are influenced by the actions of living things. Active foraging for food, for example, benefits an animal if the work allows it to eat, but the presence of enemies may make foraging so risky that the animal is better off resting where it won't be found or recognized.

Trade-offs are everywhere. Military tanks and heavily armored animals such as turtles, armadillos, or thick-shelled snails move slowly. Greater speed requires either a larger budget—a bigger, more powerful engine or a greater mass devoted to muscles in the organs of propulsion—or a reduction in mass devoted to armor. Plants protected against enemies by energetically expensive, permanent chemical defenses such as tannins, resins, and lignins generally grow slowly. Fast growth, which would improve competitive ability, can only be had if expensive defenses are replaced by alkaloids, terpenes, and other less permanent chemical agents that can be mobilized or manufactured as needed. Many plants, including some fast-growing weedy trees such as the tropical American *Cecropia*, rely on ants to remove herbivores and strangling vines. This defense is effective but costly, for the plant must house the ants and feed them with sugary compounds. Some branching corals employ a mercenary army of trapeziid crabs, which nip off the tube feet of attacking crown-of-thorns seastars. Again, such protection is not free to the host.[29]

In the realm of reproduction, trade-offs affect every aspect from investment in the young to the timing and spacing of breeding. Investing a seed or fertilized egg with a large store of food for the developing embryo protects the young offspring against starvation and allows it to grow rapidly and to a large size through its most vulnerable stages, but it means that few such embryos can be supported. Long-term parental care, as is found in humans, means foregoing the immediate benefits of reproduction in favor of larger returns later. Less investment by parents and earlier independence of offspring, however, allow individuals to capitalize on favorable circumstances faster. Allocation of resources to offspring is affected by costs and benefits, which change with age.[30]

Some trade-offs are built into the fundamental machinery of metabolism. The enzyme Rubisco (ribulose-l,5-biphosphate carboxylase-oxidase) is centrally involved both in oxygen-producing photosynthesis and in light-assisted respiration, the reverse process in which oxygen is consumed and carbon dioxide is liberated. The active site of the enzyme binds either oxygen or carbon dioxide, which are thus in competition for that site. The role of the enzyme in both photosynthesis and light-assisted respiration in plants thus sets rather strict limits on the efficiency with which carbon is fixed and released. Improvement in either function would probably benefit plants, but Rubisco's conservative structure evidently precludes this possibility.[31]

The idea that imperfection is the expected state of adaptation flies in the face of common intuition and a long intellectual tradition. In the pre-Darwinian world view, the productions of nature were thought of as God's perfect designs. Even for many post-Darwinians, the expectation that organisms are optimally adapted has held a prominent place. Proponents of this view defend it by arguing that optimality provides specific, testable predictions about function, and that natural selection molds the best possible adaptations given initial conditions, trade-offs, and circumstances. Economists, too, have constructed an elaborate theory of prices and markets based on optimal, well-informed behavior by producers and consumers. The weakness in this approach is its unfalsifiability. If an organism falls short of perfection, we can always appeal to some unknown circumstances or limitations that we had not considered in the original hypothesis. An evolutionarily more tenable position is that organisms are sufficient, that they work well enough to survive and reproduce. Good design in this sense does not have to mean optimal design. The standard of allowable imperfection varies according to competition and environmental uncertainty. Under weak competition, economic success demands nothing more than the equivalent of a passing grade, a low standard of performance. But when competition is intense and the stakes are high, as measured by absolute costs and benefits, the bar of acceptable performance is set very high, and success demands a close approximation to the ideal. Adaptation is as good as it has to be; it need not be the best that could be designed. Adaptation depends on context.[32]

Specialization and Diversity

A partial solution to, and cause of, the problems of trade-offs and imperfection is to specialize, to divide tasks among entities or parts, each devoted to just one function. Being good at chasing a fast prey animal for long distances is incompatible with protective armor. Adaptations effective for life on land are often at odds with those enabling an aquatic existence. In very many cases, a single organism could do all of these things, but it would do none of them well. As Robert MacArthur emphasized in his writings, the jack-of-all-trades is master of none. The generalist method is adequate if stakes are low, but increasing specialization is often mandated when the stakes—the standards of performance in competition—are high. Environmental variation that has little effect on performance in a world of weak competition becomes important to performance as competition intensifies. Where the total budget of energy and material available to an individual is limited, trade-offs favor specialization among parts that are loosely integrated into the whole. Specialization of this kind occurs among the cells or tissues of the body, different life stages of an individual's life cycle, species within ecosystems, and occupations in human society. Higher stakes thus impose new trade-offs, new specializations, and a finer division of the environment into domains that can be exploited profitably.[33]

Specialization and the division of labor are manifestations of diversity, or variety. The atoms and molecules that form the basis of our physical universe and of the biosphere can combine and be separated into an effectively infinite number of different structures. In the biosphere, these different structures typically have different functions. Ecologists are accustomed to think about diversity in terms of species, units composed of genetically compatible individuals that typically do not exchange genes with members of other species. Diversity within organisms is expressed in cell types, tissues, and organs; and within cells, diversity is seen among organelles and among macromolecules. Occupations—ways of making a living—reveal diversity in human societies, as do products, languages, cultural practices, institutions, and machines. The details of how diversity arises and where it is expressed vary, but, as I shall emphasize in chapter 10, diversity is an inescapable and universal attribute of economic systems, an attribute that on average builds on itself as economies develop.

Feedback

An adaptation is not merely the effect of a cause; it also enables economic entities to modify their environment, often in a way that is favorable to the entities' survival and reproduction. The extraction of raw materials and the harnessing of energy inevitably change the surroundings of a metabolizing entity. Adaptation and environment thus affect each other; they are linked by feedback.

If feedback is positive, or reinforcing, cause and effect together unleash a runaway process with all the characteristics of an arms race. If negative, or stabilizing and self-limiting, feedback acts as a brake, muting change and damping fluctuations. On the one hand, feedback can magnify small initial differences into overwhelming contrasts; on the other, it can yield constancy in the face of buffeting disruption.

An example of how feedback propels change and raises the standards of performance comes from trees competing for light in a forest. Access to light is essential for photosynthesis and growth, but for plants growing under a closed canopy of tree leaves, light is limiting. At ground level, the light available to seedlings is only 0.5 percent of that reaching the canopy in tropical rain forests, and 1–2 percent of that in the canopy of north-temperate deciduous woods. In the shade of taller trees, there is little advantage for a small plant to grow rapidly, because there is no gain in the interception of light. But if a small tree has reached a certain critical height, or if it happens to grow in a light gap where a hole in the canopy lets undiminished sunlight through, rapid growth is essential, and profitable. The key is to reach a greater height than one's growing neighbors. The faster the taller trees grow, the higher is the competitive standard that neighbors must meet to prevent being shaded. Selection involving positive feedback between competitors will operate until some limit in height has been reached. This limit could be set by strong winds or by frequent fires, which topple tall trunks, or by insufficient water or nutrients. Higher concentrations of carbon dioxide in the atmosphere, and structural features of the root system and the wood, may enable trees to reach a greater stature, meaning that the limit of stature is not absolute. At this limit, the cost of still greater stature exceeds the benefit, and negative feedback sets in. If there were no competition, a plant could simply satisfy its demand for light by remaining close to the ground and saving a great deal of energy in the process. The positive feedback, expressed both within the lifetimes of neighboring trees and over evolutionary time among species, is thus driven by competition. The negative feedback—the stalemate—is imposed by the physical forces of the environment and the technological limitations of the trees.[34]

All sorts of feedbacks pervade mate choice, the all-important activity in which all sexually reproducing animals and plants with internal fertilization engage. In many species including humans, females appear to choose male mates on the basis of their apparent fitness. Such fitness is often expressed as conspicuous healthiness, an energetically costly display of wealth or activity indicating not only that the male's ability to secure food is adequate (indeed, superior), but also that he has a wide energetic margin of safety, one great enough to afford the luxury of ostentation. Continued selection for such wealthy males by discriminating females accounts for bizarre and wasteful expenditures such as the peacock's outrageously long tail feathers, the extinct Irish elk's oversized antlers, fiddler crabs' gigantic master claws, the loud virtuoso singing in songbirds, elaborate horns in beetles, and fast sports cars. Males may likewise discriminate among females, leading in humans to extravagant displays of jewelry, makeup, fancy and seductive clothes, and elaborate marriage rituals. These displays serve no function other than to attract and advertise fitness to the opposite sex, or to compete with members of the same sex. Plants are by no means immune to these feedbacks. Energetically expensive large, nectar-rich, showy flowers evolved because they attract faithful pollinators, which ferry male pollen among plants that themselves cannot search out individuals of their own kind.

Arms races between competitors exemplify another important class of positive feedbacks. Expensive weapons and defenses developed by one competitor sometimes compel rivals to increase their own offensive and defensive arsenals, even if by doing so the competitors sacrifice other ways of increasing performance. An absurd and potentially lethal manifestation of such runaway elaboration is, of course, the arms race among powerful nations. Economic wealth, together with the imperative to deploy technology and physical competition for material resources and political influence, has powered many extremely costly episodes of military escalation ever since human populations became sedentary. Between 1863 and 1945, the maximum performance of individual weapons rose from a yield of twenty pounds of TNT, enough to kill a few people, to ten kilotons of TNT, enough to kill one hundred thousand. There have been even more far-reaching increases in yield since. These remarkable rises in the power of weapons are only the most recent in a long series of increases, and all are directly attributable to competitive pressures imposed or thought to be imposed by rival governments.[35]

Success and Failure

Emerging from the processes of metabolism, replication, variation, and adaptation is the distinction between success and failure. Adapted entities are successful in that they acquire and defend resources well enough to ensure continuity. Failure occurs when, either because adaptation was insufficient or because suitable environments disappear, an entity is unable to harness enough resources to perpetuate itself. The reason why evolution has proved to be such a powerful process, one that has spawned so much bewildering complexity and high levels of performance, is that every organism that has ever lived on Earth and every organism that lives on Earth today is the product of an unbroken chain of successful entities, stretching all the way back to the origins of life. Every entity in this chain was a going concern long enough for that entity to leave offspring. Somehow, every change—including the rare great catastrophes that precipitated extinctions among so many contemporaries—was successfully negotiated by a link in that continuous chain from ancestor to descendant.[36]

It is difficult to overstate the importance of continuity in the face of change as a manifestation of success. Each entity in the lineage was adequate hypothesis of its environment. Subsequent links successfully adapted to their environments, which were not necessarily the same as those in which their ancestors succeeded. By luck or good design, entities in this continuous lineage encountered and successfully dealt with extremely rare events, circumstances so rare that the average individual would not have experienced them or had a hypothesis about them. At least part of the hypotheses of the past accumulated, so that descendants incorporated hypotheses about rare events or at least a way of generating such hypotheses through noninherited means. The new traits of organisms are built on or modified from old traits. Every organism in effect has two histories, one of ontogeny—the process in which a single fertilized egg or other beginning structure is transformed into a reproducing adult—and the other of phylogeny—the evolution from ancestor to descendant. The new traits represent new hypotheses, or modifications of the old hypotheses as represented by traits of early embryological stages and of evolutionary ancestors. Some of the ontogenetic and phylogenetic history as revealed by the development and anatomy of organisms represents vestiges of past hypotheses that no longer apply, but much else represents adaptations that endure because they remain successful hypotheses despite their origins long ago. In short, adaptations and economies build on the past.

If an entity fails to adapt to new conditions, it can persist only if the old conditions—the environment to which the entity was adapted—continue to exist somewhere in the economy. Without such refuges, the entity will face extinction or, in terms more familiar to the human-centered economist, unemployment or bankruptcy.

The distinction between success and failure will strike many readers as too stark. Surely some entities are more "successful" than others in having more power, wielding more influence, controlling more territory, reaching larger size, or simply lasting longer. Today, the United States appears to be rather more successful than, say, Somalia. But this intuition equates success with performance and confuses short-term persistence with long-term continuity. In the strict definition of success I have adopted, the United States and Somalia are both successful as nations because they continue to exist, at least for the time being; but by almost all measures of performance—individual and national annual income, population size, political influence, average life expectancy, and so on—the United States, and indeed most other nations, surpass Somalia. In 1910, most people would have said that the Habsburg empire was more successful than Mongolia; yet the former became extinct as a political entity in the aftermath of World War I, whereas Mongolia still exists. People living in much of the former Habsburg empire enjoy higher living standards than their counterparts in Mongolia, and in that sense they perform better economically, but it would be hard to argue that they are more successful if we use persistence from generation to generation as a criterion. From an evolutionary and economic perspective, success and performance are obviously both important, but they do not measure the same thing. Many powerful, high-performing entities do well in the short run but may falter over longer intervals; other, less influential entities may persist in the long run, and in that sense they are highly successful. Natural selection and most economic policies, whether fashioned by the state or by private interests, affect short-term success. Persistence over the long run is success resulting from a cumulative process of adaptation, but cannot be assured by any short-term process such as natural selection or intentional decision-making.

Inequality

A good candidate for one of the simplest, yet most far-reaching, principles in the realm of economics is that, when two parties interact, one almost always gains more or loses fewer resources than the other. The outcome of the interaction is, in other words, unequal with respect to the commodity that is being exchanged. The significance of this inequality becomes apparent when we examine the consequences of unequal performance. One party in effect possesses more power, a greater store of resources that it can deploy, or more information—a better hypothesis—than the other. As a result, one party wields disproportionate influence and exercises a greater degree of control over its environment. This control, measured in units of power, affects not only the subordinate party—its activity, adaptive characteristics, and distribution—but indirectly also the attributes of many other members of the larger economy which the dominant party occupies and modifies. Most economic interactions, competitive as well as cooperative, from scales of inclusion ranging from organelles in a cell to biotas interacting evolutionarily across geographic barriers, are characterized by inequality. Likewise, human societies and institutions are pervaded by disproportionate influence and by unequal distributions of costs and benefits, risks and rewards, and levels of performance.[37]

Unequal gains in power are not confined to antagonistic encounters. In mutually beneficial trade, one commodity or service is exchanged for another, but the terms of trade are dictated by, and therefore are tilted in favor of, the more powerful party. I shall revisit this kind of inequality in chapter 4.

Even at the molecular level, inequality seems to be an all-important phenomenon. Chemical reactions that proceed quickly—by catalysis, for example—may generally be favored over those that proceed slowly, and these reactions may therefore come to play decisive roles in the construction of cells at temperatures at which macromolecules retain enough stability to carry out work and enough flexibility to have their bonds stretched and bent by enzymes and other catalysts. Catalysis provides power and control; it favors cooperation, and in the end it produces the molecular machinery than enables an integrated molecular economy to take on the tasks of life.[38]

Dominance and subordination are not synonymous with winning and losing, nor with success and failure. Except in the case of successful predation or a fight to the death, in which the subordinate does indeed fail by losing all control of material resources, subordinates typically do not forfeit everything.

They may, in fact, be eminently successful as survivors, even if their subordinate status has forced them into limited activity in constrained circumstances imposed by dominants. Success is measured simply by persistence or continuity, whereas dominance refers to the direct or indirect control over others. By that criterion, subordinate groups of organisms—amoebae, brachiopods, and mosses, for example—have done exceptionally well.

Economic systems are, of course, complex structures, in which the pattern of interactions resembles a web. This means that the dominant party in one interaction may well be the subordinate in another. Control thus diffuses through the ecosystem, generally from entities with high energy demands to those with more modest requirements. The ability to adapt to, or at least to accommodate, the power structure remains the ticket to success for all players—dominants and subordinates alike—regardless of how much influence an entity wields.

A Comment on Good and Evil

The last concepts I have treated—success, failure, and inequality—raise important philosophical and ethical matters that, if left unattended, could seriously derail and mislead further inquiry into economic principles and their historical consequences. We tend to associate these three concepts in everyday speech with values. Success is good and makes right; failure connotes bad and evil. Something that works—something successful—is, again in everyday parlance, somehow true; failure means falsehood. Notions of value and judgment, however, apply in very particular personal or social contexts. Something is good if, in our judgment, it benefits us or our group. Something is right if, in our judgment, it coincides with our sense of what is good. Other individuals, or other groups, might make the judgment that those same things are bad, evil, and wrong. Most importantly, success and failure refer to what works and what doesn't in a specific context. What works is not necessarily true or right or good. The myths and doctrines that humans invent and promote for the establishment of their own power and for cohesion of the society in which they wield that power can be highly effective, but effectiveness does not make these ideologies true in the scientific sense of a hypothesis being verified by experiment or observation. Inside the society, these myths and doctrines work in that they forge a whole capable of collective action to great effect, and in that sense they can be thought of as an adaptive hypothesis; but that same hypothesis may be very far off the mark of a reasonable hypothesis when applied to a broader context or if it should compete with another internal hypothesis.

The world's religions, for example, offer guidance to the behavior of their adherents, and fulfill emotional wants in many people for an abstract security and a promise of a better life to come. The monotheistic religions have been highly successful in uniting disparate peoples into societies capable of expansion and political domination over other, often polytheistic societies. The moral codes embodied in religions discourage behaviors that are destructive to society and condone those that benefit the society's members, all too often at the expense of outsiders. Religion is therefore a powerful, highly effective, and successful social institution, an adequate hypothesis about the society in which the religion originated or spread. As truth—in the scientific sense—religion fails. Miracles, supernatural deities, and explanations for the creation of the world and of humanity as religious narratives wither away as unverifiable myth in the face of unified bodies of scientific theory founded on observation, experiment, and deduction. Even the moral codes by which societies try to live can be recast in scientific terms consistent with evolutionary and economic theory, as I shall discuss in chapter 10. The point is that success and truth are not the same thing, unless one defines truth in terms other than verifiable observation or inference.[39]

Whatever our judgments and opinions about religion, pornography, war, Bach's music, or the scientific method may be, it is our task as evolutionary biologists and economists to understand how the phenotypes and phenomena they study work and arise. We must endeavor to understand success and failure and to probe the sources and consequences of inequality, and we must reserve moral judgment and political opinion to our sensibilities as ethical and responsible members not just of human society, but of Earth's entire biosphere. It is in every individual's interest to act in the common interest of as large a scale of economic inclusion as possible. What these common interests are and how adaptation can promote such common interests are questions to which I return in later chapters.

The Effect of Scale

All economic life takes place in space and time. The way in which economic units perceive and respond to their environment depends critically on their scale—their sizes, numbers, and lifespans—relative to the scale of their surroundings. By scale I mean five attributes: (1) physical size—length, area, volume, or mass of a unit or its environment; (2) reach—the distance, area, or volume traversed by a mobile unit; (3) number—the number of units, events, or environments, or some related measure of a population; (5) duration—the length of time, or lifespan, of economic units, events, or environments; and (5) frequency—the number of units, events, or environments per unit time or per lifespan. These attributes, it will be noticed, can be combined to yield measures of power, the fundamental measure of economic performance.

The range of economic scales is vast. Individual organisms, for example, span a linear size range of a few tenths of micrometers (1 micrometer = 10^{-6} m), the size of a small bacterium, to several hundred meters, as in a large tree, kelp, or fungus. This range encompasses nine orders of magnitude in linear dimensions and at least twenty-three orders of magnitude in volume or mass. There is a comparably huge range in lifespans, from about twenty minutes (4×10^{-5} yr) in a rapidly dividing bacterium to six thousand years or more for a bristlecone pine or a large coral colony. These durations span eight orders of magnitude of time. More inclusive units, of course, encompass still larger spatial scales and longer time scales. The biosphere as a whole has a continuous history of perhaps 3.8×10^9 yr (or 3.8 Ga).

The environment, too, varies dramatically in scale. Some conditions change spatially on the scale of millimeters or less—the concentration of ions across a membrane, for example—whereas others are expressed as major geographic patterns in temperature of solar radiation, spanning thousands of kilometers. Some events are so frequent that they occur on the scale of seconds or less; others—the great extinction-causing catastrophes, for example—come just every 10 million to 100 million years.

How these variations are perceived, and the extent to which economic entities can respond adaptively to them, depend on how commensurate the sizes and lifespans of economic units are with the scales of spatial and temporal variation of the environment. New circumstances tend to be calamitous for an economic unit if they are either very large or very rare relative to the size and lifespan of that unit. They can be incorporated into the adaptive hypothesis of that unit if their spatial and temporal scales more or less match those of the entity in question.

The key to economic control is to affect, and to respond adaptively to, conditions on the longest and largest scales of time and space possible. In effect, this means a fundamental transformation from the perception of change as disturbance to the incorporation of change as adaptive response. How this transformation is brought about is the subject of much of the rest of this book.

Chapter 3

HUMAN AND NONHUMAN ECONOMIES COMPARED

If we are to learn anything about the relationships between humanity and the world's resources from natural history, we must look to fundamental similarities between the economic and political structures that have emerged in human societies on the one hand, and the organization of life as it has unfolded over the course of biological evolution on the other. In chapter 2 I cited examples from both the natural and the human economy to illustrate unifying principles, but now the time has come to confront head-on the question of whether the similarities reflect the same underlying processes or a merely superficial resemblance arising from casual analogy.

Although some human-oriented observers—Jane Jacobs, Robert Wright, Steven Johnson, and many others—have enthusiastically embraced the lessons taught by nonhuman natural order, my discussions with economists and biologists overwhelmingly reveal deep-seated skepticism about the truth and wisdom of turning to principles in one domain and applying them to others. Most skeptics not only view the resemblances as metaphors and analogies rather than as expressions of the same principles operating in different contexts, but they also emphasize the uniqueness of human capacities and institutions relative to the powers and structures that have emerged among other life forms. Moreover, just as I am hobbled by an imperfect grasp of the human economy, scholars dealing with human society and its history have often failed to appreciate the full reach and subtleties of the ecological and evolutionary perspective. Perhaps most importantly, there is the ever-present danger that imperfectly understood ideas from one discipline will be deliberately or unwittingly misused in another. Agendas of racism, ethnic superiority, and eugenics were cloaked in scientific legitimacy by appeals to Darwin's "struggle for life" and Herbert Spencer's unfortunate and misleading phrase "survival of the fittest."

Other skeptics simply fear the power of the principles themselves. Many politically conservative and religious businessmen and free-marketeers who reject and despise biological evolution would be appalled to discover that the economic tenets underlying their belief in the capitalist economy rest firmly on the same foundations as do the principles of evolution and the organization of nonhuman life forms. Conservationists who revile human-economic domination would not be amused to learn that the same imperatives of competition that propel people to expand their exploitation of the world's resources are also responsible for the prolific diversity of life and for the power and beauty of wild nature, which they revere. In his book *Darwin's Dangerous Idea*, Daniel Dennett refers to evolutionary principles as "universal acid" in order to emphasize their all-encompassing explanatory power. One might equally employ the same metaphor to express the power and reach of the economic perspective, which I see as fundamentally identical to the evolutionary world view.[1]

My objective in this book is to expose common ground between human and nonhuman economies, and to build a historical framework by applying shared principles to the empirical evidence of conditions in the present and the past. Accordingly, this chapter argues for the essential similarity between human and nonhuman economic systems. I begin with a brief historical interpretation of why this similarity has been largely ignored until now. I then proceed to an evaluation of the chief reasons for considering human-economic structures and attainments as fundamentally different from those in nature. Yes, we are very powerful, and we can do things no other living things on Earth can do; but, I shall argue, our enormous power and reach have not allowed us to conduct economic life with rules fundamentally different from those that apply to living things generally. The differences are real, but they are differences in amount and scale, not in kind. In effect, economic ideas apply to an enormous diversity of systems, of which ours lies at one extreme of a continuum. In so many words, this chapter is an invitation to human-centered readers to learn about the rest of life, and to naturalists to think about human society and history in biological terms.

Uneconomic Traditions

Doubt about a sweeping synthesis that encompasses human economic affairs as well as the affairs of nature resides within each of the disciplines—ecology, evolutionary biology, paleontology, and economics—from which I draw insights and data in this book. In biology, the predominant way of thinking about organisms has been reductionist and abstract. An ever-increasing proportion of biologists has grown up in suburbs and cities, where highly artificial venues—gardens, zoos, petri dishes, and laboratory cages—are the chief meeting places between people and other living things. Most of us encounter organisms and their remains far from the places in which those creatures lived and evolved, with the inevitable result that we cannot easily envision the problems and opportunities that living things face on an everyday basis. Before 1850 or so, field observations and museum-centered investigations of comparative anatomy and development were largely in the hands of a few wealthy amateurs and of curious clergymen anxious to celebrate the marvelous handiwork of the Creator. Even then, however, field work was something farmers and common laborers did, not gentlemen with the means to engage in the study of natural history. For a brief time, it became fashionable in the halls of the great universities to study organisms and ecosystems in the wild, but in the latter half of the twentieth century these efforts were once again relegated to the smaller, less affluent institutions and to dedicated private enthusiasts. Molecular biologists routinely dismissed classification and anatomy as stamp collecting, as the descriptive accumulation of trivia. Controlled experiments, inspired by Karl Popper's contention that the only good science is hypothesis-driven experimental science, crowded out investigations that were based on observation, comparative studies, and historical reconstruction. Experiments by their nature are executed at small spatial scales and on very short time scales, meaning that phenomena of larger magnitudes were often off limits to those who took Popper too seriously.

An eagerness to emulate physics may have motivated many biologists to view organisms and their parts as abstract entities to be counted, atomized, and measured. Robert MacArthur, an accomplished naturalist as well as an eloquent and forceful proponent of the use of mathematics in biology, expressed this abstraction powerfully to me one day in 1966 after class at Princeton. It would be better, he assured me, not to know the names and characteristics of species if we wanted to identify and explain the deep regularities evident in the distribution of species in nature. Those details would divert attention away from such patterns as the equatorward increase in species numbers, the tendency for most species to be rare and for just a few species to be truly abundant, the simple relationships between habitat area and the number of species, and the general conditions that favor habitat specialization over a jack-of-all-trades strategy.

Population geneticists, meanwhile, thought of genes as abstract independent units, whose behavior could be expressed mathematically in order to assess the roles of selection, small population size, and mutation in evolution. Not to be outdone, scientists in the late 1960s embraced highly abstract protocols for inferring evolutionary relationships in order to make sense of the diversity of life in terms of ancestry and descent. The traits of species were treated as abstract markers, coded as present or absent, and collected in a table along with markers of other species, so that a program could then be applied that would indicate the order of evolutionary branching of the species represented in the table. The characters used as markers tell of shared ancestry, not of the context in which they work or in which their bearers evolved. The organism simply became a collection of markers.

In paleontology, a conscious attempt was initiated in the early 1970s by David Raup, Stephen Jay Gould, and their associates and students to create what they called "nomothetic paleontology," a discipline that was supposed to discover principles applicable regardless of time or place. To them, clades wax and wane not because of changes in circumstances, not because of the particular characteristics of the times and places in which they live and evolve, but because of intrinsic emergent properties such as the probability that a given evolutionary lineage will persist unchanged, divide into two independent lineages, or become extinct. Context in this view is unimportant; indeed, it is a hindrance to the recognition of deeper truths and patterns. Discovering how fossil creatures made a living and how their bodies functioned in ancient circumstances was condemned as irrelevant to the larger pursuit of macroevolutionary patterns of global diversification and extinction.

This intellectual climate was created by, and in turn further stimulated, the collation and analysis of huge databases on the times of first and last appearance of taxa (families and genera) and clades of fossil organisms. In an effort to discover truly global events and trends, analysts combined data from a wide diversity of regions and habitats. In many studies, tropical and temperate groups were undifferentiated; high-level predators were given the same weight and treated in the same way as suspension-feeders and herbivores. Context—environment, lifespan, functional role, adaptive type, and geographic origin—was again sacrificed in order to discern a global signal in the data. Sophisticated statistical protocols were employed to eliminate problems of inconsistent and incomplete preservation of fossils, differences in sampling effort among sites and habitats and regions, and many of the other inconveniences that potentially compromise the reading of the fossil record; but oddly missing from these analyses was the question of whether the results held biological meaning. The global scale is simply too large to be relevant to the fates of ecosystems and biotas, to say nothing of the lives of individual organisms. The diversity and geological longevity of taxa in the tundra have essentially no bearing on economic life in the rain forest or the deep sea. If events truly are global, and if long-term trends really do exist, then analyses specific to habitats or regions or clades should generally reveal the same signals. By ignoring such contexts, abstract lists and their statistical analysis at inappropriately large scales mask the economic processes that account for the events and trends we are trying to describe and explain.[2]

In the belief that current conditions suffice to explain the present-day distribution of species and resources, ecologists have felt little need to consult the past. Although this outlook is now changing, thanks in part to the realization that even apparently pristine ecosystems have been greatly modified by human activity, it is fair to say that environmental biology remains a largely ahistorical discipline.

The field of economics has similarly turned away from history and toward constructing abstract models to describe behavior in markets. Adam Smith and many German and British economists of the late nineteenth and early twentieth centuries keenly understood that history and cultural context explain how trade works, but, in the ahistorical view, the great prestige of mathematics and the appeal of discovering general laws render history obsolete. The emphasis on charting policy further encourages a short-term perspective suitable to the time scale of elections, five-year plans, or the day-to-day fluctuation of the stock market.[3]

Now, there is nothing wrong with databases, abstraction, and modeling. Indeed, these approaches have forced a much more systematic investigation than was fashionable among natural historians, who were (and regrettably often still are) content to argue from anecdotes and examples rather than through systematic inquiry. However, the scientific culture has increasingly frowned on world views that are founded on observations of organisms and their lives, and it has tended to consider history as simple narrative, untestable and without principle, full of particulars but empty of generalities. It is these attitudes that have isolated historians from economists and from many biologists. Yet it is also these attitudes that have allowed specialists to emphasize the unique properties of the systems they study and thus to avoid seeing common features.

Finally, and perhaps most subtly, Western thought—steeped in Judeo-Christian traditions—has generally elevated man above the other creatures, and created a climate against ideas linking human characteristics to those of lesser beings. True, Aristotle recognized humans as animals and felt that we could learn much about ourselves from the other species that live among us. But, whereas European cultures learned a great deal from other societies with which they came into contact beginning in the late eleventh century, the world of animals and plants was one to be overcome and exploited, not one to be studied. In fact, popular literature and children's culture often portrayed animals as having humanlike qualities of intention and emotion. Physics, astronomy, and human anatomy preceded the scientific study of ecology, animal behavior, and even history by hundreds of years.

Economic thought has, of course, not been entirely absent from biological discourse. Robert MacArthur and his associates introduced ideas from microeconomics to explain how animals allocate resources among such competing functions as growth, reproduction, and defense. Thomas Schoener considered costs and benefits when asking under which circumstances animals looking for food should maximize energy intake or should minimize search time. Michael Ghiselin interpreted sex as a mechanism for intensifying competition and as a means to speed up economic improvement. Meanwhile, systems theory and the macroeconomics of supply and demand attracted attention from scientists who were measuring movements of nutrients and energy through such ecosystems as salt marshes, mangrove swamps, and coral reefs.[4]

Oddly enough, these research programs were never properly integrated, and economics remained on the sidelines of biological thought. Ecologists tracking the economic decisions of individual animals were little interested in the systems theory and the macroeconomics of supply and demand that preoccupied ecosystems researchers, who for their part remained oblivious to the particulars of who was doing the producing, the consuming, and the recycling. Evolutionary biologists and paleontologists largely ignored both ecological camps except on rare occasions. David Raup and Adolf Seilacher, for example, applied optimality principles to argue that the makers of ancient trails systematically and efficiently searched for food on the seafloor, and Richard Cowen ingeniously showed that certain Devonian crinoids collected and transported suspended food particles to the mouth along a maximally efficient transport system of grooves in the animals' arms. These attempts, however, did not catch on; they were overwhelmed in a storm of abstraction.[5]

By the same token, economists have by no means ignored biology. From Darwin's time onward, they have been drawn to evolutionary ideas, and have often emphasized parallels between capitalism and the evolutionary struggle for life. In the past twenty years, economists have published at least five books with variations of the words economics and evolution appearing in the title, and many others treat both subjects seriously and at length. Environmental concerns have spawned dozens of books in which economics are discussed in terms of sustainability, although for the most part the authors of these works have essentially no long-term evolutionary perspective on the problem, and the solutions they advocate are couched in explicitly human terms. I shall have more to say on this topic in the final chapter.

In the end, a substantial gulf remains between the social and the natural sciences, not necessarily because scholars don't wish to sail across it, but because they don't know quite how to do it. Each side needs to be convinced that the other has something worth sailing for, and that the promised riches are really there. We want to ensure that the sailors are not wrecked on the hardships that inevitably attend risky intellectual ventures. We need a common language and common principles, and we need to think about trade if we are going to engage in it.

The Uniqueness of Humanity

Doubts about the truth and wisdom of treating the human economy and nonhuman nature as variations on a single common theme crystallize around three overlapping arguments. First, the entities of the human and nonhuman worlds are not analogous. Units in the economies of nature differ from human-economic institutions in composition, function, mechanisms of receiving and transmitting information, resources, currency of exchange, and the nature (and for some observers even the existence) of trade. Second, humans have achieved a capacity to alter environments and to cause destruction more rapidly and on a vastly larger scale than have any other life forms, thanks largely to the accumulation of technology that is not part of the human body. We have profoundly affected every ecosystem on Earth, altered the chemistry of Earth's surface, and globalized economic life. We have therefore broken constraints in place since life's beginning, and can expect to break more as we streak toward ever greater technological complexity. Third, as sentient creatures capable of prediction and of deliberate action, humanity has forged economic systems for which there are no historical precedents and to which new principles apply. The obstacles to direct comparison are therefore unique structures, unparalleled power and reach, and intentionality.[6]

Although the uniqueness of humanity and its institutions is indisputable, I shall argue that our economic system is only one of many variations on a simple common theme: an array of adaptable entities competing for resources. Despite appearances to the contrary, there is nothing in the human sphere that has fundamentally altered the principles common to all other economies of nature.

A crucial step in the unification of the economic and evolutionary perspectives is the establishment of analogies between structures or processes found in the "natural world" and those familiar to us in the human realm. We cannot understand the origins, fates, and performance of economic units without knowing what those units are; and we cannot easily transfer insights from one discipline to another without some confidence that the units we deal with are comparable in their roles. Knowing how units of different scale and inclusiveness respond to change, modify their environments, interact, evolve, and succumb to disruption is essential for the construction of a theory of economic history.

Economists have struggled mightily to pin down the human-economic analogs of such biological concepts as organisms and species. Firms (the units of production in the human economy) and households (the units of production) have been likened to organisms. Joel Mokyr and others see technologies as the human-economic analogs of species in nature, and technological change as speciation (the formation of species). Competition, so the argument goes, is between technologies (or species), not among firms (individuals). In this view, ideas are like mutations, most being unsuccessful as genetic errors are in nature. Selection in the human economy would act at the level of technologies and firms.[7]

To my mind, these attempts at analogy have met with little success and have led to few revealing insights. I would argue instead that a more precise analog of species is human occupations, or ways of making a living. Occupations in this sense combine time and place—the "niche"—with the methods and circumstances of work, the behavior and physiology employed in accomplishing the tasks of life. Among the circumstances of work we might include the enemies and allies with which we interact, the culture of employers, the markets in which we buy or sell, and the resources and rewards that make it possible for a particular way of life to exist.

To biologists who think of species as evolutionary or genealogical units, comprising individuals that share a common gene pool by virtue of common ancestry, this analogy may seem opaque. The key to the analogy between species and occupations lies in the roles that these entities play in the systems in which they live. The conditions of life create a regime of selection that yields adapted individuals whose phenotype reflects not just ancestry and the means whereby one species can be told apart from another, but also the ways of life that members are able to lead thanks to heritable adaptations. Similarly in human society, occupations have phenotypes that allow us to identify and classify individuals economically. Like species, therefore, occupations are economic units that carry out functions.

The analogy is not perfect. Human individuals can change occupations, whereas nonhuman individuals cannot transform from one species to another. The genetic cohesion that defines species evolutionarily and that provides much of the adaptive phenotype is absent in human occupations, membership in which is determined by cultural rather than genetic factors. Human unemployment seems to have no convincing equivalent among species, although it could be argued that species are locally eliminated ("become unemployed") when their ways of life are no longer viable under changed conditions. Generally speaking, membership in all human institutions, including occupations, is far more fluid—less constrained by genealogy—than is membership in species. However, the composition of ecosystems is little constrained. Species are to a significant extent functionally interchangeable, and workable ecosystems are often assembled by a heterogeneous assortment of species that may not share a long history of mutual interaction. The important theme underlying the ecological concept of a species and the economic concept of an occupation is the division of labor that enables more or less specialized, autonomous entities to carry out functions in the larger communities, ecosystems, societies, or nations to which they belong.

In human society, an occupation may entail a wide variety of tasks. A farmer sows seed, harvests crops, and sells at markets, but also repairs equipment, invests profits, and combats weeds and pests. Similarly in nature, species can occupy quite different niches and do quite different things depending on the stage of their life cycle or on individual variation. Just as species overlap greatly in what they do and how they do it, so occupations often differ in very subtle ways; yet both are economic units composed of individuals whose performance and precise tasks vary.

Consider an example. The seastar *Pisaster ochraceus* is a well-known predator in the rocky-shore ecosystem of the northeastern Pacific from Alaska to Baja California. As Robert Paine points out, however, many northern populations feed predominantly on mussels and barnacles, the competitive dominants among space-holders on the midshore, and in so doing hold a key position in that they enable subordinate competitors to settle in the patches the seastars clear. In the southern part of its range, the seastar is more apt to feed on snails. Its ecological role—or "job"—is a little different, but its occupation remains more or less the same: the species is a predator of shore animals. Now, in a system in which the division of labor is finer than in the northeastern Pacific, the occupation of predator might be divided among several species; there could be a specialist seastar feeding only on mussels and barnacles, and another targeting only snails. Specialists may even coexist with generalists.[8]

We encounter similar variation in human occupations. A salesman can sell very different things depending on where he works. In our occupationally highly diverse economy, salesmen tend to specialize on a very narrow range of products. Some people sell just Ford cars and trucks, others Toyotas, others tuxedos, and so on. In less differentiated economies, however, the occupation of salesman covers a lot of merchandise. The precise nature of the job may vary among individuals as well as through an individual's lifetime, just as it does in species in nature; and occasionally (nowadays rather frequently), people change occupations, much as many animals do. Caterpillars have different ecological roles—mainly as herbivores—than do adults, which either feed on fluids (blood and nectar, for example) or pollen, or not at all. Juvenile fish may eat worms and other small animals, while the adults eat large molluscs, crustaceans, or fish. Occupations, like species, are fluid, flexible units whose economic roles are determined by competition and its effects on the division of labor.

In ecosystems, occupations and diversity are expressed as species, whereas in the human economy they emerge through the division of labor among individuals or institutions belonging to a single species. In this sense, the human species differs from all other species, in which diversity is expressed within a single body or among the bodies of very close relatives, as in insect societies. Diversity of the kind we humans have created within a single species no doubt benefits us individually and collectively in promoting exchange, but the competition and division of labor that underpin the emergence of diversity are the same processes that are at work in all the rest of the biosphere.[9]

Production, consumption, and trade are the three activities, or categories of occupation, we most often associate with economic life. In nature, primary producers—plants and other living things that make organic matter from inorganic sources—are the foundation of a food web in which consumers—herbivores, predators, parasites, and decomposers—eat other organisms and their products. Consumers are themselves producers in that other species may eat them or the things they make. In human-economic terms, we tend to think of consumers as buyers and users of the goods and services others produce or supply. Producers grow, make, raise, gather, or prepare goods ranging from foods to fuels, manufactured items, fabrics, and building materials. In addition, many people in modern economies make their living as traders, buying from producers and selling to consumers; while still others render services of every description, ranging from education to fixing cars, governing, preaching, prostitution, construction, nursing, editing, consulting, entertaining, policing, and street cleaning. Traders and most service providers have no real equivalents in nature, although both trade and the rendering of services have certainly evolved as special functions carried out by species or individuals in nonhuman nature. Examples include pollination and seed dispersal by insects and birds, defense by soldier castes in ants and termites and by specialized zooids in the Portuguese man-of-war, and the feeding of larvae by ant workers. As before, we must remember that the various functions and occupations in which people engage are very fluid arrangements that are subject to change within the lifetimes of individuals, whereas in nature the different economic tasks are genetically more prescribed and constrained, being carried out by distinct species or by castes whose fates are determined early in life. At the species level, *Homo sapiens* began as a consumer but has culturally evolved to take on every role that nonhuman species have ever adopted.

The analogy for human technology in the nonhuman domain of life is morphology and physiology, or organic architecture. It is technology in this broad sense that enables people, societies, cells, organisms, and indeed ecosystems to convert energy and raw materials to biological (or economic) work. In organisms, technology is part of the body; in people, it is an extension—mechanical, intellectual, and cultural—that we design and that, at least figuratively speaking, takes on a life of its own. In both cases, technology evolves; in organisms it does so largely through natural selection, in humans by engineering and market forces. Some animals do, of course, use tools—sea otters breaking abalone shells on rocks, some woodpeckers using twigs to pry for insects—and others build nests, burrows, hives, and other external structures. The move toward external technology by people is therefore a matter of degree, albeit a very considerable one.[10]

Analogs are increasingly difficult to identify in the more inclusive categories in the economic hierarchy. All the units at these more inclusive scales of organization are characterized by a degree of cooperation among parts, and thus by suppression of competition within the larger units. The partnerships that emerge are the unique and contingent products of particular conditions that bring units with common interests together. It doesn't matter that a mutualistic relationship between a plant and its pollinator has no obvious equivalent in the human domain, or that a stock market where investors buy and sell shares in a venture has no equal in nature. What matters is that partnerships develop according to how effectively tasks are accomplished. What those tasks are, and who gets together with whom to perform them, are specific to each system.

Nevertheless, skeptics might argue that some human institutions behave and function in ways that no collective enterprise in the nonhuman domain does. Geoffrey Hodgson, for example, worries that "the natural selection of institutions is not like that in the biotic world, partly because of the way in which institutions become locked in by conformist and other self-reinforcing mechanisms." No observer would deny that institutions, once founded, take on a life of their own, incorporating a culture that often becomes conservative and risk-averse. The Roman Catholic Church instituted a policy of celibacy for its clergy and has steadfastly maintained it for almost eight hundred years in the face of a changing social milieu and mounting evidence of sexual misconduct. Moreover, the Church has, like many other social institutions, become an agency generally opposed to change. But how different is this from what goes on in nature? Many aspects of form in plants and animals remain highly conservative, strongly channeling change to just a few pathways. The conservative nature of human institutions and of development in organisms emerges as the inevitable consequence of the effectiveness of the underlying structures. Whether fashioned by natural selection or by intentional design, these structures arise and survive because they work well. Other structures and other body plans might work much better, but as long as the current systems function well and prototypes of potentially better systems must compete with them, the incumbent entities are secure despite their imperfections and their conservative tendencies. Natural selection, like selection in the human economy, reflects adaptation and constraint as imposed by the market forces of competition, cooperation, supply, and demand.[11]

In most human societies, individuals and groups are paid for providing goods and services. This form of exchange might appear to be a unique type of trade, without analogy in nature, but payment is in fact very widespread in other species. Plants feed nectar to the birds and insects that render the service of cross-pollination. Many plants offer nutritional rewards to ants for protecting them against herbivores. Payment may not be in the form of money, which can be converted in subsequent trade for other goods or services, but payment for one commodity with another is hardly unique to human society.

Information and its transfer provide another dimension of human uniqueness. Most of our information comes through language and learning. It is transmitted culturally, circumventing traditional gene-based inheritance. Transmission is rapid, and involves amounts of information vastly greater than what can be encoded in any set of instructions, genetic or otherwise. The flow of information is not limited to streaming from one generation to the next, as it is in biological reproduction and in evolutionary lines of descent.

This contrast is much less sharp than it might at first appear to be. In all biological systems, the linear sequence of genes is greatly enriched by the ability of living things to communicate, to learn, and to respond directly to their surroundings without the slow process of genetically based adaptation. If organisms were constrained to adapt only through genetic change, adaptation to events at time scales much shorter than the generation time of the individual would be impossible. Methods of transmitting information vary greatly and are fascinating in their own right, but they do not materially affect the nature of economic activity except in speed and volume. No matter how information is acquired, replicated, and modified, its users are still subject to competition, cooperation, trade-offs, adaptation, feedbacks, and inequality. Humans have simply and dramatically expanded the range and scale of the economy of information, but the application and modification of knowledge in which we excel have been part of life since its beginning.

There can be no doubt that humans, individually and collectively, have developed an unprecedented capacity to change and control the world. Humanity exploits as much as 40 percent of the annually produced biomass on land and perhaps 25 percent of that produced in the ocean. These estimates, which are subject to large errors, likely vary among ecosystems, being especially high for productive grasslands and the zones of upwelling in coastal marine waters, but they illustrate our global reach and power. Through intensive agriculture, made possible by the invention of artificial fertilizers and pesticides, we now fix more nitrogen than all other terrestrial organisms combined, and we export vast amounts of nitrate and phosphates to lakes, rivers, and marine coastal waters, where the nutritional environment for resident organisms has changed in nearly all parts of the world. The burning of fossil fuels—an activity of the last five hundred years at most—has created vast industry and rapid transport, but, together with the burning of terrestrial biomass to clear land for pasture and agriculture, it has also raised the concentration of the greenhouse gas carbon dioxide by 30 percent over preindustrial levels prevailing just three hundred years ago. Furthermore, we have pumped huge amounts of nitrogen oxides, sulfur dioxide, hydrofluorocarbons, and other pollutants into the atmosphere, and complemented the natural dust content of the atmosphere with particulate matter sufficiently abundant to alter patterns of cloud formation and rainfall. Nearly a quarter of the land surface is devoted to urban, agricultural, or grazing use. Human-set fires have been large and numerous enough during the last ten thousand years to leave an indelible mark on the carbon content of ocean sediments off the west coast of Africa. Humans appear to be responsible for most of the global warming of the past century.[12]

Commerce, another human economic activity that has blossomed during the last five hundred years, has resulted in the accidental or purposeful transport of thousands of species to regions where the existing biotas were not adapted to the newcomers. At least twenty of these species, including animals as well as some disease agents, have caused extinction among native species, especially on small isolated islands and in lakes. By hunting and through environmental modification, we have directly brought about the demise of many other species, especially warm-blooded mammals and birds and some large trees. Few if any species besides our own can claim such awesome ecological power.[13]

The progressive harnessing of domesticated animals, flowing water, wind, fossil fuel, and nuclear energy chronicles an extraordinary increase in individual and collective human power. Inanimate per capita energy use climbed from 0.9 megawatt hours in 1860 to 8.2 megawatt hours by 1950. From 1900 to 1990, the average per capita consumption of energy rose by a factor of five, and the human population increased threefold. The last one hundred years have therefore witnessed a global rise in the use of energy by humanity by a factor of fifteen or more.[14]

Humans command more power than has any species in the previous three and a half billion years of Earth history. This power is expressed both in the short term, by modifying old ecosystems and creating new ones, and in the long run, affecting limitations and opportunities for evolution of other life forms. Very few species are able to expand their populations in our presence, meaning that most species have been forced into an economic status quo. Only a minority of species—weeds, pathogens, and species that live in small temporary habitats typical of cities and suburbs—have any real chance of continued evolution and adaptation. We have, in other words, achieved a degree of control that ecological theory, founded on species other than our own, did not predict. We have greatly expanded opportunities for ourselves and for the species we cultivate even as we have imposed stifling limitations on the population sizes and evolutionary potential of most others.

Tempting as it is to view this human achievement as a fundamental departure from the natural order, we must not forget that humans and their technology, like other life forms and their technologies before us, are the products of evolution, a process of adaptation by economic entities in an environment of limitations and opportunities. Indeed the economic innovations introduced by people are only the latest in a long series of revolutions, which extend back to the very beginnings of life. Long before our species arrived on the scene and began to invent nongenetic technology, organisms were individually and collectively modifying their environments, learning from their surroundings, and influencing the course of history. I shall tell this history in chapter 9, but for now a few points illustrating the deep roots of our capacities are worth emphasizing.

If the simple life forms of the Archean—the era in Earth history before 2.5 billion years ago (Ga)—could have passed laws against pollution, their principal target would surely have been oxygen. Molecular oxygen (O_2), the product of oxidative photosynthesis in Cyanobacteria, was a lethal toxin, and in its reactive forms (free radicals such as O_2^-) it remains so in all organisms today. For most lineages of life up to that time, survival meant living away from the oxygen-producing culprits—in dark places, or deep within microbial mats—but a minority of others eventually adapted to the new circumstances by exploiting oxygen in respiration, a new and very powerful form of metabolism, which allowed the extraction of more than ten times as much energy from organic molecules as was possible in the older forms of metabolism that did not use oxygen.

Pollution has remained a pervasive phenomenon ever since. The plants living 320 to 290 million years ago (Ma) in Carboniferous equatorial forests of Europe and North America were so laden with lignins, tannins, and other decay-resistant compounds that their remains accumulated like so much unbiodegradable plastic debris. As a result, most of the carbon in these plants was lost to the living ecosystem, to be recycled back to the biosphere only when humans retrieved it and burned it as coal. Redwoods, eucalyptus, and pines drop toxic leaves and bark, which prevent the establishment of seedlings of potential competitors belonging to other tree species. Marine suspension-feeding animals that lay motionless on soft muddy seafloors surely did not welcome the evolution of deeply and rapidly burrowing animals, which destabilized the sediment surface and caused the immobile surface-dwellers to sink and be smothered. This immobile mode of life, so common in the Paleozoic era among brachiopods, bryozoans, corals, and snails, became obsolete in most places during later ages as sediments were increasingly disturbed by animals in search of food or safety.[15]

Were it not for the photosynthesis and the chemical breakdown of rock that plants engage in, the Earth's temperature would on average be some 18°C warmer than it is today. It is even possible that the unparalleled rise in abundance and size of land plants during the Devonian and Carboniferous periods, beginning about 400 Ma, decreased the concentration of carbon dioxide tenfold, cooling Earth to the point where large parts of the southern hemisphere became ice covered.[16]

The evolution of organisms capable of secreting mineralized skeletons beginning about 565 Ma during the latest Proterozoic eon, ushered in a fundamental increase in the control of silicon and calcium by life. Minerals of these elements formed with little or no intervention by organisms through crystal growth on the seafloor before organisms put them to defensive use in skeletons. Those mineralizing creatures that built reefs of accumulated calcium carbonate skeletons unwittingly created a complex topography suitable for thousands of species to find shelter. When silica-dependent planktonic diatoms became abundant during the Cretaceous period, about 80 Ma, their efficient extraction of this mineral from seawater robbed other silica-dependent organisms—notably sponges and radiolarians—of their silica supply. Siliceous sponges became restricted to deep waters, where the mineral remained abundant, whereas the shells of radiolarians became thinner through time. Mineralization was thus favorable for some organisms and unfavorable for others.[17]

These examples are not meant to excuse or to legitimize the pollution and destruction that humans are causing, but they do illustrate that these regrettable activities, rapid and unprecedentedly intense as they are, follow in a very long line of environment-altering activities of living entities. What requires an explanation, and a solution, is not the fact that humans affect the biosphere, but the rapidity and extent of that modification.

A hallmark of the modern human economy that might be considered unique is globalization. Long-distance trade in the human economy has been with us for thousands of years, but it involved mainly goods and slaves. In the twentieth century, the scale of globalization vastly increased, especially after 1960, and expanded to include free labor and finance capital. Wages, prices, investment, production, and consumption have all become global phenomena, over which political entities and even the great corporations have only limited control. Economic conditions in one part of the world are closely tied to those elsewhere, and the ideal of self-sufficiency embraced by those who fear the economic unsustainability of current practices has for the most part given way to mutual dependence in a swirling economic chaos of rapid change. Plentiful supplies and cheap fuel have lowered transport costs so much that efficient high-wage farm production in developed countries cannot compete with lower-cost production based on rock-bottom wages in economically less developed parts of the world. Supply and demand are now global notions, not local or regional ones. At the same time, however, weather remains an essentially local to regional phenomenon, as do at least some outbreaks of pests and diseases. Globalization has therefore led to an incongruity between the scale of economic activity, which is global, and the scale of weather-sensitive production, which in many cases has remained regional.[18]

Although the volume and spatial scale of the present human economy have no parallels in nature, globalization is by no means unknown in the prehuman biosphere. From time to time, species have expanded their ranges across previously unbreachable land or ocean barriers. Asian mammals flooded into North America across Beringia (Alaska and eastern Siberia) at various times beginning 55 Ma. When, during late Miocene time (5.3 Ma), the Beringian land bridge was severed by the opening of Bering Strait, hundreds of marine species invaded from the North Pacific to the Arctic and Atlantic Oceans. A little later, about 3.1 to 3.6 Ma, the emergence of a continuous isthmus in Central America enabled North American species to spread into South America, and South American plants and animals to spread into Central and North America.[19]

More dramatic still was the advent of long-distance migration by birds, mammals, fish, sea turtles, and even insects. Some birds annually trek between the north and south polar regions; albatrosses regularly circle the Antarctic; gray whales feed in the Bering Sea and breed in the lagoons of Baja California. Monarch butterflies migrate each year from breeding grounds in the northern United States and Canada, where they lay their eggs and raise their first larvae, to Mexico, where they overwinter. Tuna, green turtles, and the larvae of thousands of marine species make long-distance migrations, either under their own steam or passively swept along in ocean currents, across great stretches of ocean. Long-distance migration at such scales means that ecosystems in one part of the world become economically tied to ecosystems elsewhere. This kind of dependence means that the correspondence between local conditions of supply and production on the one hand, and patterns of consumption and adaptation on the other, has become increasingly disrupted and blurred as top-level consumer species traversed greater distances while producers stayed put. The northern coniferous forest, where surplus-creating production is brief thanks to cool summers and a short frost-free season, sustain summer densities of consumers and levels of consumption that are highly inflated thanks to the mass migration of summer-breeding birds from the more productive ecosystems at lower latitudes, where the birds spend the winter. Imbalances between production and consumption arise in this way; and, just as in the modern human economy, the spatial scales of production and consumption are no longer the same.[20]

Perhaps the most serious objection to the idea that the economics of humans and of nature obey the same principles is that people, unlike other life forms, can make plans and intentionally engineer elaborate technology. The self-organized units of the rest of nature evolve largely through modification of previously established structures; they build on precedence, they elaborate on themes that stabilized long ago. For example, the pattern of development of animal embryos comprises a series of interactions among cells and among groups of cells through predictable stages. Much of morphological evolution is achieved by changing the time, durations, and locations of these interactions; but much of the staggering variety we see among living as well as fossil animals is the result of endless variation on a very few fundamental themes. Sometimes these variations stray very far indeed from the theme. Octopuses look very little like the tiny, slow-moving mollusc, whose cap-shaped shell covers the soft parts of the body as the animals clamps onto a rock surface, that was its very distant ancestor. It stretches the imagination to think that arthropods, characterized by an external skeleton that molts periodically as the animal grows, should include among their ranks acorn barnacles, which have converged on the shell-bearing molluscan body plan by evolving a permanent exoskeleton whose growth, like that of the molluscan shell, occurs along the margins of its component plates.

Human engineers, by contrast, can start with completely novel structures, or employ a previously unused principle to create structure. They may well use older, tested technology as a point of departure—wheels of carts became wheels of gasoline-powered automobiles, for example—but much human technological innovation represents a break with established form. The same intentionality applies to the creation of novel social institutions or political systems. Moreover, with our capacity to accumulate knowledge of the world through scientific observation, experimentation, and inference, we can predict change and sometimes plan for it. The slow trial-and-error process that characterizes natural selection is made vastly more efficient; it is speeded up, even circumvented, when we allow theoretical predictions to dictate solutions to problems.[21]

Economists find it difficult to bridge the gap between the economies of nature, in which intentional regulation and market preferences appear to be absent, and the modern market economies of the last three and a half centuries, where intentional preferences by consumers, advertisement by sellers, deliberate regulation, and conscious business decisions play decisive roles. It is gaps like these—between ancestral and highly derived conditions—that nudge the skeptics of evolution toward the view that such complex and high-precision structures as the eye or the brain cannot have evolved from simpler versions without guidance from an intelligent god. Preference by human consumers is often conscious, whereas that of other species presumably is not; but it is preference, or at least selective consumption, just the same, and in both cases, preferences can change, through advertisement and social pressure in modern civilization or by unintentional ecological and evolutionary agencies in nature.

Activity in the modern economy proceeds largely through informed, deliberate decisions made by individuals and groups in the private and public sectors. Laws governing property rights and patents, which evolved through government action from the late seventeenth century onward, crucially stabilized and regulated ownership and its transfer, providing a more predictable context for investors and entrepreneurs. Laws, contracts, business plans, trade agreements, tax policies, price setting, subsidies, and tinkering with interest rates are other expressions of intentional regulation and economic decision-making that would seem to have no biological equivalents. And yet, nature is full of examples of regulation, of economic restraint, of subsidies, and of "prices" being set; but all of these activities came about not through deliberate policy-making in government and business, but through a self-organized accommodation of species sorting out their various ecological roles and coexisting with each other. The transition between the natural and the human economy largely involves a shift from bottom-up self-organization to top-down intentional policy. As a result, the rate and reach of economic activity, and the range of economic institutions, have increased; but nothing has fundamentally changed. Regulation remains regulation, prices remain prices, the market of supply and demand remains just that, regardless of who the actors and decision-makers are, regardless of how and where the market forces are generated.

Without question, our inventiveness and anticipatory capacity have allowed the human species to take effective control of the biosphere as no species has done before us. Yet, we should not look upon these powers as qualitatively different from those already evident during prehuman times. The greatest evolutionary innovations involved the combination and integration of previously independent life forms in a way that violates the principle of modification of established themes. The eukaryotic cell arose through the integration of separate bacterial and archaeal lineages. This cell does things that none of its components could do; it is not merely another variation on the theme of a bacterial or an archaeal cell. The emergence of complex sociality in insects, mammals, and even a few shrimps represents another organizational innovation that transcends previously established norms. With such advancement, organisms gain unprecedented control of their environment, as well as the technology to expand the range of exploitable resources.

In its full-blown expression, the ability to make and execute plans and intentionally to design structure seems fundamentally different from the passive evolution of emergent biological order as represented by a eukaryotic cell or a beehive. But I see a seamless transition. The vertebrate immune system is so flexibly constructed that it can neutralize a whole range of disease agents, including many whose precise character cannot be anticipated. Through a process of diffuse "learning," it produces a vast array of antibodies, far too many to be specified by the genetic instructions common to the immune system's cells. Similarly, the nervous system of animals deals flexibly and quickly with a very large range of circumstances, from everyday requirements and tasks to the less common but crucially important challenges posed by enemies and mates. Adaptation in general is the formation, and continued testing, of hypotheses about the environment. With more information available, improvement of the hypotheses allows an increasingly broad range of conditions to be incorporated. Learning in the way the immune system and the nervous system do it—with parallel processing using a vast multitude of simple components each endowed with a few simple instructions—is an emergent capacity that produces what we would call intelligence. The result is not very different from intentional planning, except that planning is faster, more directed, and top down, being executed by a central authority—the brain, a legislative body, an engineer, a teacher, a corporation, a homemaker—rather than by a diffuse process emanating from the coordination of simple components. In fact, central authority is itself the product of such coordination of diffuse elements. Intelligent design such as that practiced by humans, or such as that ascribed by some humans to the higher authority of an omnipotent and omniscient god, is a derived adaptation that has made innovation more rapid and more wide ranging but whose roots extend all the way back to the origins of life.

Ethics and Morality

Besides the capacity for intelligent design, humans have culturally evolved an elaborate system of ethics and morality, a code of individual and collective conduct that is often expressed in religious terms in which the rules are laid down and enforced by supernatural powers and by their human spokesmen. What sets human codes of ethics and morality apart from the behavior of other animals is that, at their best, they allow people who are not closely related to each other to cooperate and to live in peace. In most species, the degree of genetic relatedness determines the extent of trade, cooperation, and altruism among individuals. An individual is more likely to transmit its genes to the next generation if it helps its kin, which share many of the same genes, to meet the challenges of life. The more distant two individuals are genetically, the less likely they are to act altruistically toward each other. This principle operates also in human society—we feel more compelled to help immediate family members than strangers—but human cultures have, in addition, built social contracts that extend altruism more broadly, sometimes even to nonhuman creatures. There are laws enforced by governments; there are religious laws and customs originating with a few holy men but enforced by an extensive network of local clergy. The doctrines that emerge create a more or less cohesive society, which because of its common culture acts as a potent competitive unit relative to other societies.

Whether the unifying rules are right or wrong does not matter, and in any case this depends on who is doing the judging. What matters is that the rules work, or that they are perceived to work. They comprise the collective glue that allows societies to act as units competing for resources and power.[22]

Although the existence of codes of social behavior has long been thought to be unique to humans, and indeed to place humans above all other forms of life, Frans de Waal has made a compelling case from his detailed observations of nonhuman primates and other animals that the prototypes for our codes existed long before humans evolved. Birds and mammals, but not fish or most invertebrates, exhibit individual attachments to each other. Some apes have the capacity to empathize, to understand the handicaps of other individuals and to respond accordingly, as humans do. Assisting one another during crises, an element of sacrificial altruism, occurs in dolphins, elephants, lemurs, and chimpanzees. Sharing food, a condition again requiring unselfish behavior, occurs in chimpanzees, wolves, mongooses, brown hyenas, vampire bats, marmoset monkeys, and gibbons. This allows the exploitation of foods that would otherwise be too large, too rapidly dispersed, too rapidly spoiling, or too unpredictably available for one individual to find, transport, and consume. Altruism like this flourishes only if cheating is costly. Conflict resolution, punishment, and retaliation are therefore necessary ingredients for the elaboration of moral systems that bind animals together in groups in which selfishness is to some extent suppressed in favor of the common good. Monkeys, gorillas, and chimpanzees are capable of conflict resolution by a social leader, an impartial but powerful arbiter.

In de Waal's view, the seeds of human morality were sown in primates living in a highly dangerous world in which cooperation paid huge dividends. Not only did it allow for food sharing, but it also enabled individually relatively defenseless animals to mount an effective collective defense against an array of dangerous predators, with which Africa—the place of origin for the hominid line and for the modern human species—is amply populated. Other animals, notably birds and mammals, evolved many of the ingredients with which moral codes are constructed, but apparently only the human line fashioned the body of rules and regulations, laws and customs, that with considerable variation among cultures characterizes human society.[23]

The social contracts represented by legal codes and by systems of ethical and moral behavior enable humans to make decisions and policies with long-range benefits and not simply to favor short-term gains. Having such systems in place undoubtedly makes this process of engineering long-term benefits more efficient, but it does not introduce an entirely new capacity. Such a capacity in fact can be seen in economies of the past as the manifestation of emergent aspects of natural selection. As Timothy Lenton has pointed out, an economic activity that promotes growth will be favored in the short term over one that does not, much as a catalyzed reaction will be favored over an uncatalyzed one. Growth-enhancing activity by one entity has long-term advantages as well, in that it promotes growth and expansion in many other entities. By fertilizing the soil or preventing the loss of nutrients from the soil, for example, a tree will benefit not only its own growth and that of its offspring, but also the well-being of its neighbors, which in many forests are apt to belong to different species. By contrast, attributes that create monopolies or that destroy neighborhoods will not last long simply because the entities that work this way will themselves not persist.

In other words, actions and codes that look as if they are designed for the common good emerge because they confer on their bearers or creators not only long-term advantages, but the kinds of short-term advantages which natural selection can enhance. The long-term benefits cannot be selected for directly; natural selection is very much about advantages here and now, not in the distant future. Traits conferring long-term advantages emerge because they also work well in the lives of individuals and produce positive feedbacks that enhance the economic well-being for a large number of other participants in the economy. By creating a shared common interest, selfish benefits become traits for the common good.[24]

Self-awareness and free will—perhaps the most unusual among human beings' unique traits—empower us to act knowingly, autonomously, and directedly to enhance our own survival and reproduction as well as that of the groups to which we belong. And yet these traits, too, connect with the capacities that our ancestors are likely to have had and that other animals possess. Life in a social context may have set the stage for the evolution of self-awareness. As Steven Johnson speculates, "It may turn out that the brain gets to that self-awareness by first predicting the behavior of neurons residing in other brains."

Successfully "reading" the actions and intentions of others may thus predispose us to develop the ability to evaluate ourselves and to imbue our actions with conscious deliberation.[25]

The conclusion I draw from these arguments is that the characteristics making humans and their societies unique have emerged by adaptive processes from prototypes existing long before we ever came on the economic scene. Humans have uniquely great powers of communication, external technology, anticipation, and sociality, but none of these powers represents a fundamental break with precedence. They are elaborations along directions discernible throughout economic history, allowing us to speed up economic life dramatically and to expand into every part of the biosphere and well beyond. Importantly, all the processes that characterize other economies of life remain untouched. We are as susceptible to competition, trade-offs, feedbacks, and economic inequality as any of our contemporary or antecedent species. Whatever our economic future is to be, it will certainly be based on these realities. This is the rationale for studying how present and past economic systems work and how they respond to opportunities and constraints.

Chapter 4

THE ECONOMICS OF EVERYDAY: CONSUMPTION AND THE ROLE OF ENEMIES IN NATURE

ECONOMICS IS ALL ABOUT MAKING A LIVING. It is the business of every day: getting enough to eat, paying for it, competing with neighbors, investing in the future, and, above all, staying alive. We do these things by interacting with other life forms in a world replete with risks, rewards, costs, benefits, opportunities, limitations, and uncertainties. I can't think of a better way to explore the nature and consequences of economic interactions than to begin with consumption. Every organism is potentially exposed to it, and many life forms engage in it as herbivores, parasites, predators, scavengers, filter-feeders, detritus-feeders, and decomposers. Consumption illustrates the simple rules by which all economic entities live as well as the subtle and complex consequences that interactions have for consumers, producers, and the biosphere of which we are all part.

I shall use the term consumption in a broad sense to mean the eating of a living or dead organism, or part of an organism, by another living thing. Predation encompasses instances where a single large perpetrator—usually an animal, but sometimes a fungus, single-celled protistan, or even a plant—eats part or all of another animal. If the victim is a living plant, the perpetrator is classified as a herbivore. In parasitism, the consumer is typically small, and leaves its larger victim alive as it slowly eats away at the body or withdraws fluids from the host. If victims are strained from water passing through an animal's filtering apparatus or caught on collecting organs, the consumer is called a filter-feeder or suspension-feeder. Decomposers, detritus-feeders, and deposit-feeders glean bacteria or organic matter from soil, sediments, and carcasses. Scavengers are animals that eat large dead bodies.

Dividing lines among these categories are often blurred and artificial. Many predators and herbivores eat only part of their victims' bodies—a limb, tail, siphon, leaf, or fruiting body—yet we do not normally think of them as parasites. Many suspension-feeders, such as corals, ensnare and immobilize their tiny food organisms, and could be classified as micropredators. Tiny clams dwelling on mud surfaces pick up particles one by one from the sediment surface or from just beneath that surface. Here, the distinction among filter-feeding, deposit-feeding, and detritivory ceases to have any meaning. Ultimately, all these methods come down to the same thing. Whether you tap into a host's digestive system, graze plants in a meadow, browse a colony of corals, or gorge on a rat you have just killed and dismembered, you are a consumer of living things. You depend on life for your own livelihood.

I begin by considering how both consumer and victim are adapted to the risks and realities of eating. The consumer clearly has the nutritional upper hand in an episode of consumption in that it gains energy, but does this remain true on the longer economic time scale of evolution? I shall argue that it does. Consumers in particular, and enemies in general—that is, a category including predators, competitors, and parasites—exercise disproportionate evolutionary control over other species. They also affect where individuals carry out the tasks of life and, by extension, how ecosystems function. Enemy-controlled evolution, or escalation, is therefore a form of positive feedback that implies pervasive consumer control of economic life and economic change. Coevolution—adaptation of consumer and victim to each other—also occurs, but even here one party—typically the more powerful, energetically more demanding consumer—usually gains more (or loses less) than the other. In effect, the economic world is constructed by and for the rich and the powerful.

In human society, the rich and powerful are concentrated among the ranks of producers and traders rather than among consumers, largely because the former have greater knowledge about, and a greater ability to manipulate, markets. Within the human species, the distinction between producer and consumer is blurred, but the human species as a whole exemplifies perfectly the pattern that consumers have the upper economic hand over producers, for humanity is certainly the most powerful economic entity yet to have evolved on Earth.

Three attributes of consumption make this economic activity an important motor of biological evolution and economic history. First, every species, and potentially every individual, is subject to consumption in one or more of its life stages. For all intents and purposes, therefore, consumption is universal in ecosystems. Even bacteria can penetrate and partially consume other bacteria, suggesting that simple forms of consumption of one organism by another have been possible ever since the earliest forms of life evolved some 3.5 billion years ago. Second, the average victim is likely to encounter a potential consumer at least once in its lifetime. By an encounter I mean either the recognition by the victim that a consumer is nearby or an awareness by the consumer that an item of food is at hand. Consumption is thus within the reach of adaptation of populations, and not a rare event that remains beyond the realm of easy prediction on the short time scale of individuals. Third, attempts at consumption of living victims are often unsuccessful. If every attack were to result in the victim's death, then the individuals in the victim population that would produce descendants would be only those that had had no brushes with the agents eliciting anticonsumer adaptation. The consumer's imperfection as seen from the victim's point of view is thus key to the evolution of defense and to the evolutionary importance of consumption.[1]

The relationship between consumer and victim is quite literally a power struggle. Among consumers, competition is keen, with the result that an increase in the rate and reach of recognizing, pursuing, and subduing victims will be potentially advantageous. Among living victims, the threat of consumption is intense, with the result that any attribute that decreases a consumer's power and reach will be beneficial. It is to the benefit of consumers to increase power and the distance over which they exercise it; it is to the benefit of victims to limit that power and to increase their own. The pathways adopted toward these ends have important long-term consequences for both parties and for the economy they both inhabit and create. I shall look at this struggle, and at adaptations to it, first from the perspective of the victims, then from the viewpoint of the consumers.

Adaptations of the Victims

We live in a less than perfect world, and for most of us this is a very good thing indeed. If predators were generally successful in the act of predation, prey would survive mainly by good luck rather than by good traits; but in fact predators very often fail in their attempts to find, catch, and kill their victims, meaning that the prey can survive attacks by virtue of antipredatory adaptation. In fact, antipredatory adaptation requires that, at one or another stage of a predatory attack, predators fall short of perfection, and survival depends on the possession of traits consistently associated with success in remaining undetected, appearing undesirable, fleeing, confusing, or resisting.[2]

The first line of defense is to remain undetected or to look undesirable. The adaptations to thwart predators during this early phase of predation are extremely varied. A moth may look like an inedible leaf on the ground. A butterfly may display bright colors that a predator associates with aggressive or poisonous prey. So-called Batesian mimics are tasty prey that closely resemble obnoxious ones. Some passion-vines of the family Passifloraceae have a pattern of spots on the leaves that to a female *Heliconius* butterfly hunting for a place to lay her eggs look like eggs that another female has already deposited. Thus fooled, she will look elsewhere, sparing the vines from attack by the caterpillars that will hatch from real eggs. When they detect the presence of a swimming predaceous copepod crustacean, small planktonic cladoceran crustaceans (water fleas) stop swimming—stop making "noise" that the copepods are able to detect with their antennae—and sink silently through the water to safety. Many planktonic animals are transparent, so that visually hunting predators cannot see them. Lycaenid butterflies have an adornment on the abdomen that resembles a head. Visual predators like birds would therefore be tempted to attack the butterfly at the abdomen rather than at the true head. The shells of many reef-dwelling snails, which live on open rock surfaces where a host of crabs and fish patrol constantly for victims, are covered by a thick coating of rocklike coralline algae. As a result, the snails blend in chemically and visually with their rocky surroundings, which are also heavily coated by these wave-resistant and grazing-resistant seaweeds. Even for predators like me, who rely on tactile cues to find prey, these encrusted snails often elude detection. A similar effect is obtained by some sea urchins and crabs, which cover their exposed surfaces with stones, shells, bits of seaweed, and other debris. Many small snails and clams hermetically seal the vital organs in an impermeable shell, so that potential attackers are unable to detect them chemically. Even when swallowed whole by seastars or fish, these molluscs can survive for periods of two weeks or more in the digestive system of their predators, and resume a normal existence when voided.[3]

Simply remaining still without camouflage protects many animals. In trials with the tree frog *Hyla cinerea* in the laboratory, 70.5 percent of attacks by the frogs on flies occurred when flies were walking, an activity which engaged flies 53 percent of their time. Some 3 percent of attacks occurred when the flies were flying, which they did 2.25 percent of the time. Mosquitoes flew for only 4.77 percent of the time during which they were observed, but 33.8 percent of predation attacks by *H. cinerea* occurred during this activity.[4]

Remaining undetectable is a potentially effective defense, but it does not do much for high economic status. In most cases, it is linked to a motionless existence. For animals, this means inactivity—no feeding, no fighting, no mating—a passive defense rather than activity in the presence of danger. Inactivity translates into powerlessness. For long-term survival, this works very well, because the biosphere is replete with habitats and ways of life that demand, or at least do not discourage, a low-energy existence. Inactivity indeed characterizes one or more life stages of almost all organisms—eggs, pupae, seeds, cysts, and spores—and the daily cycle of light and dark enables or requires individuals to remain inactive for long periods. Seasonal contrasts impose similar rhythms on somewhat longer time scales. As long as periods of quiescence are punctuated by activity, however, individuals can engage in active competition with others for resources. The longer the inactivity, the more restricted is the power to compete, and the more an individual is forced into the interstices of the biosphere.

There is a curious exception to this generalization. The crab-spider *Thomisus onustus* in Europe preys on bees that visit flowers. Its success hinges on its color, which matches that of the flower on which it sits. The spider is thus invisible to both its predators—birds, for example—and its prey. There are likely to be more such examples in nature, but the crab-spider illustrates nicely how restricted the conditions are for an animal relying on motionlessness to be an effective consumer.[5]

Animals advertising their obnoxiousness or their resemblance to undesirable elements are less constrained in activity than are other species relying primarily on first-line defenses. The warnings are not, however, maintained only by discouraging predators at the discovery phase of an attack. Work on distasteful larval and adult butterflies that display warning coloration has demonstrated that predators like birds often deliver trial bites to these victims, and then let go in disgust or pain. The long-distance warning to the predator is therefore an effective means for the prey to reduce injury or the probability of death. Not surprisingly, insects with warning coloration tend to have robust bodies that are quite resistant to such trial attacks. In Sweden, for example, bites by the great tit on brightly colored and noxious caterpillars of the swallowtail butterfly *Papilio machaon* rarely result in injury.[6]

Many warnings, especially those in a social environment, advertise to the predator that the prey is aware of danger and that an attack would therefore most likely be fruitless. The individual broadcasting the warning receives a certain degree of protection, but so do the other prey individuals who detect the warning. This type of warning system has the effect of allowing potential victims to remain active.[7]

If prey are detected and desirable, predators will try to catch them. Fleeing—the second line of defense—is obviously not an option for plants and for many marine animals that lead sedentary lives attached to objects on the seafloor, but for many animals it offers the most effective means of avoiding predation. Potential victims crawl, run, jump, swim, fly, climb, and burrow away from predators once they have sensed danger. Good sense organs to provide early warning, an ability to move quickly or erratically, and for most animals the presence of nearby refuges where predators cannot or will not go are essential to effective escape. Moreover, escape must be done with a minimum of further sensory cues to the predator or, failing that, with a maximum of confusion.

Fleeing can be very effective. At running speeds of 70 km/h and 100 km/h, zebra and gazelle respectively can evade hyenas, lions, and even cheetahs, as long as they are far enough away from the predators when detected. Working in the Ngorongoro Crater of Tanzania, Hans Kruuk reported that Thompson's gazelle and wildebeest successfully fled from spotted hyenas during 50–75 percent of chases, depending on circumstances. These hoofed mammals have long legs, a long stride, and considerable endurance. These same animals have comparably high rates of success when being stalked and chased by lions and cheetahs. Escape is just as effective in many other predator-prey interactions. On Isle Royale in Lake Huron, moose outrun wolves in 88 percent of chases observed by Mech.[8]

Some types of escape are passive. The Chilean seastar *Meyenaster gelatinosus* manages to capture only about 1 percent of the snail and sea urchin prey it detects. The most common response by the prey is not to flee, but simply to fall off the rocks on which they were crawling. Once in crevices, the prey cannot be caught by the frustrated seastar. Many stream insects release their hold on hard surfaces when they detect predators, and escape by virtue of the current that carries them away. Cladocerans stop swimming, fold the carapace over the limbs, and passively sink through the water when they detect freshwater copepods in the freshwater plankton.[9]

In most molluscs with well-developed escape responses to individual predatory snails and seastars, the prey's reaction is elicited only upon contact with the attacker. Besides fleeing, the victims often twist their bodies vigorously when grasped, or cover the shells with a slippery lobe of the mantle. Clams and most snails simply lack the appropriate directional sensory technology to detect and locate individual predators chemically at a distance. The chemical sense organs—located on the mantle, tentacles, and foot in snails—allow for at most a short-distance detection. Chemically sensing danger at a distance enables even rather unsophisticated marine snails to avoid predators, but the avoidance—elicited by chemicals released by snails, seastars, and crabs, for example—is of predators in general, not of individuals. Long-distance, directional chemical detection typifies two major clades of snails, the Sorbeoconcha and Opisthobranchia, which include among their ranks many predators.[10]

Prey not vigilant or fast enough to detect and evade their attackers must rely on direct confrontation, the third line of defense. The methods are numerous and varied, ranging from passive resistance to retaliation, always having the effect of slowing down, disabling, or defeating the predator. The main themes revolve around preventing successful subjugation, limiting consumption, exacting a high cost by causing injury, and limiting the deleterious effects of damage and injury.

An important class of passive defenses comprises protective coverings, hard or soft structures that enclose more delicate, and therefore more vulnerable, parts of the body. Despite their extraordinary variety, these devices can be divided into five categories depending on how they grow and function: (1) egg shells and seed coats; (2) shells growing at the rim of their opening or openings; (3) armor plating; (4) molting skeletons; and (5) houses assembled or collected from the surroundings.

The egg shell or seed coat completely encloses an unfertilized or fertilized gamete or an embryo, which communicates with the outside at most by pores that allow for gas exchange. This category includes egg cases, the external covers of plant fruits, and galls, structures that the occupying parasite induces its host to make. The coverings not only serve to protect the contents from outside agencies, but commonly also play a role in spreading spores, seeds, and fruits away from the parent. Accordingly, there are modifications for floating, attachment to vagrant animal carriers, and ingestion by birds and other animals who then defecate surviving enclosed seeds.

A shell is a hollow, permanent structure, attached to and housing the rest of the animal. Organs for locomotion, feeding, respiration, and reproduction can be extended out of the shell's opening or openings. Growth of the shell takes place along part or all of the rim of the opening, meaning that the newest parts of the shell are formed at the rim. One-piece shells characterize snails and cephalopods, as well as certain segmented worms (serpulids), although these structures often have a separate plate (operculum) that can close off the opening. Two-part or bivalved shells characterize clams and brachiopods, and even one peculiar group of snails (the juliid sacoglossan opisthobranchs). Shells of large size, with small openings, reinforced rims, walls strengthened by ribs and pleats, effective closing mechanisms, architectures permitting deep withdrawal of the vulnerable organs, wall structure that prevents cracks from propagating far, and features that make the shell awkward for a predator to manipulate are some of the defensive themes that crop up again and again in shell-bearing animals.[11]

Armor plating is a very widespread form of passive defense, in which many plates, each of which grows along part or all of its margin, lie adjacent to each other or overlap as they cover the other organs of the body. It characterizes many fish, reptiles, some mammals (armadillos, for example), and echinoderms. The skulls of vertebrates also have this type of so-called dermal-bone armor. The skeleton can be rigid or flexible. Barnacles—peculiar crustaceans, in which the body is enclosed in a housing of marginally growing plates—are remarkable in achieving a skeletal form like that in molluscs, but doing so from ancestors in which the skeleton molts. On the other hand, chitons are eight-valved molluscs that in some ways converge on the flexibly armored armadillos. Extensive overlap and reduction in the number and length of sutures between plates enhance the protective function of armor plating.[12]

A potential drawback of the shell armor plating is that the vulnerable organs must, from time to time, protrude from their protective housing. The molted external skeleton of arthropods effectively solves this problem by enclosing every organ—limbs, gills, feeding appendages, and so on—in skeletal material. The skeleton is extensively jointed, making all kinds of movement possible. From time to time, however, this wonderfully flexible yet strong casing is replaced with a new, larger skeleton that forms beneath the old one and that hardens once the old skeleton is shed. It is during and immediately after molting that arthropods are highly vulnerable to attack, for movement is impossible until the new skeleton hardens sufficiently to absorb the strong muscular forces needed for escaping from or retaliating against predators.

Finally, animals in many clades construct protective houses from objects or debris collected in the surrounding habitat. The freshwater larvae of caddisflies, for example, build cases from sand grains, leaf fragments, and other materials that are stitched together with silk. Some of these cases closely resemble the coiled shells of snails, and like them have a demonstrable protective function against predators. Many annelid worms and amphipod and tanaid crustaceans build tubes by cementing grains of sediment either with a protein matrix or occasionally with calcareous cement. Some sedentary clams augment their shell defenses with a thick outer coat of mucus-impregnated sand, which again adds to the shell's protective function. Carrier shells—snails of the family Xenophoridae—and a few other gastropods adorn their shells with other shells, pieces of coral, and other items by cementing these to the outside, possibly as camouflage but perhaps also as a cheap means of enlarging the original habitation.[13]

A variation on this theme is the use of empty shells. When snails die, they leave behind empty shells, which turn out to be highly effective fortresses for many crustaceans (hermit crabs, amphipods, and tanaids), some worms (sipunculans), some octopuses, and even a few fish. Fossil cephalopods (Paleozoic nautiloids and some Mesozoic ammonoids) may also have provided secondary housing for trilobites and crustaceans respectively. The peculiar hermit crab genus *Porcellanopagurus* typically uses deeply cupped clam valves as protective housing. In typical cases, the secondary dwellers act more like the human renters of apartments than like homeowners. They cannot repair their houses, and when moving to a new larger dwelling, they typically leave their old abodes in worse conditions than when they first moved in. Some remarkable symbioses, including dwellings of Ordovician age, have been discovered in which the occupants live inside structures that began as shells but which are perpetuated and enlarged by growth of an encrusting colony of bryozoans, hydroids, or sea anemones. Whereas most hermit crabs must seek larger quarters as they grow, these symbiotic shell-dwellers gain a permanent residence thanks to the colonial architects living with them. Again we see a most unlikely convergence to the molluscan shell, a structure growing at its rim, from ancestors that opportunistically occupied empty shells, and ultimately from ancestors that simply molted the skeleton.[14]

Passive defense provided by an enclosing device can be highly effective. *Tayassu pecari*, the more powerful of two specialized seed-cracking South American peccary species, can crush the hard fruit of the palm *Mauritia flexuosa*, which requires an average force of 3,800 N to break, but not that of the next more resistant fruit, belonging to a species of *Scheelea* and requiring a force of about 4,400 N to crush. The 12,350 N of force required to crush the 28 mm fruit of the palm *Phytelephas microcarpa* far exceeds the capacity of any peccary.[15]

The shells of many adult molluscs, especially those found on reefs and in other shallow-water tropical marine habitats, are so thick or otherwise well armored that their builders appear to be immune from shell-breaking predation. At an estimated maximum strength of 10 kN, the thick, 40 mm long adult shell of the snail *Drupa morum* from the western Pacific and Indian Oceans can successfully withstand attacks by such powerful shell-crushers as the spiny puffer (*Diodon hystrix*), the crabs *Carpilius maculatus* and *Daldorfia horrida*, and spiny lobsters of the genus *Panulirus*. Many large, thick-shelled adult clams such as the eastern American quahog (*Mercenaria mercenaria*), the Californian pismo clam (*Tivela stultorum*), the tropical Pacific giant clams of the genus *Tridacna*, and the Mexican *Chionista fluctifraga* enjoy freedom from predation thanks to their massive armor, which protects against breakage and against predators that drill through the valve walls. Loggerhead turtles (*Caretta caretta*), whose biting strength has unfortunately not been measured, may be the most powerful living marine shell crushers, but they are probably unable to take large adult conchs (*Strombus gigas*) and helmet shells (genus *Cassis*) in the Caribbean. Even in cold waters, where shell strength is markedly lower, snails and clams everywhere seem to exceed the crushing capacities of the most powerful predators.[16]

In a very insightful paper, Richard Cowen identified the conditions that must be met in order for passive armor to evolve as a defense. The principal difficulty with armor, Cowen points out, is that it interferes with gas exchange. When animals metabolize, they take up oxygen and void carbon dioxide. The movement of gases into and out of the body occurs across membranes by a process of passive diffusion. When there is little oxygen in the environment, the surface available for gas exchange must be very large relative to the body's volume. When the respiring body is enclosed in a shell, this condition cannot be met, for only a small fraction of the body's surface will be in contact with the outside. Passive armor, therefore, is compatible either with a complete shutdown of the metabolic machinery, or with life in an environment in which oxygen is plentiful and gas exchange occurs over a relatively small exposed surface.[17]

A second major category of passive defenses comprises a chemical arsenal of toxins and repellants. The surfaces of flesh of countless plants, animals, and fungi contain substances that are poisonous or unattractive to attackers. Slugs and even some shell-bearing snails exude copious amounts of sticky mucus, which disables the organs of beetles, snakes, and other would-be predators. Many fish are slippery and thus difficult for predators to grasp because of the presence of mucuslike substances. Many flowering plants exude milky substances when disturbed, which disable or poison herbivorous insects. Squids, sea cucumbers, snakes, beetles, and spittle bugs discourage attackers by spraying noxious substances. Many seaweeds, including the coralline algae mentioned earlier, contain minerals such as calcium carbonate in or between cell walls, making these plants unprofitable and slow to consume. Cellulose and lignin in land plants, and silica in grasses, achieve the same effects of slowing down herbivores and causing significant wear on teeth or other organs with which the plants are grasped and torn apart.[18]

The chemical composition of the toxins apparently reflects the nutritional environment of the bearer. Plants and animals with access to a readily tapped supply of nitrogen rely on nitrogen-rich compounds, such as alkaloids, neurotoxins, and unusual amino acids; whereas those for which carbon is vastly more available than nitrogen depend on such nitrogen-free, carbon-rich compounds as tannins, terpenes, lignin, resins, cellulose, and chitin. In the presence of excess carbohydrate, such carbon-based defenses may be relatively inexpensive to produce. Interestingly, there appear to be no phosphate-based chemical defenses among plants, perhaps reflecting the scarcity of phosphate and the high energetic costs of constructing such defenses. Phyllis Coley and her colleagues have pointed out that this absence of phosphate-based chemical defense likely explains the extraordinary success of phosphate-based insecticides manufactured by humans.[19]

It is curious that, although chemical sprays are known from many mammals and birds, chemical defense in the skin or flesh are very rare in these warm-blooded animals. I am aware of only a handful of toxic birds from New Guinea, all belonging to the genus *Pitohui*. If this rarity proves to be real, it will require an explanation. Perhaps a bird or mammal, once accosted by a predator, dies even if it remains uneaten. Partial predation of the type commonly observed in amphibians, reptiles, and many other animals is conducive to the evolution of toxins, because the victim survives an attack. Warm-blooded animals may be unable to survive wounding because their injuries could quickly become infected by pathogens, which multiply rapidly in the high-temperature environment of their host.[20]

Many third-line defenses function not to prevent damage, but to limit it. One widespread form of this is autotomy, the deliberate severing by the victim of part of its body. While the predator chases and collects the severed body part, the rest of the victim escapes. Sea cucumbers, when caught by the large predatory snail *Tonna perdix*, wriggle out of their skin and bound away. Other sea cucumbers relieve themselves of internal respiratory organs. The surprised human attacker comes away with the unforgettable experience of having the fingers temporarily encumbered by the copious sticky fibers. Whether predatory fish are similarly deterred by this peculiar tribute is unknown. Fireworms of the family Amphinomidae leave their human attacker with a painful collection of glasslike needles, specialized setae that the worm autotomizes and that are almost impossible to extract from the fingers that sought to grasp the worm from beneath a rock. Urticating caterpillars, and indeed the nettles (*Urtica*) after which they are named, leave similar annoyances behind in their attackers on land. Butterflies and moths caught in spiders' webs often get away, leaving only wing scales behind for the spiders to eat. In many clams buried in sand and mud, the siphons that draw water into the animal or that carry waste away easily break off when fish or crabs attack these relatively sedentary molluscs. Sea lilies (crinoids) easily and willingly sever their arms while under attack by a human or by a browsing fish. Many crabs readily autotomize their claws. There is a famous cover photograph of an issue of *Science* from 1970 in which a surprised otter sports the autotomized claw of a crab on its nose, the result of a failed attack by the otter. I have on more than one occasion been distracted by the claw that a desperate crab had left tightly and painfully pinching my thumb as the crab swam away. Many lizards and salamanders give up their easily autotomized tails to predators. It is better to lose part of the body to the predator than to lose one's life, even though the loss of the tail imposes costs in reduced ability to feed as well as in social status. Snails of the genera *Stomatella* and *Harpa* readily sever part of the foot as they escape an attack, an action that likely diminishes the speed of subsequent movements by these animals.[21]

Plants show a variety of features that appear to limit the negative effects of damage. I would interpret the network pattern of venation in leaves as an adaptation that preserves the water-conducting and food-producing functions of leaves in the face of local vein cutting by herbivorous insects. The network allows materials to move by alternative routes in case one conduit is severed.

There is a seamless gradation of antipredatory adaptation from passive resistance to outright aggression. Sea urchin spines, the barbs on the fins of fish, and the thorns of rose bushes and acacia trees can potentially inflict injury on attackers, but they are mostly passive defenses. Venomous snakes and cone snails, stinging scorpions and bees, kicking cassowaries and horses, biting ants and moray eels, and scratching squirrels employ more aggressive forms of defense that can injure or even kill would-be predators. A widespread variant of such aggression is to employ mercenaries. Many plants are defended by ants, which kill and eat herbivorous insects and are subsidized with carbohydrate-rich liquid secretions by the plant. Pinching trapeziid crabs defend their coral hosts in the tropical Pacific and Indian Oceans against seastars and shrimps, presumably in return for shelter against their own enemies. Corals and hydroids offer protection to their bryozoan hosts by stinging attackers with their nematocysts. Sea anemones similarly protect hermit crabs against octopus predation when the shells occupied by the crabs have an outer shield of stinging symbiotic anemones.[22]

Group defense is yet another highly effective and important means of protecting animals against predators. Many ungulates and primates, including humans, employ this form of defense against large carnivores, as do schooling fish, flocking birds, and social insects. Individual defenders need not themselves be powerful, large, or well equipped with weapons (though they often are), but collectively the group provides diffuse, pervasive protection spread over a large area or volume of habitat, and incorporates cooperation and communication among threatened individuals. The advantages of such collective defense may very largely be responsible for the evolution of sociality in animals.[23]

In social animals, a division of labor among individuals makes possible the specialization of some members of the group to aggressive defense. Ants and termites have evolved soldier castes of female workers that are "all mandible." In larger human societies, there is a professional military establishment. Even without such specialization, social animals benefit in defense from individuals in the group who detect or confront enemies. The defenders alert others through warning signals, reducing the predator's chances of success. In clonal animals, too, there is defensive specialization. The Portuguese man-of-war colony develops specialized polyps with attack tentacles, and cheilostome bryozoans have evolved avicularia, unusual jawed zooids that guard the colony against marauding predatory nudibranchs and the settling larvae of potential competitors.[24]

Each type of antipredatory defense comes with costs. These are direct costs—loss of a body part, or high permanent material investment in defensive armor or toxins—or opportunity costs. Animals on the run from predators cannot feed; plants well defended against attack by energetically expensive tannins and resins are limited in potential growth rate and therefore in their ability to compete for light with plants whose chemical defenses are cheaper or more easily transported from place to place as needed. Potentially mobile animals cowering behind the closed door of a heavily armored shell, or passing the time motionless so as not to be noticed by searching predators, forgo opportunities to feed and to mate. Defenses that advertise the prey's potential danger to a predator enable the prey to be economically active, but they require substantial metabolic investments not just in a relatively tough body that would not yield when caught, but also in weapons of aggression and the musculature that powers them. Such high expenses might be more justified if structures used in aggressive defense also function in such other contexts as feeding or competition for mates. The claws of crabs, as anyone handling living crabs knows all too well, function in defense, as well as in feeding, in fighting over mates and territory, and in visual and occasionally acoustic display.

On the Pacific coast of Panama, the common intertidal predatory snail *Vasula melones* has an immensely thick, strong shell, comprising more than 95 percent of its total weight. An adult 45 mm long, with a shell wall some 5 mm thick, requires a compressive force of 8 kN to break lethally. Although the snail moves slowly, the adult is nevertheless able to forage for its food during times of the day when potential predators—fish and crabs in particular—are also active. Moreover, its relative invulnerability to attack allows *Vasula* to subdue and consume well-armored prey, which require extended handling times. Where predatory snails with thinner, smaller, and weaker shells would do well to minimize their feeding time in exposed environments, *Vasula* can expect to be uninterrupted for long periods as it drills a hole through its victim's shell or slowly enters and consumes its prey by way of a natural opening.[25]

A more common resistance defense that enables animals to carry out such activities as feeding and courting under dangerous conditions is toxicity or potential aggression. In a detailed comparative study of palatable and unpalatable butterflies in Central America, Marden and Chai showed that tasty species achieve higher acceleration and higher flight speeds, and have a relatively larger mass devoted to flight muscles (30–39 percent in females) than do species that birds avoid (muscle mass 19–30 percent). With a lower investment in expensive flight muscle, unpalatable butterflies often devote more resources to egg production and to structures that enable them to feed as adults. Moreover, adults of unpalatable species live much longer than those of edible species. Chemical defense likely has its own costs, much as does the development of heavy armor in snail shells, but investment in it enables individuals to carry on the activities of life without always having to shut down metabolism or flee.[26]

These examples illustrate the important point that antipredatory adaptation does not occur only in the context of predators, but instead proceeds in an environment of many other demands and opportunities. Adaptive solutions to predation may coincide with those to other forms of competition, or they may conflict.

An example of conflict is that between defense and attracting mates. All animals with internal fertilization must locate and approach mates. Flowering plants, too, must mate, but they often employ animal intermediaries—pollinating insects, birds, mammals, and sometimes even reptiles and snails—to carry male pollen to female ovules. Finding mates is not easy, especially for species whose populations consist of widely scattered individuals. The typical solution is long-range advertisement. Bright colors, distinctive sounds (as in frogs, crickets, cicadas, birds, and even fish), conspicuous behavioral repertoires, and unique chemical signals (pheromones) have evolved in staggering diversity as beacons to attract mates. The problem is, of course, that these same enticements also call in potential enemies, including predators and parasites. Plants and animals that rely on long-range advertisement must therefore possess the resistance defenses necessary to deal with enemies, or they must depend on sense organs sophisticated enough to distinguish between friend and foe at a distance large enough to take evasive action if necessary. Sexual selection as reflected in patterns of mate choice is therefore a potentially very powerful amplifier to selection imposed by predators and by other competitors for food.[27]

The adaptive possibilities available to species are limited by the many demands placed upon individuals. Success of individuals depends on evolutionarily solving all problems adequately if imperfectly. Sometimes success coincides with economic dominance; sometimes it does not, as in the case of many first-line passive defenses. In either case, success is measured by persistence, not perfection or economic dominance.

The Flow of Evolutionary Information

When predators interact with prey, benefits flow from one party to the other. The direction of this flow, however, depends both on the time scale over which the benefit is evaluated and on the circumstances in which the interaction takes place. From the short-term nutritional standpoint, the interaction is decidedly in favor of the predator when the attack succeeds. Food flows from victim to consumer—from prey to predator, host to guest, plant to herbivore. But not all attacks succeed, and when attempts at predation fail, the flow of benefits reverses. When the prey fails during an attack, it loses life or limb; when the predator fails, it merely loses dinner. Richard Dawkins and John Krebs dubbed this the "life-dinner" principle. The evolutionary benefit of success to the prey is that the victim has been "tested" and found adequate, giving the prey the possibility of transmitting its "good" antipredatory traits to the next generation. Success from the predator's perspective means food, but the predator may learn little from its victim on the evolutionary time scale from its interaction, even if it should fail. Unless the predator's energy reserves are critically low or its prey are scarce, the predator can secure another victim, often belonging to a different species. The prey "learns" more in an evolutionary sense from encounters with predators than the predator does from encounters with prey. Moreover, the number of prey individuals will typically exceed the number of predators that consume whole prey. This means that a predator will have more encounters with prey than a prey individual will have with predators. Opportunities for predators to cause selection among prey therefore outnumber those for prey to influence predators, at least when the two parties are interacting in isolation.[28]

But things are never as simple as they seem. Even in isolation, the evolutionary tug-of-war between predator and prey often violates the life-dinner principle, and gives a much less clear advantage to the prey. More importantly, the interaction never takes place in evolutionary isolation. With interesting exceptions, consumers influence victims far more than victims influence consumers. Consumers experience intense competition and predation from their own enemies. They respond to these enemies evolutionarily, much as victims respond to theirs. The result is *escalation*, or enemy-controlled evolution.

Consider first the circumstances under which the life-dinner principle is violated. Losing dinner is sometimes a matter of life and death for a predator. Animals in the wild with empty stomachs are common, raising the possibility that failed attempts at predation place predators in danger of starvation or of becoming so weak that they are more vulnerable to their own enemies. In his survey of predatory snails in waters around Hong Kong, John Taylor found that only 5 percent had detectable food in their digestive systems. Brian Morton and Joseph Britton found the same percentage in their survey of *Dicathais orbita*, a common predatory snail in temperate Australia. Raymond Huey and colleagues conducted detailed surveys of lizards around the world, and discovered that some species, such as those active during daylight hours, almost always have full stomachs, whereas others (especially nocturnal species) were almost always empty. Whether empty individuals suffered reduced performance is unknown. Working in coastal Maine, Robert Fritz and Douglas Morse showed that 20 percent of crab-spiders (*Misumena vatia*) pouncing on flower-pollinating insects were unable to capture enough food to grow and reproduce.[29]

The exchange of evolutionary information between predator and prey in isolation might be more even handed if the prey were able to inflict significant harm on its attacker during a failed attempt at consumption. Both parties have much to lose if they fail. Through adaptation, both can "learn" from success.

Potential victims have evolutionarily devised a wondrous array of methods to inflict injury to those who would eat them. Prey can bite, kick, poison, scratch, squirt, sting, electrically stun, trap, or throw off their predators. Bombardier beetles (*Trachinus* spp.) employ a potent binary chemical defense by squirting two liquids that, when combined outside the insect's body, react to produce a boiling-hot fluid. The copious sticky mucus exuded by many slugs (snails that have forsaken the protective armor of an external shell) disables the mandibles of predaceous beetles and the jaws of snakes. Mussels ensnare predaceous snails by attaching proteinaceous threads (byssal fibers) to their victims' shells. The snails are thus immobilized to death if they cannot wrench themselves free. Mussels and other bivalves can even use their shells to injure attackers. Rails and other birds have been known to get their toes caught between mussel valves, and predatory snails may lose part of their shell or their feeding proboscis as bivalves thwart attempts by the predators to enter between their valves. Limpets—snails with cap-shaped shells that cling tightly and snugly to rocks—can clamp down on the foot of intruding predatory snails. Some desperate snails in the clutches of the arms of a seastar bite off the tube feet of their attacker.[30]

That predators do indeed adapt to dangerous prey is abundantly evident from a multitude of studies. Horsefly larvae (*Tabanus punctifer*) and antlion larvae (*Myrmeleon crudelis*) avoid a bombardier beetle's hot stream of liquid by keeping their heads low against the ground. The grasshopper mouse (*Onychomys torridus*) handles scorpions so that the venomous sting cannot reach it. The squash beetle *Epilachna tredecemnota* and the alder sawfly *Eriocampa ovata* avoid leaf toxicity in their hosts squash and alder, respectively, by severing most of the veins in the part of the leaf they are about to consume, preventing the plant from mobilizing a chemical response. Web-building spiders may be able to avoid toxic sprays of beetles and bugs by waiting to pounce on noxious prey until the latter have released their arsenal during attempts by the prey to escape from the sticky web. Success rates of killing noxious prey are as high as 80 percent for some European web-builders.[31]

There is an even more general way in which predator and prey influence each other. Experiments on a very wide variety of plants and animals have shown that the prey's phenotype can change in an apparently adaptive direction because of the presence of a predator. Similarly, predators can respond directly to changes in prey. These organisms display what is formally known as phenotypic plasticity, the ability to change in form, chemical composition, or behavior in direct response to other organisms as well as to nonbiological aspects of the environment. Predator and prey, in other words, engage in reciprocal, nongenetic, adaptation.[32]

Consider some shell-bearing molluscs and their predators. When dog-whelks of the genus *Nucella* are in the presence of predatory crabs, chemical signals emanating from the predators cause the shells of prey snails to become thicker. Mussels (*Mytilus edulis*) exposed to chemicals released by crabs develop thicker shells and a stronger attachment of byssal threads to the rock on which they live than do mussels not thus exposed. These induced defenses decrease the prey's vulnerability to attack by the crabs. On the predator side, the presence of hard-shelled prey induces crabs' claws to become larger and more robust, with the result that the crabs can apply greater force compared to those for whom the only available food is soft.[33]

Induced weaponry of this kind also occurs in vertebrates. When eating a diet of shell-encased molluscs, some cichlid fish develop robust bones in the throat that break the shell after the prey is ingested. On a soft diet, these bones remain delicate. Something similar happens in redknots, small wading birds that fly immense distances during migration between summer and winter grounds. A diet rich in molluscs causes the gizzard, the organ in which the food is ground up, to enlarge. The gizzard shrinks prior to migration as the birds take softer food.[34]

These effects are not limited to predators eating animals. Grass-eating caterpillars and grasshoppers on a diet of tough grasses develop heads and mandibles roughly twice the mass of those fed on a diet of softer species. A similar effect has been noted in nut-cracking monkeys in comparison to monkeys fed softer diets. The sea urchin *Diadema antillarum* living in crowded conditions has a larger jaw apparatus with which to bite and scrape algal food than do individuals living at a lower population density and with a larger per capita food supply.[35]

Despite these important ways in which predator and prey influence each other's phenotypes, predators on the whole maintain the evolutionary upper hand of control. This is so because predator and prey do not interact and evolve in isolation. Their struggles take place in the presence of many other threats, and their fates cannot be understood by looking only at how they respond to each other. Adaptations of predators to locate, catch, and subjugate prey are enforced to a significant degree by the predators' own large enemies, by their competitors for food and mates, and by those who would eat them.

Observations on competition among top predators illustrate how struggles over food can influence the development of devices used in predation. Of the grazing mammals successfully caught and killed by cheetahs in Tanzania's Ngorongoro Crater, 12 percent were lost to leopards, lions, and hyenas. Bats and nighthawks hunting insects sonically on the wing at night interfere with each other, with the result that the subordinate nighthawks must catch prey in less favorable locations. The place and technique of hunting are dictated more by competition between these consumers than by characteristics of the flying prey.[36]

On sand flats and seagrass beds on the west coast of Florida, two large predatory snails aggressively compete for prey clams. The pear whelk (*Fulguropsis spiratus*), reaching a shell length of about 100 mm, has a relatively large foot with which it can push away the much larger lightning whelk (*Sinistrofulgur sinistrum*), which can reach a length of 400 mm. The pear whelk also rasps its opponent with its radula and bites with the jaw. As a result, the small but aggressive pear whelk typically wins fights with the lightning whelk, and gains access to thin-shelled clams that are quickly opened and consumed. The larger lightning whelk must make do with thick-shelled, tightly closing clams, which take eight to ten hours instead of thirty to sixty minutes to subdue and eat. Its shell lip is used to wedge open these fortified clams, and often breaks in the process. The pear whelk's radula and jaw function in feeding as well as in fighting, so these two activities are intimately linked.[37]

Blaire Van Valkenburgh and Fritz Hertel have documented a particularly interesting case involving mammalian carnivores that crush the bones of their victims. Lions, leopards, cheetahs, jaguars, pumas, spotted and brown hyenas, gray wolves, and wild dogs on average stand a 20 percent chance of breaking one or more of their teeth during their adult lifetimes. In the spotted hyena (*Crocuta crocuta*), this number is as high as 40 percent. The three large American carnivores (puma, jaguar, and gray wolf) have a per tooth breakage frequency ranging from 0.009 to 0.010 (not counting incisors) or from 0.011 to 0.019 (incisors included). For the remaining 28 species, all African, the range of frequencies is 0.005–0.020 (without incisors) and 0.008–0.027 (with incisors). These frequencies of tooth breakage generally lie well below those recorded for the fossil carnivores recovered in the tar pits of Pleistocene age at La Brea in Los Angeles. American lion, saber-tooth tiger, and dire wolf had frequencies ranging from 0.08 to 0.11 (without incisors) and 0.055 to 0.17 (with incisors). Even the much smaller coyote had much higher frequencies of tooth breakage at La Brea (0.05 without incisors, 0.068 with incisors) than in North America today (0.27 and 0.035, respectively). Alone among the species studied, the puma had less damage—in fact, no damage—during the Pleistocene than in living populations (0.10 and 0.19 without and with incisors).

One could argue that selection imposed by the predator on the prey in favor of stronger bones was responsible for the high Pleistocene frequencies of breakage in the predators' teeth, and that selection since that time has subsided. The problem is that captured prey rarely escape from bone crushing predators, so that there would be no prey that passed the bone-crushing test and no prey to pass on more resistant bones to the next generation. Van Valkenburgh and Hertel instead favor the interpretation that tooth breakage is a measure of competitive intensity among the carnivores. The predators engage in tooth-damaging bone crushing in order to compete more effectively with rivals for food. During the Pleistocene, fifteen species of carnivore at La Brea were the size of a coyote or larger; in North America today there are only seven. Africa also supports seven carnivores in this size range. These numbers, together with the data on tooth breakage, suggest to Van Valkenburgh and Hertel and to me that competition among large carnivores was more intense at Pleistocene La Brea than in North America and Africa today. Most importantly, teeth are effective weapons in dismembering prey as well as in competition among fellow predators.[38]

There are many other examples. The sharp claws of lions help to prevent rivals—hyenas, crocodiles, and other lions—from reaping the benefits of a hard-won catch. They enable the predators to quickly tear apart the prey into smaller, ingestible pieces. The jaws of the Caribbean sea urchin *Echinometra lucunter*, which scrape and gouge rock surfaces for encrusting algal food, also serve as weapons in territorial fights. Sperm whales and snapping shrimps stun prey with bursts of sound at short range. The shrimps do it by shutting a large claw in just 600 microseconds. The sounds produced are also used by the animals to communicate with each other.[39]

In crabs, the scissorlike claws—modified limbs, whose right member often differs in form and function from the one on the left—hold, manipulate, and dismember prey. They are also effective in defense against predators, in fighting, in displaying to competitors, and (in some hermit crabs and calappid box crabs) in covering other organs like a shield. The great forces that massive-clawed crabs such as *Carpilius maculatus*, *Daldorfia horrida*, and *Menippe mercenaria* apply to crush shell-bearing prey offer benefits in combat and defense as well.[40]

All species are potential victims. It would be astonishing if they lacked adaptations to—or had failed evolutionarily to "learn" from—predators or from rivals, including rivals of their own kind. My thesis is that enemies collectively exercise more intense selection on victims than victims do on their rivals. Evolutionary information, in other words, tends to flow from victims to consumers. Top-down evolutionary effects of enemies are stronger than the bottom-up effects of food, including prey. Specialization on particular foods or habitats is dictated less by selection emanating from the consumed than selection from consumers and competitors.[41]

The Characteristics of Enemies

The top-down evolutionary control exercised by predators over their prey is reflected by a previously unappreciated generalization about the relative performance levels of these animals. In every community with which I am familiar, at least some predators have keener sense organs, greater locomotor ability, and superior subjugation strength than any of the co-occurring mobile prey species. Victims, on the other hand, tend to excel in such passive defenses as armor, toxicity, and large size.[42]

In principle, prey species could be equally defended in all three of the phases of a predatory attack. They could be well adapted to avoid being recognized, caught, and subdued; but incompatibilities among the defenses against each of the stages of an attack generally rule out such an equitable distribution. Generally, prey adaptation is concentrated in only one phase of the interaction. The phase to which the prey is best adapted corresponds to the stage of the attack in which the predator is least successful.

Predators rely on sense organs to detect prey, and prey must either make themselves unrecognizable or undesirable or detect the attacker first. Predators invariably seem to possess the sense organs with the greatest acuity. Discrimination of visual images, for example, is most detailed among predators—cephalopods (squid and octopus), cubozoan jellyfish, stomatopods (mantis shrimps), crabs, insects (predatory beetles and mantids), fish, raptors (hawks and eagles), spiders, and fossil trilobites. Light perception and the formation of crude images are very widespread among animals, and have evolved at least forty to sixty times independently (albeit with apparently homologous genetic machinery); but clear, long-distance, directional vision and detailed images appear to be restricted to predators searching in well-lit environments or with the aid of in-house light-producing organs in dark places.[43]

Among snails, directional and long-distance chemical perception is possible only in marine clades with predatory and scavenging members. These snails sample water entering the front of the animal by way of the siphon, a tubular organ that in shell-bearing species is housed in a channel-like extension of the front of the shell. In other snails, water is sampled from all around the body, and chemical perception is not directional. Chemical identification of individual threats typically requires contact and cannot be accomplished with waterborne cues at a distance.[44]

Predatory crustaceans—lobsters and crabs, for example—as well as seastars also have the ability to recognize potential prey at a distance. As in gastropods, this ability requires not simply the recognition that something edible is nearby, but also where the item is. Water must be sampled and a chemical concentration gradient must be detected in order for the predator to move toward its victim.[45]

Prey that rely chiefly on rapid escape attain speeds that fall well below those of the fastest predators with which they coexist. In surveys of burrowing speeds of sand-dwelling shallow-water marine snails in Guam and Panama, we found to our surprise that the fastest species are predatory olive snails (olivids) and moon snails (naticids). All snails are slower than the many crabs, fish, octopuses, and mantis shrimps whose diets consist mostly or entirely of molluscs. Wedge clams (Donacidae) may be the fastest molluscan burrowers on sandy beaches, even outdoing olives and moon snails and various other predatory gastropods, but they are still slower than sandy-beach crabs such as *Ovalipes* and the ghost crab *Ocypode*. On the colder shores of the American Pacific Northwest, many shore snails have effective escape responses to relatively slow seastars such as *Pycnopodia* (the sunflower star) and *Pisaster* (the ochre seastar), but not against much faster crabs (*Cancer*), fish, or birds. On the plains of East Africa, predatory cheetahs attain greater speeds (112 km/h) than any co-occurring mammalian prey, including gazelles (105 km/h). In the modern North American mammal fauna, the pronghorn (100 km/h) is the fastest living mammal; but during the Pleistocene, the American cheetah was likely faster until it became extinct with the arrival of humans. Peregrine falcons, diving at speeds exceeding 200 km/h, are the fastest animals known. Like a few other raptors, these birds can kill other birds in flight. Herbivorous birds and insects tend to be weak and slow fliers compared to their more carnivorous counterparts, perhaps in part because plant food is bulky, slow to digest, and requires heavy machinery—teeth, gizzard stones, and a long digestive system—to process. Geese and herbivorous ducks, which undertake long migrations, switch to a more energy-rich diet before beginning their long flights. Burst speeds among fish are higher for top predators than for other species. The highest burst speeds recorded are for barracuda (12.2 m/s) and two species of tuna, *Thunnus albacares* and *Acanthocybium solanderi* (21 m/s). In the world of freshwater zooplankton, the fastest animals by far are predatory copepods, which at speeds up to 35 cm/s outperform their prey cladocerans (2 cm/s).[46]

The superiority of top predators over other kinds of animals in speed has existed since at least Early Cambrian time. Swimming anomalocarid arthropods of the Early and Middle Cambrian were the top predators and the fastest animals of their day. Their place was taken by shell-bearing nautiloid cephalopods in the Ordovician, acanthodian fish in the Silurian, placoderms and perhaps early sharks in the Devonian, sharks in the Late Paleozoic, and various marine reptiles in the Mesozoic. Predatory dinosaurs, most of which walked and ran on two legs, were anatomically more specialized for running than were herbivores. Their speeds, estimated to be 29.2 km/h for a theropod dinosaur trackway of Middle Jurassic age and 11 m/s (39.6 km/h) for the Late Cretaceous *Tyrannosaurus rex*, are modest by mammalian carnivore standards, but probably exceeded those of contemporary prey herbivores.[47]

A potentially contradictory picture emerges for fossil mammals, but here too it is likely that the predators of Cenozoic mammals were faster than the mammals themselves. Robert Bakker, Christine Janis, Blaire Van Valkenburgh, and other students of mammalian locomotion have long held that predatory long-distance running mammals were rare or absent for much of Cenozoic time, beginning 65 Ma. Only with the evolution of wolves in North America and cheetahs in the Old World during the Pliocene, not more than 5 Ma, did mammalian predators begin to outrun their mammalian prey. Many of the herbivorous mammals of earlier Cenozoic epochs were faster than their predators, which were typically themselves mammals that ambushed their prey from a relatively short distance. In the Eocene of North America, however, there may have been distance runners not among predatory mammals, but among ground-dwelling predatory birds. Lawrence Witmer and Kenneth Rose point to the 2 m tall bird *Diatryma* as a fearsome runner, with a 100 mm gape in its bill and a likely capacity to run after large mammalian victims. South America, an island continent for much of the Cenozoic era, was home to ground-dwelling phorusrhacoid birds, which perhaps were also long-distance runners. These examples show that we must be open to the possibility that the top predators in the Cenozoic—the age of mammals—were often birds. Even today, the top predators for tree-dwelling mammals such as monkeys in Africa and South America, and for many vertebrates on islands, are eagles or large hawks.[48]

Aspects of locomotion other than absolute speed may be the deciding factors in predation, and here too the predators apparently outdo their prey. Large African predatory cats can accelerate faster than their herbivorous prey, and therefore are able to catch them at relatively close quarters. Zebras, gazelles, and other antelopes attain top speeds in excess of those of lions and leopards, but it is acceleration that matters if the distance between the two parties is small when one becomes aware of the other. Some predators rely on the extremely rapid deployment of the grasping organ. Predatory fish can rapidly extend their protrusible jaws; cone snails shoot an envenomated radular tooth into their prey from a position at rest. Frogs and salamanders are extremely fast with their tongues in catching insects. The prey-capturing feet of the goshawk (*Accipiter gentilis*) move at a rate of 22.5 m/s, 15 percent faster than the pelvis.[49]

Predators exert greater force and mechanical power than do other animals. The American alligator (*Alligator mississippiensis*), with a bite force of 13,300 N, is the most powerful animal in which force has so far been measured. This value compares to 4,168 N in lions, 1,412 N in wolves, and 749 N in human jaws. Calculations based on bite marks on the bones of prey *Triceratops* suggest that the famous Late Cretaceous *Tyrannosaurus rex* exerted bite forces as high as 13,400 N. Based on the extraordinary shell strength of the tropical Pacific snail *Drupa morum*, which adult crabs (*Carpilius maculatus*) and pufferfish

(*Diodon hystrix*) can crush, bite forces of these predators may lie between 5,000 and 10,000 N. Predatory crabs exceed omnivorous or herbivorous ones in bite forces, and have crusher claws with a higher mechanical advantage and more specialized molarlike teeth. Extremely high power is used by some whales and snapping shrimps to stun prey with sound.[50]

Envenomation—the injection of poison—likely has defensive functions in many animals including some sea urchins (*Toxopneustes*), the seastar *Acanthaster planci*, fang-toothed blennies (Uranoscopidae), scorpionfish (Scorpaenidae), sting-bearing Hymenoptera (ants, bees, and wasps), and the Australian freshwater monotreme mammal *Ornithorhynchus* (the platypus). This method, however, evolved almost entirely among predators, and it may often have the additional or primary function of quickly disabling prey. Predators using envenomation in this way include jellyfishes (Scyphozoa and the dreaded Cubozoa, which can kill a human within minutes), cone snails and other toxoglossan gastropods, octopuses (some also lethal to humans), centipedes, scorpions, spiders, ants, deep-sea monognathid eels, snakes, and two mammals, the West Indian insectivoran *Solenodon* and the North American short-tailed shrew *Blarina brevicauda*. Many predaceous snails hasten the demise of their molluscan victims by anesthetizing them with toxins released onto the soft parts early during the subjugation phase of the attack.[51]

The ways in which the highest-performing predators excel over the most sophisticated prey—long-distance detection, speed and agility, and the application of force—reflect superior power. Finding widely dispersed prey, getting to the prey first, killing and consuming the victim before competitors do, and struggling with rivals are critical to the success of predators as they compete with one another. The larger the prey, the more these attributes of power come into play. Not only are large victims' resources worth defending, but they tend to be sparsely distributed, meaning that exploitation of such resources entails wide-ranging search. Predatory mammals have on average a home range four times the size of the range of herbivores.[52]

Arguably the most powerful animals of all are omnivores rather than dedicated predators. Individually and collectively, humans have an exceptionally broad diet. Other social primates such as chimpanzees are also omnivorous. Less well known is the fact that some of the most abundant and competitively dominant ants also have diets consisting of both plant and animal food. Descended from more strictly predatory species, these tree-dwelling tropical ants substantially supplement their animal-based diet with fluids exuded by plants. What makes these ants exceptionally powerful is not their individual large size or power, but their collective numbers. Workers in ant colonies can quickly locate food, alert nest mates to the location of resources or enemies, and effectively harvest resources too small or too widely scattered to be exploited by large individual vertebrates.[53]

Interestingly, it is the passive manifestations of power—the mechanisms that reduce or resist adversarial power without expending mechanical energy—in which prey of highest performance gain decided advantages over the most sophisticated consumers. Toxicity and body armor exemplify this conclusion well, as I pointed out in my treatment of prey defenses. Perhaps more surprisingly, the same holds for large body size, at least in productive ecosystems on land. Without exception, the largest animals on land are herbivores, not top predators. The largest herbivorous vertebrates are on average fourteen to thirty-three times more massive than the largest co-occurring predatory vertebrates, the factor being greater when the animals involved are cold blooded rather than warm blooded. The most massive warm-blooded terrestrial predator is the tiger, which with a mass up to 272 kg is small compared to elephants (5,000 kg) or extinct ground sloths (the Pleistocene *Eremotherium mirabile* weighing about 3,000 kg).[54]

Passive defense, the kind of defense where the victims seem to have a distinct edge over their enemies, can be thought of in another way. It comprises stored energy, an accumulation of resources. The great significance of this stored energy lies in its providing a potentially reliable source of food for consumers. I shall argue in the next chapter that passive defense, evolutionarily elicited by enemies, holds the key to economic expansion into environments where passive protection has the additional benefit of shielding individuals against prolonged unfavorable physical and chemical circumstances.

Although much of what I have said applies to large consumers and the control they exercise over victims, parasites too have had far-reaching effects. Were it not for the legions of pathogens and parasites living on the insides and outsides of warm bodies, mammals and birds would not have evolved their highly sophisticated immune systems, in which mobile cells carried in the circulatory system interdict harmful agents.

Wolfgang Sterrer has intriguingly and plausibly suggested that sex is ultimately an adaptive response of host genomes to infection by parasitic genes. In prokaryotes—bacteria and Archaea, the components of eukaryotic cells—transfers of genes among cells occur more or less unilaterally, with the host being the recipient and the parasite—usually a plasmid or virus—being the donor. These infections will normally be harmful to the host. In some cases, the host may be able to "domesticate" the intruding genes by tying them to its own genes and forcing the intruder to replicate with the same rhythm as the host's own genes. The genetically integrated parasite, meanwhile, can help the host by limiting subsequent infection of less well-integrated genes. In eukaryotic sex, gene transfers occur only between compatible genomes and are thus no longer parasitic. They involve both an even-handed splitting of the genome and a fusion through recombination, with both parties to the transfer contributing about equally. Sterrer sees this mechanism of even-handed exchange of genes as a means of protecting the genome against infection by pathogens capable of inserting themselves into the genome. It is also a mechanism for keeping the genes that have become integrated from selfishly replicating. This is accomplished by transferring some of the genetic intruder's genes into the host nucleus, and by breaking apart the intruder's genome through splitting and recombination. Eukaryotic sex thus evolved as a means of protecting the genomes against parasites. It is a defense against genetic consumers, a defense with extraordinary long-term consequences.[55]

A related hypothesis, offered by Richard Grosberg and Richard Strathmann, explains why all multicellular organisms go through a single-celled stage at some point in the life cycle. The single cell can purge the organism of parasites, and limits competition among the parasites that do infect. In multicellular stages, some cells acquire mutations that may cause these cells to replicate selfishly. Such mutations are largely eliminated at the one-celled stage, which therefore enforces a regime of cooperation among the cells of later multicellular stages. The eukaryotic life cycle can therefore also be understood as a means of reducing, and limiting the effects of, parasitism and error. The important point is that parasites have been important selective agents, and that many aspects of life cycles and of genetic architecture are interpretable as responses to them.[56]

Enemies and the Distribution of Vanquished Victims

Alas, many species are unable to respond adaptively to their enemies. For many, control by enemies means economic restriction. It means that the tasks of life must be carried out at times and in places where exposure to those enemies is brief or infrequent. In effect, these predation-vulnerable species live in enemy-free refuges, and in the margins and in the interstices of ecosystems.[57]

Consider a few examples. The soft, young caterpillars of many moths and butterflies in temperate deciduous forests occur mainly on the undersides of leaves, where birds foraging from above might not see them. The adult insects, which can fly and therefore have an effective escape response to predators, are much less restricted in their activities than are the larvae. Young, fast-growing snails with delicate shells typically shelter beneath large boulders or deep in crevices away from predatory fish and crabs, but thick-shelled adults forage actively for food on open surfaces, seemingly oblivious to potential enemies. Panamanian armored catfishes are vulnerable as adults to predation by wading birds. The adults live only in the deeper parts of streams and pools, where most of the birds cannot reach them. Brittle stars (ophiuroids) typically lurk beneath rocks in shallow waters, but in a few fish-free lagoons they occur in great numbers on all available surfaces.[58]

Another manifestation of control, well known to ecologists, is the regulation of victims' population sizes and densities by enemies. On many rocky shores of the western United States, the ochre seastar is a voracious predator of mussels, which the seastar subdues by pulling the valves apart with suckers on the arms. The California mussel (*Mytilus californianus*) forms a dense bed or zone on the upper middle shore, at a level where exposure to the air during ebb tide is often too long for the seastar. The lower limit of the mussel bed thus more or less coincides with the zone in which the ochre seastar becomes abundant. Ironically, the mussels grow faster and to a larger size at low levels of the shore than they do in the upper zone where they are most abundant. In fact, where the seastar is locally absent, mussels abound in the lower zone, and even well below the low-tide mark. Sea urchins, which voraciously consume kelps and other seaweeds, are a favorite prey of sea otters in California. Where sea otters are abundant, most sea urchins occur at depths below the typical diving limit of the otters, but food at these depths is scarcer than in the shallow waters frequented by otters. The Panamanian armored catfish mentioned earlier also eat algae, but adults are constrained by their would-be predators to live below the zone of highest algal abundance and productivity. Local restriction of species to suboptimal habitats by enemies is a well-documented and very widespread phenomenon.[59]

If species cannot maintain populations in any part of the range where the conditions for growth are favorable, selection ultimately drives them to specialize to these conditions. Desert plants, cave animals, deep-sea species, alpine herbs, tundra mosses, splash-zone snails in the no-man's-land between sea and land, and the initial Early Paleozoic colonists of the dry land all inhabit, and became well adapted to, places where enemies are (or were) weak and scarce. Enemies may likewise be responsible for the occupation of deep sediment layers, cavities and tunnels in rocks, and the plankton by organisms specialized to these environments.

Marginalization can take subtle forms. The evolution of fish-eating flying vertebrates—pterosaurs and birds, perhaps beginning in the Jurassic—may be responsible for the transformation of lungs to buoyancy-controlling air bladders in teleost fishes. The lung is evidently primitive among fish as an air-breathing organ. Its continued function in breathing air became untenable, so the argument goes, when aerial predators became too much of a threat for some shallow-water fish. Life in deeper water means that the lung ceased to function in breathing air, but is consistent with a buoyancy-regulating function of that organ.[60]

I have already brought up another class of examples of a retreat to safe havens and subsequent adaptation to them, namely, among pathogens and parasites. The bodies of hosts, many of which are equipped with potent passive or power-demanding defenses, are less accessible to predators of small prey than are the surrounding environments. Accordingly, living hosts offer refuge to small, defenseless creatures, which with time can become specialized as parasites, mutualists, or unobtrusive commensals as they adapt to the special conditions on or in hosts. The Coralliophilidae is a group of warm-water snails that feed on, and usually live in association with, corals and other nematocyst-bearing cnidarians. Within this specialized group, many species are more or less mobile, moving among hosts and often remaining highly exposed to potential enemies while they feed. Their shells tend to be thick and highly sculptured with ribs, spines, and other features that stiffen and strengthen the shell. A few species, however, have become sedentary, some going so far as to excavate permanent burrows in the host. Their shells are thin, delicate, and unremarkably sculptured. They may well be susceptible to predation, especially by parrotfishes, triggerfishes, and other predators that can break up coral skeletons, but their invulnerability is due largely to protection offered by the host.

This kind of specialization to initially unfavorable habitats is the expected consequence of a curious type of inequality operating at the level of populations. If enemies keep the number of individuals down in parts of the species range where resources are most abundant, individuals in that part of the population will contribute fewer offspring to the next generation than will the individuals confined to a less bountiful environment. This inequality means that most of the selection to which individuals are exposed occurs in enemy-free circumstances. The part of the population where resources are abundant and enemies wipe out most individuals has become a sink rather than a source of variants that are subsequently selected. Adaptation, in other words, operates by majority rule, and in these cases the majority live in resource-poor but relatively safe environments.[61]

Reciprocity

The idea that the predominant evolutionary and ecological control on organisms is exercised by enemies appears to conflict with the view, held by many biologists, that reciprocal adaptation of interacting parties is widespread in the economy of nature. Reciprocal adaptation, or coevolution, means that two interacting entities are responding evolutionarily to each other. As party A adapts to party B, party B accommodates to party A, which responds to B, and so forth. Plants and their animal pollinators, plants and their specialized herbivores, animals and their specialized parasites, prey and their specialized predators, mutualistic symbionts and their hosts, and antagonistic males and females of the same species have all been cited as possible examples of coevolution. If coevolution were indeed a widespread phenomenon in nature, control would be much less one sided than if enemies dominated as agents of selection and restraint.

Coevolution is most likely in cases where two parties are uniquely associated with each other, especially those that mutually benefit or become dependent on one another. As the examples below will show, these reciprocal relationships are often complex affairs, in which antagonisms and conflicts of interest play important if sometimes hidden roles. Mutual benefit—or, perhaps more often, mutual exploitation—sometimes tilts in favor of one of the parties, so that even when there is reciprocity in the interaction, one party tends to dominate in an evolutionary and economic sense.

Leaf-cutter ants in tropical America cultivate fungi in underground chambers for food. These domesticated food sources, which for the most part cannot thrive outside the ants' domain, depend for food on fresh leaves, leaf fragments, or dead material that the ants bring them. The fungi are parasitized by other fungi, which are also host specialized, The ants, cultivated fungi, and parasitic fungi thus comprise a mutually interdependent system, whose members show precisely congruent patterns of evolution. The system therefore displays coevolution involving both cooperation and antagonism. Comparable agriculture has evolved once in fungus-growing Old World termites and at least seven times in scolytine ambrosia beetles, a highly diverse group of weevils that grow fungi on the walls of galleries in the bark and wood of trees worldwide. The mutual dependence between exploiter and exploited displayed in the relationships between insects and cultivated fungi closely parallels that between humans and certain food crops. Maize, a Mexican grass, has been so modified by human cultivators that its reproduction has become completely dependent on people. Selective breeding by farmers produced plants in which the seeds do not separate from the plant and therefore do not disperse or germinate by themselves. Agriculture is thus a form of mutualism and interdependence that can be interpreted as coevolutionary cooperation, but there can be little doubt that the domesticaters gain a competitive advantage, and that the evolutionary arrangement involves unmitigated exploitation.[62]

The world's 750 species of figs (trees of the genus *Ficus*) are pollinated by tiny agaonid wasps, which are mostly specialized to a single species of fig. The figs are completely dependent on the wasps for pollination, but they pay a high price for the benefit by serving as the food source for the insects. The flowers and seeds develop on the inside of the syconium, the unique fig fruit. After mating, female wasps lay eggs in some of the seeds, which are then eaten by the newly hatched wasps. The wingless males stay in the fruit to mate, but mature females exit the syconium, taking with them pollen as they go in search of a new faraway fig host. In the new host, the females enter a syconium, deposit pollen on receptive female flowers, and lay eggs in seeds.

Both the wasp and the fig have to contend with many species besides each other. Vertebrates, especially bats and birds, disperse the seeds by eating the highly nutritious, calcium-rich, and energetically expensive ripe fruits. To gain this benefit, the fig sacrifices some seeds and makes the expensive fruit available nearly year-round. The figs must also bear the cost of parasitic wasps, which lay eggs in the syconium and eat the seeds but do not pollinate the flowers. The pollinating wasps, too, have their problems. They are parasitized by nematode worms and by several species of wasp. With the exception of the vertebrate dispersers, all of these relationships are highly specialized, with each fig having its complement of unique insect and nematode associates, and each associate occurring on only one species of fig.

This remarkable story, worked out by Alan Herre and many others, is a complex tale of intertwining antagonism and cooperation, of exploitation and mutual dependence, of reciprocal adaptation and top-down control. Despite all the ways in which this system could have been destabilized, figs have been with us for at least 80 million years, and presumably so have many of their associates. Throughout, the fig has been the dominant partner, the one that calls the shots.[63]

In the northwestern United States, the moth *Greya politella* almost exclusively pollinates the flowers, and feeds on the developing seeds, of the saxifrageaceous herb *Lithophragma parviflorum*. In many populations, the herb is completely dependent on the moth for pollination, and thus the interaction between these two species is mutualistic, albeit with a component of exploitation by the moth. In other populations, however, bees and flies may also pollinate *L. parviflorum*, and then the relationships between the moth and the herb is either neutral or antagonistic. Reciprocity thus depends on whether other pollinators are present, and always has the potential to dissolve into antagonism in favor of the moth.[64]

Every specialized host-parasite relationship is a candidate for coevolution, but only a few cases appear to show the evolutionary concordance between the two parties that would be expected if reciprocal adaptation were the rule. A long-term association between the host and guest lineages would be consistent with coevolution if the pattern and order of branching in the host were precisely duplicated by the guest. Parallels like this have been documented for many associations involving parasitic insects. Chewing lice have in this sense coevolved with host pocket gophers, although the rate of evolution in the parasites, as inferred from rates of molecular substitution, have been three to ten times faster than in the pocket gophers, the factor depending on whether the substitutions have functional consequences or are silent. The generation time in the chewing lice is correspondingly shorter (forty days) than that of the pocket gophers (one year).[65]

The much more common evolutionary pattern is host switching, in which evolution in the parasite coincides with a switch from one host to another. For example, many insects expand or switch to such agricultural crops as sugar cane when these are grown in regions where the crops did not originate. Host switching in the wild has been observed for a population of the checkerspot butterfly (*Euphydryas editha*) in the foothills of the Sierra Nevada. Over a ten-year interval, the population transferred from the native wild host *Collinsia parviflora* to the introduced weed *Plantago lanceolata*. Even among fig wasps, there have been some switches over evolutionary time from one host fig species to another. Drude Molbo and her colleagues used molecular markers to show that pairs of sister species among fig wasps inhabit host figs that are not closely related, implying that the strict coevolution between guest and host in this complex system has sometimes been violated.[66]

Perhaps a majority of animal species are parasites on one or a few host species. Peter Price estimates that some 80 percent of insects are parasitic on animals and plants. No comparable estimates are available for marine animals, but the diversity of flatworms, copepods, and snails on larger animal hosts is extraordinary. In contrast to the situation on land, however, notably few marine species parasitize single host plant species. In the rich marine floras of the North Pacific, I could find only perhaps a dozen species of snail and chiton that are specialized to live and feed on particular seaweeds or seagrasses. Coevolution between host and guest appears not to have occurred in any of these cases.[67]

Third parties may in very many cases be responsible for the apparent coevolution between host and guest. Experimental studies by Mark Hay and his colleagues have demonstrated that chemically noxious seaweeds offer safe havens for small, relatively poorly defended amphipod crustaceans and snails that would fall easy prey to fish if they were away from their hosts. In California, the limpet *Lottia paleacea* incorporates diagnostic chemicals from its food host plant, the seagrass *Phyllospadix menziesii*, into its shell, with the result that the small predatory seastar *Leptasterias hexactis* is unable chemically to distinguish the limpet from its host. The resemblance and possible but unproven coevolution between limpet and grass is thus enforced by the limpet's predator.[68]

The intimate associations between marine animals and single-celled photosynthesizing organisms benefit both parties, and so might qualify as examples of coevolution. The symbionts have a safe place to live—the bodies of well-defended animals like corals and giant clams—and the host gains nutritional benefits that translate into a competitive advantage. Symbiont-bearing corals, sponges, clams, and other animals reach larger sizes and typically grow faster than their relatives lacking symbionts, and are more effectively defended than the seaweeds whose ecological role as primary producers they have coopted. As in many cases of parasitism, however, host fidelity is evolutionarily weak and ephemeral. In a comparative study of ten strains of the symbiont *Symbiodinium microadriaticum* among twenty-two host species, Robert Rowan and Dennis Powers found no fidelity on the part of either the host or the guest. Five species of the coral genus *Madracis* in the Caribbean all share a single strain of *Symbiodinium* as guests. This flexibility and generality may result in part from the method by which symbionts are transmitted from host to host. Coral larvae (planulae) must be reinfected by symbionts while they live in the plankton before settling to the bottom to initiate new coral colonies. The apparently mutualistic association, moreover, is not without antagonistic overtones. The host can expel the symbionts during times of stress, leading to the phenomenon of bleaching in which the host loses color and the ability to photosynthesize. Colonies can become reoccupied by symbionts from the plankton, and this reinfection may again be flexible in terms of which strains are available.[69]

In the mutualism between land plants and root-associated fungi (mycorrhizae), economically analyzed by Mark Schwartz and Jason Hoeksema, both parties are able to remove carbon, phosphorus, and nitrogen from the soil, but the plant typically collects more of each element than does the fungus. The fungus, however, acquires more phosphorus than carbon. In this system, it is advantageous for both parties to trade, the plant trading away carbon to receive phosphorus from the fungus. Each party in the trade benefits from specializing to the acquisition of the elements for which it holds an advantage in extraction from the soil relative to other elements. And yet, despite this mutually favorable trade, it is the plant that exercises the greater control over the resources and the interaction, and there is no strong evidence for coevolution.[70]

A final example of symbiosis, that between the bacterium *Buchnera* and aphids of the genus *Uroleucon*, evidently does represent lock-step reciprocal evolution. The bacterium is strictly transmitted from one generation of aphids to the next, with horizontal transmission being very rare or absent. Mutual fidelity is therefore assured.[71]

Although coevolution has historically been studied chiefly in interactions between species, it may turn out to be most important as a powerful competitive process in the evolutionary interaction between males and females in sexually reproducing species. Males and females often have divergent interests. In the fruit fly *Drosophila melanogaster*, for example, competition among males has made the seminal fluid of one male toxic to the sperm of another. This toxicity, however, also reduces the survival of females, and is therefore a telling example of conflict between the sexes. Such conflicts also affect sexual displays. If, for example, a male is unusually attractive to a female by virtue of a particularly eye-catching display, the female may mate too often or in the wrong places and therefore reduce her own fitness. Traits making the female less attracted to males would be favored under such circumstances, but the males can then respond by making themselves even more attractive. Reciprocal rounds of this kind of sexual selection are thus powered by competition between the sexes. Still other examples come from the energy cost and injury to mating. Among water striders—insects of the family Gerridae that live near or on the water surface—mating for females is costly because it involves struggling with the males and because mating occurs often. Males attempt to clasp the female, whereas the females resist being clasped. This antagonism has fueled an escalation toward greater development of clasping organs in the males and features such as abdominal spines with which females prevent excessive mating. Despite this antagonism, mating is of course essential if the two contestants are to leave offspring. This community of interest ensures that the two parties exercise intense selection on each other in a reciprocal, though not always equal, manner. In the cases where armament is unequally developed between the sexes, meeting are more frequent and take longer when males have the advantage and rare when females have the upper hand.[72]

Sexual conflict and coevolution have figured prominently in the evolution of all plants and animals that practice internal fertilization. Charles Darwin was among the first to observe that the genitalia of insects, land snails, and many other animals are extraordinarily complex, and that they are among the most reliable distinguishing features that set closely related species apart. Although these distinctive traits have typically been interpreted as means by which individuals of the same species recognize each other, the elaborate courtship behaviors and other rituals prior to and during mating strongly imply that mating-related structures are the products of intense reciprocal selection in which cooperation and conflict have uneasily coexisted.[73]

That this special intersexual reciprocity has not merely been an evolutionary sideline but a major evolutionary mechanism affecting overall performance is clearly illustrated by sexual selection in our own species. In a bold departure from customary views about our brain—that its intelligence aids survival—Geoffrey Miller suggests that the human brain's stepwise evolutionary enlargement, especially involving the frontal cortex, reflects sexual selection and coevolution between males and females. Both sexes choose mates for their creativity and language, and men and women compete among themselves in different ways to impress potential mates. Using 25 percent of the body's metabolic energy but taking up just 2 percent of the body's volume, the brain is an expensive tool or "sexual ornament." With our language, we convey information about our health, our fitness, our desirability as providers, and our social status. Much human courtship is verbal, and the seat of language is in the brain. Humans, according to Miller, embarked on a path of sexual coevolution in which symbols of fitness emanate from our brains: complex thought, abstract problem solving, spoken language, creative storytelling, the projection of humor and kindness, and the like. These attributes likely aided survival as well, and in the long run were responsible for our unparalleled expansion; but the primary motor, Miller suggests, is sexual competition with the brain as the chief weapon.[74]

I interpret these and other examples of apparent reciprocity, whether species specific or diffuse, as cases of cooperation enforced either by underlying antagonisms in the relationship or by a common greater enemy. The more energy-intensive participant is the party with the greater control and the one reaping the greater competitive, reproductive, or defensive benefits. Third parties—themselves high-level consumers—are often decisive in determining the nature and outcome of the evolutionary interaction between hosts and guests. Competition between the sexes is likely the most powerful and pervasive form of reciprocal evolution, one that primes and amplifies survival-related forms of competition.

Control in Human Society

The reader may by now be willing to grant that enemies exert control in very many economic interactions, both within the lifetimes of individuals and over the long run of evolution. But does this enemy-related control carry over to the inequalities observable in human society?

When we think of the word "consumer" in human-economic terms, we think of an individual or a household. A producer, too may be an individual—a farmer, a laborer, someone who makes something or who renders a service—but in organized society, production is often in the hands of groups—cooperatives, corporations, and the like—which are often substantially larger, and which wield correspondingly more power, than a consuming individual or household. In this semantic frame of reference, therefore, it would appear to be the producers that have the upper hand. It is they who know more about the market—about prices, about supply and demand—and it is they who, through advertising and promotion, can affect market forces to their advantage. The human perspective on control raises an important and unappreciated point about economic structure. The difference between producers and consumers becomes evident mainly at the level of the species, or the occupation. Every individual of every species is both a producer—of offspring, and as a resource for others—and a consumer. There is apt to be little difference in economic power between the individual as producer and the same individual as consumer. Among species in nature, power is concentrated in individuals and populations of large predators and other animals that eat other organisms, and this power emanates from keen competition. Power in human society is likewise tied to occupations, although the source of that power has changed over the course of history. Force, whether applied by individuals or groups, has traditionally been the source of much power, and clearly remains so today. But knowledge is perhaps becoming increasingly important as a source of power and influence. The axis along which power and influence are measured has shifted from the producer-consumer continuum in natural ecosystems to a less well-defined distinction in the human economy. With the ability of individuals and groups to combine roles in single entities—production and consumption, for example—we have for the first time broken the hegemony of the consumer, at least within our own species. At the larger scale, of course, this hegemony has not been violated at all, for we are the most powerful consumer and the most powerful economic force in the biosphere.

Consumers may enjoy top-down control, but they depend utterly on what they control, namely, producers. In the next chapter, I shall look at the producer side of the economy and show how far this interdependence between rivals has come in creating a self-perpetuating economy.

Chapter 5

THE ECONOMICS OF EVERYDAY: PRODUCTION AND THE ROLE OF RESOURCES

WHY IS THE WORLD GREEN? Nelson Hairston, Frederick Smith, and Lawrence Slobodkin asked this seemingly simple question in a classic paper published in 1960. The most obvious answer is that we humans, and all the other animals and fungi of this planet, wouldn't be here without the constant toil of green plants and those other, not necessarily green, creatures that make organic matter from inorganic components. Consumption is impossible without production, and if there is too much consumption for the productive machinery to absorb, all the green things will die, along with their overzealous dependents.[1]

Every organism is, of course, a producer of sorts—growing, reproducing, and dying—potentially a resource for some other creature. It is in the collective interest of consumers to stimulate, or at least not excessively to interfere with, production. It is thus not surprising that the waste products and the dead remains of consumers often enhance or restock the supply of raw materials for plants, and that top predators directly aid primary production by consuming animals that feed directly on the organisms that make biomass from nonliving matter. Viewed from this economic perspective, an interdependence exists between producers and consumers. The system may appear to us to be remarkably well regulated and tuned, but it cannot be otherwise, for else it would not be here.

Interdependence between production and consumption holds even within organisms. Photosynthesis in plants takes place in leaves and sometimes in stems, but not in the roots, which are net consumers of oxygen just as animals are. In individual cells, photosynthesis takes place in plastids, organelles where all the requisite enzymes are located, and not in such other parts as mitochondria, chromosomes, or the nucleus. Thus, regardless of the scale at which we examine production, there is an intimate coordination among the parts of an economy that produce and the parts that consume. Small parts of an economy—the thin veneer of Earth's surface bathed in sunlight needed for photosynthesis, the leaves of plants, the zones of growth in plants and animals, and the organs of reproduction—subsidize all the rest.

The question Hairston and his colleagues posed is not as naive as it might appear to be at first. Here is the problem: plants produce enough biomass not just to perpetuate themselves, but to support a complex, multilevel economy as well, including powerful consumers. We can imagine an alternative economy in which there is just enough plant biomass left over to sustain a single tier of modest consumers, a subsistence community whose collective activities keep the primary producers down to a bare minimum, a thin soup of floating phytoplankters or a low stubble of green photosynthesizers. Indeed, such economies exist. In the eon before multicellular animals began to dominate the biosphere economically, the phytoplankton was grazed by a community of single-celled bacteria and protists, which recycled nutrients back to the phytoplankton in what has come to be called the microbial loop. This microbial loop still exists in the open ocean. On many rocky seashores around the world, grazing snails and fish keep the cover of seaweeds down to a film or a low turf, which to the human eye or hand appears as little more than barren rock. No members of these economies accumulate any power or wealth; the participants remain small because there is essentially no surplus of production. Producer biomass does not accumulate; it cycles rapidly to consumers of both living and dead organic matter.

In a precisely analogous way, agricultural production by humans for the most part remained so low until the nineteenth century that a large fraction of the human population—20 percent in Britain and the Netherlands in the early nineteenth century—could be classified as paupers. Not until agricultural production increased dramatically after 1850 was there enough of a surplus to power enormous growth in the per capita and collective use of energy. What aspects of production and consumption made this transition from a subsistence economy to a surplus economy, in which both producers and consumers accumulate wealth and power, possible?

The key to production lies in collecting and storing energy and raw materials, regulating and increasing the predictability and rate of supply, and distributing or allocating necessities to perform the tasks require to maintain and propagate the producer. These imperatives apply whether the producing entity is a cell, a redwood tree, an entire ecosystem, or an industrialized human economy. In order to grow and multiply—and, from the larger perspective, to support all those consumers—producers must acquire and maintain a surplus. In chapter 4 I concentrated on the methods that living things have devised to keep gained resources from falling into the clutches of others, and I underscored the role of consumers as agents of top-down evolutionary and economic control. In the present chapter, my focus shifts to the acquisition of raw materials, to competition among producers, and to the roles that storage, surplus, and regulation play in sustainably supporting consumers. I argue that the ability to store resources offers an entity a competitive advantage as well as a passive means of protecting the entity from potential consumers and from temporarily unfavorable conditions. The sequestration of resources for future use is, I believe, a key to the evolution of surplus and to the development of a consumer-driven economy.

Competition and Energy

David Landes speaks for many historians and economists when he writes, "All economic . . . revolutions have at their core an enhancement of the supply of energy, because this feeds and changes all aspects of human activity." Although he was clearly thinking mainly about human history and about energy-driven advances in civilization, his assertion applies with equal force to the economy of nature, where all activity depends on production. Everything hinges on how primary producers acquire energy and transform it into biological work sufficient to maintain and propagate themselves and all life forms that depend on them.[2]

The first energy to be exploited by life may have been that in the chemical bond between the two atoms of molecular hydrogen (H_2). This hydrogen, released by interactions between fluids and igneous rocks produced by geological activity deep inside the Earth, is combined with volcanically generated carbon dioxide (CO_2) by microbes (Archaea) to yield methane (CH_4) and water (H_2O). Bonds in such other simple compounds as hydrogen sulfide (H_2S) and carbon dioxide also yield energy that early microbes could exploit and that simple forms of life deep below Earth's surface still use. The released energy can power life processes directly, or is trapped and stored in such compounds as ATP (adenosine triphosphate) to be used later. Energy not thus transformed is released in degraded form as heat.[3]

In our world, most food—carbohydrates (sugars, starch, and other compounds of hydrogen, oxygen, and carbon), fats or lipids, and proteins (compounds containing nitrogen-based amino acids)—comes directly or indirectly from photosynthesis, a process requiring a restricted part of the visible spectrum of radiation, accomplished by primary producers, organisms that capture (or fix) carbon from inorganic sources. For most land plants, carbon dioxide in the atmosphere is the source of inorganic carbon, but for some mosses and for single-celled photosynthesizers in water (the phytoplankton), this source is generally replaced by the bicarbonate ion (HCO_3^-), which constitutes about 90 percent of the dissolved inorganic carbon in water.

The Earth intercepts about 1.8×10^{24} erg/s, or 1.8×10^{17} W, of solar radiation, of which 70 percent penetrates the atmosphere. According to Eric Chaisson's calculations, the entire biosphere uses energy at a rate of almost 10^{21} erg/s (10^{14} W), or 0.1 percent of the total energy reaching the Earth as radiation. Only about 31 percent of the sun's photosynthetically active radiation reaching land that is not permanently ice covered is absorbed by land plants, much of it as heat. In the ocean, which today covers 71 percent of the Earth's surface, and in which light penetrates to a water depth of about 100 m, phytoplankters absorb only about 7 percent of this relevant radiation. More than half the energy that plants absorb powers the movement and evaporation of liquid water. Evaporation takes place through small pores (stomates) on the surface of the plant, and pulls up liquid water and dissolved nutrients from the soil through the roots and stem up to the leaves, where photosynthesis and other processes transform water, nutrients, and inorganic carbon into organic molecules. Even the most efficient plants reach a maximum rate of photosynthesis at just 20 percent of the visible light that reaches them in full sunlight. In most tropical and temperate forests, 1–3 percent of sunlight reaches the forest floor. According to Egbert Leigh's calculations for the rain forest at Pasoh, Malaysia, the power output of plants per square meter of forest floor (or, equivalently, per eight square meters of sun-exposed leaf area) is not more than about 5.6 W, given an average energy input of 164 W per square meter of ground. Globally, net primary productivity—the difference between the total amount of carbon fixed by photosynthesis and the carbon respired by these producers—is about 1.33×10^8 g/s.[4]

If there is so much energy available globally, why is there such intense competition for it? The answer, as I emphasized in chapter 2, is simple: it is not the global amount of energy that matters, but the amount available to interacting units. As an understory shrub with its leaves arranged in horizontal sprays intercepts light from above, what matters is that there are canopy trees overhead, which prevent most sunlight from reaching the shrub. The amount of energy reaching a part of the forest a kilometer away is, in essence, irrelevant to these interacting individuals. It is the local availability of energy that determines what kind of competition there is, how intense it is, and what its consequences are. Availability, in turn, depends on whether and how resources are replenished, on processes affected by competitors, on diffusion, and on forces generated by the user itself.

Methods of Competition

Although phytoplankton may be among the oldest known photosynthesizers using light and producing oxygen, light may not be the limiting resource for this class of primary producers in the open ocean. Victor Smetacek has noted that phytoplanktonic organisms—Cyanobacteria, coccolithophorid green algae, dinoflagellates, diatoms, and partnerships between these and such single-celled animal-like protists as foraminifers and radiolarians—overwhelmingly occur as small, separated, single cells. Some diatoms such as *Rhizosolenia* and *Hemiaulus* form filaments or mats, but multicellular primary producers are rare in the plankton, especially in the open ocean. The floating brown seaweed *Sargassum*, which lends its name to the Sargasso Sea in the subtropical western North Atlantic Ocean, is an apparent exception, but the floating individuals all derive from large attached populations in Bermuda. Free-floating plants are common in nutrient-rich quiet waters of freshwater lakes, ponds, and canals, where they likely compete for light and shade out attached vegetation. By my count, there are at least twelve independent acquisitions of the free-floating habit among plants of terrestrial origin, ranging from liverworts such as *Riccia* to duckweed (*Lemna*), insectivorous bladderworts (*Utricularia*), the ferns *Azolla* and *Salvinia*, and hornworts (*Ceratophyllum*). These larger floating plants presumably do compete for light, as well as for dissolved nutrients, with phytoplankton and with attached submerged plants.[5]

Competition for light is important not only for attached plants—seagrasses and algae growing submerged or between tidemarks, and plants in soil or as epiphytes on land—but also for dense populations of phytoplankton in highly productive coastal waters. Place-holding stationary plants and abundant floating phytoplankton intercept light from above and from the side, depending on the orientation of the sun and on the orientation of light-gathering surfaces in winds and currents. Importantly from the perspective of competitors beneath, they also cast shade. Plant architectures emphasizing light gathering from the side include conifers, whose cone-shaped crowns expose branches and needles to the side and permit the shedding of heavy snow; and vertically disposed blades of grass and the similarly oriented leaves of sea grape (*Coccolobis uvifera*), a small tree familiar from tropical seasides. Side illumination also occurs in *Marcgravia*, a peculiar vine that plasters its leaves on the vertical trunks of host trees. The parasols of *Trillium*, the large vine *Monstera*, *Tropaeolum* (the "nasturtium" of gardens), and many other forest ground herbs, as well as the floating leaves of water lilies and the horizontally disposed sprays of leaves on branches of maples (*Acer*), intercept light chiefly from above, and cast deep shade beneath. Desert shrubs and many vines such as greenbriar (*Smilax*) have green stems as well as leaves, making the entire plant a light-absorbing surface. Kelps (Laminariales) and many other seaweeds form a dense light-absorbing canopy, but their blades, often crinkly and with wavy margins, also sway back and forth, enabling these competitively dominant nearshore plants to intercept light from many angles. Some land plants, notably vines with special junctions between blade and petiole, can change the orientation of their broad leaves under their own power so that they can take full advantage of the sun's rays as the angle of the incoming light changes with the hour and with the season.

Plants compete with other plants by casting shade, but they must in turn avoid being shaded themselves. The distance over which a leaf casts shade varies directly with its surface area: the larger the leaf, the greater is the vertical distance below over which it casts shade. Large light-gathering surfaces should therefore be deployed above competitors, but these surfaces must be sparsely distributed around the plant's stem lest the leaves overlap and cast shade on each other. In full sunlight, however, plants cannot absorb all the incident light, and therefore let through a substantial amount of sunlight for lower leaves to absorb. A land plant in dim light, whose intensity is less than 20 percent of full sunlight, should intercept as much light as possible. In the sea, submerged seagrasses can operate at just 11 percent of surface light; most attached seaweeds can live at water depths where they receive only 0.12–1.5 percent of surface light, and encrusting corallines can photosynthesize at depths where the available light is only 0.003–0.05 percent of that at the water surface.[6]

For attached organisms like plants and plant-animal partnerships, the ability to hold and gain space is essential. Shading out competitors is one mechanism; monopolizing space by preventing recruitment of competitors is another. Many primary producers do this by spreading underground or on the surface of attachment by vegetative propagation, in effect producing a colonial organism with many photosynthesizing modules that look like individual plants. Among seaweeds, vegetative propagation appears to be especially common in situations where grazing by scraping sea urchins, snails, chitons, and fish is intense, but studies comparing the photosynthesis of modules in a "colony" with single plants of the same species show that the protection offered by vegetative turf formation against grazing and against recruitment by other seaweeds comes at the cost of 33–51 percent less carbon fixed. The most productive seaweeds—kelps, which are highly susceptible to sea-urchin grazing—do not propagate vegetatively, relying instead on recruitment from spores. Among corals, the colonial form also protects against consumption. Single polyps typically die when predators attack, whereas large colonies tend to suffer partial consumption. Land plants with underground propagation or other mechanisms of vegetative propagation derive benefits in competition for space as well as in protection from trampling by large herbivores.[7]

Competition for Materials

Life depends not only on energy and its transformation, but also on material resources. After all, it is composed of matter, and all the work done by life—construction, growth, activity, activity, propagation, and competition—involves the gain or loss of substances. Almost any element or compound that is involved in some life function can be locally limited in its abundance or supply, and like energy it can thus become the object of evolutionarily significant competition.

All life forms are surrounded by walls—membranes, cell walls, and sometimes a skeleton—that impede the entry and exit of substances. Diffusion—the random movement of molecules from areas of high concentration to areas of lower concentration—fundamentally controls the rate and extent of transfer of matter into and out of a living body. For small bodies, whose volume (and therefore material demand) is small compared to the surface area available for importation of raw materials, diffusion is sufficiently rapid, and does not limit metabolic activity. Moreover, the region immediately outside the body is not depleted of the resource, because the body is so small that no envelope (or boundary layer) of fluid or gas surrounding the body is developed. In larger bodies, however, demand can easily outstrip the surface-dependent transport of substances, and a boundary layer of stagnant medium forms around the body in which the concentration of raw materials becomes depleted relative to the concentration outside the envelope.[8]

Plants surrounded by water absorb nutrients, as well as the carbon dioxide needed in photosynthesis, by diffusion across the body surface. Material exchange between producer and environment is inhibited when the fluid adjacent to the organism is stationary, that is, when there is a distinct boundary layer. It is enhanced when the medium surrounding the absorbing surface is continually being replaced. Such is the case, for example, in the turbulent open waters of lakes and oceans, and on wave-swept shores where luxuriant stands of attached seaweeds clothe the rocks. In stagnant water, however, the local nutrient supply or carbon dioxide concentration is quickly depleted, so that the primary producer resides for extended periods in an envelope of scarcity surrounded by a sea of plenty.

Two solutions to this problem are potentially available to organisms, both involving the replenishment of needed substances in the boundary layer. One solution is to move, or to grow into, environments not yet tapped. Such movement, of course, requires work on the part of the living entity. The second solution is to induce turbulent flow of the medium surrounding the body. This flow, in which the paths of molecules in the medium follow an erratic course instead of a smooth trajectory as in laminar flow, peels away the boundary layer and therefore enhances diffusion of substances by replenishing the external supply. Seaweeds flapping in the waves, and leaves rustling in a breeze, are manifestations of turbulence, which can be induced by flexibility of the absorbing surface or of its attachment to the rest of the plant. A three-dimensionally complex topography can also induce turbulence. For example, the rough surface of coral colonies can promote exchange of gases or nutrients between the photosynthesizing polyps and the surrounding water. Perhaps the crinkly blades of many kelps and the wavy-edged leaves of such familiar land plants as kale, bay laurel, holly, and lettuce perform similar functions with respect to gas exchange with the atmosphere.[9]

Another possible way in which many aquatic photosynthesizers have eased the limitation of diffusion of carbon dioxide is to induce the formation of that gas by laying down a skeleton of calcium carbonate. Ted McConnaughey has argued that the production of calcium carbonate by such algae as the stonewort (*Chara*) releases carbon dioxide, which is then taken up during photosynthesis. The link between calcification and photosynthesis evidently also exists in corals, which contain photosynthesizing dinoflagellates in their surface tissues. These organisms can therefore fix carbon even under relatively stagnant conditions, especially if they possess means of extracting nutrients from sources other than water. Interestingly, many of the calcifying green seaweeds—*Halimeda*, for example—that live in lagoons and other quiet-water conditions can take up dissolved nutrients from the sediments in which their rootlike rhizoids are attached; and corals snag small zooplankters as sources of food with their nematocysts.[10]

In land plants, carbon dioxide diffuses through stomates in the stem and leaves, but nutrients normally enter through the roots. Only desert plants and epiphytes are capable of extracting nutrients from water and dust collected on leaves. Sunlight-driven transpiration—the loss of water through stomates—and metabolism-powered processes enable plants to lift minerals from the soil. Transpiration and the uptake of carbon dioxide cease when the air's relative humidity hovers near 100 percent, but plants can still mine the minerals needed for growth.[11]

Nutrient concentrations in shallow-water sediments are ten to one hundred times higher than in the water above, and would thus provide a rich resource for primary producers living in water. It is therefore highly surprising that plants in only a few clades have become able to extract nutrients from sediments. Among primarily marine plants—that is, those plants that have always lived in the sea throughout their evolutionary history—only siphonaceous green algae appear to have this capability. Their rootlike rhizoids absorb nutrients from the sediment, and these nutrients are subsequently transported through the rest of the plant, which consists effectively of a single gigantic cell—a coenocyte—with many nuclei. Siphonaceous green algae, belonging to such genera as *Caulerpa*, *Penicillus*, *Halimeda*, *Avrainvillea*, and *Udotea*, tend to occur in tropical waters in which the concentrations of dissolved nutrients are low.[12]

Most land plants have roots, which mine water and dissolved nutrients from soil, often in close association with root-dwelling fungi and bacteria. Without plants, which secrete organic acids and other agents that help dissolve minerals, there would be no soil, and nutrients buried in sediments would remain out of reach of members of the biosphere for extended periods.

Roots developed in land plants during the Devonian period, but once they evolved, they also proved useful in aquatic habitats. C.D.K. Cook estimates that between 222 and 271 clades of living land plants have independently invaded freshwater habitats, where they often numerically and competitively dominate algae. Plants from almost every major group—liverworts, mosses, ferns, club mosses, horsetails, and flowering plants—have made the transition, and can be found in habitats ranging from swift streams to quiet muddy ponds and lakes. Only the gymnosperms—pines, cypresses, and their relatives—have no aquatic representatives. At least five clades of monocotyledonous flowering plants (including three major ones, together comprising about sixty species today) have successfully invaded the sea, where they form extremely productive ecosystems on sandy and muddy bottoms that were never effectively exploited by seaweeds. Although rooted plants have probably lived in freshwater for 350 my or more, marine flowering plants have existed for only about 70 my. Members of some forty additional clades of land plants have penetrated the zone between high and low tidemarks, where they form productive mangrove and salt-marsh vegetations. The high productivity and great economic importance of these secondarily freshwater and marine plants is due in large measure to the mining activities of their roots, an activity that has enabled these plants to push back the algal constraint of being unable to take up nutrients from food-rich, quiet waters.[13]

Scientists have argued passionately and needlessly over which elements limit the rate of production in ecosystems. Life processes critically depend on more than twenty elements, any one of which may be in short supply at scales ranging from the distance between competing phytoplankters, each 0.2 mm in diameter, to the global, and on time scales from seconds to many millions of years. The problem for organisms is that many of the elements arrive in forms that are not usable, or biologically available; rather, the elements must be transformed—usually through the efforts of other living things—into compounds or ions to be incorporated for the manufacture of more complex organic compounds. Scarcity, or the extent of limitation, therefore depends not only on the rate of supply of inorganic components, but also on how quickly and in which ways organisms recycle elements by converting them into usable forms.[14]

Consider the case of nitrogen, for example. It is hard to imagine that nitrogen could be a limiting nutrient for any organism given that it comprises about 78 percent of the gas by volume in Earth's atmosphere. In its inorganic (or dinitrogen) form of two nitrogen atoms bound into a single molecule (N_2), nitrogen is inaccessible to organisms save for a few microbes, which in several independent lineages—some Cyanobacteria and spirochaetes, for example—have evolved biochemical pathways to convert N_2 into ammonia (NH_3) in the absence of oxygen. This so-called nitrogen fixation is an energy-intensive process catalyzed by an iron-rich enzyme, nitrogenase. The NH_3 is then oxidized to nitrate (NO_3), which is usable by other organisms. On average, a kilogram of seawater contains 1,000 millimoles of N_2, but only 0 to 40 millimoles of nitrogen in the form of nitrate, and 0 to 1 millimoles of nitrogen in the form of ammonium (NH_4^+).[15]

The importance of nitrogen limitation to attached and floating larger primary producers is underscored by the success of plants that harbor nitrogen-fixing microorganisms. The partnership between nitrogen-fixing bacteria and plants is best known in legumes, with 17,500 species in the Fabaceae, including beans, clovers, acacias, and their relatives; but it also occurs in floating ferns of the genus *Azolla*, as well as in some elms (Ulmaceae), the large Chilean herb *Gunnera*, cycads, and even some bryophytes. Within the Fabaceae, the association with nitrogen-fixing rhizobia evidently evolved many times. The nitrogen content of leaves depends on the availability of nitrogen to the plant and affects the desirability of the plant as food for herbivores. Plants living in bogs and acid waters, where biologically usable nitrogen is in very short supply, sometimes resort to prey insects as a source of nitrogen. This kind of carnivory is familiar in bog-dwelling sundews (*Drosera*), pitcher plants (*Sarracenia*), tropical *Nepenthes* vines, and freshwater bladderworts (*Utricularia*), all of which have leaves modified for catching and digesting insects. Perhaps even more strangely, pines in northern forests obtain nitrogen from soil-dwelling arthropods, which are killed by root-associated fungi and whose nitrogen content is then taken up by the tree through the offices of these fungi.[16]

In some parts of the open ocean, especially in waters near the coasts of large land masses, copious nitrate supplied by rivers, and now also by farm fields, supports a high rate of production of the plankton. With the addition of nitrate, production increases—indeed, it may have doubled in many coastal regions once farmers began to use artificial fertilizer on a large scale—indicating that nitrate is a nutrient limiting the productivity and size of the phytoplankton community. In other parts of the ocean, especially in regions far from continental sources of dust and river runoff, there is enough nitrate but not enough iron or silicon. This appears to be the situation in the southern ocean around Antarctica, as well as in parts of the North Pacific and the eastern tropical Pacific. The addition of iron stimulates production by Cyanobacteria and such other phytoplankters as dinoflagellates, diatoms, and coccolithophores, whereas higher concentrations of silica ($Si(OH_4)$), the biologically usable form of silicon) are associated with higher rates of production of diatoms, comparatively large phytoplankters that build a boxlike, predation-resistant skeleton of silica. When iron is added, nitrogenase-carrying Cyanobacteria convert more of the superabundant molecular nitrogen into biologically available nitrate. Iron and silicon enter the ocean as windborne dust from continents, as dissolved ions in river water, and as part of the chemical brine released by hydrothermal processes associated with undersea volcanic activity. Currents carry these nutrients far and wide, and upwelling assisted by winds or by rising hot plumes of mineral-rich water fertilizes surface waters with these nutrients. The availability of iron and silicon thus depends on how much dust, runoff, and hydrothermal fluid enter the ocean, and on how effectively currents and convection disperse them to sites of photosynthesis.[17]

The burning of fossil fuels and the manufacture of synthetic fertilizers have dramatically increased the rate of supply of fixed nitrogen to the biosphere. Peter Vitousek and colleagues estimate that $(90–130) \times 10^{12}$ g of nitrogen was fixed on land, and perhaps a similar amount in the sea, in preindustrial times. By 1990, industrial manufacture of nitrogen fertilizer produced 90×10^{12} g nitrogen per year, a figure that may rise to as much as 135×10^{12} g per year by the year 2030. The cultivation of nitrogen-fixing crops such as alfalfa and other legumes adds another 40×10^{12} g per year, and the combustion of fossil fuels increases the input by another 20×10^{12} g per year. As Vitousek and his team point out, human activity in terrestrial ecosystems has doubled the supply rate of fixed nitrogen over all natural sources combined. Agricultural and nearshore marine productivity has risen by leaps and bounds around the world as a consequence. The transformation from forest and grassland to pasture, croplands, cities, and industrial zones may add another 50×10^{12} g per year of nitrogen. These extraordinary increases in the input of nitrogen may constitute the most profound change in the nitrogen cycle in the last 2 billion years or more, resulting in a huge rise in productivity, largely wrought by and for the benefit of humanity, but also with many unintended and often deleterious effects on rivers, lakes, and coastal marine ecosystems.[18]

Silicon is as abundant as sand on an ocean beach, but there does appear to be competition for it among primary producers as well as among other life forms. Sponges that lay down a skeleton containing silica grow much more robust spicules and a more rigid framework at high silica concentrations than when the concentration is low. Diatoms are evidently better at extracting silica from the water—that is, they can concentrate the mineral when the water around them contains smaller amounts of the dissolved silicic acid—than are such other silicon-using organisms as sponges and radiolarians. With their rise in abundance during the last 40 my or so, diatoms have reduced silica concentrations in the ocean worldwide, and perhaps contributed not only to a reduction in the mass of silica skeletons in radiolarians and shallow-water sponges, but also to the progressive restriction of siliceous sponges to deep and higher-latitude waters from a more cosmopolitan and shallower-water distribution during the Mesozoic era.[19]

Even carbon, that most central of elements in the domain of life, can be limiting under certain circumstances. Only about 0.3 percent of the carbon in the Earth's crust is available to organisms. Experimentally induced increases in the local concentration of carbon dioxide in the air in temperate forests have a fertilizing effect on the trees, and raise the primary productivity of the forest as a whole. Land plants take up carbon in the form of carbon dioxide gas, which diffuses through stomates. Those same stomates let in oxygen, which competes for space on the active site of the enzyme Rubisco, involved in standard photosynthesis, and allow water to escape through transpiration. These relationships among carbon dioxide, oxygen, and water create fundamental constraints on photosynthesis. When oxygen concentrations are high and carbon dioxide concentrations are low, as in the modern preindustrial atmosphere, as much as 40 percent of the carbon the plant takes up is lost because of respiration in the light, the process that occurs when oxygen instead of carbon dioxide attaches to the active site of Rubisco. Moreover, if the plant's environment is warm and dry, the stomates must often close to prevent excessive water loss, but this closure also prevents the uptake of carbon dioxide. In an atmosphere rich in carbon dioxide, therefore, plant photosynthesis is potentially rapid even under rather dry conditions, provided the soil contains sufficient nutrients, with the likely consequence that plant biomass would be substantially higher in water-starved regions than it is today.

Land plants have invented several clever mechanisms to alleviate (though never to eliminate) these constraints. In so-called C_4 plants, carbon is initially fixed as a four-carbon molecule (malic acid) by the enzyme PEP carboxylase (phosphoenol pyruvate carboxylase), which is located in large, thick-walled cells (kranz cells) that surround bundles of vessels and other mesophyll cells where the more standard enzyme Rubisco is situated. PEP carboxylase has a higher affinity for carbon dioxide and a low affinity for oxygen. The four-carbon molecule made by the kranz cells is transported into the mesophyll, where Rubisco transforms it into the more standard three-carbon form (phosphoglyceric acid), well away from high concentrations of oxygen. This type of C_4 photosynthesis has evolved at least thirty-one times in eighteen families and 7,500 species of land plants, including many warm-adapted grasses. It is especially characteristic of plants in warm, dry environments. Another form of photosynthesis, the so-called crassulacean acid metabolism (CAM) also employs a C_4 pathway, but it evolved separately from standard C_4 metabolism and today occurs in some thirty thousand species, mainly characteristic of hot dry places. In CAM plants, carbon dioxide is taken up during the cool night, when water loss through open stomates is less than during hot daylight hours, and is fixed in a four-carbon form. During the day, with the stomates closed, the four-carbon compound is converted into the standard three-carbon form. CAM plants thus reduce the incompatibility between water retention and carbon uptake by carrying out stages of photosynthesis at different times of the day.[20]

Recent experiments have demonstrated that tobacco and celery, both C_3 plants, can take up significant amounts of carbon dioxide from the soil or as the product of respiration from the roots. This carbon dioxide is dissolved in sap that is transported as malate in the xylem tubes to cells surrounding the xylem in stems and in leaf petioles, where it is converted into three-carbon compounds and carbon dioxide. This ability to take up carbon dioxide from roots and soil may again increase the carbon supply as well as alleviate functional tradeoffs in the management of water, carbon dioxide, and oxygen.[21]

Under conditions of plentiful light, nitrate, and phosphate, diatoms in the open sea are sometimes limited in their growth by the concentration of dissolved carbon dioxide. In the ocean, dissolved CO_2 accounts for less than 1 percent of the total dissolved inorganic carbon that is potentially available to the phytoplankton. Limitations imposed by low concentrations of dissolved CO_2 occur briefly during so-called diatom blooms, times when diatoms grow at maximum rates in the spring at high latitudes when nutrient-rich water from the deep well up to the surface. Recent German work, moreover, indicates that at least some diatoms can temporarily store CO_2 in a four-carbon malate form, perhaps in the organism's cytoplasm. This malate is then released within the cell, to be taken up by Rubisco in the chloroplasts. This type of C_4 photosynthesis may be advantageous to diatoms in environments where light availability is reduced by intense turbulence.[22]

Globally, water is another ubiquitous material that could hardly be considered a limiting resource; yet, for many plants and animals, it is. In the first place, land organisms generally do not tolerate water with a high salt (NaCl) content, such as seawater. In the second place, water is very unequally distributed on land, with many parts of the polar regions and at mid-latitudes receiving less than 200 mm of rainfall per year. Because photosynthesis depends on the entry of CO_2 through stomates, the loss of water through evaporation is an inevitable consequence of photosynthesis. Indeed, as I mentioned earlier, evaporation is also essential because it pulls up water and nutrients. At the other extreme, plants living in air saturated with water cannot engage in transpiration, meaning that nutrients cannot reach the leaves from the soil. This circumstance, too, limits primary production.[23]

For plants and animals of marine origin, the typical way of coping with limited water is to reduce the surface area from which evaporation takes place. Snails—periwinkles (Littorinidae), nerites, and some limpets—living high on the shore, where they are exposed to air for days at a time, often close their tissues in a hermetically sealed shell, and passively wait until waves wet the rocks and feeding can proceed for a brief time. Activity is restricted to periods when these creatures or their habitats are wet or submerged. The union of egg and sperm typically occurs in a water medium in primitively aquatic plants and animals, as well as in such early land-adapted groups as bryophytes and amphibians. Independence from water in reproduction through the evolution of the seed in plants and the enclosure of the embryo in amniotic membranes in vertebrates enabled land creatures to penetrate much further from sources of water.[24]

Reducing the surface area of evaporation is a feasible solution to the shortage of water only if it does not compromise the surface-related uptake of other essential materials. For early land animals, which depend on oxygen, such a reduction may have restricted them to very small sizes, and thus to a low oxygen demand. Only when oxygen became abundant enough in the atmosphere could a larger body's demand for oxygen be satisfied with a relatively small surface compatible with minimal water loss. The limitation on size was probably gradually eased, first in the Devonian and more so in the succeeding Carboniferous period, as land plants and modifications in the carbon cycle to be discussed later pumped oxygen into the atmosphere.[25]

Limitations in the supply of water have been overcome by humans through irrigation. By introducing a year-round water supply to fertile soils in river valleys, the Sumerians beginning about four thousand years ago raised agricultural production to a point where the surpluses could support urban centers. By 1990, irrigated land comprised about 270 million hectares, or 15 percent of total cultivated land, and accounted for some 30 percent of world agricultural output.[26]

And then there is the interesting case of oxygen. Initially, photosynthesis was accomplished in the absence of free oxygen (O_2) by various microbes using pigments (bacteriochlorophylls) that absorb photons (particles of light). Purple bacteria (Chromataceae) absorb light at the low end of the visual spectrum in the near-infrared range of wavelengths, 900 to more than 1,000 nanometers (nm). The pigments of green sulfur bacteria (Chlorobiaceae) absorb light of wavelength around 750 nm. Cyanobacteria were likely the first microbes in which photosynthesis liberated free oxygen, an element that to all other forms of life was a toxin. When the metabolic process of respiration evolved, oxygen became a valuable resource for organisms other than plants. In respiration, or aerobic metabolism, organic molecules combine with oxygen to yield carbon dioxide and water. Compared to pathways of metabolism such as fermentation, in which free oxygen is not used, respiration yields sixteen to eighteen times as much ATP and other energy-rich compounds. Interestingly, fermentation produces ATP at a higher rate than does respiration, meaning that this oxygen-free metabolism is favored when entities are competing for abundant external resources; whereas respiration produces a larger quantity, or yield, of ATP, meaning that this oxygen-requiring process is favored when resources are internal to the entity. The latter circumstance endows the entity with more control than does competition for resources that must enter the entity via diffusion, and moreover enables the entity to maintain high use of energy regardless of the ups and downs of external supply.[27]

Today's atmosphere contains about 20.9 percent free oxygen by volume, but concentrations of oxygen vary greatly among habitats in the biosphere. A given amount of water saturated with oxygen, for example, contains only 5 percent of the oxygen that a similar volume of air holds. The amount of oxygen in water also depends on temperature and salinity. A cubic meter of fresh water at 0°C contains 10.22 liters of oxygen, whereas a cubic meter of fresh water at 20°C contains only 6.35 liters. Seawater at 20°C (NaCl content 35 parts per thousand) contains just 5.17 liters of oxygen per cubic meter. Animals living deep beneath the surface of sediments, in the guts of animal hosts, and in decaying carcasses, among other habitats, have little or no free oxygen available to them.

As I noted earlier, low oxygen concentrations constrain respiring animals to be small or, equivalently, to have a body shape in which the surface area is very large compared to the volume. In such wormlike or sheetlike animals, moreover, no part of the interior is very distant from the surface, through which oxygen must diffuse. Features that cover the surface of diffusion, such as an exoskeleton, may not have been feasible below a concentration of oxygen of about 1 percent by volume in the atmosphere. Warm-bloodedness, powered flight, and other highly energy-intensive work may have been possible only when active ventilation in living muscles in an oxygen-rich medium supplemented passive diffusion. Although passive diffusion remains the only way for oxygen to enter cells, the key to high metabolic activity lies in maintaining high concentrations of oxygen just outside the cell so that diffusion can proceed rapidly. Above a concentration of about 30 percent by volume, however, spontaneous fires easily start, and organisms may gain little additional benefit from a higher oxygen level.[28]

Phosphorus, however, may be the ultimate limiting nutrient in the long run. Unlike nitrogen, it lacks an atmospheric source, entering the biosphere only from the deep crust through the eruption of volcanoes and by the metal-rich fluids from hot vents in the ocean floor, or by the weathering of rocks on land. It is useful to organisms only as phosphate (PO_4^{3-}) in which form it plays a crucial role in storing chemical energy and as a component of nucleic acids. For every atom of phosphorus in the oceanic plankton, there are 106 atoms of carbon and sixteen of nitrogen. When biologically available nitrogen is scarce relative to phosphate—that is, when there are fewer than sixteen atoms of nitrogen for every atom of phosphorus in the environment—microbes capable of fixing nitrogen from the atmosphere become more numerous. As consumers and decomposers in the plankton eat these nitrogen-fixers, the nitrogen enters the system as dissolved ammonium (NH_4^+) and nitrate, so that the ratio of available nitrogen to available phosphorus increases and the limitation of nitrogen is temporarily eased. Because there is no atmospheric source of phosphorus, no similar mechanism is available to eliminate the scarcity of that crucial element. Only increases in the supply of phosphorus through geological processes can stimulate productivity when phosphate is the limiting nutrient. Thus, many elements—iron, silicon, nitrogen, and even zinc—can limit productivity regionally or locally on medium to short time scales, but only phosphorus limits the rate of production of biomass on very large spatial and very long temporal scales. In short, limitation depends on the size and the time span considered.[29]

Surprisingly, there are sources of energy that forms of life other than humans do not harness directly. Prominent among these are winds, currents, and waves. Organisms have not discovered windmills, water wheels, or hydroelectric power. Nevertheless, plants and animals do benefit from the movement of the air and water that surround them, and in fact often depend on such movement to acquire the light and food necessary to sustain themselves. Consider life on wave-swept rocky shores. Egbert Leigh and his colleagues calculated that the heavily surf-exposed west-facing shore at Tatoosh Island, Washington, receives an average of 0.3 W per square centimeter of shoreline. This rate of delivery of kinetic energy is twelve to eighteen times higher than the energy delivered by solar radiation during the sunniest month of the year, calculated to be between 0.017 and 0.025 W/cm^2. This wave-battered, low-sunlight North Pacific coast supports large brown seaweeds with an enormous productive capacity, estimated to be as high as 14.6 kg per square meter per year for the sea palm (*Postelsia palmaeformis*), which exposes 24 square meters of frond surface per square meter of rock surface. By comparison, a highly productive Hawaiian grassland produces only 8.5 kg per square meter per year. Leigh and his colleagues believe that the waves not only stir the fronds so that all of their surfaces are exposed to at least some light, but also renew the supply of nutrients on which photosynthesis in the fronds depends. The plants have no mechanism to draw nutrients up from the holdfast, and therefore require the nutrients to enter by diffusion across the surface of the fronds. In calm waters, the supply of nutrients would be depleted rapidly, setting an upper limit to the plant's productivity. Turbulence therefore increases the plant's exposure to light and nutrients, the essentials for photosynthesis.[30]

Similar arguments help to explain why many suspension-feeding animals achieve larger body sizes and large population densities in environments characterized by heavy surf and strong currents. Although many suspension feeders create their own currents, many others—corals, some worms, and sea lilies, for example—rely on currents to bring food particles to their organs of collection, and even the current-producing filterers often benefit from the movement of water around them. As long as the animals are not dislodged from their moorings, they can collect more food when the supply is continually replenished through the action of outside forces. Many mobile consumers are strongly inhibited by fast currents and intense wave turbulence, because the lift exerted on them while moving will sweep them away. There is food in abundance, but the strong forces acting on consumers prevent it from being exploited. The kinetic energy on wave-battered shores and in swiftly flowing streams therefore directly and indirectly benefits those primary producers and suspension-feeders that are securely fastened to the shore or streambed.[31]

On land, wind limits the height of trees by toppling those that cannot resist gusts during storms, and also interferes with flying birds and insects. Indirectly, however, many stationary plants rely on the wind to disperse their spores, pollen, and seeds. Air movements also disperse the fragrances of flowers and the pheromones of animals, so facilitating the location of or food.

Humans have supplemented food with energy from sources that other organisms have exploited little or not at all. We are the only organisms that burn other organisms or their remains in order to cook our food and to fuel machines for industrial production and transport. Nuclear fuel is entirely unused by other forms of life on Earth. Wind and water power were critically important in shipping and in driving machines before the Industrial Revolution of the late eighteenth century.

Productivity and Supply

If energy and raw materials locally limit primary producers, it would stand to reason that an increase in supply of the necessities of life would stimulate increased primary production. Up to a point, this is indeed so; but an increase in supply beyond the capacity of prevailing technology to absorb and use the resources does not yield more production, and may in fact disrupt the system. Most importantly, it temporarily favors fast-growing, more ephemeral producers at the expense of slower-growing ones. In the long run, if the increase in supply persists, adaptation should produce a community of producers adapted to that higher supply, and these would be producers with greater competitive power.

The life forms best able to take advantage of an increased nutrient supply are opportunistic weeds whose *modus operandi* is rapid exploitation. They are characterized by rapid growth, poorly defended bodies, short visible lifespans, and early and prodigious reproduction. When the supply of nutrients is cut off, or when other conditions of life for the actively growing opportunists become unfavorable, the weed enters a potentially lengthy inert phase, spent as a seed, overwintering egg, spore, or cyst. In short, weeds live a roller-coaster existence between all-out growth and reproduction during an active phase amidst plenty, powered by rapid metabolism, and an inert resting state amidst inclemency and danger.[32]

In many aquatic ecosystems, three more or less distinct categories of primary producers exist, which differ in their competitive method. At the weedy extreme are phytoplankton, which can efficiently and rapidly take up nutrients and compete for light with attached seaweeds and plant-animal partnerships. The addition of nutrients through upwelling or through human agency typically results in plankton blooms. The second level of weediness is represented by annual attached seaweeds, which in the absence of grazers quickly overgrow, shade out, and inhibit recruitment of perennial species. Experiments by Boris Worm in the Baltic Sea have demonstrated that the nutrient enrichment that has characterized the marine ecosystems of this region and of many other coastal waters around the world has favored annual seaweeds at the expense of longer-lived rockweeds, kelps, and red algae. On reefs, the addition of nutrients and the removal of grazers similarly favors ephemeral fleshy algae over photosynthesizing corals and encrusting coralline algae. Perennials, the third and least weedy category of primary producers, persist in the face of intense grazing by virtue of sophisticated chemical and architectural defenses, which tend to be incompatible with rapid growth. Where grazers are present under a regime of high nutrient supply, as in reefs along continental coastlines and on many surf-swept shores around the world, all these types of primary producer coexist, the weeds being held in check by grazers; but where grazers are removed, the perennials are imperiled, and the weedy species take over.[33]

Very similar relationships exist on land. Weedy plants grow quickly, lose high percentages of production to consumers, channel less biomass to decomposers, and store fewer nutrients than more permanent, better defended producers. They characterize light gaps in forests, where full sunlight reaches the forest floor after a treefall; as well as the more familiar but less "natural" abandoned lots in towns and suburbs. Many flowering plants have adopted a combination of traits, in which a perennial tree body bears ephemeral deciduous foliage. Deciduous leaves are thinner, grow faster, lose a higher proportion to herbivores, and have a higher nutrient content than more persistent leaves. By deploying deciduous organs of photosynthesis, plants are able to capitalize on briefly favorable conditions while avoiding the costs of maintenance and of potential carbon loss during unfavorably cold or dry seasons.[34]

The distinction between opportunists and more permanent entities seems to be very general. Parasites that kill their host must find another before the death of their host kills them as well. These parasites are perceived to be virulent because they multiply and spread rapidly without regard to the longer-term health of their immediate surroundings. Other parasites track their host from generation to generation, and are therefore longer-lived guests whose selfish interests lie in keeping their host alive.[35]

Storage and Investment

Perhaps from the very beginning of metabolism, production and other forms of economic work have depended on the ability of entities—molecules and unicells at first, multicellular organisms and societies later—to store energy and material resources for subsequent use. If a resource is used as soon as it is acquired, any interruption in supply of that resource means serious economic disruption for the entity in question. Such disruption can occur when competitors or the risk of consumption prevent access, or when some circumstance beyond the entity's control interferes with the source of supply. By storing materials that would otherwise be usable in production of new tissues or of offspring, entities forgo immediate advantages in favor of a longer-term benefit. There are many examples of storage in nature. The organic molecule ATP stores energy that can later be tapped under controlled conditions to carry out the tasks of life. In CAM plants, as discussed earlier, carbon dioxide taken up at night is held until daylight hours for use in photosynthesis. Temperate and desert plants store huge amounts of food or water in the stem and in underground roots, tubers, or bulbs. The nutrients and water accumulate in times of plenty, and are tapped when growth conditions are not yet ideal, such as in early temperate spring when daylight hours are short and temperatures are low. Seeds and yolky eggs store food for the developing embryo, which can grow rapidly without having to photosynthesize or feed during its most vulnerable and dependent phase. Kelps in Nova Scotia accumulate nitrogen in the fall, store it in winter, and tap it in the early spring when the water is still close to freezing and daylight remains short, but waterborne nutrients are abundant. In the summer, these large algae store carbohydrates, which are tapped for rapid growth during the still relatively warm autumn when abundant nitrogen returns. Migrating birds and whales store fat for the long voyages they undertake during migration. In these and many other examples, storage expands the range of conditions under which an entity can do work, and often increases power, the rate at which work is done.[36]

Plants and most animals store food in their own bodies, but some birds and mammals have broken this constraint by maintaining provisions in caches underground or in some other relatively safe place. Jays and rodents often depend on a seasonal food supply of nuts and seeds, but by burying their food during times of plenty, they have enough to eat during lean times. In the process, because not every hidden item is recovered, these caching animals are instrumental in dispersing the plant producers of the nuts and seeds to environments where germination is likely.

Humans have carried extracorporal storage even further. Smoking and salting fish and meat are very ancient techniques for conserving food. Cooking and then vacuum-packing food was invented in 1795, and tin cans came in 1812. Mechanical refrigeration was perfected in the mid-nineteenth century, and after 1870 meat could be shipped frozen from New Zealand to Europe. The storage of water was essential to the development of agriculture.[37]

Perhaps the greatest economic innovation in human society was the invention of money, an exchangeable commodity that in effect stores the power to purchase and sell goods. Because in the monetized economy money earns interest and thus accumulates, it buys goods and services in ways that simple exchange of those commodities and of labor could not. By accumulating, investing, and then spending money, we create opportunities for production, consumption, and ways of making a living that could not exist without mechanisms for storing wealth. Like other reserves, money that is stored rather than invested or spent can do no economic work; but money invested or spent translates into economic power.

Storage means control, and control means competitive advantage. To store resources for later use is to function under a wider variety of circumstances, to possess things that, for a time at least, others cannot have, and to accumulate wealth, so that possibilities can be realized that previously would have remained beyond economic reach. But inevitably storage also means defense, the ability not simply to accumulate for later use, but to keep others from those accumulated reserves. Such an imperative would be unnecessary without competitors and consumers, and so there is always an imperative to defend resources, as I discussed at length in chapter 4.

It is in passive accumulation that producers hold a key advantage over consumers. Recall from the last chapter that predators tend to outperform their victims in sensation, locomotion, and the use of force, but typically not in such passive attributes as large size, toxicity, and skeletal strength. Competition-inspired storage, serving as well for defense, further ratchets up competition and defense, and creates potentially reliable resources for yet more determined consumers. Here, then, we have a potent example of escalation, of the evolutionary interdependence, of producer and consumer.

Regulation and Control

A fundamental problem faced by organisms that compete for any kind of resource is that, at least initially, competitors have little control over supply. Resources can arrive at times when, because of low temperatures or some other reason, they cannot easily be taken up or used. Their supply may be very erratic, times of scarcity being punctuated by episodes of overabundance, with the result that resources go unused. Desert plants must live in a regime of scarce water, but when the rains do come, they are often prolific and violent. Suspension-feeding brachiopods, which have low metabolic rates and thus filter particles slowly, cannot capitalize on a rapid influx of food.

I shall illustrate this predicament with an example of resources for hermit crabs. As discussed in chapter 4, shells built by molluscs offer effective protection against outside forces, including predators. When the builder dies, its shell often remains intact. Several groups of animals, notably hermit crabs, have become specialized to live in, and carry around, abandoned shells. The soft vulnerable abdomen and the crab's eggs are hidden within the shell, and in many cases the entire crab can withdraw, closing the shell's opening with one of the crab's claws. As the hermit crab grows by periodically molting, it must find larger quarters, a considerable challenge in most places because most shells are occupied either by their original builders or by secondary occupants such as other hermit crabs. The supply of shells is therefore often limiting both for individuals, who often fight over shells, and for the population. What controls the supply of shells? In some marine environments, and especially in temperate lakes and streams, shell-bearing snails often die during extreme weather—floods, storms, heat waves, and the like—which by their nature occur only once in a while. Shell supply is therefore low during most intervals, but there is a superabundance at a few times. In such environments, shells might well provide good protection, but they cannot feasibly be employed as a resource because the normal supply is too low to accommodate the molting frequencies and body-growth rates of the hermit crabs. When shells do become available, the supply is suddenly so large that the population of hermit crabs cannot capitalize effectively on the shells. As a result, most of the shells stay empty and eventually disappear into the ground or are destroyed.

But now the predators of molluscs enter into this economic picture. Predators exert a more or less consistent, if not constant, consumer demand on their prey. If they do not destroy the shells of their molluscan victims, predators produce a reliable supply of empty shells, a supply regime predictable enough for shells to be effectively exploited by secondary shell-dwellers. In Florida and probably elsewhere, hermit crabs have acquired the behavior of lurking near predatory snails, which do not destroy their prey's shells, so that these crabs are in a good position to acquire housing that is in excellent condition. Predators thus indirectly aid hermit crabs in stabilizing, and in effect making available, a resource that the hermit crabs do not themselves control.[38]

There is at least one other elegant solution to the problem of an unpredictable resource supply, a solution that in one way or another has been widely applied by organisms. The secondary occupant can let its shell grow, much as a mollusc does. But hermit crabs, unlike molluscs, do not have the ability to secrete calcium carbonate at the shell's rim to make it bigger. They accomplish the growth of their house by going into partnership with a sea anemone, or another organism that encrusts and enlarges the shell. Often, these encrusting animals continue growth in the same spiral direction as did the original builder before the latter died. In other cases, the shell is extended as an expanding tube, still encasing the hermit crab but growing straight or in a curve instead of in a tight spiral. When the encruster is a cnidarian, it provides the occupant with the additional benefit of discouraging would-be predators by the presence of nematocysts. However it arises, this kind of partnership solves the problem of an erratic shell supply, and essentially eliminates competition among crabs for this resource.[39]

Interestingly, there are no freshwater hermit crabs. In fact, with the exception of a few African cichlid fishes that lay eggs in empty snail shells, shells appear not to have been used as housing by any freshwater animals. One possible explanation is that predation is rare relative to other more unpredictable causes of mortality, effectively making the shell resource unavailable to would-be secondary occupants. It is also possible that the shells don't provide a sufficient protective benefit, although this would seem unlikely in many tropical rivers and lakes, where snails can be relatively thick shelled. Encrustation of the type frequently observed on the shells of both living snails and secondary occupants in the sea is also unknown in freshwater.[40]

Consumers probably play a very important general role in smoothing out the rate of supply of resources for the organisms on which these consumers depend. Herbivorous mammals fertilize the soil on a daily basis as they eat, digest, and defecate the remains of their food. Such a regular regime of fertilization makes it much easier for plants to exploit soil nutrients than if those nutrients entered chiefly from erratic instances of flooding, volcanic eruption, or some other, physically controlled, external input. There is, in other words, a sort of unintended cooperation that benefits both parties because it regulates resource supply to a more predictable regime. Some of this cooperation takes truly bizarre turns. Karen Porter showed that the freshwater phytoplankters ingested by tiny cladoceran crustaceans reside in a highly enriched environment while in the crustaceans' digestive systems. Although many of the individual cells are digested and therefore killed, others pass through unscathed and nutritionally enriched. Algae ingested by copepods can even photosynthesize while in the gut. It is likely, in fact, that many symbioses between animals and single-celled photosynthesizers arise as casual cases of ingestion.[41]

Some producers exploit the inabilty of consumers to take up temporarily superabundant resources. They do this by confining their production to a very brief interval, during which consumers are overwhelmed. In the rain forests of southeast Asia, there is a large diversity of tall, emergent trees belonging to the family Dipterocarpaceae. These trees engage in a practice known as mast fruiting, in which the fruits of all individuals and all species ripen every few years at the same time. The edible fruits are so numerous that fruit-eating mammals and birds eat them all, ensuring that at least some of the seeds in some of the fruits will germinate.[42]

On the Atlantic coast of Panama, seventeen species of green algae of the order Bryopsidales—all highly productive, abundant seaweeds on reefs, including members of such genera as *Udotea*, *Penicillus*, and *Halimeda*—become fertile during a single night. The next morning, nine species release the contents of their cells, including motile unfertilized eggs and sperm, and promptly die. The precise time of release varies slightly among the species, enabling fertilization (union between egg and sperm) to be mainly between gametes of the same species. From the perspective of consumers—plankton-eaters in this case—the release is highly concentrated to just a single morning. This remarkable pattern increases the likelihood of successful fertilization, but it is also an effective—and in this case cooperative—method of reducing losses of gametes to consumers. Once fertilized, the eggs sink to the bottom and, as long as they are not grazed, can germinate into new plants.[43]

A similar case of cooperative spawning has been documented for various species of coral, including members of the well-known branching genus *Acropora*, both in Australia and in the West Indies. Like the Panamanian seaweeds, these corals practice external fertilization by means of gametes that would be very tasty morsels for the many plankton-eaters that frequent reefs.[44]

A particularly curious case is that of periodical cicadas. In the central and eastern parts of North America, three species each mature en masse every thirteen years, and another three species do so every seventeen years. The larvae live underground, where they feed on tree roots. The adults that crawl out of holes in the ground are slow fliers, have thin cuticles, and in general show no discernible behaviors or any structural or chemical defenses that could be interpreted as antipredatory adaptations. They call incessantly and are easy even for me to catch and hold. Red-winged blackbirds love to eat them. For most populations and individuals of predatory species, the mass appearances of such vulnerable prey are probably difficult to track. The capacity of predators to exploit these items is therefore utterly overwhelmed by the overabundance of periodical cicadas, which lasts for about one month in late spring.[45]

There are many other potential examples. Plankton spring blooms precede, and may well be a cue for the reproduction of, zooplankton consumers. Intertidal crabs release swimming larvae en masse during periods of maximum difference between high and low tide. The flush of new leaves after the dry season or after the winter temporarily overwhelms herbivorous caterpillars, which peak in abundance later. The mass maturation of Arctic mosquitoes may for a time be too much for populations of hungry summer-breeding birds to consume.

What all these examples have in common is that a particularly vulnerable phase of life stage, with few or no consumer-oriented adaptations, must be passed through in the presence of powerful enemies. Overwhelming one's foes with numbers requires very high production rates for a short time, to say nothing of remarkably precise timing. Legitimate questions remain about whether this behavior is an adaptation or has the happy but fortuitous consequence of protecting vulnerable stages against complete eradication, but it certainly illustrates how the unpredictability of resources can be both a constraint (for the entity relying on the resource) and a potential benefit (for the entity controlling the resource).

Producers as Resources

Producers provide the necessary, though not sufficient, conditions for consumption. A resource becomes increasingly likely to be exploited by a consuming entity as the supply of that resource becomes more plentiful and reliable. Storage and regulation of supply, both of which are evolutionarily favored in the presence of consumers, create circumstances that are in turn favorable to the enhancement of the consumer class. A series of positive feedbacks thus exists between those entities that produce and those that consume. But these feedbacks are in some ways indirect. It is competition among the consumers that sustains the effects of consumers on producers. Without such competition, consumers might affect producers only in a negative way.

In the sea, where about 95 percent of primary production is accomplished by unicellular floating plankton, consumption of this resource is chiefly in the hands of either single-celled protists, which together with the smallest fraction of the phytoplankton comprises the microbial loop, or larger suspension-feeders, animals that strain or catch food particles from the water surrounding or flowing through them. Some suspension-feeders, notably sponges and brachiopods, are highly effective at straining out bacteria-sized particles, whereas others concentrate on larger phytoplankton and on zooplankton. Freshwaters, too, support their share of suspension-feeders.

Experiments in which suspension-feeding clams are exposed to varying concentrations of food particles have demonstrated that the body sizes and growth rates of individual consumers both increase as the concentration of planktonic food rises. Geographic comparisons between planktonically productive regions, such as the west coast of South Africa and the tropical shores of southeast Asia, and less productive waters, such as those of eastern South Africa and isolated islands in the Pacific, point to the same conclusion. Many scallops in the productive northern Indian Ocean and in southeast Asia, for example, reach a linear dimension greater than 100 mm, whereas species in Micronesia and Polynesia do not exceed 50 mm. As Rodrigo Bustamante and his South African colleagues note in their studies of the relationship among nutrients, production, and consumption, the measure that most closely predicts the biomass of consumers is not simply the concentration of dissolved raw materials or of planktonic cells, but their effective availability. This measure combines the concentration of nutrients in a given volume of water with the rate at which water passes through a suspension-feeder's filter. Effective availability of nutrients is therefore determined by the density of biomass and by the velocity of flow created by either currents or waves imposed by outside forces, or by the suspension-feeder's own metabolism. Food-rich waters in areas of fast currents and intense wave action support the highest densities, the most rapid growth rates, and the largest body sizes of suspension-feeders. Shallow subtidal bankss of mussels (*Mytilus californianus*) up to 200 mm long and of barnacles exceeding a height of 150 mm offer classic examples along the productive, current-swept, and wave-washed shores of California.[46]

In the world's coastal zones, primary production is overwhelmingly accomplished by attached plants or by animal-plant partnerships. Whereas net primary productivity in the unproductive open ocean is just 50 g carbon per square meter per year, and in upwelling regions is some six times higher, my rough calculations show that average worldwide productivity of attached marine plants is about 500 g carbon per square meter per year. In Nova Scotia, dense beds of seaweeds (mainly brown kelps) have an average net primary productivity of 1 kg carbon per square meter per year, about ten times higher than that of phytoplankton in the same region. Californian beds of the kelp *Macrocystis* produce as much as 1.4 kg carbon per square meter per year. Incredibly high rates of algal production are achieved by shore seaweeds in the Baltic Sea, where Boris Worm reports values of 4.9 kg carbon per square meter per year.[47]

This prodigious production by nearshore attached plants not only supports a cadre of herbivores, but also adds particulate food to the water and therefore helps to support suspension-feeders. Generally speaking, the higher the productivity, the higher is the biomass of herbivorous consumers. Predators may eat herbivores and keep grazers down, but significant levels of herbivory cannot be maintained if primary production is slow. Sometimes, intense herbivory occurs when plant biomass is high, as on western South African shores where extraordinarily high densities of herbivorous limpets (up to 200 individuals per square meter, with individuals up to 100 mm in length, with a combined whole wet weight of up to 13 kg per square meter) are subsidized by dense canopies of kelp. This was also the situation in the North Pacific, where ten-ton Steller's seacows grazed luxuriant forests of kelps and other seaweeds. Elsewhere, however, standing biomass is kept low by relentless cropping. This is the case in coral reefs, where armies of parrotfish (Scaridae), surgeonfish (Acanthuridae), damselfish (Pomacentridae), sea urchins, crabs, and other herbivores consume algae and corals at rates close to those in which carbon is fixed by these primary producers. Similarly, manatees, dugongs, sea turtles, and large herbivorous fish kept seagrass meadows down to a low stubble before these warm-water herbivores were exterminated or severely depleted beginning in the seventeenth century.[48]

Consumption of plants on land is also stimulated by high plant productivity and by the availability of raw materials. Comparisons between a mangrove community in Florida in which nitrogen availability is high by virtue of roosting pelicans, with a nearby mangrove with lower nitrogen levels and without guano-producing roosting birds, revealed that herbivory by moth larvae and beetles removed about 2.5 times more buds and leaf area in the enriched forest. Vegetation in the enriched forest had a higher nitrogen content and therefore constituted a nutritionally better forage for the insects. This example illustrates a general trend for productive vegetation growing on fertile soils to support higher levels of loss to herbivores than less productive plant assemblages growing on poorer soils."[49]

There is abundant evidence that consumers are limited in abundance and body size by food. Grasshopper populations explode after particularly heavy rainfall because their plant food grows luxuriantly. With more nutrients available, plant foods become richer in nitrogen, with the result that many herbivores—especially actively growing young individuals—can survive. Phytoplankton blooms after episodes of extensive runoff from high islands enable large numbers of larvae of the crown-of-thorns seastar (*Acanthaster planci*) to survive. These larvae feed on phytoplankton, and normally die, or at least fail to settle. When the phytoplankton supply is large enough for many larvae to settle, an outbreak of adults follows. The adults, which eat reef corals, then cause widespread mortality among corals.[50]

Ecologists have vigorously debated whether and how productivity affects diversity. John Chase and Matthew Leibold, for example, find that comparisons of plankton communities among ponds in a single watershed reveal a pattern in which productivity and the number of species both increase when productivity is relatively low, but above a rather high rate of production, diversity falls. When similar comparisons are made among ponds from several watersheds, however, the relation between the number of species and the rate of production remains linear even at very high levels of productivity. Thus there is a paradox. Highly productive local communities and ecosystems tend to support fewer species than do less productive ones, but on the larger scale of regions and biotas, higher production rates are associated with a greater richness in species.[51]

Agriculture and the Rise of Civilization

Throughout this chapter, I have emphasized that production is necessary for consumption, and that regulation of the supply of resources makes increases in productivity possible. Perhaps the ultimate form of this regulation is agriculture, the cultivation of food organisms by consumers. Many groups of animals have evolved means of managing the producers which sustain them, either within their own bodies or in their immediate vicinity. Many sponges, corals, sea whips, sea squirts, bivalves, and even some snails maintain photosynthesizing microbes in their tissues and derive substantial competitive benefits from the association. The microbes typically have free-living stages or states that are picked up by the larvae of each new generation of host animals. Ghost shrimps culture bacteria on organic debris that these crustaceans bring from the sediment surface to galleries deep beneath the sediment surface. Damselfish on tropical reefs, nereid worms on mudflats on temperate shores, and large limpets in South Africa and California maintain aggressively defended territories in which they "garden" food seaweeds.[52]

These associations, however, fall short of fully developed agriculture, in which cultivators protect food crops against enemies, sow new recruits to replace harvested food, improve the conditions of growth for crops by fertilizing and weeding sites of cultivation, breed domesticated species selectively for higher yields and nutritional quality, and remove wastes. To my knowledge, full-blown agriculture has evolved only ten times. Nine instances involve fungus-farming insects (see chapter 4), which grow their crops in galleries, mounds, or chambers well hidden from the insects' competitors and predators. Humans are unique in practicing agriculture with primary producers, which together with domesticated animals and even fungi are managed by collectively or individually owned plots of land.

Human agriculture began about eleven thousand years ago, shortly after the end of the last ice age, and as much as forty thousand years after the emergence of technologically and culturally sophisticated modern humans. Peter Richerson and his colleagues have suggested that this late emergence is attributable to the harsh climates of the glacial Pleistocene, which were not only cold and dry in many places but also highly variable over time. Only after the Pleistocene, Richerson and colleagues argue, did climates become predictable enough to ensure reliable yields of cultivated plant crops. I am more inclined to the view that postglacial ecosystems in the Middle East and China, where agriculture first originated, became more productive as growing seasons lengthened, summer temperatures rose, and rainfall increased. Whichever interpretation is correct, yields of grain per hectare may have risen by a factor of one hundred or more during the transition from gathering to cultivation. Agriculture may have been a response to a rise in the size of the human population associated with the new high-productivity regime in grass-dominated postglacial lands. In turn, it allowed the human population to grow still further.[53]

But it may be the invention of irrigation in the Fertile Crescent that propelled humanity into civilization. By regulating the supply of water to crops, humans enabled yields to rise to a point where a consumer class of people not directly engaged in food production could emerge. The first cities arose with the emergence of the Sumerian civilization about six thousand years ago in southern Mesopotamia, where rainfall was relatively plentiful, soils were fertile, and the nearby Persian Gulf teemed with fish. Relative to the more northerly parts of Mesopotamia, the productive south gained a trading advantage, and during Uruk times (5,800 to 5,300 years ago) developed a society of government officials, artisans, and traders, whose livelihoods were made possible by the surplus production of the land and the nearby sea.[54]

Over the centuries, increases in agricultural production and regulation of food supply have been concentrated in eight directions: (1) the addition of new crops; (2) methods to prevent soil erosion, nutrient loss, and salt accumulation; (3) the development of higher-yield crops and livestock; (4) the application of artificial fertilizers; (5) control of pests and diseases either through biological means or through the use of manufactured chemical agents; (6) the invention of machinery to increase the rate and spatial reach of planting, maintaining, and harvesting crops; (7) preservation and storage of food; and (8) rapid and more equitable distribution of food. Against all odds, and in a trend that almost certainly cannot be sustained, food production has kept ahead of human population growth globally (though with many regional exceptions), with the result that per capita wealth has on average risen, especially during the twentieth century.

Among the factors stimulating agricultural production, the addition of crops and the expansion of production to new lands preceded the others and perhaps made such factors as the introduction of fertilizers and pesticides possible. In Europe, the human population grew by a factor of 3.5 between 1750 and 1950, yet living standards generally rose, and the continent was launched into the Industrial Revolution early in the period. In China over the same two hundred years, the population grew by a factor of 2.6, and although China was able to feed itself, its average living standard at most kept even with the population and in the first half or more of the twentieth century it actually declined.

The reason Europe did not suffer a similar fate, Kenneth Pomeranz suggests, is that Europe—and later North America as well—relied increasingly on land outside Europe for the production of food, fuel, fiber, and building materials. Until 1839 in British colonies, and 1863 in the United States, this enormous land subsidy was amplified by the "free" labor of slaves. The subsidy of land and labor freed Europe to convert land formerly dedicated to agricultural or forest production to other uses, and enabled a large proportion of the population to work in commerce, government, scientific exploration, technological development, research, and the military. Agricultural land itself became more productive, first thanks to the spread of the South American potato beginning in the eighteenth century, and later through the introduction of natural and artificial fertilizers and pesticides. Fossil fuels, especially coal, began to substitute for charcoal and wood as fuel, and freed even more land. New World sugar, cotton, and wood imported into Britain could have covered 30 million acres had they been grown in Europe, an area even greater than the area of forest needed if wood had been substituted for coal. This reliance on land in the New World skyrocketed through the rest of the nineteenth century. According to Pomeranz, coal output rose fourteenfold, sugar imports rose elevenfold, and cotton imports jumped twentyfold in Britain from 1815 to 1900.

China, on the other hand, was not materially subsidized by foreign lands. The rich coastal regions of the Yangzi Delta did trade with the less productive interior, but trade like this was insufficient to lift the land constraint, just as trade within Europe was insufficient to alleviate Europe's internal pressure on land. Peanuts, sweet potatoes, and eventually maize were introduced to China by Europeans after 1500, but these crops did not make inroads into the diet as potatoes did in northern and western Europe.

The substitution of labor and capital for land in Europe, thanks to the colonial subsidy of land and labor in the New World, propelled Europe into the Industrial Revolution and led to further vast increases in agricultural production. China might have enjoyed such fruits as well had it not aborted overseas exploration in the early fifteenth century. Ultimately, then, the technological success of Europe and North America owes its origins to a cultural predilection toward expansion and innovation. Europe's expansionist tendencies, spearheaded by military aggression and conquest, began with the First

Crusade in 1096. From that time onward, Europe took in new crops and new technology from the Arab world and from the East. Europe kept pushing its sphere of influence outward, especially during and after the fifteenth century, in a way that few other peoples had done. The Arabs had expanded westward and eastward from Arabia beginning in the seventh century, but most of the lands they conquered were dry and thus agriculturally unrewarding. Polynesians spread far and wide in the Pacific, but they too encountered few new crops and no significant land that would have subsidized their economies. Through mutually beneficial trade and even more through wanton compelled exploitation, Europe bought freedom from its land limitation by subordinating much of the rest of the world.[55]

The key to consumption is producing a surplus. Wherever this has been achieved, often in collusion with the consumers themselves, the rise in production reflects increased power and reach. Competition among producers and among consumers ultimately drives these increases, but although competition is ubiquitous, the ability of economic entities to respond adaptively to it is not. In the next chapter, I turn to the ways in which power is enhanced and to the conditions under which opportunities arise that allow economic power driven by competition to increase.

Chapter 6

THE INGREDIENTS OF POWER AND OPPORTUNITY: TECHNOLOGY AND ORGANIZATION

STANDING ON A SPRINGY carpet of fallen twigs and needles on the forest floor of a redwood grove in coastal northern California, I strain to hear the haunting, minor-key song of a lone hermit thrush. Against a constant background of the low, restful sound of the wind blowing through branches a hundred meters over my head, the song emanates from somewhere so high that I am unable to discern the acoustic boundary between forest and sky. The day above is bright and sunny, but here among the massive trunks and the still blooming rhododendrons and the sparse ground cover of giant *Oxalis* and spreading leathery ferns, I stand in deep, cool shade, the air still and moist on this June morning. I am in a city of sorts, a place so utterly dominated and constructed by one species that every living thing, every activity, every local environment is there by the grace of that one plant architect. The redwood gives this forest its shape and nurtures its life. Its trunk is home to mosses and lichens, to spiders and diminutive herbs growing in holes. Its fallen parts—logs, twigs, needles, and cones—provide livelihoods for a multitude of microbes, fungi, slugs, and tiny insects. The shade-loving ground herbs and understory shrubs would wither away if the trees that cast the deep shade were to be felled. Without design or forethought, generations of giant redwoods built, maintained, and wielded power over this forest in the face of fierce winter storms, rainless summer days, unforeseeable earthquakes, and the mainly frost-free cool maritime climate of fog and sun that the great North Pacific imposes from the west.

Just a few kilometers away, wild surf crashes on a rock-bound shore. Long fronds of lithe kelps sweep back and forth as water rushes in and out. Sea urchins, foraging from their excavations in the seafloor, gnaw and scrape what they can beneath the kelp canopy. Higher on the rocks, it is a dense cover of filter-feeding animals—tightly adhering mussels, acorn barnacles, and gooseneck barnacles—that create the biological landscape. Here and there this continuous animal cover is interrupted by a huge seastar, humped over a clump of its favorite mussel prey. Gulls above and sea otters below hunt for food, and once upon a time—before humans exterminated them centuries ago—there would have been Steller's seacows, ten-ton behemoths lazily browsing seaweeds at the water surface. And all this wealth of production and consumption, dominated by a forest of kelps and by large consumers, is made possible by the same cold plankton-rich North Pacific waters that define the weather in the nearby redwood forest.

These displays of natural power now coexist alongside some of humanity's most spectacular manifestations of the use and application of power. In every town and city there is a steady background white noise created collectively by motors and engines of all sorts, constantly consuming fuel and doing humanity's work. Universities, private firms, and government have created a remarkable system of directed invention and innovation, nowhere better exemplified than in California's Silicon Valley. Bridges, highways, airports, and the agencies that keep the entire transportation system running smoothly lubricate trade in a display of awesome collective power and organization.

Great works of nature and of humanity all have something in common: they exemplify power. The competitively dominant producers provide food, create structure, offer shelter and living space for others, modify the environment of life, and even in death nourish their surroundings, chiefly to their own advantage but also to the benefit of many other members of their economy. The dominant consumers regulate when, where, and how the economic units with which they interact make their livings, and determine the adaptive responses that producers and fellow consumers deploy to defend themselves and the resources they control. We cannot hope to understand how economies work and develop without knowing what power is, who has it, how it is maintained and augmented, and what happens when powerful agents are displaced or eliminated.[1]

Historians and sociologists have dissected power in exclusively human-centered terms. In his excellent book *The Anatomy of Power*, John Kenneth Galbraith cites personality, property, and organization as the chief ingredients of power in human society. Those in power seek to bring about the obedience and submission of others through various forms of persuasion, including the threat or reality of punishment and retaliation, aggression, the promise of rewards or money or prestige, and the more subtle methods familiar as advertising, propaganda, indoctrination, and education. Persuasion is in the hands of individuals who are knowledgeable, inspiring, charismatic, rich, ruthless, attractive, or physically strong and courageous. It also rests with organizations—governments, religious institutions, corporations, political parties, special-interest groups, and centers of learning—whose collective capacities and resources far exceed those of any single individual. Wealth, might, knowledge, and control are needed to acquire more of the same. Power is driven by competition among self-interested parties. The attributes that make nations rich and powerful—greater trade and more investment, military might, technological innovation, knowledge, and permissive social and political attitudes favoring enterprise and individual freedoms—may seem highly specific to human affairs, and might appear to apply mainly to that very short period of human history in which nation-states and large institutions rose to cultural and political prominence, but they are merely manifestations of more general phenomena that have affected power from life's beginning.[2]

My focus in this chapter is on power. The attributes of power belong partly to economic units themselves—their metabolism, size, and organization—and partly to the environments that support and are created by living things. The environmental component of power—temperature, habitat size, productivity, and the competitive context—interacts with the technological component of power in individual economic units to produce an economy in which the distribution of power depends on the costs and benefits of improvement and the risks and limitations of performing economic work. Economic opportunity—the extent to which entities can adapt to and modify their surroundings—thus depends on the degree to which demand for increased power is realized. The technological and organizational qualities that confer or permit power require high, predictable, and well-regulated supplies of energy and material resources. Adaptive scope—the range of adaptive possibilities available to economic entities, or the number of pathways available for the deployment of power—is dictated by how effectively entities can extract, transform, regulate, store, and apply resources. Conditions under which entities grow in size or in numbers provide opportunities for adaptation by reducing the costs of failure, increasing the benefits of success, and relaxing the material and technological constraints. If constraints predominate over opportunities, economic conditions perpetuate the adaptive status quo.

I begin this chapter by characterizing the technology and organization of power. Increases in an entity's temperature, metabolic rate, size, and structural complexity are the chief mechanisms by which entities can acquire and deploy more power and doing so under a wider range of circumstances. Then, in chapter 7, after specifying what I mean by opportunity and constraint in economic and evolutionary terms, I turn to the environmental component of power. I ask where the energy and raw materials essential to life come from, how they are distributed and dispersed through the biosphere, and what becomes of them. This account draws heavily on the particulars of our own planet's atmosphere, ocean, crust, and mantle, and on the intimate and far-reaching interplay between biological and geological processes. I argue that conditions most conducive to growth and adaptation, and thus providing the greatest adaptive scope and economic opportunity, occur where the rate of supply of raw materials produced by weathering and eruptions from Earth's interior is predictably high and where life maximally modifies, enhances, and regularizes these rates. These conditions prevail in warm, shallow, well-lit waters and moist warm lowlands in or near mountainous terrain with plentiful runoff.

By elaborating on these requisites for power, I am not endorsing the economic dominance that emanates from power as a good or a bad thing. My position throughout this book is that we need to understand phenomena dispassionately before we construct policies based on them. Economic power affects the well-being of every living thing on Earth. What it is, how living entities acquire and nurture it, and what its long-term consequences are for those who have it and for those who do not are questions fundamental to all interpretations of history and to all scenarios for our collective future.

Power and Efficiency

The physical dimensions of power offer a useful guide to the ways in which power is expressed in the economic domain. Recall that power is defined as energy per unit time. Energy, in turn, is expressed as force times distance, or mass times the square of velocity. Mass, distance, force, speed, energy, and duration are therefore all potential components of power. An entity's power increases if any or all of the first four of these components increases, or if the sixth (duration) is reduced. Ecologically, this means that powerful entities are large, fast, wide-ranging, rapidly metabolizing units capable of exerting strong forces, storing and regulating resources, and responding appropriately to a wide variety of circumstances. Power makes for prolific producers and demanding consumers with a wide reach.

Another way of looking at power is to think in terms of costs and benefits. Power can be increased by lowering costs as a way to increase productivity, or raising benefits, or both. Economists have tended to emphasize reducing costs as a way to maximize profits, that is, increasing efficiency. By efficiency I mean the quantity of output relative to the quantity of input. A given absolute expenditure of energy or material resource thus yields a higher absolute output if efficiency increased; or, by saving costs on labor or on materials, a given amount of product can be made for less.

I believe that this emphasis on efficiency is misplaced. Economic success depends on absolute performance, and very often—in human-economic contexts as well as in the evolutionary marketplace—high levels of performance go hand in hand with reduced efficiency. The most efficient animals in nature are those with chronically low metabolic rates—brachiopods and sea lilies (crinoids), for example—and those at rest. The shell-bearing cephalopod *Nautilus pompilius* extracts as much as 20 percent of the oxygen from normal seawater when at rest, and up to 50 percent when oxygen concentration in the water surrounding the resting animal is close to the animal's minimum threshold. When swimming, however, *Nautilus* removes just 5 percent or less of the oxygen from the water that streams over its respiratory surfaces. Warm-blooded animals are notoriously inefficient when compared to cold-blooded ones. Cold-blooded amphibians and reptiles channel 15–30 percent of the energy they assimilate into production, whereas for warm-blooded birds and mammals only 0.5–3.0 percent of assimilated energy can be transformed to production. The daily cost of locomotion in mammals ranges from 0.5 percent of daily metabolism in a 10 g animal to about 6 percent for an elephant. Faster locomotion increases these costs, but the advantages in food gained, time saved, and risk reduced are so great that these added costs are insignificant compared to the benefits. In fact, a problem for active warm-blooded animals is disposing of excess heat produced by an inefficient engine. This is why we sweat, dogs and birds pant, and bees and termites ventilate their nests. In our technological world, internal combustion engines and atomic power plants give off vast amounts of unused heat, but their power yield is so great and provides such clear economic advantages that their inefficiency is tolerated, much as it is in warm-blooded animals.[3]

In all economies, I suggest, efficiency becomes important when power is low and output cannot be increased in absolute terms. This occurs when energy or raw materials are sufficiently scarce that reducing the cost of acquiring them is the only way of not losing ground. Increases in power, however, are sufficiently beneficial that considerations of efficiency are secondary, especially if productivity also benefits the supply of raw necessities. In such cases, absolute performance is far more important than efficiency. Thus it pays to be efficient for subordinate members of an economy, and it pays to increase in performance for those in power.

Temperature

The energy state of an economy—that is, its temperature—largely determines what its members can do and how fast they can do it. Temperature—the average kinetic energy of the moving molecules in a gas—affects every chemical process and every physical property associated with life. It influences not only the cost of doing business, but the speed at which tasks can be accomplished, and perhaps most importantly the range of adaptive options available. Temperature is, in other words, the crucial link between energy and time, the two components of power.[4]

As a rule of thumb, biological activity and metabolic rates increase exponentially with temperature over a range of 0°C, below which metabolism ceases because of the phase change of water from liquid to solid, to 40°C, above which many molecules deform or break apart. The precise relationship between activity and temperature varies according to organism and environment. In their comparative survey of bivalved molluscs adapted to various regimes of temperature, Peck and Conway found that activity increases by a factor of 2.29 for every rise of 10°C over a thermal range of 10 to 30°C. Between 0 and 5°C, this factor rises to 3.64, meaning that thermal dependence is much greater in species adapted to the cold conditions in polar regions than it is for temperate and tropical species. Studies on fish reveal similar strong dependencies of metabolic rate on environmental temperature.[5]

Physiologists have discovered a great variety of mechanisms by which strict thermal dependence is lessened. Enzymes, which enhance chemical reactions by reducing the activation energy or "hurdle" that must be overcome to transform reacting molecules into a new configuration, permit biochemical processes to proceed at temperatures typical of the biosphere, far below the temperatures that would be required in the absence of these catalysts. This effect can be amplified with the production of more enzymes at low temperatures, although such a solution entails considerable expense to the organism. Despite these and other molecular accommodations, a degree of thermal dependence remains. Organisms whose body temperatures conform to those of their surroundings are slower in the cold than in thermal regimes characteristics of the tropics or of the temperate zones in summer. For them, lower temperatures mean reduced power.[6]

For this reason, temperature greatly affects the scope of adaptation, the range of adaptive possibilities open to organisms. With little power available, and with resources limited in supply and reliability, the range of options is small. In the deep sea and the polar oceans, where water temperatures hover just above the freezing point of seawater, all cold-blooded organisms have low metabolic rates, imposed in part by the frigid conditions and in part by a seasonal or chronic dearth of food. All the clams there have small, thin, slow-growing, unornamented shells. They lead sedentary lives, slowly burrowing into the fine-grained sediment or permanently attached to stones, sunken wood, or whale bones. Slow and sedentary bivalves with unobtrusive, slowly growing shells occur prolifically in the warm waters of tropical seas as well, but they are joined there by large, fast-growing species, species with frilled and spinose shells, and clams specialized for rapid burrowing, swimming, and even leaping. Brachiopods, whose metabolic rates are typically just one-third to one-tenth those of bivalves of comparable size, mainly live in cold, dark, and oxygen-poor waters, where their adaptive range is severely limited. No brachiopods have become specialized as rapid burrowers, swimmers, and leapers. Even in the tropics, they are tiny animals confined to deep crevices, caves, and cavities. The important point is that low metabolic rate, whether imposed by the cold or by technical limitations of physiology, constrains adaptive scope. Conditions or technologies permitting high metabolic rates allow for a very much greater range of adaptive choice.[7]

The permissive effects of higher temperatures are manifested in many other ways related to performance. Flight among insects near freezing can be accomplished under very limited conditions of very low wing loading, whereas at higher temperatures the range of morphologies suitable for sustained flight is much greater, and notably includes heavy-bodied insects. Fish, lizards, snails, clams, and crabs have a very limited range of top speeds in the cold, none being fast; whereas at high temperatures, all these groups contain sedentary species as well as extremely fast ones. In the cold, plants can achieve high growth rates only if they can tap large amounts of stored energy in roots or underground stems. In warm moist conditions—tropical climates and the summers in temperate East Asia and eastern North America—plants show a great range of growth rates, from very slow (oaks and many understory plants) to as much as 15 cm per day in some vines. These rapid growth rates are sustained by photosynthesis, not by food reserves. In short, constraints on specialization to particular extremes are much more severe in the cold than at the temperatures of typical summers and tropical climates, generally 20°C and above. All the options available in the cold are still open in warm conditions, but in addition there are options based on high power output that are accessible only at higher temperatures.[8]

The physical properties of materials reinforce the dependence of life processes on temperature. For example, the viscosity of water—its stickiness, or the resistance a body encounters when moving through it—decreases by 45 percent from 1.79×10^{-3} Nsm^{-2} (newtons times seconds per square meter) at 0°C to 0.80×10^{-3} Nsm^{-2} at 30°C. Even a relatively small rise in temperature from 10 to 20°C is associated with a decrease in the viscosity of seawater from 1.39×10^{-3} Nsm^{-2} to 0.87×1010^{-3} Nsm^{-2} at 30°C. This dependence of viscosity on environmental temperature has important implications for all organisms living in water, whose viscosity at 20°C is sixty times greater than that of air. Work must be done to overcome the resistance encountered by a body moving through a fluid due to viscosity. For example, animals that filter food particles out of the water by drawing currents through a filtering system encounter viscosity-related resistance because the current and the beating cilia creating the current are subject to viscosity. Animals moving through water or through some other viscous medium are also impeded. They and their moving parts must overcome viscosity-related resistance in order to swim, burrow, crawl, or sink. For all these functions, the work needed to overcome viscosity increases as water temperature falls. The result is reduced power at lower water temperatures.[9]

In air, by contrast, viscosity rises as temperature increases. Air's viscosity increases by a factor of 1.1 from 1.718×10^{-5} Nsm^{-2} at 0°C to 1.919×10^{-5} Nsm^{-2} at 40°C. This effect, magnified by the higher water-vapor content of air as temperature rises, provides greater lift for animals flying in warm air than for those at low temperatures. Whether winds of a given velocity are more damaging to tall trees at high temperatures than in the cold has not to my knowledge been investigated.

The cost of production of skeletal material also varies with temperature. Many skeletons, such as the shells of molluscs and brachiopods and the articulated internal skeleton of plates in echinoderms, are made up chiefly of the mineral calcium carbonate. This mineral becomes more soluble, and is therefore energetically more costly to deposit, at lower temperatures. This has implications not only for the potential growth rates of individual animals, and for marine plants like *Halimeda* that also precipitate calcium carbonate, but also for repairing damage to the skeleton. In effect, the cost of a calcium carbonate skeletal defense rises as temperatures fall. That such costs are biologically significant is clear from A. Richard Palmer's finding that the growth rates of "soft" tissues in snails are limited by the growth rate of the enclosing shell. In the cold waters of the Antarctic, in fact, shell-bearing animals tend to be very small, no bivalves exceeding a linear dimension of 100 mm and most being much smaller, whereas animals without a skeleton and those that support structures made of silica can reach very large sizes.[10]

The beneficial effects of higher temperatures reverse when temperatures exceed a threshold, whose absolute value depends on both the organism and the temperature range to which it is accustomed. Although some microbes such as the nitrate-reducing archaean *Pyrolobus fumarii* can grow at a temperature of 113°C, most proteins and nucleic acids begin to lose their all-important three-dimensional structure at temperatures approaching 100°C and membranes across which substances pass into and out of the cell become fluid. Chlorophyll used in photosynthesis degrades above a temperature of 75°C. Some algae, fungi, and single-celled protists can withstand temperatures as high as 70°C. Mosses can tolerate 60°C, vascular land plants can manage at 48°C, and some desert insects remain active at 51°C. Some high-intertidal snails tolerate rock temperatures of 50°C or more, and fish succumb above 48°C. Mammals and birds, which maintain body temperatures of 42°C or less, live near the upper end of their thermal tolerances, and cannot normally survive body temperatures of more than 43°C. Of course, many cold-adapted species succumb at much lower temperatures, and their optimal performance lies well below that of tropical species. In the sea and in fresh water, very few species remain metabolically active above 40°C, and high-level predators shut down above 30 or 31°C. Because intense activity itself produces heat, it may be limited at high environmental temperatures, so that excess heat can be lost through convection, evaporation, panting, or some other means. Environmental temperatures of 25 to 30°C seem to be optimal for temperature-conforming plants and animals in the tropics both in the sea and on land. For animals capable of regulating their temperature, optimal performance is achieved when the body is somewhere between 30 and 42°C, depending on the species.[11]

Some costs are lower in the cold. In organisms whose body temperature rises and falls according to their surroundings, the basal metabolic rate—the rate of energy used for body maintenance—is significantly lowered as conditions cool. Experiments by Andrew Clarke show that Antarctic invertebrates experience a 55 percent decrease in the rate of metabolism when the temperature drops from 10°C to 0°C. Growth rate and the energy annually devoted to reproduction also decrease, but by only 40 percent and 30 percent, respectively. This means that, although total energy intake is less at 0°C than at 10°C, energy expended in simple maintenance constitutes a smaller fraction of the total energy budget at the lower temperature. By the same token, a rise in temperature increases maintenance costs relative to other functions. Whether this increased expense is biologically meaningful is debatable. The criterion by which organisms are "judged" is absolute performance, and it remains to be seen if the gains in performance with a rising temperature outweigh the increased basal expenditures. In other words, the energetic inefficiency associated with life and higher temperatures may be economically acceptable as long as absolute levels of performance are sufficient to assure continuity in the next generation. For animals that maintain a high constant body temperature, such as mammals and birds, maintenance costs rise dramatically as the temperature of their environment decreases. For this reason, warm-blooded animals must have thick insulation to live in polar waters, and there is a severe lower size limit on body size for these animals.[12]

Another potential advantage of the cold is the lower vulnerability to disease. Microbes that parasitize hosts and cause disease are highly temperature sensitive, and therefore grow and reproduce faster in warm bodies than in cold ones, at least up to a point. It is for this reason that warm-blooded animals must close open wounds very rapidly, a necessity that appears to preclude widespread regeneration of lost body parts. It also accounts for the enormous elaboration of the immune system, which identifies and eliminates foreign intruders. These costs of vigilance against enemies are therefore apt to be substantially lower in cold environments and in the bodies of cold hosts.

Metabolism

Perhaps the most obvious way to regulate temperature and to gain power is to increase the rate of metabolism, the rate at which energy is taken up and used for economic work. There are essentially two ways of doing this. The first is passive: live in a warm place, where the tasks of life proceed at high rates and where the scope of adaptive possibilities is large. For the lazy, however, economic performance entirely depends on the energy-related permissiveness of the environment. It is the environment that provides the power, not the economic entity itself. The second way is active, and involves significant investments in power-generating machinery. Rapid metabolism is achieved by the entity's own physiology. It makes the metabolizing individual less dependent on the current energy state of its environment, but more dependent on a consistently high supply of food.

Let us look briefly at an example that does not involve temperature. Many freshwater and marine animals make their living by suspension-feeding, collecting tiny food particles from the surrounding water. Passive suspension-feeders—corals, many sea anemones, crinoids, and some brittle stars (ophiuroids)—simply extend their collecting equipment—tentacles, arms, mucus-coated surfaces—to intercept food particles arriving on currents of water. In food-rich, turbulent waters, this way of feeding works quite well, as exemplified by the predominance of passive suspension-feeders on reefs and wave-swept shores, and in swift streams. Passive suspension-feeding, however, poses two problems and implies two corresponding constraints. One is that animals must be exposed to currents. Water movement is fastest well above the hard surface, and slow or negligible in the thin film or boundary layer adjacent to a solid surface. This means that feeding structures must be deployed well above the seafloor, but this exposure places these organs, and the animal as a whole, in positions where predators can easily find them. It is therefore not surprising that most passive suspension-feeders rely heavily on either of two types of defense: toxicity or protection by stinging cells, as in corals and other cnidarians, and easy breakage and rapid replacement of feeding structures, as in crinoids and other echinoderms. The second problem is that passive suspension-feeders are at a clear disadvantage in calm waters. Without currents to bring them food, the passive suspension-feeders quickly exhaust their local food supply. Calm waters prevail not only in such common environments as ponds, lagoons, and sheltered bays, but also in many environments where large predators are physically limited in their ability to locate and catch prey. These habitats include crevices, caves, the undersides of boulders, and dense vegetation.

Active suspension-feeders, by contrast, create their own currents, drawing food particles to them and transporting the catch to the organs of digestion. Most bivalved molluscs (clams) create feeding currents by adding a feeding function to the beating cilia in the gills, which also serve for gas exchange. This enables bivalves to occupy sandy and muddy bottoms throughout the world. Some bivalves can burrow very deeply into the sediment, sometimes as deep as 60 cm, maintaining contact with the water above through long pipes (siphons) through which water must pass before reaching the food-sorting areas in the gills. The beating cilia in the gill must overcome the very considerable resistance that the siphons offer to the flow of water, a capacity made possible by modification to the gill structure and by a high metabolic rate. Many crustaceans, suspension-feeding fish, and baleen whales actively collect suspended food by swimming while water passes through the extended collecting gear.[13]

Strong feeding currents may exact a high energy cost, but they provide clear competitive advantages. Bryozoans—sedentary colonial invertebrates that grow over rocks, shells, and other surfaces—illustrate this advantage well. In his studies of bryozoans in the northern Adriatic Sea, Frank McKinney has found that one group (the cheilostomes) tends to overgrow another (the cyclostomes) in about two-thirds of the interactions he observed. The median velocity of feeding currents produced by cheilostomes (1.14 mm/s) is considerably higher than that of cyclostomes (0.82 mm/s), an ability made possible by the larger size of the lophophore, the anatomical structure in each member (zooid) of the colony where the currents are generated. Cheilostomes grow more rapidly, grow to a larger colony size, and are able to take larger food particles. Moreover, filtered water depleted of food is expelled at the edges of the colony, whereas in cyclostomes the colony edge is the location from which food-bearing water is drawn. This means that, when a cheilostome and cyclostome interact along a colony border, the cyclostome is at a distinct disadvantage not only because it has less power to draw in food-laden water, but also because its neighboring cheilostome diverts food. Other attributes, notably of organization and specialization within the colony, contribute to the competitive dominance of cheilostomes, but the competitive outcome with neighboring suspension-feeders may chiefly be decided by the power of the current-generating machinery.[14]

The competitive superiority of cheilostomes over most cyclostomes has had long-term ramifications. Cyclostomes, which originated as early as the Middle Ordovician (about 460 Ma), comprised a small minority of bryozoans during the Paleozoic era (543–250 Ma), but thereafter diversified in the Mesozoic era, especially from the Middle Jurassic to the middle Cretaceous (170–100 Ma). Many of these cyclostomes lived as crusts on surfaces. Cheilostomes arose in the latest Middle Jurassic or early Late Jurassic (Oxfordian to Kimmeridgian epochs), but underwent a major diversification beginning in the Late Albian part of the late Early Cretaceous, about 105 Ma. Most of the Mesozoic cyclostomes had been encrusters, but when cheilostomes began to diversify, erect species—those with branching colonies rising like tiny shrubs or trees above the seafloor—increasingly replaced encrusters in this group. Unlike encrusters, which often compete with neighbors, erect bryozoans frequently occur in isolation, and rarely interact with other bryozoans. The competitive superiority of cheilostomes therefore manifested itself in several ways: (1) increasing diversity relative to cyclostomes; (2) restricting cyclostomes to erect growth forms that interacted little, and to deep-water habitats where food is less abundant; and (3) disproportionately overgrowing neighboring cyclostomes, whose growth and power to reproduce are therefore reduced. Interestingly, the cheilostomes originated and later diversified in nearshore environments, where competition from cyclostomes and other suspension-feeders was intense.[15]

Given that higher body temperatures enable living things to work more rapidly, it is not surprising that many lineages, including animals with dominant economic roles, have evolved various means of generating and maintaining heat. By doing so, they can be economically active under a great variety of conditions, and are less at the mercy of cold than are those forms of life whose body temperature more or less matches that of their surroundings. They can sustain activity longer, exert stronger forces for more extended periods, and grow faster as long as food remains plentiful and reliable.

Particularly high rates of metabolism in animals are associated with the production of internal body heat. In birds and mammals, internal organs—heart, liver, intestines, kidneys, and brain—are largely responsible for the production of heat while the animal is at rest. Comprising 8 percent of the human body's mass, they produce 70 percent of human body heat. Shivering—the contraction of muscles without doing work—contributes to heat production, especially in the cold. The brain and retina supply heat to such fast-swimming fish as mackerel, tuna, and billfishes, but red muscles situated deep inside the body also contribute heat in tuna. In bees, moths, and many other insects, heat is generated by shivering of the flight muscles. Animals that do not produce internal heat nonetheless gain some benefits from operating at high temperatures by basking in the sun. This habit is very widespread among lizards, snakes, and such insects as butterflies, grasshoppers, and dragonflies.[16]

Even some plants achieve metabolic rates high enough to heat their flowers by some 15 to 35°C above the temperatures of their surroundings for periods up to fourteen days. Skunk cabbage (*Symplocarpus foetidus*), which in the eastern United States flowers in late winter, is an aroid in which metabolic activity heats the flowers by burning starch reserves in the roots, presumably to disperse the foul-smelling odors to attract pollinating flies or perhaps to keep the flies warm while the flowers are being pollinated. The water lily *Nelumbo nucifera* at 10°C expends 1 W of power to heat its night-blooming flowers. Even in tropical forests, flowers of some six plant families are 3.4 to 4.0°C warmer at night than the surrounding air. This high energetic investment benefits the large beetles that feed and mate in the warm interior of the flowers, where these insects can remain active without the risk of being detected by their enemies.[17]

The benefits of high metabolic rates and of high and constant body temperatures are dramatic. Brian McNab, for example, found that higher rates of metabolism are associated with higher rates of potential population increase in mammals. Warm-blooded marine mammals, tunas, and sharks maintain faster average swimming speeds and probably have higher maximum speeds than do fish whose body temperatures match those of their surroundings. Among land animals, too, average walking or running speeds and the duration over which very rapid locomotion can be maintained are six to seven times greater in warm-blooded species than in reptiles and amphibians of similar mass. This enables warm-blooded animals to range more widely and to encounter food more often. Burst speeds do not differ much between warm-blooded and cold-blooded vertebrates of the same mass on land, but in cold-blooded amphibians and reptiles such maximal performance is mainly done through unsustainable anaerobic metabolism, which draws on stored resources. It takes hours to replenish the reserves. Among vertebrates, flapping flight (flight involving movement of lift-generating wings) is invariably associated with high body temperatures. Even in insects, flight and other energy-guzzling activities require a hot body. Some moths can fly at temperatures near 0°C, because they raise the temperature of their flight muscles by 20 to 30°C. In dung beetles (Scarabaeidae), internally generated warmth provides a competitive edge in locating food (dung) and rolling dungballs to safe sites for consumption. In short, the thermal independence and greater power provided by a hot engine enable animals to compete effectively and to consume prolifically.[18]

Thermal independence and high body temperatures exact great costs. Absolute energy costs of maintaining a warm-blooded songbird at body temperatures ranging from 37 to 42°C average 6.247 W × $kg^{3/4}$ of body mass, a value more than one thousand times higher than that of typical marine invertebrates (0.0056 W × $kg^{3/4}$). The power required to maintain the average warm-blooded mammal at body temperatures of 37 to 39°C is 3.39 W per $kg^{3/4}$, about half that in songbirds but ten times that of the average lizard and land plant (0.378 W per $kg^{3/4}$). The power budget of lizards is about 2.5 percent that of birds. Warm animals provide ideal conditions for temperature-sensitive pathogens. Richard Goss argues that infections under these conditions spread so quickly that wounds must close quickly.[19]

The highest power output known in nonhuman animals is 160 W per kg of muscle mass in euglossine orchid bees flying in heliox, a gas mixture with the normal 20 percent oxygen content and with the rest composed of helium. This mixture has a density of 0.4 kg/m^3, one-third that of air at 20°C at sea level. Hummingbirds flying in heliox can achieve a power output of 133 W/kg of flight-muscle mass. In nature, performance levels are normally lower, but flying animals still exceed all others in power output. The extremes of power output are approached during mate selection and in episodes of escape from pursuing predators.[20]

At the level of the cell, rapid metabolism is associated with high concentrations of mitochrondria and of enzymes. Enzymes such as cytochrome oxidase, which is responsible for oxygen-consuming (aerobic) metabolism, are housed on membranes of the mitochondrion. According to John Ruben, total mitochrondrial membrane surface area is four times greater in the liver, kidney, heart, and brain tissues of warm-blooded rats than in the comparable organs of lizards, which do not produce significant internal heat. This difference arises in part from the larger mass of the heat-producing internal organs, and in part from the higher density of mitochondria in the tissues. The mitochondria in skeletal muscles generate about twice as much metabolic power as those in the internal organs of mammals, and thus differ both in density and in individual metabolism from their counterparts in the organs where heat is produced at rest.[21]

As part of the machinery of rapid metabolism and heat production in birds and mammals, many internal organs have become modified. Oxygen can be delivered at higher rates thanks to increased densities of capillaries and higher ventilation rates in the lungs, increased capacity of red blood cells to carry and deliver oxygen, and high-pressure pumping of the heart. Skeletal muscle, which produces heat during exercise, has three to four times as much mass in warm-blooded birds and mammals as in reptiles. In short, the entire system is geared to receive and use large amounts of energy, oxygen, and ATP.

High speed, rapid acceleration, and keen vision are among the capacities that in animals are made possible by the dedication of a large proportion of body mass to muscles and other organs rich in mitochrondria and very high in energy demand. Some 30 percent of the body mass of the South American flightless bird *Rhea americana* comprises leg muscles, which enable this animal to maintain a speed of 4 m/s while running on a treadmill. While running, the rhea consumes 2.85 ml oxygen per kilogram of body mass, a rate thirty-six times higher than that when the bird is metabolizing at rest. The shell-bearing cephalopod *Nautilus pompilius* with a top speed of 0.4 m/s, devotes 4.5 percent of its total mass (shell together with internal organs) to propulsive musculature. The size of the muscles, and the volume of the water that is expelled as the animal moves by jet propulsion, are constrained by the volume of the shell's living chamber, in which all of the organs and the ejectable water must be packed. Evolutionary loss of the external shell relieves this severe constraint by enabling muscle mass and the volume of expelled water to increase. Squids, which lack external shells, attain a speed of 2 m/s, and devote some 35 percent of their body mass to muscles. Fish such as tuna, which create forward thrust by side-to-side movements of the streamlined tail powered by internal red muscles, are still faster (up to 21 m/s in extreme cases), accelerate more water as they swim, and devote almost 50 percent of their body to propulsive musculature. In fish, power output scales as the cube (third power) of top speed.[22]

The application of force during feeding is similarly expensive. In the leaf-cutter ant *Atta sexdens rubropilosa*, which cuts leaves from plants and then transports them to underground gardens where the leaves serve as substrate for domesticated fungi eaten by the ants, the energy demanding muscles of the mandible comprise more than 50 percent of the mass of the ant's head capsule and more than 25 percent of the mass of the entire body. While cutting leaves, these ants sustain metabolic rates thirty-one times higher than the rates of oxygen consumed while the animals are at rest.[23]

As the center for the processing of sensory inputs and of mechanical responses to them, the brain of warm-blooded animals—and the visual cortex in particular—is one of the most rapidly metabolizing organs of the body. The human brain comprises 2 percent of body mass but utilizes 15 percent of the oxygen used. Sensory acuity, rapid and coordinated response, learning, and memory are key functions of the brain that require metabolic work and that enable an animal to compete under a wide variety of circumstances.[24]

Low metabolic rates, then, greatly limit adaptive options, where high rates enlarge the scope of adaptation. I shall illustrate this point with one final example from animals in nature, an example drawn from my own work on shell shape in molluscs.

The shell of a mollusc is essentially a hollow conical tube, closed at one end (the apex, or oldest part of the shell) and open at the other (the aperture). The shell arose as an external skeleton, enveloping all the internal organs, and attached to the rest of the body by one or more muscles. Growth of the shell takes place at the rim surrounding the aperture, and involves the deposition of an organic matrix and a mineral component of calcium carbonate, both laid down by the edge of the mantle. In its simplest configuration, the shell can be thought of as a cone with a circular base (the aperture) and the apex situated directly above the center of the base. The rate of growth can be measured along a line from the apex to the rim of the base, or from one point along that line to another, a distance representing some known increment of time. The shape of the cone is determined by its rate of expansion, or apical angle. A tall narrow cone expands slowly, whereas a low cap-shaped cone expands rapidly. Most real shells are more complex, and are typically coiled in a spiral configuration. The coiling approximates the curve known as the logarithmic spiral, in which adjacent coils move farther apart as one travels away from the point of origin of the spiral. The logarithmic spiral has the property its shape remains the same as material is added to its open end. Shells conforming precisely to logarithmic-spiral growth are therefore coiled cones that do not change shape as they enlarge through growth at the rim.

No real shell actually conforms exactly to this logarithmic pattern of growth. There are two reasons for this. First, shell shape depends on growth rate. Second, growth rate varies through time as well as from place to place along the rim of the aperture. All else being equal, the rate of expansion of the cone is greater when growth rate (and thus metabolic rate) is higher. A low, flat, conical shell expands rapidly because it grows rapidly. Parts of the rim located far from the apex will tend to expand more rapidly than parts closer to the rim, and correspondingly grow faster. The enormous variety of shell shapes observed in nature therefore reflects the relationship between growth rate and cone expansion on the one hand, and the shape of the rim on the other. The shape of the rim depends on many factors, some genetic, others related to the forces that the environment imposes on the shell-bearer and that the mollusc itself exerts.

The two-part shell of a clam represents a variation on the theme of the one-piece conical shell characteristic of primitive shelled molluscs and of snails. In the bivalved shell, slow growth is associated with an inflated, highly convex form, whereas rapid growth is reflected in a flattened valve. Brachiopods, which evolved the bivalved shell independently of molluscan clams, show the same relationship between growth rate and shell shape.

The early molluscs in which the conical external shell arose seem to have been sluggish creatures crawling on hard surfaces on the seafloor. Their shells reflect passive resistance to predators, and likely fit completely over the vulnerable internal organs. Metabolic rates in these early shell-bearers were probably low, because the shell cones were generally narrow and tall, and bivalves had inflated shells. Many later groups show notable increases in the rate of metabolism. Not only are these increases reflected faster and more frequent locomotion, but also by the much greater range of shell shapes, including many in which the rate of shell expansion is high.[25]

Thanks to the harnessing of power from outside sources, humans have effectively increased their metabolic rates by at least an order of magnitude. For short periods of intense activity, human muscles deliver 800 W of power, but for intervals of a day or more they operate at an average of about 250 W. Power increased in steps with the domestication of pulling and drafting animals, the harnessing of wind and water with such inventions as the water wheel, windmill, and sailing ship, the exploitation of fossil fuels with the use of steam engines and internal combustion engines, and the controlled use of nuclear energy. By 1990, the per capita average use of fossil fuel amounted to about nineteen full-time human equivalents of all-day, year-round labor. Eric Chaisson's calculations indicate that an average person in the United States consumes 25 kW per day, whereas globally the figure is 3 kW.[26]

Although supply of resources dictates what level of metabolism can be achieved, it is demand—imposed by consumption and by competition—which drives some entities toward higher metabolic rates. Those who win fights over food, territory, or mates benefit from, and are dependent on, rapid metabolism. The same applies to those who can afford expensive symbols of health and fitness, individuals able to grow rapidly, and agents capable of exerting great force. Supply makes demand possible, but competition and consumption are the processes that select for increased power and control among winners.

Size

Few attributes are more reliable indicators of absolute power than absolute size. All else being equal, the power demand of a living entity is proportional to the entity's mass. Large size—expressed as mass, volume, territory, or numbers—is widely associated with competitive dominance. Large opponents tend to win fights with small ones; large corporations control markets more than small ones, and large nations set policies and define terms by which smaller governments must abide. Although exceptions are known, large organisms are less susceptible to lethal predation than small ones. Not only can they produce a larger number of progeny of a given size, but large organisms have more evolutionary flexibility in whether they produce many small young or a few large, well-nourished ones. A large body can carry more energy reserves than a small one, and is able to isolate its interior parts better from external assault. On average, the lifespan of an individual scales with the one-quarter power of body mass, meaning that larger organisms are able to adapt to rarer circumstances. With a greater number of parts, or with a large population size, increased size allows for more redundancy—a greater margin of safety—and a finer division of labor. Parts can become specialized to fulfill functions that in a smaller body or population are accomplished by all-purpose units. For example, whereas large animals have distinct systems for digestion, respiration, circulation, and ventilation, tiny animals often lack these, the various functions being adequately executed through passive diffusion across the body surface.

Much of the internal structural diversity we observe in organisms is related to transportation networks. In animals the separate systems for circulation, ventilation, excretion, and nerve communication are operated by energy-expending pumps or other motors—heart, lungs (and diaphragm in mammals), cilia (in gills), kidneys, and neurons (which produce electrical discharges)—keeping diffusion-limited surfaces well supplied with necessities and taking away wastes. In plants, transport tends to be powered by outside sources, such as the sun's heat causing evaporation, which pulls up water and nutrients through the roots. There is probably active transport of organic substances through phloem and of soluble defenses such as alkaloids and terpenes through latex tubes, as in poppies (Papaveraceae), milkweeds (Asclepiadaceae), and dandelions and their relatives (Asteraceae).[27]

Most such networks—circulatory systems, the vascular system (xylem and phloem) of land plants, digestive systems, and road networks—all have in common a hierarchical structure, in which a few large primary conduits branch into increasingly smaller tubes, ultimately terminating in tiny capillaries, vesicles, leaf veinlets, tracheoles, and other units where exchange of materials mainly takes place. Transport systems of this kind have a fractal structure, meaning that the architecture of small scale parts resembles the architecture of the system at larger scales. They service and therefore penetrate a given volume or area more or less completely, enabling all parts of the body to receive and void the raw materials and products of metabolism. The power drawn by cells scales with mass, but the delivery of supply and of power scales as the three-quarters power of mass in space-filling fractal transport systems. This is so because the mitochondria, enzymes, and other power-generating structures are situated on surfaces, membranes where exchange—the interaction between supply and demand—takes place.[28]

Economists have long understood the importance of economies of scale, meaning that many costs rise less rapidly than do the benefits of greater mass or greater numbers. The production of internal heat, for example, enables a large vertebrate to achieve and maintain a higher temperature for a given power output because the surface area of skin across which heat is lost increases only with the two-thirds power of mass, whereas heat production scales with mass or volume. Dinosaurs would not have had to produce as much heat as mammals do in order to achieve high and relatively constant interior temperatures, because their huge volumes were encased in a relatively small area of surface. Economies of scale do not apply when the investments required to support a larger body scale with a power of mass greater than unity.

As usual, there are limitations. Besides the same kinds of costs that entities with high metabolic rates face, large-bodied organisms usually belong to small populations, which through inevitable fluctuation are more susceptible to extinction than populations of many individuals. As I discuss more fully in chapter 7, small environments such as islands cannot support large individuals because the population in a small area is unsustainable.

Organization

Most members of an economy, be they human or nonhuman, deal with a vast number of challenges and opportunities. Organization—a physical partitioning of an entity into separate, semiautonomous, communicating parts—enables tasks to be divided among submits or to be carried out at different times. The parts—modules, chapters, departments, compartments, domains, cells, and elements—comprise a whole that exercises a degree of central control.

Organization arises spontaneously first through chance encounters and interactions in conformance with the laws of physics and chemistry, and is then codified through genetic or cultural regulation, allowing it to evolve through adaptive modification. Components arrange themselves according to prevailing forces, chemical gradients, and other external cues. The double membrane, a ubiquitous feature of cells, self-assembles because the ends of component molecules interacting with the external liquid medium of water point in one direction while the ends that do not interact with water point in the opposite direction and bond with like ends of other molecules. Cells comprising connective tissue in the vertebrate skeleton arrange themselves along stress fields imposed by muscular forces and by the environment. The direction and magnitude of growth of body parts are influenced by everyday forces, as illustrated by the shape changes wrought by a diet of hard food in claws and jaws of crabs, fish, insects, and monkeys (see chapter 4). Cell movements in the developing animal embryo cause tissues to encounter other tissues, and these interactions in turn result in yet more movement and more interaction. A very great deal of the complexity we observe in living things arises from responses of parts to internally and externally imposed forces and chemical conditions.[29]

Two complementary means of achieving a division of tasks exist in all economies. One is differentiation, the process whereby initially similar, more or less redundant units acquire different functions or specializations. The other is unification, in which previously independent parts come together to form a larger, integrated economic whole. No matter how the whole is organized, its parts must cooperate at least to some extent. Parts must exchange information and materials cheaply and quickly. Integration must not be so all-encompassing that every part and every function becomes interlocked with every other; parts and tasks must all retain some degree of autonomy in order to preserve flexibility. In short, the differentiated or unified parts should be cooperative, specialized, communicative, and semiautonomous.

The power of organization derives from five overlapping advantages: (1) greater redundancy, meaning that the system becomes more forgiving of error and disruption, and that variants arising within the system are not automatically crippling to it; (2) the elimination of incompatibilities and functional trade-offs by increasing the autonomy of parts and reducing or diffusing central controls; (3) increased generation and testing of variation, making the "search" and selection for adaptations faster and more directed and allowing for better prediction and anticipation; (4) exploitation of resources not previously tapped; and (5) more effective use of resources, waste products from one process being put to use in another. An economic entity's organization thus confers flexibility, protection against disruption, responsiveness to change, and enhanced performance on many fronts.

If only one part engages in a given task, the loss or injury of that part could compromise the whole. In terms of genes, a mutation to a single copy might be lethal, but if there are multiple copies in the genome's sequence, the error will not jeopardize the entire genome or the body that the genome specifies. No innovation comes into existence perfectly hewn. Error is thus necessary for the generation of variation. Redundancy as multiple copies, each subject to subsequent variation, ensures that variants are generated without disastrous consequences for their bearer.

Consider the advantages of redundancy in colonial animals. By a colonial (or clonal) animal I mean one in which genetically identical units are produced by simple division or budding without sexual reproduction, that is, without the shuffling of genes from two parents. Corals, bryozoans, and tunicates are familiar examples of marine colonial animals, which live sedentary lives attached to rocks or to living substrates. Coloniality offers protection against predators by turning a lethal attack into a case of partial predation. In colonies of social insects, where individual modules are not physically connected as they are in the marine invertebrates, redundancy among workers makes possible the exploitation of highly dispersed resources such as nectar, pollen, and tiny prey, which come in small bits distributed over a wide area. Failure by one or even many workers to find food does not compromise the colony as a whole. Redundancy in colonies also allows the deployment of effective diffuse defense by workers, soldiers, specialized polyps, or jawed zooids. Presumably, advantages like these would apply also to organisms that achieved a multicellular grade of organization from a single-celled state.[30]

The elimination of functional conflicts, made possible by increasing autonomy of interacting parts, expands the range of conditions under which an entity can compete. Autonomy is achieved when parts grow, act, and respond more or less independently of each other. To be sure, there is always some kind of central authority—a brain, central government, dictator, board of directors, spiritual leader, captain, and nucleus, for example—which coordinates activities and provides direction as well as some globally necessary functions, but diffuse and local controls ensure that many functions can be carried out simultaneously and that adaptive responses by one part do not jeopardize or conflict with responses by others. A system of connected yet semiautonomous parts is versatile and responsive. The parts of such a system are separated by porous walls or their metaphorical equivalents, which provide both a degree of independence and communication.

It is no accident that this general description of a versatile system applies to democratic government. As the history of highly centralized forms of government makes all too clear, rigid allegiance of society and its parts to central authority exposes everyone to errors or foolish policies generated at the seat of power. Napoleon's and Hitler's decisions to invade Russia proved diastrous for them both and for their nations, France and Germany. The Cultural Revolution of the late 1960s in China destroyed a whole generation. Theocratic and dictatorial governments, in which leaders demand obedience rather than approval, squelch competition and independent inquiry, propound an inflexible orthodoxy, and hamper initiative on the part of individuals and groups under their control. These conditions are not absent even in representative democracies—decisions to go to war are made by a powerful oligarchy, and dissent is ruthlessly discouraged—but accountability does constrain and spread out power.[31]

Versatility in organic form is often achieved by decoupling one structure from another or by breaking dependencies of one function on another. Consider some examples from the architecture of vertebrates. In early dinosaurs, including the predatory theropods, the hindlimbs and tail formed a single unit of locomotion. There was extensive muscular connection between the femur (thighbone) and the long tail. In birds, this single locomotor unit evolved into two modules, the hindlimbs and the tail, with few or no muscular connections between the two. In conjunction with the evolution of flight, which involved modification of the forelimbs into wings, the tail was able to take over a new function of stability, augmenting the wings in flight. The acquisition of a new function would have been impossible had the tail remained anatomically part of the hindlimb system.[32]

Early land vertebrates and their amphibian and reptile descendants cannot run and breathe air at the same time. Their limbs are splayed out to the side, so that the animal's underside rests on the ground. During locomotion, the body flexes from side to side and the limbs on the left, and then those on the right, step forward. As the body flexes, the lung inside compresses, and the animal is unable to breathe. As a result, breathing and locomotion are incompatible; the animal must do either one or the other but not both at the same time. The evolution of erect posture, in which the belly is held off the ground and the limbs move back and forth in vertical planes beneath the body, eliminated this constraint. Erect posture evolved in therapsids ("mammal-like reptiles") during the Early Permian period, as well as in dinosaurs and in several lines of crocodiles. An important consequence was that, whereas early land vertebrates were limited as predators to ambushing their prey over short distances, the evolutionarily more derived predators with erect posture acquired the option—by no means always realized—of pursuing their quarry over long distances. The functional dissociation between breathing and running therefore widened the range of adaptive possibilities among predatory land vertebrates.[33]

Intuition might lead us to expect that greater functional specialization and autonomy of parts should be associated with a large number of parts, but that intuition seems in general to be wrong. If there is strict central control, it does not matter how many parts there are, because none of the parts has the independence to specialize to any one function. All that this kind of organization can achieve is safety in numbers, a kind of unimaginative redundancy. The walking legs of centipedes and millipedes are numerous, but they all do pretty much the same thing. So do the limbs behind the mouth of a trilobite. Many crustaceans, on the other hand, have limbs variously modified for swimming, walking, grasping prey, crushing or dismembering prey, and sheltering eggs, all in the same individual. The ability to carry out multiple functions is thus made possible by subjecting the same basic structural element—the crustacean limb, serially repeated from front to back along the animal's body—to different and semi-independent regional regimes of regulation and growth. Successively younger major clades of arthropods—from basal groups of trilobites, crustaceans, and myriapods (millipedes and centipedes) to more derived decapod and other crustaceans, arachnids (mites, spiders, scorpions, and harvestmen), and insects—show a general increase in the number of the number of semiautonomous domains in which the parts are fashioned to carry out distinct functions, and a decrease in the absolute number of limbs and body segments.[34]

According to David Jacobs and his colleagues, molluscs originally had multiple zones of skeletal formation, in each of which more or less identical spicules of calcium carbonate are laid down. Chitons retain this organization in modified form, having eight plates from front to back. In other molluscs, the number of skeletal units has been reduced to one or two. In early univalved snails, the shell was probably a single domain.

A mutation affecting one sector of the shell-secreting mantle margin would affect the entire structure. In more derived clades such as the Sorbeoconcha, the single shell has been divided into two or more domains. Modifications of the anterior end to accommodate a canal for the siphon, for example, do not affect more posterior sectors. Differentiation like this not only increased the functional versatility of the shell, but also greatly increased the range of shell shapes.[35]

Vertebrates, too, show trends toward increased regional differentiation and semiautonomy accompanied by a reduction in the number of parts. Teeth, bones of the skull, the terminal bones of limbs, bones along the vertebral column, and elements of the jaw have all decreased in number and undergone increasing functional specialization in successively derived major vertebrate clades. Mammal teeth are distinguished as incisors, canines, premolars, and molars, and their functions range from grasping and gnawing to puncturing and grinding. Fish, amphibian, and reptile jaws typically contain more teeth, which are less functionally specialized, not least because grinding is possible only in the more advanced members of just a few clades of dinosaurs and therapsids.[36]

As these examples make clear, increased autonomy allows parts to take on new, often specialized, functions. A similar benefit is achieved when previously independent components come together through symbiosis. Through internal trade between the now semiautonomous units, the whole is able to exploit resources and to use waste products that none of the components in isolation would have been able to exploit. Stinging cells (nematocysts) likely descended from independent protistan ancestors, and provide defensive and feeding capabilities to jellyfishes, corals, and other cnidarian hosts. Carbon dioxide, a waste product of animal respiration, is put to good use by photosynthesizing symbionts in many marine animals. As I emphasized in chapters 2 and 3, cooperation among symbionts is a crucially important way to improve competitive performance.

In principle, many potential adaptive solutions exist to meet any given challenge, but finding solutions presents a formidable problem. Finding solutions is like looking for a needle in a haystack. Vast numbers of states must be searched before arriving at even a single potential solution, to say nothing of finding a good solution. Moreover, the pathway of change to a desirable solution may be so long and contorted that there simply isn't enough time in the universe to complete it or even to discover the pathway. Rigid specialization—

by a genetic code, for example—is not feasible, simply because the code would be excessively large, prone to breakdown, and inadequate for anticipating the many challenges and opportunities an economic entity is likely to encounter during its lifetime.

Organizational structures promoting the exploration of possibilities emphasize two attributes of semiautonomous interacting parts. One is the ability to combine and recombine. Parts can dissociate and reassociate in new combinations, generating a very large diversity of possibilities quickly. This is the great long-term (though not the short-term) benefit of sex, in which the genetic codes of two individuals combine to form genetically distinct progeny. The second attribute is the ability to generate new structures and new properties through interactions of parts. This is how many structures are formed during the development of animals. The interactions generating the novelty arise when parts grow or move, so that they come into contact with other parts. If any of the interactions or combinations prove to be useful, selection can stabilize them by bringing them into genetic control.[37]

The process of selection itself, of identifying a good solution among many possibilities, can be made more efficient by reducing the role of random events or, to put it differently, of the benefits of being first. I would argue, for example, that internal fertilization is a means of exposing eggs and sperm and their bearers to more efficient selection, with the result that offspring will be fitter on average.

Here is how the argument works. If individuals release eggs and sperm into the surrounding medium (especially water), encounters that could result in union (fertilization) are essentially random. The probability depends almost entirely on the abundance of the gametes, and thus on the density of individuals producing them and on the quantity of eggs and sperm the individuals of the two sexes release. To be sure, the vigor with which a male gamete (sperm) propels itself will increase its chances of encountering an egg, and this vigor may correlate with the progeny's performance; but randomness remains important. Internal fertilization (or one-to-one mating), by contrast, involves selection by one or both mates for fitness-related characteristics that are correlated with the fitness of progeny. The union is between the gametes of two specific individuals, whose identities are determined through a prior selective process. There is still a role of chance in terms of which sperm arrives at the egg first, and there can still be competition among the sperm of different males mating with the same female, but some of the element of randomness has been removed. The key is mate choice before fertilization takes place. In animals, this typically takes place through competition among males, competition among females, or a system whereby one mate literally makes a choice among several members of the other sex. In many land plants, mating is done vicariously through third parties, pollinators that can move among individuals that entice these go-betweens with nutritional rewards. As I pointed out in chapters 2 and 4, the sexual selection related to mate choice powerfully amplifies top-down consumer-related selection. My point here is that it also makes the process of selection more effective in that it reduces the chance, and increases the role of competition with respect to offspring performance.[38]

Perhaps the most sweeping manifestation of concentrated, coordinated power to emerge from simple, locally communicating, semiautonomous components is intelligence. When many components acting in parallel use the same few rules to accept or reject available choices, the whole adapts, or learns; it develops a better hypothesis of its environment. The emergent collective intelligence is thus a largely reactionary capacity, an ability to predict, to organize information in ways that benefit the whole.

Learning is one thing; control is another. The kind of intelligence we think of as human does more than predict the environment; it exerts influence, it controls, and it modifies circumstances according to what has been learned. This ability to make decisions and to act in accordance with prediction distinguishes what we might call central intelligence from collective intelligence. Although self-organization of multiple components can produce collectively intelligent systems such as cities, economies, ant societies, and certain kinds of computer software, it cannot act alone to produce evolutionarily derived forms of central intelligence. There must be a positive feedback between the bottom-up, diffuse learning that characterizes some self-organizing systems, and the emergent, top-down structures imparting central control. The result is a highly responsive, integrated system in which learning and communication are diffuse but control and reaction are under central authority.[39]

Greater power brings not merely immediate competitive advantages, but in the long run also a greater potential to adapt and to modify as well as a faster, more encompassing means of exploring and taking advantage of possibilities. Communication, flexibility, and semiautonomy appear to be the essential properties of organization that make these capabilities possible.

Power, opportunity, and resources interact in a series of feedbacks. They stimulate each other, yet they mutually impose limits. In the next chapter, I shall look at the role that resources play in these feedbacks, examine the processes and places where resources come from and disappear, and when and where economic opportunities for growth and expansion are expected to occur.

Chapter 7

THE INGREDIENTS OF POWER AND OPPORTUNITY: THE ENVIRONMENT

ECONOMIC POWER DERIVES NOT just from the technology and organization of the players, but also from the environment. In chapter 6, I emphasized the properties of the participants in an economy; here I delve into the characteristics of the environment that provide opportunity and impose constraint.

By *opportunity* I mean the potential for improvement in the performance of economic tasks. When an economic unit expands by growing in size or increasing in numbers, resources must be sufficiently plentiful for it to maintain itself and for a surplus devoted to the additional spending necessary to support that growth. Expansion implies that constraints are eased or lifted. By constraint I mean conditions or events that either maintain the status quo or that bring about loss in performance. Under constraint, the possibility of improvement is blunted by a constellation of internal and external controls despite the ubiquity of imperfection. Improvement under constraint is possible, but only in some directions, and only at the direct cost of reduced performance in other functions.

Like all other phenomena, opportunities and constraints occur at many scales of time and space. They can be temporary, frequent, and local; or chronic, rare, and global. Whether a circumstance is an opportunity or a constraint depends on whether the affected unit can capitalize on it. If the unit can respond, the circumstance is an opportunity; if it cannot, the condition is a constraint.

These rather abstract ideas are perhaps best communicated in a metaphor. In what I shall call the metaphor of motorized balls in a bag, think of a flexible bag filled with balls of various sizes, each ball being provided with an internal motor that enables the ball to move. When closely packed in the bag so that overall expansion is impossible, all the balls are severely constrained in their movements. The system is maintained in a crowded status quo in which nothing much happens. Even the most powerful balls stay in position, being prevented by those around them from interacting with balls far away. But now suppose that there is some form of disturbance. Perhaps external pressure is relieved, allowing the bag as a whole to expand. Perhaps the system collides with an object, destroying or disabling some of the balls. Or perhaps the temperature of the whole system increases, providing more power to the balls.

Whatever the cause, the disturbance brings about change. Individual balls, especially the surviving ones with the larger motors, will travel farther, affect a larger number of their neighbors, and move faster. Not every ball will have such opportunities; it is the powerful ones, those with the faster metabolism, that will move under their own steam rather than being moved by others. The less powerful ones will either be little affected, staying in the spaces between, or be pushed around and sometimes even crushed. With sufficient opportunity, the more energetic balls may even enlarge the bag from within. The larger, more powerful balls may grow, filling up the spaces and preventing expansion of their more lethargic counterparts. Sooner or later, as wriggle room for even the most potent balls diminishes, the system returns to a constrained status quo, and an overall rigidity will once again ensue.

Metaphors can be at once illuminating and misleading. We must tread carefully in extracting insights from them lest these revelations emanate from parts of the metaphor that do not accurately represent the abstract phenomena we are trying to understand. The point of the metaphor is that opportunity can be likened to freedom, to a realization of potential. Constraint literally means an inability to move, to be tied down, being confined or encumbered. It means that the status quo, however far from perfect it may be, is maintained. Opportunity means growth, movement, improvement, adaptation, flexibility, freedom, and change. It is, of course, not universal; opportunity for some means constraint and disruption for others. It is very much in the eye of the beholder.

The attainment of power is obviously contingent on the availability of abundant energy and raw materials, but it also requires reliability. An increase in free energy or material supply beyond the technological capacity of an entity does not benefit the entity and will likely harm it. The increase will be a stimulus only if it is reliable—that is, if it can be incorporated into the adaptive hypothesis of the entity—and if the entity possesses the equipment to capitalize on the windfall.

At any given time, a living thing has available to it a certain amount of energy, or at least the capacity to take up energy. This amount, commonly referred to as the energy budget, is not strictly energy, because it typically accumulates with time. It is thus more properly a power budget. Units with a large budget can devote more resources to any given task than can those with a small budget, and they can gain a competitive advantage even if the pattern of allocation—the relative amounts of energy and time allotted to each task—remains invariant. Just as importantly, larger power budgets permit extreme patterns of allocation, or extreme specialization to some particular task, as long as minimal requirements of other functions are satisfied. With increasing power, living things ease the constraining effects of trade-offs; they can do more things better.

Consider two hypothetical examples of spending money—or energy—on defense. One small nation has an annual budget of 1 billion dollars, half of which is allocated to military personnel, weaponry, and related expenditures. Given that only 500 million dollars remain to cover all the other functions of government, this level of investment in defense is very high indeed. A second, much larger nation has an annual budget of 5 billion dollars. If it chose to spend the same amount of money as the smaller nation in absolute terms, defense would account for a mere 10 percent of the total. Under normal circumstances this would not be much of a burden. Even if the larger nation were to spend half of its revenues on defense—2.5 billion dollars, in other words—it would still have 2.5 billion dollars left to cover all the other costs. This level of investment is also very high, but because the total budget is five times that of the small nation and because the larger nation can realize some economies of scale, the perceived burden of a 50 percent allocation to defense is lighter. In fact, under extreme conditions, the larger nation might choose to allocate 60 percent of its budget, or 3 billion dollars, to military matters, and still have an absolutely larger amount left over to fund other necessities than the entire budget of the smaller nation. This example should not be taken to plead for or to excuse high allocations to bellicose pursuits. It is instead meant to illustrate that adaptive options are more numerous with a larger budget, and that governments can more easily persuade their citizens to accept high and increasing outlays if their budgets are large and growing.

When the total budget grows, presumably because of overall economic growth, small changes in allocation are made with relatively little pain or opposition, because most sectors can experience some absolute growth. In a static or declining budget, one funded function will suffer as another receives a more generous allocation. The options for change, to say nothing of overall improvement, are limited even if wants compete more fiercely and as demand for services multiplies.

This view of opportunity and competition contradicts the widely held belief that increased competition will by itself create opportunity and adaptation. If improvement in competitive ability or defense entails reduced performance in other essential tasks, an entity gains little; it is then best to keep doing the same things as before, because these things, however imperfect, worked. Increased competition thus reinforces the adaptive status quo; the adaptive hypothesis remains unchanged. Intense competition under conditions permitting overall expansion allows entities to improve without compromising other functions too much. The adaptive hypothesis is accordingly modified. Competition by itself is insufficient to bring about adaptation or improvement; it must occur in an environment that permits growth.

Economists and historians have long appreciated the connection between economic growth and opportunity. A.G.B. Fisher aptly captures the prevailing view when he writes, "If economic progress were to cease to-day, we should find it necessary to submit to rigid and ossified social stratification, which is rightly abhorrent to large sections of modern public opinion." Shortly after World War II, economic growth in the West was seen not only as a means to ensure full employment, but as the only feasible solution to rid the world of poverty and, what was more important at the time as the Cold War was getting under way, as the mechanism for increasing allocations to military defense. The economist W. Arthur Lewis similarly argued that "The advantage of economic growth is not that wealth increases happiness, but that it increases the range of human choice." Presciently, Lewis also perceived the role that economic growth plays in solidifying top-down control. Speaking of economic growth, he writes, "It increases opportunity for dictators to control minds, through mass communication, and men's bodies, through highly organized police services." He might have added that it has also provided corporations with increasing possibilities to create and channel people's wants through advertising.[1]

We can think of an economy as a three-dimensional structure, whose base represents production and whose top represents consumption. Using this metaphor, I find it useful to classify the agencies that permit or constrain growth or movement in three categories: (1) bottom-up agencies, external conditions that influence the availability of energy and raw materials; (2) top-down agents, economic entities that consume other entities; and (3) technology, the machinery—structural, biochemical, and organizational—that makes extraction and retention of resources a reality. Bottom-up agencies act most directly on all other dependent entities; they provide supply-side control. Consumers exercise top-down control by affecting the distribution and adaptations of entities with which they compete or which they consume; they also often affect supply. In ecosystems, consumers stifle innovation and limit the adaptive options for victims if bottom-up supply is stable or declining, but they provide "incentives" for adaptation and expansion if the base of the economy can grow. The technology required to increase extraction and to protect resources is almost always costly. It therefore becomes established only if resources are abundant, accessible, and reliable; and those technologies that make resources so will achieve the best performance.

Many readers may find this view of opportunity too narrowly focused on external conditions. They might instead point to the inventions of technology—the evolution of complex genetic architecture in animals, the origin of sex in eukaryotes, the development of the internal combustion engine, the creation of the stock market—as breakthroughs leading directly to further innovation and to economic growth. Without denying the stimulatory importance of these inventions, I would argue that they are most likely to enhance power and growth only if external conditions permit. If the technology should happen to arise at economically unfavorable times, it might fizzle; there would be little market for it. In other words, circumstances permit—and technology enables—economic growth to take place and some entities to gain power.[2]

In short, I perceive two fundamental constraints on economic power: (1) technology and the organization required for its development are energy intensive, and thus costly to implement; and (2) new technology is not immediately superior to the old. Two circumstances appear to be necessary to overcome these constraints: (1) there must be economic disruption or a permissive environment of sufficient free energy and material resources to provide the capital and the room for error that are necessary to allow innovations to arise and improve; and (2) there must be sufficient competition—demand—to serve as the selective agency favoring the large investments needed. These circumstances depend on the size of the economy in which an entity finds itself, and on where and how resources become available.

The Size of Units of Space

Big economies are powerful economies, and they support powerful competitive dominants. Populations of entities with high requirements for energy can be sustained only in economies of large size, because the disturbances that inevitably affect economies merely diminish population size when entities are spread over a large area, whereas they might well doom populations when habitats are small. Moreover, selection and adaptation can be more precise, more finely tuned, and lead to greater refinement in larger populations, because incremental changes in the frequencies of genes that encode adaptations can be much smaller when the numbers of individuals subjected to birth and death are large. If, for example, we have a population of just ten individuals, the addition or subtraction of one individual changes the size of the population—and, potentially, the frequency of an allele—by about 10 percent; whereas if the population consists of a thousand individuals—still rather small by the standards of many species—sensitivity increases to 0.1 percent, allowing for a much more precise honing of adaptation. Finally, a large population—a large economy in general—is more likely to contain individuals that possess or that generate beneficial traits or solutions than a small one. Julian Simon, an ardent progrowth economist, passionately argued that human population growth is beneficial in the long run because one or two individuals in a large population would be clever enough to conceive of answers to seemingly intractable problems, many of which would of course have been the direct consequence of burgeoning growth in human numbers.[3]

The size of an economy depends not just on the size of the place in which it operates, but also on the size of the environment as perceived by the participants. Traditionally, the notion of size simply refers to the area of islands, continents, ocean basins, watersheds, lakes, biomes, nations, or other geographic entities. Population size, species numbers, income levels, and measures of productivity might be added as indications of the size of an economy. Yet, from the point of view of individual participants, the size of the economy they inhabit may differ radically from one member to the next. To a small shopkeeper selling to a local market, the size of the economy is tiny compared to the size perceived by the local franchise of a corporation with a global reach.

Consider the small archipelago of Fernando de Noronha, off the northeast coast of Brazil. This tiny group of islands, with a total area of about 16 km^2, is home to a distinctive and unique biota of land animals and plants, most of which are found nowhere else. For these land creatures, Fernando de Noronha is an unambiguously definable place. For shore animals, however, the unity of the archipelago is far less secure. A few shore snails—the periwinkle *Nodilittorina vermeiji* and the limpet *Patelloida noronhensis*, for example—are unique to the archipelago, and for them Fernando de Noronha's volcanic coast represents the totality of their economic space. But most other marine species living in the islands have broad distributions extending to mainland Brazil and even including the entire Caribbean region. For them, the effective economy reaches widely into regions where these species interact with many entities unknown in the biota of Fernando de Noronha.[4]

Dispersal by means of planktonic larvae across deep-water tracts of ocean inhospitable to adults stitches together many small island economies into a larger whole. This effect is most strikingly portrayed by species in the Indo-West Pacific region. Some snails, crabs, sea urchins, and seastars extend from the Red Sea and East Africa as far east as Easter Island, halfway around the world. These species occupy an effectively gigantic ecosystem, where they interact with a multitude of other species.

The important point is that the size of an economy is only partly indicated by the sizes of suitable environments. It is affected also by the extent to which patches of suitable circumstances are interconnected, or held together, by economic communication and trade; and this interconnection is to a large degree determined by the properties of the economic players themselves, by their technology of dispersal and trade, by the ability to move under their own power or by taking advantage of winds and currents. The greater the mobility of players, and the greater the extent of contiguous "suitable" circumstances, the larger are the economic systems of which the players are a part, and the more powerful those systems and their members are. The sizes of ecosystems, the distances between them, and the ability of individuals and populations to travel from one patch to another thus all contribute to the effective sizes of economies.

The most extreme condition of interconnectedness has been achieved by our own species, which has effectively globalized economic activity. Regional and local distinctions still abound, of course, but high-speed communication and extensive trade have for the first time made global measures of such economic attributes as productivity, diversity, and carbon storage biologically and economically meaningful. There might once have been a time when species, including ours, could externalize costs by throwing things away and allowing wastes to accumulate in places of no interest to us. Now that there is no place away from us, that time is long gone.

With these caveats about size in mind, we can now consider how and why economies of large effective size are more productive and spawn more powerful dominants than smaller economies operating under similar physical circumstances. All else being equal, large economies can support entities with high energy demands and high productivity, whereas small economies cannot sustainably support entities possessing such high power.

Areas Large and Small

When Charles Darwin came to the relatively unspoiled islands of the Galápagos, he was astonished at the docility and fearlessness of native lizards and birds. It was obvious to him that many of the predators with which these animals had to cope on continents were absent on isolated islands like the Galápagos. With this observation, Darwin initiated a long affair between biologists and oceanic islands, a fascination that has yielded important insights into the study of the evolution and distribution of species.

To Darwin, always the careful observer, isolation from continents was not the sole difference between islands and other land masses:

> Although I do not doubt that isolation is of considerable importance in the production of new species, on the whole I am inclined to believe that largeness of area is of more importance, more especially in the production of species, which will prove capable of enduring for a long period, and of spreading widely. Throughout a great and open area, not only will there be a better chance of favorable mutations arising from the large number of individuals of the same species there supported, but the conditions of life are infinitely complex from the large number of existing species; and if some of these species become modified and improved, others will have to be improved in a corresponding degree or they will be exterminated. Each new form, also, as soon as it has been much improved, will be able to spread over the open and continuous area, and will thus come into competition with many others. Hence more new places will be formed, and the competition to fill them will be more severe, on a large than on a small and isolated area. . . . (T)he new forms produced on large areas, which have already been victorious over many competitors, will be those that will spread most widely, will give rise to most new varieties and species, and will thus play an important part in the changing history of the organic world.[5]

There, in a nutshell, is the fundamental argument linking maximal performance and area. Later authors have quantified these arguments with models and with population-genetic theory, but the basic form of the argument remains as Darwin expressed it. Absolute standards of maximal performance are low in small areas, high in large ones. Species in the two areas taken separately are successful in the economic context in which they evolved, but if a connection between the areas were to be established, the species from the small area would be at a lethal competitive and defensive disadvantage.

The difference between islands and continents that has received the most lavish attention from ecologists is the number of species. Robert MacArthur and Edward O. Wilson noted that the number of species is related to the area of a land mass. Although the precise nature of the relationship varies, and is far more complex than MacArthur and Wilson initially envisioned, the number of species S tends to rise exponentially with area A: $S = bA^z$, where the constants b and z vary widely.

MacArthur and Wilson thought that the number of species on an island is determined by two processes—immigration, which adds species, and extinction, which eliminates them. Later work, especially by Lawrence Heaney on mammals and by Jonathan Losos and Dolph Schluter on lizards, shows that these two opposing processes do indeed account adequately for the number of island species in some cases; but a third process—speciation, or the evolution of new species, in this case from immigrants—contributes importantly to diversity on most islands. The minimum size of an island where species formation can occur depends on the extent to which immigrant gene pools are isolated from those of populations on neighboring islands or continents; but, according to Losos and Schluter, speciation is rare among *Anolis* lizards on West Indian islands smaller than 3,000 km^2.

In the Galápagos Islands, today's fifteen ground-finch species of the genus *Geospiza* are all the result of repeated speciation from a single ancestral species, which could have arrived in the islands or their geological precursors as early as 14 Ma during middle Miocene time. Similarly, the Hawaiian finches are all descendants of one ancestral American immigrant species.[6]

The size of an economy sets limits not only on the diversity of the economy's members, but also on their per capita power and size. Comparative analyses of land masses of different areas have revealed that the maximal per capita mass of consumer species increases with the square root of area, and that the average herbivore reaches a maximum biomass about ten times that of carnivores. Moreover, the size limit of warm-blooded consumers, with very high resource requirements, is set much lower than that of cold-blooded consumers. Thus, for a given area, warm-blooded top carnivores are just one-fifth the mass of top cold-blooded ones. For herbivores this ratio is one to sixteen. Area, in other words, affects the total food supply available to a population, and small populations of large animals with high demands cannot persist in small environments. Therefore, the land biotas of small islands cannot and do not support individuals and populations with high energy demands and economic power.[7]

Consider the largest carnivores, for example. Australia, a continental island with an area of 7,686,884 km^2, supported as its largest carnivorous mammal the marsupial lion *Thylacoleo*, an animal weighing some 75 kg. Australia's top carnivore, however, was likely the giant lizard *Megalania prisca*, which at a length of 3 m and with an estimated body weight of 380 kg was like Flores's Komodo dragon writ large. When South America was an island continent before the Panama Isthmus connected it with North America beginning 2.8 Ma or earlier, the top warm-blooded predators were borhyaenid marsupials and phorusrhacoid ground-dwelling birds. These were largely eliminated when North American carnivorous mammals—puma, jaguar, bears, and dogs—reached South America and diversified there. The faunas of many islands—New Zealand, Hawaii, and New Caledonia, for example—were devoid of mammals except for bats. Here, the top predators were eagles and hawks. Before humans arrived in New Zealand some one thousand years ago, the top predator was a 16 kg eagle (*Harpagornis moorei*), the largest known predatory bird, which could probably kill and eat moas and which, together with a 3 kg harrier (*Circus eylesi*) and flightless adzebills of the genus *Aptornis*, enforced nocturnal habits on many of New Zealand's ground-dwelling flightless birds. In the Hawaiian Islands, the top predators before human occupation were a harrier (*Circus dossenus*), owls of the extinct genus *Grallistrix*, and the locally extinct eagle (in the *Haliaeetus leucocephalus/albicilla* species complex). The island of Crete (current area 8,260 km^2) during the Pleistocene supported an 11 kg otter, whereas nearby Cyprus (current area 9,251 km^2) hosted a 2 kg genet as its largest carnivorous mammal. Bahamian islands with an area less than 100 m^2 cannot support predatory lizards, and instead harbor very high abundances of spiders, which are the default top predators. Lizards on such small islands experience physiological stress and are eliminated from time to time by hurricanes, and therefore cannot maintain permanent populations.[8]

All these top predators are small compared to the animals evolving on large continents. Today's largest carnivore is the Siberian tiger, weighing up to 272 kg. In Africa, there were much larger carnivores: the early Miocene creodont mammal *Megistotherium osteothlastes* (estimated weight 880 kg) and the early Pliocene bear *Agriotherium africanum* (750 kg), for example. Although there is some question whether these animals were strict carnivores, nothing of comparable size is found on smaller land masses.

These same relationships apply to herbivores. Plant-eating mammals occur only on larger islands, such as Australia, where the extinct massive-bodied marsupial *Diprotodon* stood 1 m at the shoulder and reached a weight of 2,800 kg; and South America, which supported such ground sloths as the one-ton *Glyptotherium* and the 6 m long *Megatherium*. Crete and Cyprus both evolved dwarf elephants—the 90 kg *Elephas creticus* in the early Pleistocene, and 3,200 kg *E. creutzburgi* in the Late Pleistocene, both on Crete, and the dwarf Cyprus hippo of the Pleistocene weighing 200 kg. The top herbivores on many somewhat smaller islands are or were birds: elephantbirds (*Aepyornis maximus*, 275 kg) on Madagascar, moas (12 to 13 species of the order Dinornithiformes, 160 kg) in New Zealand, and moanalos—flightless ducks weighing up to 7.5 kg belonging to *Thambetochen*, and related genera, with tortoiselike beaks—in the Hawaiian Islands. On the island of Mauritius in the southern Indian Ocean, the dodo (*Raphus cucullatus*), which at an adult male weight of 28 kg is the largest known pigeon, was a huge flightless fruit-eater and likely also a leaf-eating herbivore. The solitaire (*Pezophaps solitaria*), a smaller (17 kg male adult) pigeon, probably filled a similar role on the neighboring island of Reunion.[9]

In the absence of individually and collectively powerful consumers and competitors, two opposing trends have characterized the evolution of island life, or more generally the living things in small economies. First, weedy immigrant ancestors have given rise to species that become the island's top consumers and producers. Their performance may not match that of the big players of larger ecosystems, but they evolve the same kinds of traits to the extent permitted by the island's size and fertility. Lizards disperse fairly easily to islands, and occasionally evolve to become the islands' top predators, as on the Indonesian islands of Flores and Komodo (the 54 kg Komodo dragon, *Varanus komodoensis*) and in Pleistocene Australia (*Megalania prisca*). Without herbivorous mammals, many islands host gigantic land tortoises weighing up to 1,000 kg, as on the Galápagos and the Indian Ocean island of Aldabra, or flightless birds—the elephantbirds of Madagascar, moanalos of Hawaii, the dodo and solitaire of the Mascarine Islands, and parrots (kea and kakapo) of New Zealand—all descended from smaller ancestors. Among producers, invasive weedy families gave rise to woody trees, which become the local competitive dominants. Composites—members of the Asteraceae—overwhelmingly comprise herbs, such as dandelions, sunflowers, thistles, and daisies, but on many islands—the Galápagos, St. Helena, and the Macaronesian islands of the Azores, Madeira, and the Canaries—invaders from nearby continents have given rise to impressive clades of trees.[10]

The second trend is that species native to islands where high-performance consumers are absent or different often have reduced defenses compared to their continental counterparts. In the Channel Islands off southern California, many plant species have reduced levels of chemical defense compared to their mainland ancestors, evidently in response to relaxed selection by herbivorous insects. In New Zealand, about 10 percent of the woody-plant species belonging to seventeen families display a peculiar growth form in which branches diverge at wide angles, so that most of the edible leaves are located in the interior, protected by an outer bulwark of tough twigs. This unique architecture appears to have slowed down herbivorous moas. On the other hand, spines, though present in some species of *Aciphyllum*, are not well represented in the New Zealand flora. They are effective against mammals with soft lips and noses, and are widespread in understory plants in all the major continents but not on islands. Finally, the populations of the newt *Taricha granulosa* on Vancouver Island, British Columbia, have one thousand times less tetrodotoxin in their skin than do mainland populations, and the island's garter snakes (*Thamnophis sirtalis*) are more susceptible to this toxin than are mainland snakes.[11]

Another manifestation of this trend is the widespread reduction in body size and metabolic rate of island species that are the descendants of large, powerful immigrants. On the island of Jersey, in the Channel Islands off the coast of France, Adrian Lister has documented a sixfold decrease in body mass of the European red deer (*Cervus elaphus*) over a six-thousand-year interval between 126 and 115 thousand years ago during the last warm interval of the Pleistocene before our own. The frequent assumption of flightlessness among island birds, especially rails (Rallidae), is linked to a reduction in metabolic rate and often in body size. Bradley Livezey has linked flightlessness in the Galápagos cormorant (*Nanopterum harrisi*) with a reduction in the size of the metabolically expensive pectoral muscles, which in flying birds power the wings.[12]

An important geographic attribute that distinguishes the top consumers and the competitively dominant producers on islands is their very limited range. Most of these species evolved their traits on the island, and are found nowhere else. The economy in which they live is effectively small, and the power that these dominant island species have evolved is correspondingly limited in comparison to that of the rich and the powerful on larger land masses. Among dominants, the attributes of power—large size, high metabolic rate, high growth rates, and so on—are favored once the initial colonists have adapted to the island's conditions, but the small economy of the island places severe constraints on the extent to which per capita power can increase. Islands will simply never produce the kinds of economic dominants that larger economies do.

If all of this is true, why is it that so many continental species succeed so spectacularly on islands once they have managed to cross the water barrier with human help? Rats, mongooses, rabbits, goats, house sparrows, starlings, mynah birds, mosquitoes, and giant African land snails certainly did not evolve their high reproductive and competitive vigor on islands, but they immediately assumed leading ecological roles in island ecosystems upon their arrival. The explanation for this apparent conundrum, I believe, resides firstly in the attributes of high-performing continental species, and secondly in the selective regime prevailing on isolated islands.

A necessary condition for long-distance dispersal across hostile territory is that the dispersing individuals survive. They do this either by shutting down metabolism for long periods or by carrying large energy stores for use during the crossing. Coconuts and humans exemplify the second of these solutions; and lizards, ferns, and some coastal weeds exemplify the first. But very few warm-blooded mammals, notorious for their success on islands, have the ability to suspend animation, making them uniquely poor colonizers under their own steam. By providing safe and quick passage, humans have effectively eliminated the metabolic hurdles associated with the dispersal of most species. If, in addition, dispersing individuals leave many of their diseases and parasites behind, the founding population will be especially vigorous.

For species that evolve and remain restricted in small, isolated islands, selection for high performance reaches an effective limit dictated by the island's size and productivity. There is no evolutionary incentive to perform better than this, until some foreign species with greater reproductive and competitive powers becomes established. In other words, the extra outlays needed to enhance performance levels are unwarranted in the absence of invasion, because they do not provide sufficient benefits.

The situation for marine life on islands is interestingly different. For shallow-water plants and animals living on the seafloor, stretches of open ocean pose barriers just as they do for land life, but many marine species possess larval stages that spend days to months in the plankton, with the result that they disperse far and wide among far-flung land masses. For these species, therefore, effective population size and the effective size of the economy are large, even if the species are locally rare in most places. Particularly striking is the observation that large powerful predators have huge ranges. The most dramatic evidence comes from the Indo-West Pacific region. All large reef-associated predatory fish and crabs have enormous geographic ranges, often extending from the Red Sea and the Indian Ocean coast of East Africa to eastern Polynesia. These animals include specialized fish-eaters such as a number of moray eels and groupers, as well as species specialized to eat hard-shelled prey, such as two species of puffer (*Diodon hystrix* and *D. holocanthus*), wrasses, the crabs *Carpilius maculatus* and *C. convexus*, and palinurid spiny lobsters. The puffers in fact occur throughout the marine tropics. The voracious coral-eating crown-of-thorns seastar (*Acanthaster planci*) extends from the Red Sea eastward to the Pacific coast of tropical America. Other predators of corals—the puffers *Arothron hispidus* and *A. meleagris* and various coralliophilid snails, for example—have similarly extensive distributions. No herbivorous fish, snail, or crab has such a range.[13]

Sources and Sinks

A useful way to look at opportunity and constraint and the effects of scale on them is to consider where resources come from (the source) and where they go (the sink). Inevitably, the economic transformation of resources entails a certain amount of waste. Unused energy is given off as heat, by-products of reactions are useless or even poisonous and are disposed of. Disposal, however, can take two forms. Often, disposed wastes become available for use by other members of the economy; they are, in other words, recycled. Others end up in forms unusable by any member, or they are exported to locations where they remain out of reach or out of trouble for members. Increased control—that is, increased opportunity—means that resources are created and recycled more effectively, and that sinks in which wastes are useless, lost, or harmful are reduced. Opportunity and constraint are thus about what and who controls sources and sinks.

Although a few resources that life could use come from outer space as meteorites and dust, most are of the Earth itself. Their availability depends on a complex interaction among geological processes and the metabolism of life itself. The geological processes include: (1) tectonics—upheavals of the crust, powered by internal heat, shattering and pushing apart continents, and thrusting up volcanoes and mountain ranges—adding volcanic rocks (ash, basalt, and granite) to the crust, minerals to the ocean, and carbon dioxide and other gases to the atmosphere; (2) weathering, the physical and chemical breakdown of rocks on land; and (3) processes of burial, sedimentation, and subduction (the movement of slabs of crust downward into the mantle). Once near or at Earth's surface, materials are distributed by wind and water to all parts of the biosphere by weather-related mechanisms whose intensity depends on temperature, topography, and other factors. Materials remain in the biosphere for widely varying periods of time, determined in large part by how effectively organisms can prevent them from being removed; but eventually they sink into sediments or are ferried aboard subducting slabs into the mantle, where they reside out of the reach of life for tens to hundreds of millions to years until tectonic upheavals once again cause them to surface. Nitrogen is returned to its inert molecular state by denitrifying bacteria active in oxygen-poor waters. Understanding the time dimension of supply, distribution, use, and loss of resources is therefore essential for answering questions about how living entities can avail themselves of material opportunities and overcome material constraints.[14]

Heat inside the Earth creates the forces that drive tectonic movements in the crust. A difference in temperature between deep Earth and the surface produces convective currents of heat and molten rock. When such convective systems reach the Earth's surface, volcanic eruptions and emissions from hydrothermal vents release heat, carbon dioxide, metal-rich compounds, and igneous rock—solid basalt and granite and volcanic ash (tephra), as well as smaller particles of dust. Explosive eruptions—typical of volcanoes in Indonesia, Japan, Colombia, and the Lesser Antilles, among many other places—release astonishing quantities of matter. The eruption of Mount Tambora (a volcano on the northern coast of Sumbawa, Indonesia) in 1815 is estimated to have released about 10^{15} kg, with a volume of about 400 km^3. The largest known eruption of the last one hundred thousand years was that of Mount Toba, in northern Sumatra, which 73.5 thousand years ago erupted about 10^{16} kg, or 2,800 to 4000 cubic kilometers of material.[15]

As I shall discuss more fully in chapter 9, violent eruptions of this kind cause disruptions lasting months to a few years at most, and appear to have few if any long-lasting deleterious effects on the biosphere. Less violent eruptions from fissures, such as the 1783 Laki eruption in Iceland, probably disrupt the atmosphere much more. Large-scale versions of this kind of eruption produce gigantic quantities of basalt and, as discussed in chapter 9, appear to be associated with several of the major extinction events of the geological past.

By releasing carbon dioxide and nutrients at and above Earth's surface, and by scattering sunlight, volcanoes should in the long run stimulate economic life in the biosphere. Carbon dioxide, of course, acts as a greenhouse gas and stimulates plant production. The ash and dust emanating from volcanoes on the dry land accumulate and quickly decompose to produce highly fertile soils. In most climates, forests and agriculture on volcanically derived soils are more productive than those on other soils in which soluble nutrients are less abundant. As much as 0.5 percent of the humus layer in volcanically derived soil is phosphorus, much of it in forms readily available to plants. Some tephras contain as much as 4 percent potassium, mostly as potassium oxide (K_2O). The dense human population of Java owes its persistence to the nutrient-rich soils of that volcano-studded Indonesian island. Mineral-rich particles also fall into the ocean, and may contribute to the high productivity in nearshore waters in volcanic areas. Perhaps most intriguingly, the particles hurled into the atmosphere by volcanic eruptions and carried around the world in the stratosphere stimulate photosynthesis in trees by making light available to more leaves. In a cloudless sky without dust, a few leaves are exposed to beams of direct sunlight while many more reside in deep shade. The sunlit leaves receive more light than they can use in photosynthesis, whereas the shaded leaves receive none. Dust scatters light and therefore reduces areas of shade, so that many more leaves are exposed to diffuse light on sunny days than would be the case in the absence of dust. In the year following the June, 1991, eruption of Mount Pinatubo in the Philippines, photosynthesis in a hardwood forest in Massachusetts jumped by 23 percent over the rate during the previous year. Even two years after the eruption, photosynthesis was still 8 percent higher than before the eruption. Volcanoes may therefore boost worldwide productivity.[16]

During the last fifteen years, geologists have discovered past episodes, lasting hundreds of thousands to perhaps as long as 3 million years, of massive volcanism that took place beneath the surface of the ocean, where they formed gigantic undersea domes and plateaus. In producing basalts over vast areas, these volcanic episodes resemble flood-basalt eruptions on the land, but they cover larger areas and disgorge much greater volumes of rock. In so doing, they displace seawater onto low-lying parts of the continents as sea levels rise, and thus create extensive stretches of shallow, productive water and adjacent moist lowlands where terrestrial productivity tends to be high. The undersea eruptions are accompanied by the release of copious carbon dioxide, which warms the atmosphere, and by hot, mineral-rich brines at hydrothermal vents. The intense heat in the vents convects the nutrient-rich brines up and away from the source, so fertilizing the ocean from below much as the dissolved nutrients coming in from rivers fertilize the ocean from above. In chapter 8, I shall examine the historical evidence in favor of the hypothesis that these kinds of eruptions globally stimulated life.[17]

Whereas flood basalts and undersea plateaus form as precursors or accompaniments to the breakup and separation of great blocks of crust, some mountain ranges rise as continents collide. During collisions, one continental block is forced beneath the other, causing the zone of collision to be elevated into ranges and plateaus to altitudes of 1 to as much as 5 km above sea level. The resulting topography influences climate by directing wind currents and affecting patterns of rainfall and runoff. Prevailing winds—westerlies in the temperate zones, tradewinds from the northeast and southeast in the northern and southern tropics, respectively—drop large amounts of water on the windward flanks, and leave the leeward slopes in a dry rain shadow. The Asian monsoon, evidently already in existence during early Miocene time as revealed by the presence in North China of wind-blown loess deposited 22 Ma, intensified as the Himalayan ranges and the Tibetan Plateau rose during the late Miocene, about 8 Ma. In North America, the Rocky Mountains may be indirectly responsible for directing cold polar air far southward in winter and hot humid tropical air far northward periodically throughout the year along the east coast, which lies 2,000 km east of the mountains. Most importantly for the present discussion, glaciers and rivers draining newly thrust-up mountains carry physically eroded sediments down into the lowlands and into coastal waters.[18]

At the Earth's surface, many elements critical to the functioning of cells become available to organisms through the breakdown of rock or through recycling among living things. Two processes—physical erosion and chemical weathering—are involved in the breakdown of rock. Physical erosion, the breakage of rock into pieces ranging in size from boulders to silt, is an episodic process associated with storms, floods, earthquakes, landslides, underwater slumping of sediments, and other manifestations of impact and grinding as water, winds, and sediments move. This process is most intense in steep mountainous landscapes where rainfall is high, in dry unvegetated deserts subject to dust storms, on wave-exposed coasts, on land scoured by ice, and on parts of the seafloor subject to earthquake-induced mass movements of sediments. Dust further enters the atmosphere through volcanic eruptions.[19]

Zhang Peizhen and colleagues have intriguingly observed that physical erosion is most intense globally at times of large, frequent swings in climate, a regime characteristic of the last 4 million years but not of earlier epochs in recent geological history. They argue that the land under a constant climatic regime comes into a sort of balance with the prevailing agents of erosion—rivers, glaciers, and wind, for example and that, when climates shift, new styles of erosion replace the old, and tens to hundreds of thousands of years must elapse before the land once again equilibrates with the new erosion regime. Rapid and frequent swings between wet and dry, or warm and cold, regimes prevent the attainment of balance, and therefore support a regime of intense physical erosion, in which large amounts of coarse sediment are loosened and washed away.[20]

In addition to these nonbiological circumstances favoring physical erosion, living things also play a major role in fracturing or grinding rocks into fragments. Much of the sand around coral reefs, for example, is formed by the grinding activities of parrotfish, which eat mineralized seaweeds and corals by breaking off small pieces of reef and fragmenting them with pharyngeal bones after swallowing. Grazing limpets, chitons, and sea urchins excavate cavities in rocks for feeding as well as for protecting themselves against other scrapers. Some snapping shrimps literally hammer rocks as they construct burrows in limestone or basalt. The effects of these so-called bioeroders is greatest on tropical and warm-temperate shores, where protection by a cavity or burrow laboriously ground, pounded, or scraped into rock provides an effective defense against the many surface-roaming scrapers and predators. On land, tree roots are known to split rocks by exerting great pressure as they grow in length and diameter. In contrast to the physical causes of erosion, these biological causes are relentless and constant in their rock-destroying activities, and thus not episodic. Mechanical bioeroders, as well as the organisms that use chemical means to dissolve rocks, speed up the entire process of physical erosion. This amplification is certainly greater in warm climates than in cool ones, at least in the sea.[21]

By chemical weathering I mean the transformation of minerals—silicates and carbonates—in rock into soluble fractions and clay minerals through the activities of microbes and land plants. Through respiration and the secretion of acids, these organisms are especially effective in breaking down particles in young soils, those formed recently as the result of erosion and subsequent deposition of newly exposed bits of rock. Like other life processes, chemical weathering speeds up as temperature increases. The release of such biologically important nutrients as phosphorus, iron, potassium, and calcium is therefore especially rapid in warm regions with high rainfall and young soils formed by intense physical erosion.

Chemical weathering is not just some incidental or necessary by-product of metabolism. Plants and the root fungi (mycorrhizae) symbiotically associated with them secrete acids as a means to dissolve minerals, so that dissolved nutrients such as phosphorus can be taken up and transported to where they are needed. Relative to lichens and mosses, which practice dissolution of minerals on a modest scale, vascular land plants with true roots weather minerals at rates five to twenty-three times higher, depending on the mineral. Not surprisingly, these effects are most dramatic for deeply rooted forest trees, which can effectively mine nutrients to depths as much as 18 m below the soil surface in some tropical rain forests.[22]

Material resources remain effectively inaccessible to organisms if they are either tied up in the well-defended bodies of living things or if they are sequestered in rocks and sediments away from the activities of life. On the time scale of individual organisms, there is a tug-of-war between consumers that attempt to appropriate resources from the living bodies of other life forms, and individuals that attempt to prevent such consumption. Huge differences exist in the average time that matter is locked up in organisms. The average turnover time of plant biomass in the open-ocean plankton is two to six days, whereas the comparable figure for many plants on land is nineteen years, more than one thousand times longer. For attached seaweeds, the mean turnover time is about one year. Some trees store carbon for six thousand years.[23]

On the much longer time scales of tectonics, the remains of organisms and their metabolic products are lost to the biosphere for thousands to many millions of years. For example, gigantic deposits of iron, other metals, and methane, which Archaea and bacteria formed directly or helped precipitate, persist for millions or billions of years. We still mine coal that accumulated as peat produced by land plants growing in the poorly drained soils of lowland marshes, bogs, and forests 300 million years ago. Lignins, produced by land plants for antimicrobial defense, resist degradation by bacteria and fungi, and even today are carried by rivers to the oceans, where they are buried in sediments. Silica, laid down in the skeletons of various organisms—diatoms, radiolarians, silicoflagellates, tintinnids, and sponges—accumulates in cherts. The calcium carbonate produced in the skeletons of algae, corals, foraminifers, molluscs, and a host of other creatures is locked up in limestones that may persist for billions of years. Gypsum and phosphorite deposits, many of large size and millions of years old, sequester phosphorus. Even such soluble minerals as halite (seasalt, NaCl) accumulate as seas dry up, and remain out of reach of organisms for long intervals. In the central United States, for example, there are huge salt deposits of Silurian age formed 440 to 405 Ma. In the ocean, the remains of marine animals, together with fecal pellets and the waterlogged remains of land plants, sink to the bottom. The fraction of material that does not decay through microbial activity accumulates as organic carbon in sediments. Compaction of sediment expels water and eventually turns carbon-rich mud and silt into carbon-rich shale. These deposits become available to the biosphere only when they are exposed to physical erosion and chemical weathering.[24]

The burial of materials in sediments represents the failure of the ecosystems above to capture and employ these resources. To increase supply of materials is not enough for stimulating economic growth; it provides a stimulus only if organisms and more inclusive entities can effectively absorb, use, and recycle the supply. As I noted in chapter 5, a major difficulty for economies and their members is that the rate of supply varies, and even more that injections of new resources come in short, dramatic pulses, which can neither be predicted nor capitalized on. The most beneficial increases in supply therefore occur over long intervals, as predictable, regulated infusions, which enter an economy technologically capable of transforming supply to economic work. An increased supply allows adaptation, but in order to be capitalized on also requires adaptation.[25]

By recycling nutrients and mining them from places where they would otherwise remain out of life's reach, organisms greatly enhance the supply of usable materials and thus increase their productivity. When plants and animals die, their remains are consumed by decomposers and scavengers. Herbivores and predators speed up the transfer of nutrients by consuming tissues of organisms that are still alive. Animals that burrow into sediments add water and oxygen, and therefore expose organic matter to rapid decay. In this way, burrowers make nutrients that would otherwise be lost available to other forms of life. Like all other life processes, these ways of retrieving and recycling materials depend on temperature, their rate more or less increasing exponentially as temperature rises. The positive feedbacks between life and the sources on which life depends thus work most effectively when conditions are warm or when at least some members of the ecosystem have the high metabolic rates necessary for gaining access to buried resources.[26]

Just how complex the relationship between life and its potentially limiting or enabling resources can be is poignantly demonstrated by the controls on oxygen. On the time scale of thousands of years or less in the biosphere we inhabit, the concentration of oxygen in the atmosphere (about 21 percent by volume) and in surface waters of lakes and oceans is determined by the balance between two life processes: photosynthesis, in which carbon dioxide and water react in the presence of light to form the basic organic (carbohydrate) molecule CH_2O, and respiration, which combines organic-carbon compounds with oxygen to yield water and carbon dioxide. Today, more than 99.9 percent of oxygen liberated by primary producers is consumed in respiration. Sulfate-reducing bacteria, which form pyrite (FeS_2), also release molecular oxygen, which is consumed by other bacteria that produce sulfates; but the oxygen traded in these microbial pathways amounts to only a little more than 10 percent of the oxygen trade by life forms engaged in photosynthesis and respiration. Most of the tiny imbalance between consumption and production of oxygen occurs in the sea, where burial of organic matter in sediments removes 0.1 percent more oxygen than primary producers liberate. The part of the biosphere where most of the oxygen is produced—the photic zone, the part of the system on which the sun shines—constitutes a tiny fraction of the biosphere's volume. But oxygen is consumed throughout the biosphere, both within the photic zone and away from direct sunlight. Even within the photic zone, some zones of low primary productivity—the mid-latitude parts of the open ocean far from continents, known as the central gyres, for example, are net sinks of free oxygen.[27]

The long-term controls on oxygen, however, depend on two circumstances: the presence of reduced molecules with which oxygen can combine and the efficiency with which organisms capitalize on inorganic resources and recycle organic matter. As long as all the free oxygen produced in photosynthesis is consumed by organisms or combines with other molecules through other processes such as fire, there will be no net change in the global amount of this gas. Where production overwhelms the capacity to take up oxygen, organic matter accumulates unused because it is not completely oxidized. As a result, there is net accumulation of oxygen. The inability of living things over large areas to capture and retain organic matter as it falls to the sedimentary reservoir thus increases the availability of free oxygen in many surface environments, and makes possible the evolution of biological tasks and performance levels requiring high rates of oxygen-consuming metabolism. Long-term changes in the oxygen reservoir of the atmosphere, ocean, and biosphere thus depend on the primary productivity of plants and microbes—that is, the rate of production of organic carbon—and the rate at which organic carbon is entombed out of reach of the economy of the biosphere.

Imbalances leading to an increased concentration of free oxygen come about in various ways. The key, according to Timothy Lenton, is phosphorus, the nutrient that limits life on long time scales. Phosphorus availability determines primary productivity on land as well as in the surface waters of the ocean. Primary productivity, in turn, determines the rate at which organic carbon reaches the ocean floor. The rate of sedimentation on the ocean floor—a rate which itself is correlated with primary productivity, with runoff from the land and the accumulation of dust from the atmosphere—determines how much of the organic carbon reaching the bottom is buried and thus lost to the biosphere. The faster the rate at which sediments accumulate, the higher the proportion of the organic carbon that escapes being combined with oxygen. Phosphorus becomes available to organisms either through recycling or through chemical weathering or rocks on land. Chemical weathering, especially of a nutrient as desirable as phosphorus, is greatly amplified by plants. Faster weathering yields more phosphorus, which makes higher plant productivity possible, which results in the burial of more organic carbon, which in turn yields an increased concentration of oxygen. This positive feedback does not operate without limit, however. The rate of photosynthesis is reduced at increasing concentrations of oxygen in the atmosphere, leading to less burial of organic carbon and therefore to a decrease in atmospheric oxygen levels.[28]

The feedback between oxygen and phosphorus I have just described may not be ancient. Phosphorus may be a globally limiting nutrient in today's biosphere, but its average worldwide concentration in the ocean (2.3 micromoles per cubic meter) is considerably higher than that calculated for ocean waters more than 1.8 billion years old (0.15 to 0.6 micromoles per cubic meter) by Christian Bjerrum and Donald Canfield. This very low early phosphate availability, Bjerrum and Canfield hold, was directly related to the scarcity of oxygen in the deep ocean. During many intervals of the Archean (before 2.5 Ga) and the Early Proterozoic era (2.0 to 2.5 Ga), gigantic deposits of iron oxides were laid down on the seafloor by reaction of ferrous ions (Fe^{2+}), with oxygen. Phosphorus readily combines with these iron oxides, and therefore is unavailable to life. Its scarcity limited primary production and therefore the burial of organic carbon, with the result that free oxygen was unable to increase in concentration. The feedbacks among oxygen, phosphorus, and organic carbon may therefore not have arisen until oxygen had saturated the reduced minerals accumulating in the deep sea. This, according to Anbar and Knoll, may not have happened until the late Mesoproterozoic era, about 1.25 Ga. Anbar and Knoll argue that in the long interval from 1.8 to 1.25 Ga, the biosphere as a whole may have been limited not by phosphorus, but by nitrogen, owing to the abundance of oxygen-starved sulfide in the deep sea and the low rate of supply of iron, molybdenum, and other elements essential in the microbial process of nitrogen fixation and the conversion of ammonia to biologically useful nitrate.[29]

This complex web of feedbacks among primary producers, phosphorus, burial of organically produced carbon, sedimentation rate, and oxygen has a global reach, but some environments and regions play far more decisive roles in global control than do others. The burial of organic carbon, for example, is concentrated today in marshes, in estuaries, beneath upwelling coastal waters, and in the deep stagnant basins of such enclosed bodies of water as the Black Sea and the Sulu Sea, where the great currents that ventilate the deep ocean do not reach. In the distant past, the equatorial coal swamps and forests of the European and North American Carboniferous period removed huge amounts of plant-produced carbon from the biosphere. Bacterially produced methane accumulates deep in sediments offshore beneath many of the world's continental shelves and beneath polar permafrost. Oxygen is more soluble and less rapidly spent in respiration in cold than in warm waters, with the result that shallow-water sediments in the tropics and in the temperate-zone summer are more readily depleted of oxygen and more apt to sequester organic carbon than are the sediments that lie beneath the deep open ocean and those accumulating during the winter months at temperate latitudes in shallow water. High rates of sedimentation, which favor the burial of organic carbon and thus indirectly the conservation of oxygen, are concentrated in coastal waters receiving large amounts of material derived from rainy mountainous terrain.

The areas of high primary productivity are located in rainy tropical lowlands, in coastal waters near continents and large islands, in zones of upwelling, and at the mouths of rivers. Tropical rain forest covers about 13 percent of the Earth's land surface, but with an average productivity of 1,098 grams of carbon per square meter per year, it accounts for 35.9 percent of terrestrial primary productivity. By contrast, boreal coniferous forest (taiga), with an average productivity of 200 g/m^2y, covers 10 percent of the land surface but accounts for only 6.2 percent of world terrestrial productivity. Desert, polar desert, and tundra together are found over 16.7 percent of the Earth's land, but with a mean productivity of 53 g/m^2y support a meager 3.0 percent of productivity on land. Grasslands are less productive than forests; tropical grasslands cover 12.2 percent of the land and account for 11 percent of production on land, whereas the comparable figures for temperate grasslands are 7.2 and 4.5 percent respectively. About 0.1 percent of the ocean's surface lies over upwelling zones, but planktonic productivity in those zones is about 5.9 percent of the total planktonic productivity.[30]

Any geological process that yields more topography, larger shallow seas, more extensive lowlands, and more river runoff will thus lead to an increase in productive surplus available to consumers, as well as to a potential rise in oxygen. It will also yield more opportunity worldwide for entities with the equipment and know-how to capitalize on the newly available resources.

Recently, an additional mechanism has been proposed for raising the Earth's concentration of free oxygen during the early phases of the history of life. In the atmosphere before 2.4 billion years ago, oxygen was essentially absent, being consumed as quickly as oxygen-liberating photosynthesizing microbes produced it. Methane (CH_4), on the other hand, occurred at very high concentrations (100 to 1,000 parts per million by volume) compared to that of the modern atmosphere (1.7 parts per million). This methane was produced by microbes (methanogens) that consume organic matter (CH_2O) made by primary producers that liberate oxygen. Nitrogen-fixing Cyanobacteria also produce methane as a by-product. Methane, when wafted into the upper atmosphere, is decomposed into one atom of carbon and four atoms of hydrogen in the presence of ultraviolet radiation. The carbon recombines with oxygen to form carbon dioxide, but the four hydrogen atoms escape into space. The net effect of these reactions is the elimination of atoms with which oxygen can combine, and therefore an increase in the concentration of oxygen, originally produced by cyanobacterial photosynthesis. Ultimately, the gain of oxygen from methane-induced escape of hydrogen into space is the result of the splitting of water molecules during photosynthesis in the presence of a carbon source.[31]

Catling and his colleagues proposed this model mainly for the Archean eon, the period in Earth history when free oxygen was essentially absent from the atmosphere and largely before the formation of iron oxides with which phosphorus readily combines. There is another possible and very intriguing circumstance in which this mechanism might work. Gigantic eruptions of methane occurred from time to time in Earth history. Microbially produced methane is typically stored in crystalline form (methane hydrate) in sediments at high pressure and low temperature in polar regions and beneath the ocean's continental shelves. When a disturbance—a major volcanic eruption, or the heat of impact from a comet or asteroid—either decreases the pressure from the sediments above or raises the temperature, these hydrates become unstable and bubble into the atmosphere as methane gas. Although most of this methane probably combines with oxygen to form carbon dioxide, it is possible that some of it reaching the upper atmosphere breaks down into its constituent carbon and hydrogen atoms by ultraviolet radiation. If so, and if the mechanism proposed for the Archean could also work under the unusual circumstances of a catastrophic methane release into an oxygen-rich atmosphere, then the rare but dramatic expulsions of methane might be followed by an overall increase in oxygen concentration in the atmosphere.

Dispersal of Resources

Sources, sinks, and biological intermediaries are not alone in determining the availability of raw materials. Equally important are the currents of air and water that transport nutrients. Were it not for such transport, coastal waters would not receive nutrients from land, and organic matter sinking to the seafloor could not be recycled by the primary producers in the waters above.

In soils, chemical weathering releases most of the more soluble ions during the first 150 thousand years after their formation. As a soil ages, its nutrient content therefore declines, because most of the remaining nutrients are found up in insoluble compounds that are unavailable to plants. Unless the nutrients in the soil are replenished by floods or airborne dust, the ability of the soil to support high primary productivity declines with time. Indeed, experiments in Hawaii have demonstrated that dust from Asia is largely responsible for maintaining the fertility of forest soils there.[32]

Along the western coasts of North and South America, as well as off northwest and southwest Africa, prevailing winds drive relatively warm surface waters offshore. These waters are replaced from below by cold waters welling up from the depths. Because dead animals, larger plankton, and fecal pellets tend to sink, these bottom waters are greatly enriched in nutrients compared to the surface waters. When upwelled water reaches the surface, where photosynthesis takes place, the nutrients support highly productive ecosystems of plankton and of attached seaweeds, which in turn sustain huge populations of fish, suspension-feeders, and marine birds and mammals. Ocean currents diverted by underwater topographic features can also lead to upwelling. Upwelling also occurs in a narrow band on either side of the equator in the western Pacific, as well as in the Arabian Sea, Gulf of Oman, and in parts of the southern Philippines, eastern Indonesia, and the southeastern United States.[33]

Upwelling is ineffective as a mechanism for recycling nutrients if the planktonic primary producers and their consumers do not sink to the bottom after death. This is the situation for the microbial loop, the minute (0.2 to 0.3 micrometers) cyanobacterial phytoplankton and its single-celled consumers. These organisms and their waste products are simply too small to sink very far before all their contents are reclaimed by other members of the surface plankton. The microbial loop is likely to be one of the most ancient ecosystems on Earth, giving way to a system dominated by larger animal consumers only in the latest Neoproterozoic eon perhaps as early as 600 Ma. Upwelling may therefore have become an important mechanism only when the members of the plankton and their excreta became large enough and heavy enough to sink to the seafloor.[34]

Rivers are the chief means of transferring the nutrients released through erosion and weathering from the land to lakes and oceans. Estuaries are therefore highly productive sites in the marine biosphere, whereas the shores of low-lying atolls, from which very little water runs off, are very little subsidized with nutrients from the land. About 92 percent of the phosphorus and 80 percent of the silica in the sea today enters by rivers. Rivers transport about 4.1×10^{15} g of dissolved material per year, of which about 5 percent is organic matter. Eroded particles (the so-called detrital load) comprised about 12.5×10^{15} g before human-wrought changes, and 18.2×10^{15} g in today's rivers, the increase being due largely to deforestation and faster flows induced through the building of dikes. Another 2.0×10^{15} g of particles enters the sea through glaciers each year. All these materials are transported in an estimated 4×10^{19} g of water per year.[35]

Organisms, of course, have a hand in this transport of nutrients, because they have considerable control over the water cycle. The presence of forests enhances rainfall and therefore water runoff, and overall therefore increases the rate at which water cycles among the ocean, atmosphere, and the bodies of living things. Deforestation by humans temporarily increases the amount of water removed as runoff, because evaporation—water lost to the atmosphere—is greater in forests than on bare soil.

I shall revisit the increasing role that organisms have played over the course of time in controlling the sources, sinks, and transport of resources when I treat major patterns of history in chapter 10. For now, I want to make the point that organisms collectively have a large hand in creating opportunities for themselves. These opportunities are not evenly spread either in space or in time. The geography and history of opportunity are the subject of the next chapter

Chapter 8

THE GEOGRAPHY OF POWER AND INNOVATION

THE WORLD IS A thrillingly varied place. Some of that pleasing variation resides purely in the realm of physics and chemistry—the topography of landscapes, patterns of rainfall and temperature, saltiness of the water, composition of soil and sediment, power of waves and currents, sizes and locations of land and water masses—but a great deal of it is the product of life itself. A tropical rain forest in South America has much in common climatically with one in Africa or Asia, but the inhabitants differ in interesting and important ways. Gliding vertebrates—various mammals, lizards, frogs, and snakes—abound in Asian forests but are not represented in tropical America. The reverse is true for mammals with prehensile tails. These differences, in turn, reflect the presence in Asia but not America of many trees rising above the general level of the canopy, giving gliders a potential advantage over other forms of arboreal locomotion. Asian and African forests are populated by many large, destructive herbivores such as elephants and rhinoceroses, whereas plant-eaters in tropical American forests are small. Africa and Asia, but not the more islandlike land masses of South America, Madagascar, and Australia, have natively evolved mammals with horns and other head ornaments that function in mate choice and in defense.

Such contrasts are not limited to the land. Shore assemblages of snails are richer in species and in shell defenses in the tropical western Pacific and Indian Oceans than elsewhere in the marine tropics. Cold-water shells look drab and are far less precisely ornamented with ridges and spines than their tropical counterparts. Arctic marine ecosystems are populated by bottom-dwelling crabs, fish, and marine mammals such as walruses and gray whales, which depend on the rich life of the seafloor for food. Antarctic systems, by contrast, lack crabs and the large bottom-dwelling vertebrates, and harbor small fish stocks not suitable for commercial exploitation.[1]

I could expand this list of examples to fill a book or more. The point is that there exists a geography of context in which economic life takes place. Absolute levels of performance vary from place to place and over time; so do the numbers, types, and rates of supply of resources. Some environments allow for a great range of adaptive solutions; others restrict the number of available evolutionary pathways. Life amplifies and elaborates the physical variation of Earth by superimposing regionally and temporally distinct adaptations, traditions, and patterns of trade on the basic conditions created by our planet's position and rotation. Sometimes life creates division and spatial separation where none would have existed; at other times, most notably in the human-dominated world, long-distance trade tends to obscure regional variations, with the result that economic activity homogenizes the conditions of life.

These regional variations are not random. They reflect economic conditions—the type and severity of challenges, the costs of failure, the benefits of success, the affluence of one's neighbors, the power of one's peers—as well as adaptations to them. In chapters 6 and 7 I identified the circumstances that promote economic opportunity, adaptability, and growth, as well as the conditions in the environment and in the organization and architecture of life that impose constraint to adaptation. Now I turn to the geographic variation in these conditions, to a geography of opportunity and constraint.

I first show that the economic spatial divisions of the world, whether they be the forests and fields of nature or the nations of human civilization, result from competition and from the responses of living things to it. Boundaries and differentiation arise because competition for resources enforces specialization, which entails the division of labor.

Without competition, there would be little incentive for life forms to create or modify spatial structure. With competition, the variations imposed by chemical and physical conditions are amplified and altered, boundaries created and dissolved, and regional distinctions exaggerated. An economic geography of habitats and regions is superimposed on a physical topography of mountains and valleys, air and ground, lands and seas, ocean water and ocean floor, open ground and cave, brook and river.

The most favorable economic conditions in nature, especially for living things whose internal temperatures match those of the environment, occur in large, warm, sunlit, nutrient-rich, biologically diverse settings where competition is intense. These settings subsidize less productive parts of the biosphere by acting as sources of nutrients, biomass, species, and—most importantly of all—evolutionary innovations. Warm-blooded animals, and humans most of all, are less strictly tied to environmental limits of temperature, and therefore have expanded the range of settings in which innovation and adaptive improvement proceed, but even they thrive best when supplies of resources are abundant and reliable and when adaptive response to intense competition is economically possible.

It is all well and good to establish general principles such as these, but economic life and economic history are made in very particular geographic circumstances. We happen to live in a world in which the productive tropics encompass very large areas of suitable ocean and land, and in which there is a relatively large number of biogeographically distinct regions. But geography changes. Land masses break up and collide; islands emerge and disappear beneath the waves; rivers change their courses; and all these machinations influence climates, species pools, productivity, and all the other attributes of economic systems. The remainder of the chapter is therefore concerned with the application of the general principles to the particulars of geography as it exists today and as it developed in the geological past.

Creating and Dissolving Boundaries

We take it for granted that living things and the ecosystems they create are bounded in space. Some creatures live in the sea and others on land, but rarely does a single species or individual do both. The deep-sea environment is really quite unlike the seashore, and few if any animal species occupy anything like the full range of ocean depths. Parasitic insects are often so specialized that they are restricted to a single host species. Accustomed as we are to classifying our surroundings into distinct environments or regions—sea, lake, stream, land, shore, soil, island, continent, tropics, polar regions, Europe, Africa, the Caribbean, the central Pacific—we tend to think of the boundaries between species, between ecosystems, or between biomes as "natural," enforced by accidents of physics, chemistry, and geology.

Of course, there is a correspondence between the bounds as reflected in the distributions of individuals, species, and ecosystems, and the physical and chemical heterogeneity of the Earth's surface, but this correspondence is not preordained or invariant. Instead, economic geography—the spatial distribution of economic entities and their interactions—is very much affected by life itself.

Consider the world of single-celled organisms, for example. Bland Finlay has shown that most single-celled free-living protists—diatoms, flagellates, ciliates, and amoebae—have cosmopolitan distributions. The eighty-six ciliated protists known from a crater lake in Australia also occur in lakes in northern Europe. Many microbes are not only enormously abundant, but they also show a very wide range of tolerance.

G. Evelyn Hutchinson pointed out long ago that many protists are at home in both fresh- and saltwater, a distribution that is rare among more complex animals except among physiologically specialized vertebrates. In effect, these organisms observe few boundaries; for them, the world is a far more homogeneous place than is the world as perceived by most animals and plants. Whether this homogeneity also applies to the millions of genomically distinct prokaryotes—Bacteria and Archaea—is not yet known, but the smaller size of these life forms together with their extraordinary abundance strongly imply cosmopolitan distributions.[2]

Even at the local scale, very small organisms live in a different world than larger ones. Over a 10 s interval, the diffusion velocity of a molecule of oxygen in water (0.020 mm/s) is approximately the same as the swimming speed of a bacterium. Over shorter time intervals, diffusive transport is even faster than swimming, meaning that diffusion is sufficient to replace oxygen (and many other necessities) for a stationary bacterium. For larger organisms, convection—movement of the organism relative to its medium—is the predominant mechanism by which depleted resources are replaced in the organism's vicinity. Mark Denny estimates than organisms less than 1.4 mm in linear dimension in air, and less than 0.14 mm in water, live in a world dominated by passive diffusion; whereas larger organisms must move, create currents, or live in currents (bulk movement of the medium) to prevent local depletion.

For organisms that move, or that live in a flow, the way in which the world is perceived depends on the dimensionless Reynolds number Re, defined as Re = rua/m, where r is the density of the fluid medium, u is the mean velocity of the body moving through the fluid, a is the length of the body in the direction of movement, and m is the fluid's viscosity. When Re is less than one (that is, when the body has a velocity of 1 mm/s or less or a length of 1 mm or less or both), the organism lives in an environment dominated by viscous forces, and its resistance to movement is determined entirely by the viscosity of the medium. For bodies with Reynolds numbers between 1 and 1000, and for streamlined bodies with a larger Re, flow around the body is laminar, meaning that particles of fluid follow smooth paths as they stream past the moving body. Above Re = 1000, particles take irregular paths, and flow becomes turbulent. Not only is more work needed to drive a body through the medium when flow is turbulent, but movement is also "noisier," meaning that nearby observers can more easily detect the moving body. The important point is that the heterogeneity of the environment as experienced by organisms depends on the size of that organism and on the rate at which the organism and its medium move relative to each other.[3]

We as humans readily perceive the distinction between gravel and sand, and so do many clams. For smaller clams, however, there may be little difference. Clams 1 to 2 mm long experience even sand grains as relatively large objects, so that for them habitats "on the sand surface" or "beneath the surface" may appear indistinguishable. Many invertebrate larvae may likewise make little distinction between swimming with cilia just above the seafloor or moving with cilia on the seafloor. Ways of classifying spatial structure vary according to the size, movements, and sensory apparatus of organisms. Economic geography on the scale familiar to humans thus emerged as the sizes of living things increased, and as the specialization that accompanies trade-offs in competition came to be expressed at larger spatial scales. Where millimeter-scale or smaller variations might have mattered most in the unicellular economies of the Archean eon, spatial structure on larger scales became important for larger life forms in the succeeding Proterozoic and especially the Phanerozoic eons.[4]

Economic performance is sensitive to the environment. The trade-off principle dictates that a given entity performs better under some conditions than under others, and selection reinforces this difference. Within a population, selection favors traits that enhance performance in those environments where individuals are most successful, usually where individuals are abundant. Competition with other entities establishes and modifies the range of individuals and populations. The geography of life is thus a geography of competition.[5]

Limits of range expand—or break down—when species are released from competition under conditions where few entities divide up the available resources, or when technologies evolve that enable individuals and populations to function as potent competitors under a wider range of conditions. Well-adapted incumbents normally keep individuals and populations in their place, but when these incumbents disappear, or when some innovation produces a superior adaptation in the competitively restricted population, expansion can occur.

Consider some examples drawn from my own observations of marine molluscs. In the West Indies, many small snails—*Agathostoma fasciata, Cerithium varium, Columbella mercatoria,* and *Smaragdia viridis,* among others—live on the narrow blades of seagrasses, where they tend to graze the tiny plants and animals that settle on the blades. With the exception of *Smaragdia,* which I have never seen away from grass blades, these species occur in many habitats in addition to seagrass meadows, including the surfaces of stones. By contrast, snails inhabiting seagrass blades in the western Pacific and Indian Oceans tend to be specialized to that habitat, and not to occur under nearby stones. These Indo-Pacific species, in other words, observe distinctions among habitats that their West Indian counterparts do not.

Similarly, most molluscs that live high on the seashore in cool-temperate parts of the northern and southern hemispheres also occur below the tidal zone, whereas most warm-temperate to tropical high-shore species are restricted to that zone. The single high-shore periwinkle on Fernando de Noronha, a tiny (16 km) archipelago off the northeast coast of Brazil, extends over the entire vertical extent of the shore, whereas the dozen or so periwinkles of the Caribbean tend to be restricted to particular zones in the splash zone and upper intertidal belts. In the absence of competitors, species often experience ecological release, expanding into environments they would not have occupied in the face of competition from adapted entities. With the arrival of new competitors, by contrast, species often contract their distributions, becoming specialized to that part of the spectrum of habitats where their adaptations still suffice.[6]

Another nice example comes from Alan Kohn's work on the food habits of predatory cone snails, members of the family Conidae. In the western Pacific and Indian Oceans, dozens of species of this group co-occur, each one more or less specialized to a rather narrow range of prey. There are specialists on other snails, on fish, various groups of segmented worms, and even one (*Conus imperialis*) limited to fireworms. On Easter Island, however, there is only a single species (*Conus pascuensis*), whose diet contains members of seven phyla, a far greater range than is eaten by *Conus miliaris*, the widely distributed species elsewhere in the Pacific most closely allied to *C. pascuensis*.[7]

Though common, this sort of competitive release is not universal. Jonathan Losos and colleagues showed, for example, that Cuban species of the lizard genus *Anolis*, which evolved various habitat specializations under species-rich conditions, did not become more generalized when they were introduced to various islands in the Bahamas, which are occupied by at most one or two species. I suspect that cases like this will turn out to be the exception rather than the rule, but like all other exceptions, they deserve careful scrutiny.[8]

The haphazard and arbitrary nature of borders between political entities further underscores how economic and political power superimposes spatial subdivision on Earth's physical geography. The border between the Netherlands and Belgium corresponds to no "natural" feature, and entirely reflects temporarily diverging interests and political allegiances from the late sixteenth to the mid-nineteenth centuries. People on either side of the border speak a common language and practice a common religion. Boundaries between nations shift as wars are won and lost, as lands are added through conquest or shed through the dissolution of great empires. Within nation-states, provinces compete among themselves for the financial resources of the central government; so do cities. Sometimes, central authority is strong enough that these internal differences in self-interest are suppressed or unimportant. In systems with more diffuse control, competition flourishes and creates a complex political geography.

The important message is that barriers are not absolute. Competition from others limits ranges and imposes barriers for some entities, but enables others to expand their ranges and eliminate boundaries. Traditionally, biogeography—the study of the geographic distribution and history of life on Earth—has been founded on the supposition that life forms respond to the forces that shape physical geography. I claim instead that the geography of life and of economic interactions is a co-construction between physical and economic forces. Life doesn't just respond to its environment; it creates and modifies that environment, smoothing out differences here and creating or accentuating others there.

Geographic Patterns of Performance

In the last chapter, I made the case that large warm environments near sources of prolific raw materials and exposure to plentiful sunlight support high productivity, intense competition, and powerful evolutionary control by consumers. Although we can identify the factors that allow for the deployment of great power, the realization of that power depends on the particulars of space and time. For this reason, I now turn to the task of specifying and characterizing regions and environments where potential and realized economic power are greatest and where innovation is concentrated.

A temperate-zone observer visiting the tropics for the first time cannot fail to be overwhelmed by the enormous variety and vigor of life. There is an extravagance about the equatorial zone, a concert of extremes in colors, sounds, textures, poisons, aggression, fragrances, speeds, armor, camouflage, and mechanical power. The business of life being conducted here is hard-driving and brash, not soft and genteel as one might find on an Alaskan shore, in a New Zealand forest, or in the quiet tall redwood forests of coastal California. The indelible impression is that competition and predation occur faster, kill greater proportions of victim populations, and have led to greater specialization in the tropics than in climatic zones where winters are cold and summers are cool.

Objective evidence amply bears out this impression. Conservative estimates of leaf consumption in forests indicate that this kind of herbivory removes an average of 7.1 percent per year in the north-temperate zone, 11.1 percent per year in tropical rain forests, and 14.2 percent per year in tropical dry forests. Given that most plants devote only 10 percent of their resources to reproduction, these losses are significant. It is therefore not entirely surprising that tropical trees, despite their generally higher losses to herbivores, have better defended leaves than do their north-temperate counterparts. Alkaloids—nitrogen-containing compounds of low molecular weight that protect against insects and that can poison vertebrates as well—are both more common and more toxic in tropical species. About 35 percent of tropical plant species contain them, as compared to 16 percent of temperate species.[9]

Experimental evidence for greater chemical defense in warm-adapted as compared to cool-adapted plants comes from work on salt-marsh plants in the eastern United States. Steven Pennings and his colleagues offered fresh northern and southern plants (from Rhode Island northward, and Georgia southward, respectively) to various potential herbivores, and found that northern plants were preferred over southern members of the same species in 85 percent of cases.[10]

Other forms of consumption also seem to be more prevalent at lower latitudes. An average of 50 percent of birds' eggs is lost to predation in the tropics, as compared to 35 percent among temperate species. Experiments with ants and spiders as prey show that the rate of predation is consistently higher at tropical sites in Peru and Gabon than in temperate northeastern America.[11]

Among shell-bearing molluscs, which have been favorite targets of my research, the mix of architectural types provides a good indication of the kinds of predation that have had the greatest evolutionary influence at different sites. Thickenings, spines, tubercles, and knobs—external features that impart resistance to crushing and that make the shell larger and therefore more time-consuming to handle—occur in 10 to 20 percent of tropical shallow-water marine species of shell-bearing snails living on rocky bottoms, but in only 2 percent of cool-temperate and no polar species. Features such as teeth or folds that impede entry into the shell's aperture occur in 15 to 40 percent of tropical rocky-bottom species and in less than 5 percent of cool-temperate ones. Snails with a markedly narrow slitlike opening occur in 15 to 50 percent of tropical rocky-bottom species, depending on location, and in less than 11 percent of those on cold-temperate shores. Some snails are able to close off the opening with a mineralized operculum. Such protective structures occur in 5 to 12 percent of tropical and less than 2 percent of temperate snail species. In the family Turbinidae, the mineralized opercula of tropical species are significantly thicker relative to their diameter than those of species within such temperate regions as southern Africa, New Zealand, southern Australia, Chile, and California. Sand-dwelling snails show very similar geographic patterns in shell architecture. One architectural type especially well represented among sand-dwelling species is the turreted shell, a slender high-spired structure enabling the vulnerable soft parts to withdraw deeply when danger threatens. Shells of this architecture occur in 15 to 35 percent of tropical sand-dwelling snails, depending on location, and in less than 5 percent of temperate species.[12]

These architectural patterns in snails, together with similar ones in bivalves, indicate that forms of predation involving the application of force account for a higher proportion of deaths in warm-water species than in cold-water ones at comparably shallow depths. Experiments by Mark Bertness and colleagues support this conclusion. Empty shells glued to rocks in Panama were far more likely to be scraped away and broken, mainly by fish, than were similar shells set out on the cold shores of Massachusetts. Snails that drill a hole at the edge where the two valves of a clam shell come together account for higher proportions of drilled individuals in tropical samples than they do in cold-temperate ones.[13]

The predators responsible for these trends are better equipped to use force in the shallow marine tropics than at higher latitudes. Although no one has compared absolute forces, specialized shell-crushing crabs of such tropical to warm-temperate genera as *Carpilius, Daldorfia, Eriphia,* and *Ozius* have relatively larger crusher claws with a higher mechanical advantage than any temperate species. Tropical fish puffers, eagle rays, various wrasses, triggerfishes, and many others, and stomatopods also have mouthparts specialized for crushing and hammering that are less developed in their cool-temperate counterparts. High-latitude communities in places like the Bering Sea, the west European shelf, and Argentina are dominated numerically by mollusc-eaters that use slow methods of predation, such as seastars opening or swallowing clams, snails drilling or entering prey without the use of force, and fish swallowing victims whole.[14]

In at least some groups of animals, the frequency of highly toxic species is higher in the tropics than at higher latitudes. Sea cucumbers toxic enough to kill fish within fifteen minutes when force-fed to these potential predators account for more than 80 percent of species in the tropical western and eastern Pacific, whereas no species in the cold waters of Washington State are highly toxic. Species that void Cuvierian tubules, which provide a highly effective underwater glue, are predictably encountered under boulders on tropical reef flats. Off the west coast of Mexico and Central America, they account for 17 to 29 percent of local sea cucumber species, whereas in Washington species with this noxious habit do not occur. Sponges whose toxins kill fish consumers within an hour account for about 9 percent of species in Washington, 11 percent in warm-temperate southern California, and 33 percent in tropical Vera Cruz, Mexico.[15]

Intimate guest-host associations are both more common and more specialized in warm seas than along colder shores. More than 50 percent of tropical East African shrimp species live obligately with hosts, whereas no shrimps in England live this way. Shrimps and fish that "clean" hosts of parasites are known in warm-temperate and tropical waters but not in the cold-temperate zone. Cap-shaped snails (lottiid and patellid limpets, and hipponicid hoof-shells) that excavate depressions on rock surfaces and on shells are found only in warm-temperate and tropical environments, and are best interpreted as animals specialized to thwart grazers and scrapers.[16]

It is not surprising that warm-blooded animals, which maintain high constant body temperatures during times of activity in the face of wide variation in the temperature of their surroundings, show no obvious latitudinal patterns of adaptation. Martin Moynihan noted thirty years ago that temperate mammals and birds can be just as aggressive and powerful as are tropical species. Many species, in fact, are widely distributed in both temperate and tropical regions. Examples include the mountain lion (or puma) of the Americas, the Asian tiger, the Old World lion (known from Europe in historic time), and hundreds of species of migrating birds. Compilations of top running speeds reveal no tendency for tropical mammals to be faster than temperate ones. For these animals, environmental variation in temperature has become a minor factor in determining either opportunity or constraint. Instead, as I noted in chapter 7, it is the effective size of the environment that sets limits to the level of performance.[17]

For technologically sophisticated humanity, a good argument can be made that performance has historically been greatest in the temperate zones. Although modern humans and their ancestors originated in tropical Africa, and achieved agriculture-based and urbanized civilization first in warm-temperate to tropical areas such as southwest Asia, India, China, Mexico, and Peru, technological advances of the last two thousand years have increasingly given the economic edge to peoples in the cooler climates of Europe, Asia, and

North America. Much of this advantage derives from trade with warmer countries, as well as from increasingly effective clothing and housing to shield against the cold. Further advances may erase the temperate zone's advantage in the long term, especially if the debilitating effects of tropical parasites and diseases can be overcome through additional medical breakthroughs.

The trends in adaptation as one passes from the tropics to higher latitudes are duplicated along other, smaller-scale geographic gradients. As productivity or temperature or both decrease, maximum performance declines accordingly along with opportunity and the market for innovation. Examples of these gradients include those from low to high levels on the seashore, from low to high altitudes, from the surface of sediments to deeper levels within sands and muds, from well-lit open-surface habitats to dark nutrient-poor caves, and from rain forests to deserts. All these gradients are interesting and important, but I mention them here only in passing so that I can concentrate on three other gradients, those with water depth, water salinity, and water availability.

Some kinds of predation seem to be less important toward deeper water. Tatsuo Oji noted a decrease in damage to the arms of a Caribbean stalked crinoid with increasing water depth, and supported the interpretation that the present-day restriction of immobile stalked crinoids to deep-water environments is due in part to the evolution of arm-nipping fish. In the Bahamas, Sally Walker and colleagues set out trays of empty shells of two Pacific snail species, and followed the fates of these shells, some of which became temporarily occupied by hermit crabs, for several years. Shell breakage occurred at all depths, but was most frequent at the shallowest depths (15 to 30 m) and least common at depths below 200 m. In the deep North Atlantic (below 200 m), repaired shell breaks are known from the greatest depths sampled (about 4,880 m), and occur at frequencies of 0 to 20 percent depending on the species. These numbers roughly accord with frequencies observed in shallow-water temperate snails but lie below those in many tropical species. The damage to the generally thin shells of deep-sea molluscs is probably caused by predators with broad diets and unspecialized methods of subjugation, for no deep-sea predators are known to break shells with equipment dedicated to that function.[18]

Even greater contrasts exist between the biotas of shallow-water marine environments and those in fresh water. Alan Covich and I noted that freshwater snails everywhere are poorly armored compared to their marine counterparts. Only two freshwater genera (the southeast Asian *Rivomarginella* and the North American *Acella*) have a narrowly elongate aperture, a feature extremely widespread in marine snails; and no freshwater snails have the shell opening reinforced by thickenings on the inside of the rim. In great contrast to marine snails and clams, there are no fast crawlers, swimmers, fast burrowers, or jumpers among freshwater molluscs, and none has a streamlined shell, ratchet sculpture (in which the external ribs are oriented so that forward motion entails less friction than backward slippage), and deep withdrawal of vulnerable organs into the shell. Most large freshwater clams achieve stability in unstable sand or gravel by weighting the downward-pointing front end of the shell. This adaptation is rare in large marine clams, which are either thick-shelled throughout or adapted to burrow quickly if they happen to be exhumed by currents or predators. Detection of individual predators at a distance by molluscs is unknown in fresh water. Drilling, a very widespread form of predation in which snails or octopods penetrate the shell wall by excavating a small hole, is also absent in fresh water save for a report of drill holes made by unknown agents in the shells of snails from the brackish Caspian Sea.[19]

In the sea, the habit of burrowing in sand and mud and of excavating tunnels in shell walls or rocks is widespread, but in freshwater these habits are less common. Mammals—muskrats, beaver, water shrews, desman, and platypus—excavate dwelling burrows in the banks of lakes and streams. So do many beetles, mayflies, stoneflies, midges, crayfish, and crabs. The dwellings of gryllotalpid mole crickets extend down to a depth of 30 cm beneath the sediment surface. Burrowing for food in sediments is done mainly by earthworms in freshwater to a depth of not more than about 5 cm. The only freshwater organism known to tunnel into shells is the southeast Asian annelid worm *Caobangia*. No species seems to excavate rocks as so many animals do in the sea.[20]

The general decrease in performance observed with increasing latitude, increasing water depth, decreasing salinity, greater sediment depth, decreasing rainfall, higher altitude, and other gradients has evidently been stable throughout the history of life. This is not the case, however, with the gradient between sea and land. Whereas the dry land began as a less productive environment than the sea, the tables turned when land plants reached the size of trees about 370 million years ago. Not only did this mean a reversal in the gradient of top economic performance, but it also changed the pattern of evolutionary invasion between these two physically contrasting environments.

Sea and Land

The seashore is one of those boundaries between two realms that, despite great differences in the conditions of life and in the evolutionary history of the residents, represents a zone of thriving exchange between ecosystems. On one side is a biological world dominated by insects, birds, mammals, fungi, and soil-rooted land plants. On the other side are fish, snails, echinoderms, crustaceans, a host of filter-feeding and sediment-ingesting animals, and a contingent of primary producers comprising blue-green algae, phytoplankton, and attached seaweeds. The no-man's-land between, where immersion in saltwater alternates with exposure to air and rain, is home to a specialized complement of denizens, some belonging to marine lineages—a few mussels, periwinkles, seaweeds, barnacles, crabs, blennies, and gobies—and others with a terrestrial heritage such as kelp flies, some spiders, lichens, salt-marsh grasses, and mangrove trees.

Across this zone, there is a lively trade in nutrients and in organisms. Rain washes nutrients off the land, and these nutrients stimulate the growth of phytoplankton, which in turn feeds the larvae of many marine animals. Charles Birkeland showed that outbreaks of the coral-eating crown-of-thorns seastar (*Acanthaster planci*) on Pacific islands come three years after exceptionally heavy runoff brought on by intense rains following prolonged dry periods. The leaves and other debris dropped by mangroves wash into adjoining coastal waters, and also feed juvenile fish, which as adults move further offshore. These consumers, subsidized by inshore and terrestrial production, exercise intense ecological and evolutionary control on the coral reefs that occur in low-nutrient waters seaward of highly productive mangroves and seagrass meadows. Seabirds—animals ultimately of terrestrial origin—feed at sea but roost, nest, and defecate on land. Globally, according to Gary Polis and colleagues, seabirds transport some ten to one hundred thousand tons of phosphorus (a limiting nutrient for most land plants) from productive inshore waters to adjacent areas of land per year. About 87 percent of the nitrogen requirements of land plants on Marion Island in the southern Indian Ocean are provided by seabirds. On the desert islands in the Gulf of California, where some 28 kg of marine biomass wash up along each meter of coastline per year, the density of spiders just above the tideline is six times higher than that further inland. Islands with seabird colonies support 2.2 times as many land arthropods as do islands without bird colonies.[21]

The pattern of subsidy varies according to which of the realms involved in the exchange is more productive. In places like the Gulf of California, where productive ocean waters lap onto unproductive desert, the marine ecosystem heavily subsidizes the ecosystems on land. This pattern is likely to prevail along the dry west coasts of North and South America and southern Africa, as well as in the polar regions. The situation is reversed in mountainous and luxuriantly vegetated regions. Here, it is the land that subsidizes adjacent marine ecosystems with nutrients carried in river water or in tidal channels from the land and from coastal marshes. The high marine productivity of coastal marine ecosystems in southeast Asia, Atlantic North and South America, tropical West Africa, the Pacific coasts of tropical Central and South America, and many of the larger islands in the tropical Pacific and Indian Oceans is sustained by runoff from even more productive forests and savannas.

Some subsidy of the sea by the land occurs over great distances. At the height of the last ice age, some 18 to 22 ka, strong winds carried dust laced with iron and silica from dry continents to offshore areas of the southern ocean, the North Pacific, and parts of the tropical eastern Pacific and Atlantic Oceans. Productivity in the plankton in these oceanic regions far from land was much higher than it is today. With less wind-blown dust, these regions show consistently depleted levels of iron and silica, so that planktonic rates of production are low despite the ready availability of such other critical nutrients as nitrogen and phosphorus.[22]

On the evolutionary time scale, the dominance of ecosystems relative to each other can be gauged by the pattern of evolutionary transition of lineages between them. Remarkably few lineages have made the jump between the sea and the dry land, but the pattern of transition switched from a one-way invasion of the land from the sea before about 300 Ma to a lopsided pattern favoring land-to-sea expansions later.

Most of the enormous diversity of fungi, plants, and animals on land stems from initial invasion of at least ten lineages from the sea, many perhaps with freshwater intermediaries. Divergence times calculated from molecular rates of evolution would place the split between bryophytes and vascular plants at 841 Ma, implying that multicellular plants had already colonized the land by this time in the Neoproterozoic era. The earliest fossil evidence of eukaryotes on land comes from trackways, perhaps made by euthycarcinoid arthropods, discovered from rocks from the Cambrian-Ordovician boundary (505 Ma) in southern Ontario. There is evidence of fungi by the Middle Ordovician (460 Ma). By no later than 450 Ma, in the Late Middle Ordovician, spores of land plants and trails of millipedes appeared. These were followed in the Silurian or Early Devonian (420 to 390 Ma) by the ancestors of mites, springtails (colembolans), and of insects, and in the Early Carboniferous (340 Ma) by undoubted terrestrial vertebrates. Two or more groups of land snails and one group of scorpions invaded the land a little later during the Carboniferous period.[23]

The earliest evidence of a return invasion, from the land to the sea, dates from the Late Permian (260 Ma), when a mesosaur reptile became secondarily marine. From this time onward, invasions from land to sea greatly outnumber invasions in the opposite direction. Robert Dudley and I estimate that 122 insect lineages expanded from the land to the sea, sometimes by way of fresh water (seventy cases) and sometimes directly (fifty-two cases). At least sixty-two other arthropod groups invaded marine habitats. At least thirty-five lineages of fossil and living reptiles made the transition to the sea, including snakes, crocodiles, rhynchocephalians (today represented by the terrestrial tuatara in New Zealand), turtles, and lizards. Marine mammals—whales, seals, the Pliocene Peruvian sloth *Thalassocnus*, and seacows, among others—belong to at least seven distinct lineages. Among birds, marine habits have evolved dozens of times, ranging from birds that wade in the water to those that skim over water, dive, and plankton-feed. Only five lineages of land plants—three major and two minor clades—have become truly marine, as seagrasses. Mangrove trees and salt-marsh plants in another forty to fifty lineages have colonized marine intertidal habitats. We were able to identify just ten minor invasions from the sea to the land for the last 250 my, five among snails and not more than five among crustaceans. No plants or vertebrates have made this transition during the past 350 my.[24]

How is this turnaround in evolutionary subsidy to be explained, and why and how does it differ from the ecological pattern of subsidy? Four factors—productivity in the donor environment, productivity in the recipient environment, competitive pressure from species in the donor biota, and competition from incumbents in the recipient biota—control invasion of lineages, whereas differences in productivity or available biomass between adjacent environments account for the net movement of nutrients. Although the ocean and the dry land contribute about equally to global primary productivity, the larger area of the ocean (71 percent of the Earth's surface versus 29 percent) ensures that the rate of biomass production on land is, on average, higher than that in the ocean. About 25 percent of the ice-free land surface area (3.3×106 km^2) supports net primary productivity greater than 500 g carbon per square meter per year. Productivity this high or higher prevails over just 1.7 percent (5.0×10^6 km^2) of the ocean.[25]

When regional productivity happens to be higher in the sea than in the adjacent ecosystems on land, potential invaders from the land are beckoned by abundant food, and intense competition may sweeten the enticement. But most land creatures are initially ill suited to cope with the radically different medium of saltwater and with the well-adapted incumbents that already inhabit the water. Of course, potential invaders from the sea to the land face comparable opportunities and hurdles if recipient ecosystems on land produce biomass at a greater rate than the donor ecosystems in the sea. Before land plants established forests about 370 Ma in the Middle Devonian, both productivity and the intensity of competition must have been substantially less on land than in coastal marine environments. Fungi, plants, and animals with modest metabolic demands could have been pushed by higher-energy competitors and consumers into marginal habitats, such as freshwater and the dry land. Adaptation in these biological frontiers rapidly increased productivity and competition, and set in motion the same escalation that had earlier produced a biologically rather sophisticated ecosystem on shallow seafloors and seashores. No later than 300 Ma in the Late Carboniferous, terrestrial productivity and perhaps the competitive performance of economic dominants began to exceed those in the adjacent coastal ecosystems of the sea. Globally, opportunity on the land was greater than in the sea, but so was resistance from marine incumbents and competitive pressure among high-energy species. These relationships did not prevail everywhere, of course. To this day, deserts and polar regions on land remain much less productive than nearby inshore waters. Resistance from terrestrial competitors normally prevents marine species from colonizing the dry land, but on small islands they may have been weak enough to allow a few animals to colonize from the sea.

Land vertebrates have done better at colonizing the sea than all other land-derived organisms. Even though at first they were not adept at swimming or at hearing under water, they were nonetheless active even in the initial stages of colonization of nearshore habitats, and therefore could compete head to head with well-adapted but generally lower-performing incumbents. For these land vertebrates, the most important factor may have been high marine rates of production. Early marine whales, for example, evolved on the productive northern shores of the tropical Tethys Ocean during the Early Eocene, in what are now Pakistan, India, and Egypt. Desmostylians—extinct herbivorous marine mammals—and seals became marine along the productive shores of the northeastern Pacific during the Late Oligocene and earliest Miocene respectively. Otters entered the sea in the northeastern Pacific during the Pliocene, and facultatively marine otters occur on the productive shores of Chile today. The Peruvian Pliocene sloth *Thalassocnus* is yet another example of a mammal that became marine in a place where marine productivity was dramatically higher than terrestrial rates of production.[26]

Regional Inequality

The case of the sea and the land illustrates nicely how productivity and levels of maximum performance affect evolutionary patterns of invasion and spread of species. Large, warm, productive regions should produce species of high performance. When opportunities present themselves allowing species to spread, as when a barrier disintegrates, movements of species should be largely one way, from the region with high potential performance to the region with a lower potential. This is indeed what we observe in nature.

Consider, for example, movements of species between the tropics and the temperate zones. Among species that do not regulate body temperature, movement is almost entirely one way, from low to high latitudes. In a preliminary survey of molluscs and barnacles originating 20 to 25 Ma (during the Late Oligocene and Early Miocene) in the North Pacific, I found twenty-nine American and nine Asian clades in warm-temperate to tropical waters that gave rise to cold-adapted species, but no temperate clades spawning topical species. The more gradual transition between warm and cold waters in the eastern Pacific probably accounts for the difference in numbers between America and Asia. Nowhere in the world have clades with origins at cold high latitudes or in the cold deep sea penetrated the warm tropics. This is so also for land plants.[27]

Impressive as this asymmetry of invasion is, adaptation to the cold occurs in only a minority of warm-adapted lineages, and even when it does occur, cold-adapted species rarely penetrate to polar latitudes. For example, the nine East Asian lineages that became cool adapted during the Late Oligocene and

Early Miocene do not extend to Kamchatka or the Bering Sea, but occur only as far north as Hokkaido and the southern Kurile Islands. At the level of the genus, only 32 of the 105 snail lineages (30 percent) recognized by Warren Addicott from the Early and Middle Miocene of warm-temperate to subtropical central California spawned cold-adapted species, even though this time interval was one of the most favorable for the evolution of cold-water snails in the North Pacific.[28]

The one group of animals that may prove to be the exception to the one-way expansion out of the tropics are warm-blooded birds and mammals. I am unaware of any studies on this topic, but there are certainly examples of mammals that, unlike most cold-blooded animals and plants, are spread widely in the temperate zones and the tropics. The case of hominids is especially interesting. The evolutionary tree and fossil record of primates indicate that hominids, the genus *Homo*, and the modern human species *Homo sapiens* originated in tropical Africa. Unlike most other anthropoid apes, the genus *Homo* successfully penetrated the cold-temperate zone. The first to do so appears to have been *H. erectus*, which by about 1.36 Ma had reached 40° north latitude in cold-temperate North China. Technologically advanced *Homo sapiens* secondarily reinvaded the tropics on many occasions, the best-known examples being Europeans colonizing all tropical regions, native Americans extending into tropical America from the north, and East Asians expanding into the Asian and Pacific tropics. It would be interesting to know if the temperate to tropical transition was also made by technologically less sophisticated hominids, before the invention of composite weapons about 300 ka, long-distance weapons 50 ka, and the domestication of animals some 10 ka.[29]

Among today's marine tropical biotas, that in the Indo-West Pacific—stretching from the Red Sea and East Africa to Polynesia, and from southern Japan to northern Australia—stands out for its diversity and for its specialization of host-guest associations. It is the only biota harboring bivalves (mussels of the genus *Lithophaga*) and snails (the coralliophilids *Magilus* and *Leptoconchus*) that tunnel into, and gain antipredatory protection from, living stony corals. Elsewhere in the marine tropics, *Lithophaga* bores into the dead parts of coral skeletons, and coralliophilids feed on but do not tunnel into coral hosts. Almost forty species of pontoniine shrimps live in obligate association with corals in the western Pacific and Indian Oceans, whereas no obligate associations of this kind are known in the Atlantic. Clownfishes (*Amphiprion* and *Premnas*, belonging to the damselfish family Pomacentridae) are obligate associates of stichodactylid sea anemones, and all are confined to the Indo-Pacific. Many fish can live with sea anemones in the Caribbean, but none does so obligately. Carapid pearlfishes live inside sea cucumbers, both in the Indo-Pacific (seventeen species) and the Caribbean (one species), but only in the former region do they feed on the internal organs of their hosts. Coral-guarding crabs are known in the Pacific and Indian Oceans but not in the

Atlantic. Some bivalves—giant clams of the genus *Tridacna*, and smaller cardiid cockles of the genera *Corculum* and *Fragum*—contain photosynthesizing dinoflagellates in their light-exposed mantle. Today, these bivalves occur only in the Indo-West Pacific, and nothing comparable to them is found in the Atlantic. Finally, snails of the hoof-shell hipponicid genus *Sabia* usually live on the shells of snails and hermit crabs. They excavate a pit on the host shell, from which they are difficult for scrapers and people to dislodge. Today, Sabia is known only from the Indo-West Pacific.[30]

Antipredatory shell architecture also reaches a higher level of sophistication in the Indo-West Pacific than elsewhere in the tropics. Between 35 and 40 percent of shallow-water snails living on rocky bottoms in the Indo-West Pacific have the aperture of the shell occluded by thickenings along the rim. In the tropical Atlantic, only 15 to 19 percent of species have such features. Narrowly elongate apertures occur in 35 to 50 percent of Indo-West Pacific rocky-bottom snails and in only 15 to 25 percent of tropical American ones. Ratchet sculpture, which aids burrowing in sand-dwelling snails, occurs in 16 percent of Indo-West Pacific snail species on sandy bottoms, but in fewer than 10 percent of such species elsewhere in the tropics. In general, adaptations enhancing armor and locomotor capacity are best expressed and most frequent in the Indo-West Pacific, intermediate in the tropical eastern Pacific, and lowest in the tropical western and eastern Atlantic.[31]

If the Indo-West Pacific is a center of marine adaptive sophistication today, it may have been less so in the recent geological past. Architectures and intimate associations that are now found only in the Pacific and Indian Oceans also occurred in the warm Atlantic during the interval from the Oligocene to the Pliocene, from about 27 to 1.8 Ma. For example, coral-boring coralliophilids occurred in France during the Late Oligocene to Middle Miocene and in the western Atlantic during the Pliocene. *Sabia*, the pit-forming snail living on host shells, unexpectedly turned up, together with pitted host shells, in deposits of Late Miocene and Early Pliocene age in the Dominican Republic. Rocky-bottom snails with narrow or otherwise reinforced apertures suffered more than did species with wide apertures during the extinctions of the Pliocene in the western Atlantic, meaning that interoceanic contrasts were smaller during the Miocene than today.[32]

As the largest biota, and the one in which specialization and antipredatory adaptations are most frequent and maximally expressed, the Indo-West Pacific has exported many species, especially to other parts of the tropics and to the temperate southern hemisphere, but imported none. Since the Isthmus of Panama formed in the late Pliocene or Early Pleistocene, before 1.8 Ma, the tropical Atlantic gastropod fauna received at least twenty-one species from the Indo-West Pacific, probably by way of South Africa, which during the Late Pliocene was warmer than today. At least sixty-five species crossed from the eastern parts of the Indo-West Pacific to the tropical eastern Pacific during the Pleistocene. The Suez Canal, opened in 1869, has let through hundreds of species from the Red Sea (part of the Indo-West Pacific) to the eastern Mediterranean. An unknown but probably substantial number of species from the Indian Ocean extended around South Africa during warm intervals to colonize tropical West Africa. No molluscan lineage has been able to invade the western Pacific and Indian Oceans from the other tropical regions.[33]

During the same interval of about the last 2 million years, the tropical western Atlantic received at least fourteen molluscan immigrant species from the eastern Atlantic, whereas the eastern Atlantic received at least forty-one from the western Atlantic (counting eleven Indo-West Pacific species that crossed the Atlantic from west to east after they first reached the western Atlantic). Ten of the immigrants to the west became widespread, whereas only twelve of the immigrants to the east did so. Successful molluscan invaders account for only 0.5 percent of the present-day tropical eastern Atlantic fauna and about 0.3 percent of that in the western Atlantic. Widespread immigrants in the eastern Pacific account for 2 percent of living shell-bearing molluscs there. In all three regions, oceanic islands—Clipperton and Cocos in the eastern Pacific, Fernando de Noronha in the western Atlantic, and the Cape Verde Islands in the eastern Atlantic, among others—have been more affected by immigrant species than mainland coasts. In short, the pattern of invasion as revealed by tropical marine molluscs shows that biotas exporting many species import few, whereas those exporting few import many. Area appears to be the best predictor of a biota's status with respect to the export of species. As I indicated in chapter 7, it is also associated with the highest maximum performance; moreover, the proportion of the biota that has become extinct is smallest in the biotas with large area (the Indo-West Pacific) and high productivity.[34]

When the Bering Strait, separating Alaska from Siberia, became a seaway linking the North Pacific and Arctic Oceans during the latest Miocene (5.5 to 5.4 Ma), a potential route was established for invading marine species. At first, a tiny trickle of species expanded from the North Atlantic and Arctic Oceans to the North Pacific, including a few bivalves as well as cod and herring and brown algae of the genus *Fucus*. Invaders from the Atlantic or their descendants comprise at most twenty-four confirmed molluscan species. These account for at most 3 percent of the North Pacific shell-bearing molluscan fauna. Then, beginning in the Middle Pliocene (3.5 Ma), hundreds of North Pacific species expanded their range through the mainly ice-free Arctic Ocean to the Atlantic. At least 143 invading molluscan species from the North Pacific or their descendants colonized cool-temperate European shores, whereas 176 did so in eastern North America. Species of Pacific origin comprise 10 percent of the north European molluscan fauna and at least 22 percent of that on the American side of the Atlantic. On American rocky shores, they comprise the majority of common species, including periwinkles, dogwhelks, and mussels, as well as such other organisms as kelps, seastars, sea urchins, hermit crabs, barnacles, true crabs, and fish. In the northern biota, invaders and their descendants of Pacific origin outnumber those of North Atlantic origin by roughly ten to one. Once again, it is the larger region, with the smaller magnitude of extinction over the last 3 my, that contributes the great majority of invaders, and the smaller, more extinction-prone region (50 to 75 percent of species extinct since the Middle Pliocene) that receives the bulk of the colonizers.[35]

In these and all other cases of invasion from one biota to another, the species taking part comprise a minority of those in the donor biota. The Line Islands in the central Pacific—one of the easternmost outposts of the Indo-West Pacific, in the path of the eastward-flowing current that brings planktonic larvae to the eastern Pacific—support about 250 species of molluscs and 70 species of coral. Of these, 13 percent and 17 percent, respectively, have colonized habitats on the west coast of tropical America. These percentages would be much lower if I based them instead on the much richer fauna of, say, the Philippines or Fiji. Some 23 to 46 percent of molluscan species in the North Pacific, and less than 1 percent of species in the Atlantic, expanded their ranges through Bering Strait. For the majority of species, therefore, either the remaining barrier is insuperable or successful invasion is thwarted by incumbents in the recipient biota.[36]

The dominant plants and animals of the great continental land masses further exemplify the point that large economies give rise to an environment in which competition is so intense that the standards of winning are exceptionally high and the economic performance of entities is contingent on great absolute power. Studies of evolutionary branching within clades show that fast-growing, prodigiously reproducing weeds as well as more steadfast competitive dominants overwhelmingly originated on the largest continents, and it is on the large land masses where all sorts of specialization to intense competition and predation reach their maximum expression.

The most complete story comes from mammals. During the last 65 my, the largest continuous land masses have been Eurasia and Africa. Evidence accumulating from evolutionary studies of mammals indicates that these two continents have served as the places of origin for the major groups that have subsequently spread to most other continents of the world. Rodents, carnivores, the major clades of ungulates (including the derivative whales), primates, and many subgroups within these clades all have their origins in Asia during the Cretaceous period or the subsequent Paleocene epoch of the Cenozoic era, and then spread to North America and Europe, and thence to South America and Africa beginning in latest Paleocene time, when the land corridors linking the northern continents became sufficiently warm. Asia continued to function as the source for later waves of immigrants to North

America. The first humans in North America almost certainly came from Asia about twelve thousand years ago.[37]

Through most of the last 60 my, North and South America harbored continental biotas that were separated by a wide, deep stretch of tropical ocean. Beginning some 12 to 14 Ma during the Middle Miocene, movements of great tectonic plates in the Earth's crust created island arcs and shallow-water passages, which during Pliocene time (3.5 to 1.8 Ma) gave way to a continuous, twisted Central American land bridge connecting the two continents. The biological effects of the isthmus and of the geological configurations leading up to it were dramatic. Marine biotas that previously were linked through the Central American seaway became divided into a western Atlantic and an eastern Pacific component, whereas the land biotas of North and South American began to exchange species. South American armadillos, anteaters, porcupines, phorusrhacoid (large flightless predatory fossil birds), and hundreds of lineages of rain-forest plants and animals moved north. Dogs, cats, camels, weasels, raccoons, deer, mice, elephants, and rattlesnakes spread from North to South America. Although initial movements of mammals were about the same in number in the two directions, the later invasions were tilted heavily in favor of North American lineages moving south.

The South American mammal fauna was at a disadvantage for two reasons. First, a larger proportion of mammals became extinct in South America during the Pliocene than was the case in North America. This may partly have been the consequence of the second reason: the mammals invading from North America had evolved under a competitive regime much more sophisticated than that prevailing on the smaller continent of South America. The predatory marsupials of South America were evidently no match for the cats, dogs, bears, weasels, and other carnivorans that came from the north. No South American herbivorous mammal had evolved horns or antlers, devices that enabled North American hoofed mammals to engage in mating-related combat and to defend themselves against predators. No native South American snakes were venomous until rattlesnakes established themselves. The fleetest North American mammals—cheetahs, wolves, pronghorns, and rabbits—were far faster than any native South American ones were. Bone-crushing predators were plentiful in the pre-isthmian North American mammal fauna but unknown among South American carnivorous mammals. South American mammals in North America had limited success, diversifying little or not at all and mostly remaining in tropical to subtropical zones. These invaders perhaps succeeded because they had modes of life different from those of native North American mammals. Whereas no North American mammals in pre-isthmian times could be classified as armored, several invaders from South America—armadillos, ground sloths, and porcupines—had evolved heavy external armor or quills. The habit of eating ants and termites may not have evolved in any North American mammal, but South American anteaters brought it successfully to the southern parts of North America.[38]

The competitive prowess of North American mammals stems not only from the large size of North America, but also importantly from the periodic connections of North America with the even larger, still competitively more sophisticated biota of Asia. Thus, Asia is the ultimate source of the carnivorans, hoofed mammals, and rattlesnakes that were to prove so successful in South America.

A tragic modern manifestation of geographic inequality is taking place on isolated oceanic islands. Largely through human agency, many continental plant and animal species have become established on islands, where they have decimated the native biota. The Florida predatory land snail *Euglandina rosea*, originally introduced foolishly to eradicate the giant African land snail *Achatina fulica* (itself an introduced plant-eating garden pest), caused the extinction of at least five native land snails on the French Polynesian island of Moorea within a few years of its introduction. Rats, weasels, and cats doomed dozens of native bird species in New Zealand and on many tropical Pacific islands to extinction. Humans exterminated perhaps two thousand bird species in Oceania, extinguished the forest palm on Easter Island, and caused extinctions among numerous other native island birds and reptiles worldwide. Dwarf hippopotamuses on Mediterranean islands died out when humans arrived a few thousand years ago, as did giant lemurs when people settled Madagascar about one thousand years ago. Introduced aggressive grasses eliminated native plants on Ascension Island in the South Atlantic. In every case, the low-energy populations of oceanic islands, unaccustomed to high-energy consumers, were unable to cope with plants and animals that had evolved under circumstances where competition and predation by sophisticated species was the norm. The effects of the invaders are magnified when, as is almost always the case with human-caused introductions of species, the arrivals are accompanied by destruction of native island forests. Complex vegetation provides refuges, and complicates the search efforts of even the most diligent consumers. Without forest cover, native species have no more places to hide, and fall easy victim to the newcomers.[39]

Human history, too, has been profoundly influenced by geographic inequality. In his brilliant book *Guns, Germs, and Steel,* Jared Diamond meticulously lays out the argument that Asia—its southwestern parts in particular—was more favorable than any other continental region on Earth to the domestication of plants and animals by humans and to the subsequent development of metal-based technology. The combination of a very large land area favorable to large-grained grasses and edible legumes, behaviorally malleable large grazing mammals, and fertile soils made southwest Asia an ideal place for humans to domesticate plants and animals for food and animals for labor. With agriculture came a dramatic rise in human population size which was accompanied by the evolution of many potent disease organisms whose virulence and transmission depended on high densities of hosts. People in these large societies developed partial immunity to these emerging diseases and also perfected the technology of warfare. As these societies spread around the world, powered by writing and by cohesive religions, they encountered other human groups that had not previously experienced the diseases or the weaponry with which they were confronted. As Diamond points out, it is not the individual people or even their cultures of southwest Asia that were inherently superior, but the exceptionally favorable and competitive biota in which the earliest inventors of agriculture and metal technology happened to live.[40]

At an earlier stage of human evolution, the biological environment may have played an equally decisive role. Hominids, and the modern human species *Homo sapiens*, arose in Africa, home to an extraordinary array of potent predators and potential competitors. Perhaps nowhere on Earth during the Pliocene and Pleistocene was competition among high-energy consumers on land ratcheted up to levels as high as in Africa. Snakes there are fast and highly venomous. In contrast to tropical Asia, Australia, and the Americas, few birds in Africa are brightly colored, perhaps implying that the long-distance visual advertisement for mates made possible by conspicuous coloration is untenable in the face of intense predation. The Pleistocene mammal fauna of North America was in some ways comparable in competitive prowess to that of Africa, but primates and parrots, which could potentially have given rise to highly intelligent species, did not penetrate across the Central American isthmus much beyond the tropical zone. Early technological advances by humans, including stone tools by 2.5 Ma and the purposeful use of fire by 1.9 Ma, represent highly effective means of competing with and defending against enemies. The emergence of intelligence may well be connected with sexual selection among humans, but it crucially also enabled hominids to solve problems and to develop out-of-body technology rapidly in an intensely competitive world. We shall never know if these advances would eventually have occurred in such other regions as South America, Madagascar, and southeast Asia, where primates also occurred, but I would hazard the guess that a competitive environment on a par with that of Pliocene Africa provided the selectional background in which exceptional intelligence would most likely have evolved.[41]

The Geography of Innovation

Compared to many intervals of the geological past, our time is characterized by relatively dispersed continents and by a biota that is both longitudinally and latitudinally well differentiated into distinct geographic units. One might expect that exceptionally favorable conditions for the evolution of high-powered organisms existed at those times when the continents were united into one or two giant land masses, or when large parts of the ocean were occupied by a single, geographically more or less homogeneous marine biota at low latitudes. More generally, adaptive innovations that require or entail a high investment in energy should arise in warm, effectively large economies in which competition is intense and resources are sufficiently abundant and reliable to enable entities to respond to such competition adaptively.

It has proved curiously difficult to subject this hypothesis to objective scrutiny. One problem is the rarity of truly path-breaking innovations. Unique innovations, such as the evolution of flight in insects and perhaps the union among prokaryotes to forge the eukaryote cell, are hard to compare among themselves, even if we could identify the conditions under which they originated. Each rare innovation comprises a population of one, clearly insufficient for a statistical analysis. A more tractable type of innovation is one that has occurred several or many times independently, one that leaves diagnostic traces in the fossil record, and whose context of origin can be reconstructed with confidence from an analysis of the evolutionary tree and from inspection of the time and place of first geological appearance of each instance of the innovation in question. The desirability of looking to the fossil record for clues about the time and circumstances of innovation diminishes the attractiveness of many innovations that have arisen multiple times, such as well-developed eyes, social organization, orb webs in spiders, and the symbiosis between root-associated fungi and land plants. These latter types of innovation are recognizable only in living forms, and although their origins can be reconstructed from evolutionary trees, instances among extinct forms can neither be recognized nor evolutionarily traced. We should in no way shun these novelties, but the most complete picture of innovations and the conditions under which they arise emerges when there are multiple instances in extinct as well as in living clades.

A second problem is that the train of events triggered by a favorable circumstance or by a technological advance takes time. The effects may not become fully apparent for millions of years in the case of biological evolution, or decades to hundreds of years in the case of human history. Causal connections may thus be more apparent than real, more the figment of wishful thinking than of economic reality. In some ways, it is all too easy and too tempting to connect a train of events with a trigger; in other ways, it is all too problematic. Dangerous as this territory may be, I think it is worth exploring.

In a quest for a power-enhancing innovation that can be unambiguously recognized in fossils and that evolved often enough for appropriate statistical analysis to be carried out, I settled on the so-called labral tooth of snail shells. The labral tooth is a blunt or spinelike, downwardly pointing projection at the edge of the outer lip (growing margin) of the shell of some predatory marine snails. Experiments and comparative observations by other researchers show that the presence of the labral tooth speeds up the subjugation phase of predation on barnacles and mussels by a factor of two to three. Moreover, inserting the tooth between the shell wall and the operculum of a prey snail, or between the valves of a prey clam, the tooth-bearing predator may protect its feeding organ (the proboscis) from being amputated by the prey. I currently estimate that the labral tooth evolved independently sixty times during the last 80 my of predatory-snail history. Without fossil material, we would have just thirty-eight cases, and therefore a far less complete picture of the circumstances of evolution of this minor technological improvement. Although some snails with a labral tooth live in cold-temperate shores in such places as southern Chile, southeastern Alaska, and southern New Zealand, all instances of origin of the labral tooth occurred in tropical or warm-temperate waters. Of 46 origins after 34 Ma (the end of the Eocene epoch), all but one were on productive continental coastlines. The single exception is a species found today only in the Galápagos and Cocos Islands in the tropical eastern Pacific. The Galápagos are exceptional among archipelagos of oceanic islands in being bathed in highly productive upwelled waters. Tooth-bearing snails do reside on the shores of unproductive oceanic islands, but all belong to tooth-bearing clades with continental representatives.

Complementing this geographic evidence, the times of origin of clades with a labral tooth appear to have supported nearshore marine ecosystems with very high planktonic productivity, as indicated by the large size of suspension-feeding barnacles, clams, and snails. Intervals when a larger than expected number of snail clades gave rise to tooth-bearing members are the Late Cretaceous (nine cases), Early Miocene (twelve), Late Miocene (six), and Pliocene (seven). The Early Paleocene and Early Oligocene, times immediately following mass extinctions, were characterized by small suspension-feeders worldwide, and no first appearances of tooth-bearing clades are known from these intervals. The long Eocene epoch also had few origins of the labral tooth. Although some herbivorous and predatory snails of this age reached enormous sizes—the herbivorous Middle Eocene *Campanile* exceeded a meter in length, for example—suspension-feeders remained remarkably small in most parts of the world.[42]

Another minor innovation in snails—the transition from right-handed to left-handed shell coiling—presents an instructive contrast with the labral tooth. In most snails, the shell exhibits right-handed coiling, meaning that the opening appears on the observer's right as the shell is held with the apex pointing up and the aperture facing the observer. In left-handed shells, the opening appears on the observer's left side. Although handedness affects mating in snails that practice internal fertilization, and left-handed individuals of normally right-handed species often show slight deviations from regular spiral coiling, there is no evidence that left-handedness provides either a benefit or a disadvantage once the young have hatched. During the past 65 my of the Cenozoic era, left-handed coiling has arisen dozens of times among clades in fresh water and on land, but only nineteen times among marine snails. These nineteen clades originated, or at least appeared first, in circumstances in which predation risk for either the young or the adult is low. Whereas labral-tooth-bearing clades always appeared first in warm, productive regions, including the competitively rigorous Indo-West Pacific and tropical eastern Pacific regions, at least three left-handed clades originated in polar seas, and none appeared in either the Indo-Pacific or eastern Pacific. In short, the evolution of left-handed species seems to require tolerance of error, but it is inhibited by intense competition of predation because left-handedness, unlike the labral tooth, offers no obvious survival benefits.[43]

Much more important evolutionary innovations also seem to have originated in productive, warm environments. Flowering plants, for example, probably began as fast-growing herbs in disturbed, warm, lowland habitats during earliest Cretaceous time. Their combination of new or elaborated features—network venation in the leaves, effective water-conducting vessels in the stem, and nourishment of the embryo in the seed with the endosperm—is consistent with a weedy habit emphasizing rapid growth. David Jablonski noted that the tropics during the last 250 my served as the place of origin for the great majority of clades distinctive enough to be labeled orders by taxonomists. If the origins of these orders coincide with the appearance of some enabling invention, as Jablonski implies, then the productive and competitive environments of the equatorial regions were very conducive to innovations. Energy-intensive cheilostome bryozoans (see chapter 5) arose nearshore in Late Jurassic time as weedy, fast-growing, linear colonies in productive environments.[44]

Innovation and Geological History

Another way of assessing the history of innovation is to ascertain whether periods and places that meet all the criteria for economic growth do indeed register a higher than expected number of adaptive innovations and improvements. Warm, productive conditions should prevail over large areas and should support ecosystems in which competition is intense. Time intervals during which these conditions existed should be characterized by (1) high or rising sea level, associated with large contiguous tracts of productive lowland and shallow sea; (2) wide or broadening belts of warm climate, often (but probably not always) linked to high concentrations of carbon dioxide and other greenhouse gases; (3) active tectonics, introducing new resources through volcanism, seafloor spreading, and widespread erosion from large mountain belts; (4) increasing levels of oxygen in the atmosphere and ocean; and (5) large, biologically more or less homogeneous regions in which small populations of high-powered organisms can be sustained without the threat of stochastic extinction.

Several times stand out as exceptionally favorable to evolutionary innovation because episodes of massive mountain building of undersea volcanism are concentrated in them. We are far from identifying all such intervals, especially those in the very distant past, but evidence accumulating from the work of geologists makes it clear that there is a broad coincidence between the tectonically driven increases in oxygen and nutrients and the evolutionary origins of major economic advances.

According to David DesMarais and colleagues, the first assembly of Earth's land masses into a supercontinent took place during the Late Archean eon, somewhere between 3000 and 2400 Ma. Within this interval, at 2700 Ma (or 2.7 Ga), the earliest known molecular markers of eukaryotic life are preserved. Oxygen-liberating photosynthesis may likewise have appeared at about this time. Obviously, a tight link between tectonic and biological events cannot be established for early episodes of the biosphere's history, but it warrants further investigation by those probing the mysteries of far-distant Archean time.

Breakup of the first supercontinent began during Paleoproterozoic time, 2.2 to 2.1 Ga. This episode was in turn followed by large-scale mountain building. Rifting and weathering released nutrients, stimulated primary production, and led to widespread burial of organic matter, in turn enriching the atmosphere and surface ocean with oxygen. This enrichment came to an end perhaps 1.8 Ga as another supercontinent, probably of low relief, emerged. During this interval of enrichment, fossils interpretable as multicellular eukaryotes first appeared.[45]

The emergence of photosynthesizing eukaryotic seaweeds and plankton as competitively superior primary producers in productive nearshore marine environments coincides with the Grenville episode of mountain building in North America about 1.25 Ga during the Late Mesoproterozoic era. This emergence came after a 500 my period of relative economic stability, characterized by low rates of primary production in the ocean and perhaps by global limitation of production by nitrogen instead of by phosphorus. Continental relief may have been low, because the difference between the thickness of crust beneath the continents and that beneath the ocean was perhaps half that prevailing after 1.0 Ga. With the Grenville episode, according to a hypothesis advanced by Anbar and Knoll, global limitation by phosphorus was reestablished. Given that phosphorus supply is controlled by erosion and by formation of new crust, processes related to tectonics, productivity once again became tightly linked to upheavals in Earth's outer layers. As phosphorus became more abundant through erosion associated with mountain building, productivity and the burial of organic carbon increased, leading to a rise in oxygen.[46]

Although the time of origin of multicellular animals remains controversial, it is likely not to have been much earlier than the mid-Neoproterozoic, 800

Ma or later. Fossilized remains become common after 600 Ma. The formation of the equatorial supercontinent Rodinia about 900 Ma and its breakup beginning 800 Ma evidently enriched the atmosphere, as well as the shallow and deep ocean, with oxygen and nutrients. This interval coincides with a significant diversification of planktonic and bottom-dwelling marine life, a point to which I shall return in chapter 10.[47]

Perhaps the best-known interval of evolutionary opportunity is that comprising the Ediacarian period of the latest Neoproterozoic and the Early Cambrian period of the Paleozoic era, 600 to 520 Ma. This interval, the final third of which marks the so-called Cambrian explosion, witnessed the evolution of the first mineralized skeletons, planktonic consumers of phytoplankton, early suspension-feeders, animals burrowing deeply into sediments, and large predators capable of inflicting injury to their prey (see chapter 10). Geographic reconstructions show that the Ediacarian to Early Cambrian interval was marked by major reorganization of the Earth's crust. While continents in the northern hemisphere rifted and drifted apart, creating extensive seaways and large areas of shallow ocean as sea level rose, those in the southern hemisphere collided to form a single large land mass, Gondwana. Mountain building on the African part of Gondwana created a landscape of high relief, suitable as a source of nutrients loosened by erosion. The global climate warmed following several ice ages in the pre-Ediacarian Neoproterozoic, and oxygen levels probably rose as very large phosphorite and chert deposits formed in many basins around the world. Northern rifting was evidently accompanied by vigorous volcanism, releasing more nutrients as well as the greenhouse gas carbon dioxide. In short, the Ediacarian to Early Cambrian combined an abundance of nutrients with warming, a rise in sea level, increasing oxygen in the atmosphere and ocean, and the wide spread creation of productive ecosystems.[48]

The interval from the Late Cambrian to the Middle Ordovician, approximately 510 to 460 Ma, witnessed fewer path-breaking innovations, but was nonetheless a time of notable expansion. Planktonic animals, suspension-feeding animals, predators, well-defended prey, deposit feeders and other deep burrowers in sediment, and life on land all greatly expanded and diversified as members of many independent clades moved into previously marginal parts of the biosphere. These expansions took place against a backdrop of major continental rifting, sea-level rise, and the development of very large shallow seas on low-lying parts of the continents.[49]

Major evolutionary events during the Middle Paleozoic—the interval from the Silurian through the Early Carboniferous, approximately 410 to 340 Ma—include the origin of predators capable of breaking shells with specialized equipment, development of spines and many other forms of passive defense among marine animals, and above all the colonization and rapid diversification of animals and plants on land. Rates of chemical weathering increased thanks to the roots of land plants, and, beginning in the Early Devonian, oxygen levels rose while the concentration of carbon dioxide in the atmosphere fell. By the Middle Devonian, about 370 Ma, there was enough oxygen in the atmosphere to support forest fires, recorded as the earliest charcoal deposits. There were episodes of rising sea level during this interval as well, leading to the creation of large, productive, shallow-water ecosystems. Although various continents began to collide as early as the Late Ordovician, ultimately leading to the assembly of the supercontinent Pangaea in the latest Paleozoic era, there were also important episodes of rifting, especially during the Late Devonian and Early Carboniferous. The associated release of nutrients at hydrothermal vents and of carbon dioxide thus increased the supply of raw materials and may have entailed global warming. These conditions may well have provided economic opportunities for the evolution of land vertebrates and of energy-intensive insect flight during the Early Carboniferous.[50]

Geographic events of the Late Paleozoic—the Late Carboniferous and Permian periods, 320 to 250 Ma—nicely illustrate the stimulatory role of effectively large economies on evolutionary opportunity. The Late Paleozoic was a time of generally falling sea level and of continental collision, culminating in the formation late in the interval of the supercontinent Pangaea. The very large continuous areas of habitat on land may have set the stage for the evolution of more active locomotion among vertebrates during the Early Permian, the rise of herbivory by land animals in the latest Carboniferous and Early Permian, and the first indications of warm-bloodedness among therapsid mammal-like reptiles during the Late Permian and of full endothermy by the Early Triassic. The jump in herbivory apparently coincides with a rise in carbon dioxide and with a poleward expansion of warm climate. Meanwhile, the marine biosphere witnessed few innovations as it either failed to expand or actually contracted and as warm equatorial waters were diverted toward the poles through the erection of a land barrier during the Namurian stage of the Middle Carboniferous period. This barrier, formed as part of the coalescence of land masses, interrupted the flow of currents around the equator, and may have reduced the area of the warm tropcial zone while initiating ice ages at high latitudes.[51]

The influence of Pangaea continued to be felt well into the Mesozoic era. Dinosaurs, the largest land animals of all time, originated about 230 Ma during the Carnian stage of the Late Triassic, and had attained huge size by the Early Jurassic, after 200 Ma, when the world's land vertebrate fauna was still globally rather uniform. Pterosaurs—the first vertebrates to employ flapping flight—also appeared during the Late Triassic. The marine biota, too, appears to have been very widely distributed and therefore to have constituted an effectively very large economy during the Late Triassic and Early Jurassic. Late Triassic seas may have supported the largest marine predators of all time, ichthyosaurs at least 30 m in length. Whether the exceptionally large economies of the Early Mesozoic are alone responsible for the extraordinary sizes of top consumers at this time remains an intriguing and unsolved problem. Oddly enough, much of the diversification taking place during Middle and Late Triassic time among marine animals represents variation on Paleozoic themes rather than dramatic new architectural and ecological departures. The Late Triassic does record the first appearance of mineralized single-celled plankton, new groups of fish and of land-derived marine vertebrates, and photosynthesizing corals and bivalves, perhaps stimulated by early rifting beginning 230 Ma. The world of land plants, too, saw few innovations as floras worldwide were dominated by gymnosperms.[52]

The rifting that began in the Triassic period heralded a more general breakup and drifting of the continents in the Jurassic and Cretaceous periods of the Mesozoic era. As Pangaea splintered and sea-floor spreading between the fragments accelerated beginning 175 Ma in the Middle Jurassic, sea level rose, reaching a level in the Late Jurassic some 100 m above that at the beginning of the period. By the Middle Jurassic (165 Ma), marine conditions warm enough for widespread deposition of limestones extended over 85° of latitude, some 20° more than in today's oceans. Warm, shallow, productive seas therefore expanded even as land masses, especially in the southern hemisphere, remained large. These conditions were generally favorable to the evolution in many clades of deeper burrowing animals in the Early Jurassic, expansion of many insect and snail clades in the Late Jurassic, and the evolution of flight in birds in the latest Jurassic. High levels of volcanically generated carbon dioxide in the atmosphere stimulated land-plant production, creating conditions favorable to the widespread occurrence of gigantic conifer trees estimated to have reached a stature of 40 to 60 m over large expanses of the tropical and temperate zones. At the same time, huge dinosaurian herbivores evolved, remaining large on southern continents until the end of the Cretaceous.[53]

The Cretaceous period, 145 to 65 Ma, may well rival the Ediacarian to Early Cambrian interval as a time of exceptional geological upheaval and evolutionary ferment. The period began with the emplacement of some 2×10^6 km^3 of basalt on the Shatsky Rise, a submarine plateau in the northwestern Pacific Ocean formed about 145 Ma near the Jurassic-Cretaceous boundary. During the Valanginian stage of the Early Cretaceous (130 Ma), as South America and Africa began to rift apart, basalt erupted into the South Atlantic and on the nearby splintered land masses. Like later undersea episodes of volcanism, this event brought with it high rates of chemical weathering and burial of phosphorus. A much larger volcanic episode occurred during the Late Barremian and Early Aptian stages of the Early Cretaceous, 124 to 121 Ma. An estimated 55×10^6 km^3 of basalt erupted on to the ocean floor in the Indian and Pacific oceans to form the Ontong Java and Manahiki Plateaus and the Mariana and Pigafetta Basins. This volcanism initiated a 40 my period of rising sea levels, global warming, increasing levels of carbon dioxide in the atmosphere, and—given the frequent burial of large amounts of organic matter in extensive black shales—higher concentrations of oxygen in the atmosphere.

These trends were reinforced by additional episodes of undersea volcanism during the Cretaceous. From 116 to 108 Ma, an interval encompassing the Late Aptian and Early Albian stages, about 24×10^6 km^3 of basalt formed the southern part of the Kerguelen Plateau beneath the southern Indian Ocean. Still more volcanism beneath the waves took place from 95 to 87 Ma in the Late Cenomanian, Turonian, and Coniacian stages in the southern Indian Ocean (central Kerguelen Plateau and Broken Ridge), the Caribbean region (Caribbean Plate, peak formation 90 Ma), the Arctic basin, and the South Atlantic. From the Early Aptian (120.5 Ma) to the Turonian (90 Ma), sea level rose by at least 125 m to a height some 250 m above present sea level, with the result that the ocean covered 20 percent of the Earth's continental area and 77 percent of Earth's total surface. Temperatures in the tropics at various times during the Cretaceous from the Berriassian to the Turonian perhaps exceeded those of the warmest parts of the equatorial zone today by 1 to 2°C, and may have hovered between 30 and 31°C. During the Aptian stage, warm conditions favorable to the deposition of reef limestone prevailed over 95° of latitude. In short, the Cretaceous—especially the Aptian to Early Campanian interval, 120 to 80 Ma, was a time of warmth, plentiful nutrients and oxygen, and large shallow-water and lowland continental habitats made possible by undersea volcanism and the rifting and drifting apart of continents.[54]

Though perhaps less appreciated than the rifting of continents and the formation of large volcanic plateaus beneath the oceans, significant mountain building also occurred during the Cretaceous, resulting in large tracts of high elevation, up to 5 km in western North and South America during the last two stages (Campanian and Maastrichtian) of the period. Mountain regions 1 to 2 km in elevation were formed in South China and the Himalaya-Alpine region during the Late Jurassic and Cretaceous, and like the mountain-building episodes in the Americas must have yielded large amounts of eroded sediment to nearby lowlands and coastal marine habitats.[55]

A tide of innovations washed over the biosphere during the Cretaceous, especially during its first half. The earliest flowering plants are known from 140 Ma, but the major spread of the group took place during the middle Cretaceous warm interval, 120 to 90 Ma. Three groups of insects—termites, ants, and bees—achieved sociality during the Early Cretaceous. Army ants—the major group of tropical predators using coordinated attack by millions of worker from a single colony—originated about 105 Ma in Gondwana, the huge, then still largely intact southern continent comprising what is now South America, Africa, Australia, and Antarctica. Many groups of fish, gastropods, and crustaceans evolved predatory habits at this time, and groups such as brachiopods and crinoids, which had been abundant elements of shallow-water communities before the Cretaceous, became restricted to deeper water. The second half of the period witnessed a general increase in the expression of armor in shell-bearing snails and cephalopods, accompanied by increased frequencies of failed attacks by shell-damaging predators. Several groups of photosynthesizing plankton became mineralized and greatly diversified during the Aptian and Albian stages of the later Early Cretaceous.[56]

Compared to the Jurassic and Cretaceous periods of the Mesozoic era, the Cenozoic was dominated by continental collision rather than by rifting. Only two undersea volcanic episodes, both early in the era and both associated with subsequent continental breakup, are known for the Cenozoic. Approximately 8×10^6 km^3 of basalt was erupted by the Deccan Traps event between 65.6 Ma (latest Cretaceous) and 62.5 Ma (Early Paleocene epoch), of which roughly one-quarter was deposited on land in northwestern India and three-quarters covered the sea floor in nearby parts of the Indian Ocean. North Atlantic volcanism during the Paleocene (61 to 55 Ma, peaking 56 Ma) produced about 7×10^6 km^3 of rock, much of which erupted on to the ocean floor. Elsewhere, significant mountain building, associated with collision and compression of continental blocks, proceeded throughout the Cenozoic. The Himalayan Mountains and Tibetan Plateau, formed as India and Asia collided, already existed by Early Miocene time (22 Ma). The high Andes and the Altiplano of South America became prominent features during the Late Miocene and Early Pliocene (8 to 3 Ma), at the same time that the high ranges of New Guinea were being thrust up. All this activity during the Late Miocene and Early Pliocene led to widespread erosion and chemical weathering, releasing large quantities of phosphorus into the world ocean.[57]

Three intervals during the Cenozoic—the Early Eocene (55 to 49 Ma), the Late Oligocene to Early Miocene (27 to 17 Ma), and the Early Pliocene (5.3 to 3.5 Ma)—are notable for the wide latitudinal extent of warm conditions. Essentially tropical conditions existed during the Early to Middle Eocene as far north as southern Alaska and the North Sea Basin. In the later warm episodes, the present-day cool-temperate North Sea supported warm-temperate to subtropical faunas. In possible contrast to earlier ages, the warm periods of the Cenozoic do not seem to coincide with high concentrations of carbon dioxide, nor were they brought on by extensive undersea volcanic activity except perhaps during the earliest part of the Eocene. Instead, changes in ocean currents brought about by land barriers may be responsible for the Cenozoic warm spells. During much of the Mesozoic, ocean currents flowed along the equator more or less unimpeded by land masses, but continental collisions in southwest Asia and the Indonesian area during the Early to Late Miocene and the formation of the Central American isthmus during the Early Pliocene caused these currents to be diverted northward and southward away from the equator, so transporting vast amounts of warm water toward polar latitudes.[58]

Just how this physical geography and history of the Cenozoic matches the record of evolutionary innovation is not entirely clear. The warm Early to Middle Eocene records extraordinary diversification of marine life, including the first appearance of herbivorous marine vertebrates and the repeated evolution of armor among many molluscan clades. Peak warmth at the Paleocene-Eocene boundary enabled mammals and probably many other groups to spread across high-latitude Beringia from Asia to North America, precipitating wholesale reorganization of the faunas in the Americas. An apparent rise in productivity during the Late Oligocene to Early Miocene interval, recorded as an increase in body size of suspension-feeders, is accompanied by major radiations and innovations in many clades of molluscs and barnacles. Perhaps this reflects the overall warming during this interval, but other, as yet unknown physical factors may also have been at work. Finally, the Early Pliocene warmth coincides with the last major episode of species formation among marine animals and, of course, with the emergence of hominids. Further documentation of these temporal coincidences and relationships constitutes a major and intriguing challenge for researchers in the years to come.[59]

I do not pretend that the foregoing narrative constitutes a rigorous evaluation of the claim that innovations are concentrated during times of widespread warmth, high productivity, and effectively large ecosystems, but the information at hand is at least consistent with it. One problem is that dates of first appearance of adaptive traits are often not well constrained. Another more interesting point, noted by James Kirchner, is that the origin and spread of innovations and the diversification of clades carrying the new adaptations, play out over long intervals of geological time, often for millions of years. External triggers that set in motion the economic changes associated with innovation are therefore especially difficult to identify and to date. As Kirchner points out, extinctions in the geological record are also difficult to pinpoint (see chapter 9), but they tend to occur over much shorter time intervals, so that there is a tighter correspondence between trigger and consequence for extinctions than for episodes of adaptive innovation.[60]

Still another fascinating possibility is that external triggers may have become less important as economic life develops. Once warm-blooded consumers evolved, the dependence of powerful agents on temperature was at least reduced if not wholly eliminated. Temperature-related triggers still affected those life forms whose body temperatures change along with their thermal surroundings, but even the evolution of cold-blooded organisms, insofar as it is economically controlled by top consumers that often work at high constant body temperatures, may be less strictly affected by temperature than they would have been before endotherms evolved. Similarly, the ability of many top consumers, especially endotherms, to move long distances beginning in the Early Mesozoic may have kept the major geographic units of consumer-dominated economies effectively large, irrespective of the precise configuration of lands and seas. Temperature regulation, extensive movement by individuals, and other economic characteristics of advanced top consumers are ultimately made possible by favorable circumstances controlled by geological processes beyond the control of organisms, but once they have evolved, and provided they can withstand or rapidly recover from the inevitable disturbances that affect economies from time to time, they provide the economy with an increasingly strong and persistent feedback and control mechanism that increasingly generates and tests innovations and new emergent structures regardless of external conditions.[61]

In spite of these caveats, a link between adaptive evolution and warm, productive conditions globally is well supported. In the Jurassic period, for example, innovation and diversification among shell-bearing ammonoid cephalopods occurred during episodes of sea-level rise, when suitable shallow-water habitats increased in size and when conditions generally warmed. Extinctions, by contrast, were concentrated during times when sea level fell and habitats shrank. For the fossil record of the Phanerozoic eon as a whole—that is, for the last 543 my—the rate of diversification (the rate at which new genera arose) is strongly and tightly correlated with increased concentrations of carbon dioxide in the atmosphere, and therefore with temperature. All these lines of evidence therefore point to a general confirmation of the hypothesis of evolutionary opportunity: adaptive improvement is most likely when resources are abundant enough and economies large enough for many populations to respond to intense competition by evolving energy-intensive, power-enhancing characteristics of architecture and organization.[62]

The Human Advance

Most students of human history attribute the great economic advances of our species to human action and social circumstances. For them, economic expansion and greater wealth are sufficiently explained by technological innovation (especially in matters of food, medicine, transportation, communication, and manufacturing), increased trade, and social permissiveness. Factors over which people have little control—climate, volcanic eruptions, earthquakes, floods, and the like—are either dismissed as unimportant or blamed for economic disruption. Economic growth and the consequences of innovation play out over relatively long intervals of time and may not be apparent immediately following a triggering event, whereas weather-caused catastrophe is more readily identified and associated with an unusual circumstance. I do not for a moment wish to minimize the benefits of trade, technology, and tolerance in human history—in fact, I deal with these matters in the final two chapters—but I want to consider briefly how the warm, productive conditions that have stimulated adaptive evolution in nonhuman life also create circumstances favorable to the actions and social structures that historians rightly associate with economic improvement. What follows is meant to be illustrative rather than comprehensive, but it shows that human history, for most of its duration, has not escaped the age-old enabling role that permissive climate plays in economic life.[63]

During times when most people lived at subsistence levels, any circumstance that promoted the rate and reliability of supply of resources would stimulate production and create the possibility of surplus. As I noted in chapter 5, the independent origins of agriculture in the Old and New Worlds shortly after the waning of the last ice age were made possible by a new climatic regime of longer, warmer growing seasons and plentiful rainfall in regions of fertile soil and domesticable species. Later, during the warm spell of the medieval period (from 800 A.D. to the thirteenth century), widely favorable conditions enabled Europe to emerge from the economic and political woes of the centuries following the fall of Rome. Viking trade, exploration, and colonization began in earnest at the end of the eighth century, resulting in the settlement of Iceland (in 874) and southwestern Greenland (by the end of the tenth century) and in the establishment of farming in those far-northern lands. Arab civilization and scholarship thrived in southwest Asia, North Africa, and Spain at a time when rainfall was more plentiful than it was after 1300. China at this time underwent a period of extraordinary technological inventiveness and industrialization, in part powered by production-enhancing improvements in rice cultivation. In Europe, where cities of any size had disappeared after the fifth century, new urban centers of commerce and trade sprang up all across the continent after the year 1000, and in the thirteenth century achieved a measure of political and economic independence from the ruling nobility.

After the weather-related famines and epidemics of the fourteenth, and to some extent the first half of the fifteenth, century, during which the European population fell sharply, the second half of the fifteenth century witnessed relatively warm climates, good harvests, and a growing population. In the Low Countries, increased demand for food was met by a series of important agricultural advances, especially the cultivation of nitrogen-fixing crops (beans, peas, and clover) and food plants for livestock on land that in earlier times would have lain fallow for a year or more. Wide-ranging explorations also began during the fifteenth century. Basque and English fishermen harvested bountiful fish stocks on the Grand Banks of Newfoundland, while Portuguese and later Spanish explorers established links with West Africa, India, the East Indies, and eventually the New World.

The feedbacks between these economic innovations and enabling climates increasingly insulated much of human civilization from the famines and epidemics that gripped our species most of our history. The climatic instability, frequently poor harvests, and raging episodes of plague and other diseases that disrupted European economic and social life during the fourteenth, sixteenth, and seventeenth centuries became less destructive toward the latter half of the so-called Little Ice Age (the period from 1300 to about 1850), before the magnitude of the climatic fluctuations and inclement conditions began to lessen. The combination of slowly spreading agricultural improvements, new crops, and intercontinental trade gradually cushioned civilized societies against local and regional crop failures. Conditions of growth became more persistent, or could be more easily created or stimulated by human action; they became correspondingly less dependent on the geographical and climatic factors that had helped trigger episodes of human expansion before 1600. After 1850, when urbanization and a rise in the emission of greenhouse gases from human activity warmed the climate worldwide and put an end to the Little Ice Age, growth became so commonplace and so apparently independent of our surroundings that many economists unwisely jumped to the conclusion that climate-driven control on resources no longer mattered.[64]

If climate and the life it supports provide the conditions for economic growth and well-being, they can also engender catastrophe. It is to the role of disruption, both as constraint and as opportunity, that I turn next.

Chapter 9

BREAKING DOWN AND BUILDING UP: THE ROLE OF DISTURBANCE

DISRUPTION IS A FACT OF LIFE. As long as nothing upsets the status quo or interrupts the flow and regulation of resources, economic entities can continue to thrive. The problem is that no economy, no matter how sophisticated its structure or how powerful its chief architects, is immune from disturbance. Just as a well-adapted genome experiences most genetic mutations as harmful, a smoothly running economy is apt to be disrupted by a change in conditions.

And yet, disruptions eventually lead to a new order, one in which energetic upstarts from the old regime often give rise to a new hegemony of elites. If a storm causes a mature forest tree to blow down, the resulting patch of sunlight—a light gap in the parlance of ecologists—can support fast-growing weedy plants, which can mature and set seed before slower-growing, shade-making replacement trees once again close the canopy. If all the trees are destroyed regionally, surviving weeds could eventually evolve into a new cast of productive dominants, evolutionarily unrelated to the old guard. Disturbance may thus spell doom to some, but provide openings to be capitalized upon by others.

Disturbance is all about upsetting the status quo and establishing a new order. It is about destruction as well as recovery, about how a system breaks down as well as about how a system is repaired or reconstructed. To understand these processes, we must understand how disturbance acts, which parts of an economy are most susceptible to it, and how disturbance can turn into opportunity. These are my aims in this chapter.

I shall argue that elimination of top consumers may result in the equivalent of an economic recession, but by itself does not spin out of control to cause economic collapse. Interference with the machinery of production, on the other hand, does reverberate throughout the ecosystem, with disastrous consequences not only for primary producers but for the many organisms that depend on these producers directly for food and living space. All the mass extinctions of the geological past, and many of the minor ones, began with a productivity collapse. Recovery from such crises is slower for consumers than for opportunistic producers, so that postcrisis ecosystems remain in a recessionary state of low demand for extended intervals of time. Nonetheless, the high-powered weeds that do best in the early recovery phases prepare ecosystems for the successful establishment of more permanent, powerful competitors, and are in a good evolutionary position themselves to give rise to the new dominants. In the end, disturbance is at worst a temporary setback to economic systems, even if that setback is accompanied by prolonged perturbations in supply and demand. At best, it stimulates economies in directions the economies would have taken even in the absence of disruption.

In the final part of the chapter, I consider how these ideas, inspired by observations in modern ecosystems and the fossil record, apply to the human-economic realm. Here, too, ecological collapses devastate economies, but disruptions generated within the human economy itself have come to play an increasingly destructive role in modifying the imbalance of power. Again we see a parallel between natural history and human history: as adaptations by the dominant members of economies and by the larger economies themselves increasingly counter the collateral damage done by bottom-up crises, economies become more vulnerable to internally generated disruptions, which on the whole (and thus far) are less destructive than the externally driven catastrophes they replaced.

Classifying Disruptions

In order to assess how disturbances affect economies, I begin by imposing several classifications of disruptions and their consequences. One classification arrays disruptions along the supply-demand axis. Some disturbances begin by affecting demand, or consumption. When top predators are removed, they are replaced by secondary consumers, which do not create the same intense selection or exercise the same ecological controls on other species that the top predators did. This situation is like a sudden decline in the buying power of rich people. There would be less patronage of restaurants, lower attendance at concerts and plays, and a declining demand for luxury items and automobiles and houses. If that kind of consumer-driven disturbance leaves the machinery of production more or less intact, the result would be the equivalent of a recession or, in the event of a more severe downturn, a depression, an excess of productive capacity relative to demand. If, on the other hand, a disturbance begins by interfering with production or supply rather than with consumption or demand, the result could be a bottom-up collapse, ultimately leading to mass starvation or deprivation among consumers if the disruption is severe enough.

Consumers are, of course, themselves producers. This is true in ecosystems, where surplus production of consuming animals permits higher-level consumers to make a living as top predators; but it is even more true in human societies, where production and consumption are in the hands of single individuals. We can then frame the discussion about disturbance by asking the following question: is the loss of an economic unit within an ecosystem or within a society more disruptive because of the loss of a unit of consumption or because of the loss of a unit of production? I shall try to show that it is the decline in production more than the decline in consumption that causes and propels more severe economic disruption.

A second axis along which we can think about disturbance is the size of an economy. As I argued in chapters 7 and 8, an increase in size allows for more powerful components to emerge and spread, whereas a decrease imposes stringent limits on the economic power of the highest-performing members. Enlargement of an economic system by growth or by fusion of previously separate systems places incumbents at risk of being outcompeted by more powerful intruders and may further restrict the activity of subordinate members. Trimming the size of an economy through shrinkage or fragmentation also places the dominant, high-performing components at risk, this time because the most power-intensive lifestyles in the large economy become unsustainable in the smaller economy with diminished resources.

Feedbacks among economic players dramatically affect how disruptions ripple through the playground. Strong dependence of one player on another means that, if disturbance affects one, it will endanger the other as well. Disruption therefore has the potential not only to affect entities, but the interactions among entities. Competition may intensify for some and diminish for others; cooperation may be dissolved in some cases but be strengthened in others. Groups that held an advantage before the disruption may be at a disadvantage afterward. But there will also be interactions, and because of that there will always be feedbacks. The way a disturbance plays out in the larger economy depends on how patterns of interaction and feedback change.

As with all other economic phenomena, disturbances vary widely in their extent, duration, frequency, and intensity. They tend to obey a so-called power law: many mild, frequent, short-term, small-scale disruptions, few less frequent, more intense disruptions of larger extent, and very few global calamities. Garden-variety disturbances are so commonplace that individual organisms encounter them once or more often during their lifetimes, and they are thus easily incorporated into the adaptive hypotheses of individuals, especially in those with long lifespans. At the other end of the continuum, global catastrophes are so rare that adaptation to them, even by the most inclusive units of economic systems, is unlikely. Yet, because all living things belong to lineages that have passed through truly rare and large-scale catastrophes, we can expect a certain cumulative, if unintended, adaptation to rare disruptions. A disruption of a given magnitude and extent today may have less drastic consequences than a similar disruption would have had in the past, and more lineages may now be able to transform disturbance into opportunity. General syndromes that enable economic entities to weather disturbances, indeed to profit from them, may have emerged as the most important and enduring economic adaptations.[1]

Disturbances work because they reduce or wholly destroy the power of affected economies. They interfere with metabolism, and disrupt communication and organization among parts. Survival during disruptions occurs either because the unit involved happened not to be exposed to them—that is, it occupied a refuge—or because it was able to respond or resist effectively—that is, it was adapted, or at least predisposed, to cope with a crisis by virtue of successful responses to other calamities. These adaptations roughly fall into two classes, those involving passive resistance—shutting down the system and enduring the adverse circumstances until conditions improve—the pathway of suspended animation, and those involving active response—sensing and escaping poor conditions, modifying unfavorable circumstances, or making other appropriate adjustments that involve the use of power—the pathway of intervention. Elements of both classes—suspended animation and intervention—are likely to be crucial in surviving rare, grave disturbances.

These types of defense are, in the end, not very different from those that living things have against their enemies. But because the more severe disruptions are unpredictable, adaptations to them must be of a very generalized, all-purpose nature. The important point is that the agencies of everyday—competition and predation from enemies—may prepare species for the less frequent, less predictable, more disruptive events that cause whole ecosystems to break down. Shutting down the metabolic machinery works well not only for successfully passing through the digestive system of a fish or a seastar, but also for making it through a short productivity crisis. Recognizing danger and learning how to solve problems are effective means of dealing with the everyday risks and opportunities arising from economic interactions; they may also be crucially important as methods to avert annihilation during crises.

The Record of Extinctions

Few phenomena have captivated scientists in general, and paleontologists in particular, more than extinction. The modern quantitative study of extinction began when David Raup and especially John J. Sepkoski, Jr., initiated a mammoth project of compiling the first and last appearances in the fossil record of every family, and later every genus, of marine animals. Similar work was started by others on land animals and plants. Among the aims of this ambitious effort was to identify times in Earth history when the rate of extinction (the number of taxa dying out per million years) or the magnitude of extinction (the number of taxa with last appearances during a given time interval divided by the total number of taxa living during that interval) reached a peak. Dozens of paleontologists eventually became involved in collecting and interpreting the data.

What emerged was a global seesaw curve in which a measure of extinction was plotted against time. Against a backdrop of generally decreasing rates and magnitudes of extinction with time, five major peaks rose above all the rest. Raup and Sepkoski identified these as the five mass extinctions: the end-Ordovician (440 Ma), Late Devonian (360 Ma), end-Permian (251 Ma), end-Triassic (201 Ma), and end-Cretaceous (66 Ma). At least three earlier mass extinctions have been recognized since as more information has become available on the distribution of fossils and the precise ages of rocks chronicling major geological events. These are an extinction among planktonic organisms in the Late Neoproterozoic (580 Ma), an extinction at the transition between the Neoproterozoic era and the Cambrian period (543–542 Ma), and an Early Cambrian event (514 Ma).

Controversy surrounds all these events. Several—the Cambrian, end-Ordovician, Late Devonian, Late Permian, and end-Triassic events particularly—may consist of two or more closely spaced events, and there is considerable skepticism in some quarters about the reality and extent of the Late Devonian and Late Triassic extinctions. In addition to the eight or so major events, there are many "minor" ones, in which extinction was either less severe or more regional in character. All the extinctions of the last sixty-five my fall into this latter category.[2]

It is worth pausing for a moment to dwell upon the gravity of the phenomenon of extinction. The disappearance piecemeal of a few small plants, narrowly distributed in a single valley or on one mountaintop, or of a rare island bird, is lamentable but not ecologically devastating. It would never be seen in the geological record. Not only is the loss of such a small number undetectable statistically, but many of the species involved would never be preserved in the first place. The fossil record is entirely silent on the comings and goings of such life forms. The extinction events that the fossil record does chronicle—even the so-called minor events—are quite a different matter. Were we as individual humans to witness and live through one of these events, we would experience it as a devastating crisis, whose long-term consequences of instability and disruption would far exceed our lifespans. Times of crisis are like economic depressions writ very large. Extinction is, in short, a phenomenon not to be taken lightly, not to be played down as a minor inconvenience or as the minor unintended consequence of human economic activity. It is a phenomenon worth probing and worth avoiding.

Top-Down Extinction

Exactly when humans first reached the Americas remains uncertain, but one fact about their arrival is uncontestable: human arrivals found a remarkable fauna of large mammals and birds there. Cheetahs, true lions, camels, horses, mastodonts, mammoths, and ground sloths roamed Late Pleistocene North America. South America at this time boasted a fauna of glyptodonts, ground sloths, horses, giant rodents, mastodonts, and members of two native groups of hoofed mammals, the litopterns and notoungulates. By about ten thousand years ago (10 ka), North America had lost thirty-three out of forty-five genera of large mammals (those weighing 44 kg or more), whereas South America lost forty-nine out of approximately fifty-five genera of this size. In North America, there were no birds as large as these mammals, but birds of above-average weight, especially those that ate living or dead large mammals—several eagles, condors, and "Old World" vultures—became extinct.

In Australia, twenty-three of twenty-four genera of large vertebrates—mostly mammals, but also snakes, lizards, turtles, and the flightless bird *Genyornis*—died out about 45 ka, a date indistinguishable from the time of arrival of modern humans. Madagascar witnessed the loss of giant lemurs, elephantbirds, and tortoises upon the arrival of people from the east. Faunas and floras on smaller islands around the world suffered devastating losses at the hands of human invaders. Jared Diamond estimated in 1984 that 115 species and subspecies of mammals, and 171 species and subspecies of birds, had become extinct worldwide since the year 1600, all directly or indirectly through human actions. Losses in the sea have been less dramatic but are nonetheless ecologically significant. They include the extinction of the Atlantic gray whale (*Eschrichtius robustus*), the great auk (*Pinguis impennis*) of the North Atlantic, the sea mink (*Mustela macrodon*) and Labrador duck (*Camptorhynchus labradoricus*) of the eastern United States and Canada, the Caribbean monk seal (*Monachus tropicalis*), Steller's seacow (*Hydrodamalis gigas*) of the North Pacific, and the flightless ducks *Chendytes lawi* and *C. milleri* of the west coast of North America. Freshwater biotas in the southeastern United States lost dozens of clams, snails, fish, and crayfish. Countless other species in all easily accessible environments have been severely depleted to the point of ecological if not actual extinction through such human activities as hunting, damming, urbanization, introduction of foreign species, deforestation, and pollution.[3]

The biotas that suffered most are those in which humans arrived late and with technological sophistication: the Americas, Australia, isolated oceanic islands, and Madagascar. Continents such as Europe, Africa, and Asia—where humans and their ancestors have lived for hundreds of thousands of years or longer—were much less affected by extinctions during the Late Pleistocene. In Africa, fourteen Late Pleistocene genera of large mammals became globally extinct, two survive in Asia, and forty-one survive in Africa. Europe lost three of thirteen large-bodied mammal genera at the end of the Pleistocene, and one more since that time.

There is overwhelming evidence that the near or total elimination of large, powerful consumers by humans has led to dramatic changes in the ecosystems over which these animals once presided. In the tundra of Beringia—a region stretching from eastern Siberia to northwestern Canada, which shared a common biota during the Late Pleistocene when Bering Strait was dry land—large mammals including woolly mammoths, bison, and the extinct horse *Equus lambei* grazed a vegetation of Arctic sage (*Artemisia frigida*), bunchgrasses, sedges, and various smaller herbs. With its relatively high rates of transpiration, this dry-adapted vegetation was quite productive, no doubt aided by manure from the large grazers. When the large mammals disappeared, this vegetation was replaced by a much less productive flora of mosses, lichens, and low shrubs growing on largely water-logged soils. The predominant plant-eaters now are caribou, which feed on mosses and lichens rather than on grasses and Arctic sage. Extinctions in Beringia thus caused a more productive vegetation to be replaced by a less productive one. The dry-adapted species today persist on south-facing sub-Arctic slopes, but the steppe vegetation on which the Late Pleistocene mammals fed in the Arctic has no modern equivalent.[4]

At about the same time, the extinction of several large mammals eliminated the dispersal agents for many tree species in tropical American forests. These large mammals, including elephant-like gomphotheres and giant ground sloths, pass fruits and seeds with thick, hard coats through the digestive system, where the protective coats are softened, so that the seeds can germinate once they are voided as excrement. The large mammals in question were also major herbivores. With their disappearance, the forest likely changed from a relatively open vegetation with a broken canopy to the closed-canopy forest we typically associate with wet tropical lowlands today. Interestingly, the introduction of cattle and horses by the Spaniards in the sixteenth century may have given several tree species a new lease on life, for these animals have to some extent filled the role of their extinct larger Pleistocene counterparts by softening and dispersing seeds in their digestive tracts.[5]

Hunting during the eighteenth century in the North Pacific eliminated the Steller's seacow, a ten-ton behemoth, the top herbivore in kelp forests from California to northern Japan. Daryl Domning's careful historical work reveals that this mammal concentrated on canopy species, so that kelp forests were probably much more open before the seacow became extinct than they are today. At the same time, another marine mammal—the sea otter, *Enhydra lutris*—was being heavily exploited for its fur around the North Pacific rim. Careful field studies carried out by James Estes and his colleagues show that this mammal, which was exterminated from most of its range by the early twentieth century but which has since rebounded, is a major predator of bottom-dwelling invertebrates and, at least locally, of fish. In the presence of thriving sea-otter populations, grazing sea urchins (species of *Strongylocentrotus*) and abalone (species of *Haliotis*) are more or less restricted to deep crevices or to water depths below the typical diving range of the otters. When otters are absent, the grazers occupy open surfaces in shallow waters as well as at depth, reaching population densities sufficient to transform luxuriant kelp forests into unproductive barrens dominated by crustlike coralline red algae. As in the previous example, therefore, elimination of high-energy consumers brought about dramatic changes in a widely distributed, productive ecosystem. Interestingly, the system without the big predators and grazers is generally less productive, pointing to the important role that top consumers play in enhancing their own food supply.[6]

Successive elimination of major consumers has transformed Caribbean coral reefs into seaweed-dominated meadows. As painstakingly reconstructed by Jeremy Jackson on the basis of early accounts of reefs by explorers in the seventeenth and eighteenth centuries, reefs and adjacent lush seagrass meadows in the West Indies supported an abundance of large animals: seagrass-eating green turtles (*Chelonia mydas*) and manatees (*Trichechus manatus*), sponge-eating hawksbill turtles (*Eretmochelys imbricata*), large predatory fish (sharks, groupers, and wrasses), grazing conch (*Strombus gigas*), and the mollusc-eating loggerhead turtle (*Caretta caretta*), among others. Native peoples had exploited the conchs for centuries, as indicated by huge middens of their shells throughout the West Indies; but the large predators and grazers were harvested on a large scale only after 1655, the year when the English captured Jamaica. Grazing sea urchins (*Diadema antillarum*) had always been abundant on reefs, but by 1980 they, along with small fish, were the only significant consumers of seaweeds and corals in large parts of the Caribbean. The final coup came in 1984, when a rapidly spreading disease wiped out about 97 percent of the population of *D. antillarum* from Florida to Panama and Barbados. With few grazers left to keep seaweeds in check, reef surfaces were soon covered with a rapidly growing carpet of algae, which smothered adult corals and effectively prevented settlement of new coral recruits. In the seagrass meadows, turtlegrass (*Thalassia testudinum*) grew much longer blades than it had under the constant grazing attack by turtles and manatees. With longer-lived blades, many organisms including pathogens were able to settle on these plants. Thus, the exploitation by humans of large consumers brought about major changes in the shallow-water marine ecosystems of the Caribbean.[7]

An equally profound change may be transforming salt marshes in the southeastern United States into barren mudflats. In a series of clever experiments run over two years, Brian Silliman and Mark Bertness showed that, when the micrograzing salt-marsh periwinkle *Littoraria irrorata* at natural densities (1,200 individuals per square meter) is protected from predation by blue crabs (*Callinectes sapidus*) and several other predators, damage to the blades of salt-marsh cord grass (*Spartina alterniflora*) caused by the snail's scraping with the radula encourages the growth of fungi, which eventually kill the grass. Collectively, snails in the absence of predatory control therefore destroy the salt-marsh vegetation. During the 1990s, overharvesting of blue crabs by humans in parts of the southeastern United States has thus indirectly brought about the destruction of large tracts of salt marsh, and has replaced a highly productive ecosystem, with a productivity of 3,700 g carbon per square meter per year, by a mudflat in which plant productivity (or at least consumable biomass) is negligible. Silliman and Bertness's work thus shows how regulation by predators favorably affects plant production, and how elimination of top consumers potentially destabilizes both the regulation and the productive machinery of an ecosystem.[8]

No less dramatic than the ecosystem-level changes resulting from overexploitation of high-level consumers are the effects of fragmenting previously continuous habitats. Land use for agriculture, roads, and urban areas has changed large tracts of continuous forest into a patchwork of islandlike refuges surrounded by habitats unsuitable for most forest dwellers. John Terborgh and his colleagues have documented the consequences of such fragmentation on a semideciduous dry-forest ecosystem in the Caroni Valley of Venezuela. The construction of a 4,300 km^2 hydroelectric impoundment there in 1986 flooded much of the valley, leaving a series of island fragments of forest surrounded by water in the newly created Lago Guri. Compared to the large "mainland" forest, the small remnants lack such top-level consumers as predatory harpy eagles, ant-eating armadillos, and various fruit-eating mammals and birds. The remnants support very large populations of herbivorous howler monkeys, iguanas, and leaf-cutter ants. Like herbivores elsewhere that have been released from control by predators—white-tailed deer in the eastern United States, wild pigs in fragmented forest in Malaysia—these excessively abundant consumers are inhibiting recruitment of young forest trees and are preferentially favoring the survival of chemically well-defended plants at the expense of faster-growing, more palatable species. Similar changes took place when Barro Colorado and other smaller islands of rain forest became isolated from the "mainland" of Panama in Gatun Lake, formed in 1914 when a dam was constructed across the Chagres River as part of the construction of the Panama Canal. In time, the abundance of grazers is adjusted downward in the truncated ecosystems as reduced production (or supply) equilibrates with consumption (demand).[9]

These and many similar examples have led many biologists to hold that extinction due to predation or competition is the rule. Charles Darwin, Lee Van Valen, and John Maynard Smith are among the prominent biologists who have taken this type of extinction as a matter of course, as a pattern to be expected. In support of this view, Thomas W. Schoener and colleagues have shown that the presence of predators on small Bahamian islands greatly increases the likelihood of local extinction of lizard populations after the latter had been decimated by a hurricane. Habitat loss for the lizards may have eliminated refuges, and therefore increased the effectiveness of predators searching for and catching these reptiles. Of course, local extinction is not global extinction—the lizards did, after all, survive on islands that were predator-free before the hurricane—and the role of predators in bringing about the local demise of populations would have been nil without the habitat alterations wrought by the storm.[10]

The elimination of top consumers through overexploitation or habitat fragmentation by humans has profoundly changed ecosystems of all kinds, but has not yet led to a cascade of further extinctions. The replacement of steppe grassland by moss tundra through the extinction of mammoths and horses in Alaska at the end of the Pleistocene was not accompanied by known extinctions among other animals or among plants, despite the reduced productivity of the vegetation. The same conclusion holds for the other examples. In all these cases, it is likely that some specialized parasites became extinct along with their hosts, but there is no evidence that extinction cascaded very far despite the dramatic reorganization of interactions among species. The dismantling of the ecosystem occurred piece by piece, beginning at the top, disrupting the system to be sure but not bringing about outright collapse on an ecosystem-wide scale. Refuges of production remained, and controls formerly exercised by large consumers shifted to smaller, less powerful agents.

Skeptics might argue that insufficient time has passed to reveal the full consequences of disruptive effects of top-down extinction at the level of additional species loss. Plant species whose individuals live hundreds of years may look healthy, but they may be unable to reproduce and therefore succumb to extinction. However, the loss of top consumers in temperate and tropical America occurred thousands of years ago, so that any widespread species extinctions stemming from the demise of high-level consumers should have been detected. Chronic overfishing and the extinction or decimation of high-energy marine vertebrates through human agency have intensified and become global more recently, but the organisms at lower trophic levels potentially at risk of extinction have relatively short lifespans and can therefore be expected to have suffered some extinctions if top-down effects precipitated collapse of a magnitude sufficient to cause extinction. In any case, the greater the time lag, the greater is the opportunity for adaptation among surviving species. The loss of top consumers and the fragmentation of habitats result in lower productivity, but they leave the production system essentially intact, if significantly altered in species composition. Because consumers eat living organisms, their presence speeds up the cycling of nutrients to the plants in the ecosystem, whereas their absence means that material transfer is delayed, taking place mainly through decomposition once the producers or their parts have died. Production, however, continues to support a cast of consumers, and it is the rate at which biomass is created and the rate at which it is accumulated that determine the size and complexity of the community of dependents.

Other skeptics might point out that many extinctions could have gone unnoticed because we would not have known of the existence of the unlucky species in the first place. This is certainly true for animal and plant groups that are rarely or poorly preserved as fossils, such as many insects, herbs, mosses, worms, spiders, and even birds. Groups that do have a good fossil record, however, chronicle few losses since Late Pleistocene time, the interval during which the top-down disruptions wrought by humans are most widespread. I know of just two molluscs, the eastern North American whelk *Atractodon stonei* and the Moroccan shore dogwhelk *Spinucella plessisi*, that are represented as fossils in Late Pleistocene deposits and that are extinct today. All other Late Pleistocene species are still extant. Much-needed systematic examination of old museum collections may reveal more examples of species that have become extinct since these species were first collected, but the evidence at hand implies that few species have become extinct in marine and continental ecosystems where humans have tampered with the most powerful consumer species.

The loss of top consumers will bring about extinction only under somewhat unusual circumstances. In order to extinguish a victim species, a consumer species must eliminate or reproductively incapacitate all members of that species, or cause an irreversible decline in its population. The consumer must thus penetrate or destroy all of the victim's habitats and geographic refuges. Complete exploitation of this kind is unlikely when the consumer and victim species coexist and when the victim is able to respond adaptively to the consumer. Consumers seldom harvest all members of an entire population or destroy all the habitats occupied by it. Instead, they switch to more abundant victims of other species when a heavily exploited prey species becomes rare. In short, a cascade of competitively generated extinction should be expected only if species have become mutually and exclusively dependent on one another, a circumstance that appears to be rare in nature even among such groups as host-specific herbivorous and pollinating insects where such specialized relationships are most to be expected. Research by Laurel Fox and others has shown that very few insects are restricted to a single host species throughout their ranges, and very few predators are specialized to a diet of a single prey species. Interactions in nature are flexible and varied, and it is this flexibility and variation that confers a certain independence on most species even in communities where the outcomes of interactions carry high stakes. Extinction due to consumers and other competitors should, in my view, be the exception rather than the rule.[11]

A necessary condition for a consumer to cause the extinction of another species may be an external event that enables a high-intensity consumer to spread to a biota in which comparably performing consumers had been absent. Such an event might have a physical cause, or it might result from technological innovation by a consumer that permits the consumer to cross a previously impermeable barrier. For land biotas, an example of a physical event leading to the fusion of previously separate biotas might be a drop in sea level, which would transform a stretch of shallow ocean into a lowland corridor between the mainland and a small island. Whether this kind of event is important in bringing about extinction is very doubtful. Many smaller land-bridge islands may be connected and disconnected from adjacent mainlands at a frequency sufficiently high that species adapted to the island have little time to evolve. Few land-bridge islands have been separated from nearby land masses for more than ten thousand years, and although evolution can proceed rapidly, contacts between mainland and island populations from time to time may prevent the island population from adapting to the new circumstance.[12]

Humans exemplify the second mechanism by which consumers with superior technological skills can cross a previously insuperable barrier. People already possessed some of the key advances—fire, long-distance weapons, and domesticated dogs—that made rapid exploitation of vulnerable species possible and that led to far-reaching environmental modification when the human species invaded Australia, the Americas, and oceanic islands. Suffering most at the hands of the newcomers were the very slowly reproducing large mammals, birds, reptiles, and trees, which were unable to recoup their losses even if people did not kill every individual outright. The biotas of islands were especially prone to extinction under the new regime because the evolution of many island natives had proceeded in the absence of high-energy consumers and competitors, and because they produced very few offspring per year.

Apart from humans, few species are good candidates as agents of recent extinction. The most likely culprits are energy-intensive animals, especially warm-blooded mammals, characterized by broad environmental tolerances, high per capita food requirements, and dense populations. Diseases such as rinderpest carried by livestock, distemper transmitted by dogs and cats, and various blights carried on introduced trees can also decimate populations and sometimes cause extinction. Of the twenty-two species implicated in recent extinctions of species on land and in freshwater, there are eleven mammals (including cats, rats, pigs, goats, foxes, mongooses, and weasels among others), one bird, one snake (the brown tree snake *Boiga irregularis*), four fish, two snails, and three microbes. Fourteen of these villains cause extinction largely through consumption, and three do so through a combination of consumption and habitat modification. Top-down extinctions on continents have been confined to microbial pathogens; those caused by nonhuman animal consumers have been restricted to islands and lakes. There are as yet no examples of marine extinctions attributable to biological agents other than humans.[13]

If invasion is a necessary condition for top-down extinction, plausible cases of this kind of extinction in the geological past should be identifiable by the sudden appearance of species whose evolutionary roots lie outside the region in question. Though many such sudden appearances are known, very few cases qualify as probable examples of extinctions resulting from invasion. One such case is the disappearance of several groups of native hoofed mammals (litopterns and notoungulates) and of predatory marsupials in South America following the arrival of true carnivoran mammals—dogs, cats, bears, and weasels—from North America. Invasion from the north began in earnest around 3 Ma, during the middle Pliocene epoch, when Central America was transformed from a series of islands into a continuous land bridge. Whether the carnivoran newcomers in fact were responsible for the demise of native predators and herbivores can perhaps never be known, but their higher metabolic rates compared to those of the South American natives provide circumstantial evidence in favor of this hypothesis.[14]

The interval from the Late Oligocene to the Middle Miocene, about 27 to 17 Ma, witnessed several replacement events among high-level carnivoran mammals in both North America and Eurasia. In the early phase of this replacement, nimravids (saber-tooth-like predators) and hyaenodontids gave way by earliest Miocene time (23 Ma) to early dogs, bears, and bear dogs in North America, and to bears, bear dogs, and hyenas in Europe. The doglike carnivores were, in turn, partly displaced or replaced by felids (cats), which arrived in North America from the Old World in the Middle Miocene (17.5 Ma) and which increased in size through the remainder of the Miocene epoch, to about 6 Ma. Cats also increased in size in Europe, where a similar displacement of doglike predators took place. Saber-tooth cats appeared in Europe about 18 Ma, and by immigration in North America about 11 Ma, adding to the displacement of the earlier carnivores. As these cats were diversifying and increasing in size, hyaenids in Europe and borophagine dogs in North America also proliferated and became larger. Immigration of the true dog genus *Canis* from North America to Europe in the Late Miocene (7 to 8 Ma) may have contributed to the decline of European hyenas and borophagine canids. They also diversified in the Late Miocene in North America, where *Canis* also eventually replaced borophagines. Whether these replacements reflect competitively superior predators driving less sophisticated predators to extinction remains unclear, but extinction due to competition is at least a possibility.[15]

The problem with these ancient cases is that the fossil record is not detailed enough to allow us to distinguish between two contrasting, mutually incompatible scenarios: (1) invasion of superior foreign animals followed by extinction of native forms; and (2) extinction of native forms followed by invasion of foreign species. In the second scenario, one would expect the extinctions to coincide with (and in part to be caused by) a change in climate. In that scenario, invasion requires the prior removal of well-adapted incumbents, and does not imply that the successors are superior to their antecedents. In the first scenario, invasion requires a geographic opportunity, such as the geological removal of a barrier, but not necessarily a change in climate; but we would expect the invaders to be competitively superior to the native species whose extinction we are trying to explain. If, however, the invaders cause extinction among native species by the spread of new diseases instead of by their superior performance, as has often been suggested, even a comparison of the animals' performance levels will not provide a definitive distinction between the two scenarios. All we can do at present in cases of ancient extinction is to point to circumstantial evidence. That evidence seems to indicate a minor, and perhaps geologically recent, role for top-down consumer-driven extinction.

Bottom-Up Extinction

Preoccupation with our own increasingly important role in extinction has directed the attention of biologists and conservationists away from the causes that have been responsible for most of the extinction events in the geological past. These extinctions, I shall argue, emanate from a general collapse of ecosystems brought on by major interruptions in primary production—in photosynthesis—and associated disruptions in the supply of raw materials. Loss of primary production results in a cascade of extinctions by one or both of two mechanisms: (1) elimination of the food base of consumers; and (2) elimination of habitats that are formed and controlled by organisms.

Our familiarity with forests, grassland, and coral reefs can make us forget that these habitats, with their intricate three-dimensional complexity, are wholly the work of organisms. If their architects should die, through either overexploitation or starvation, there would be collateral extinction of other species on all sides. Plants, as well as suspension-feeders such as mussels, oysters, tubeworms, and fossil crinoids, create spaces in which animals can hide and in which consumers' power of search and capture are physically constrained. With fewer places to hide and a less encumbered life-generated topography, victims come under increased exploitation by consumers just as the production machinery of the ecosystem is being dismantled. Many species, in other words, are being squeezed both from above by consumers and from below by diminishing resources. Selection from consumers is intensifying while the ability of victim species to respond adaptively is declining.

Starvation as a cause of extinction might seem self-evident, but it is worth examining a little more closely, because not all species are equally vulnerable to it. Species particularly threatened by bottom-up collapse should have one or both of the following general characteristics: (1) high metabolic demand throughout life, with few mechanisms to conserve or store resources; and (2) direct dependence on primary producers or on the environments that those producers create. On a more hopeful note, resistance to supply-side disturbance is conferred by (1) the ability to shut down metabolic machinery during one or more life stages, or to store resources for seeing individuals through periods of scarcity and adversity; and (2) dependence on food sources or environments not directly connected with photosynthesis.

When populations of single-celled phytoplankton crash, either because of prolonged darkness or because crucial nutrients are not being delivered at their usual high rates, suspension-feeding animals directly dependent on this food source suffer enhanced mortality, reduced growth rates, and lower reproductive success. Large animals with high rates of metabolism would be especially vulnerable because of their high maintenance costs, which cannot be sustained when food supplies drop below a certain level for a sufficiently long time. Many marine animals have planktonically feeding larval stages, even if their adults feed as herbivores or predators on larger organisms. These larval stages do not have ways to store food, and are therefore exceptionally vulnerable to fluctuations in the supply of phytoplankton. Even if the adults can survive a protracted interval of famine, the larvae may be unable to do so. New recruits therefore cannot be added to the population of adults, and if this period without larval settlement exceeds the lifespan of adults, the population and perhaps the entire species is doomed.[16]

The importance of suspended animation in resisting extinction is well illustrated by the pattern of extinction among phytoplankters at the end of the Cretaceous. There was far more extinction among phytoplankton genera without resting stages—coccolithophores, 73 percent; radiolarians, 85 percent; and foraminifers, 92 percent—than among planktonic diatoms (23 percent), which go through a bottom-dwelling resting phase.[17]

On the consumer side, bottom-up extinction seems to hit the most powerful entities hardest. All large herbivorous and predatory vertebrates died out during the crises ending the Permian, Triassic, and Cretaceous periods, leaving smaller relatives behind to mount a recovery. Fast-growing, competitively superior marine bivalves, brachiopods, and corals with the ability to photosynthesize suffered far greater losses during these crises than did relatives without symbionts.[18]

On a very small scale, the effects of a bottom-up ecosystem collapse become dramatically evident during El Niño years, when warm unproductive surface waters replace cold, nutrient-rich upwelling waters on the Pacific coasts of temperate and tropical America. In normal years, the rich waters of the eastern Pacific support a huge ecosystem of phytoplankton, small fish (anchovies and sardines, for example), seabirds, and marine mammals. When upwelling ceases and the supply of nutrients from the bottom is interrupted, the whole edifice collapses, resulting in mass mortality and in widespread reproductive failure. During the last 2 my, the frequent reductions in productivity associated with these El Niño events have evidently not led to extinction of marine species in the eastern Pacific, perhaps because all vulnerable species had already been winnowed from the system by earlier fluctuations in the nutrient supply. They do, however, illustrate how ecosystem collapse can occur when the nutrient supply is shut off, and how a cascade of extinction could be put in motion if the disruption occurred over a large enough area for a sufficiently long time.[19]

Past changes in productivity are difficult to gauge directly because the fossil record of primary producers is both incomplete and ambiguous. For the plankton, only those large and symbiont-bearing animal-like protists that build mineralized skeletons have left us a detailed record. With relatively rare exceptions, these mineralized planktonic photosynthesizers—dinoflagellates, coccolithophorids, diatoms, silicoflagellates, radiolarians, and foraminifers—are known only from the Mesozoic and Cenozoic eras. Acritarchs—organic-walled, cystlike structures thought to be the resting stages of planktonic dinoflagellates or similar algae—appeared much earlier (1.5 Ga, during the Mesoproterozoic era). Planktonic Cyanobacteria, which account for much of the photosynthesis in the open ocean, have left tell-tale signatures since at least 2.7 Ga, but there is no interpretable record of their physical remains. Seaweeds, too, have left a highly incomplete record. Despite the enormous abundance and large size of kelps (Laminariales), only a single fossil, the Middle Miocene *Julescranea*, is known that could represent this important group of temperate primary producers.

Although land plants above microbial grade may have existed as early as the Late Cambrian, 510 Ma, a more or less continuous record of vegetative remains does not begin until the Late Silurian, about 420 Ma. Extracting estimates of productivity from these remains, however, is essentially impossible. The best we can do is to argue that productivity probably declined when many species of photosynthesizers became extinct. At the very least, the loss of enough individuals to bring about the extinction of common plants, corals, and especially phytoplankton must have represented a sharp reduction in productivity.

Many geologists have turned to the fossil record of chemical signatures of life in order to infer changes in the rate at which organic carbon is made from carbon dioxide during photosynthesis. Carbon comes in two forms, or isotopes: the heavier ^{13}C and the lighter, more abundant ^{12}C. Organic matter produced in photosynthesis with the help of the enzyme Rubisco contains carbon in which the ratio of ^{13}C to ^{12}C is 25 parts per thousand lower than that in the source carbon dioxide. In other words, the lighter ^{12}C isotope is preferentially incorporated into organic matter relative to the heavier ^{13}C form. If the source of the carbon is the bicarbonate ion (HCO_3^-), as in some phytoplankton, there is little change in isotopic composition from the source to the organic form. Respiration similarly does not alter the isotopic composition of carbon when organic carbon is transformed back to carbon dioxide. Geologists thus interpret an enrichment in ^{12}C (that is, a decline in ^{13}C to ^{12}C ratio) as indicating a reduction in the rate of photosynthesis by Rubisco.

The problem with this approach is that processes other than photosynthesis also affect the carbon-isotopic ratio. One of the most important of these is the formation of methane (CH_4) by Archaea through the interaction of carbon dioxide and hydrogen. Compared to the source carbon dioxide, the methane produced is depleted in ^{13}C by 60 to 80 parts per thousand. An enrichment in ^{12}C relative to ^{13}C in sediments could thus signify a massive release of methane as well as a decrease in photosynthesis, or both. Now, a mass release of methane—a poisonous gas, which in today's atmosphere would quickly combine with oxygen to produce carbon dioxide—might itself bring about a collapse in photosynthesis, especially if it were accompanied by the release of such other toxic substances as hydrogen sulfide. The point is that the processes resulting in a decrease in the ^{13}C to ^{12}C ratio—photosynthesis and burial of organic carbon, and the sudden release of methane—are separate but geologically difficult to distinguish. An increase in the ratio of ^{13}C to ^{12}C can be brought about by vaporizing limestone, a possibility when a celestial body strikes a limestone-rich region. Changes in the isotopic composition of carbon, in other words, allow us to infer large and important changes in how carbon is transferred among parts of the biosphere and its supporting environments, but they provide few clues about changes in productivity.[20]

As a biologist, I would supplement the isotopic approach with evidence from the animals that directly depend on plant production. In the sea, about 95 percent of photosynthesis is carried out in the plankton by single-celled organisms, which form the food base for suspension-feeding animals. A higher rate of supply of phytoplankton makes possible larger populations of faster-growing suspension-feeders, which individually grow to a larger body size. The sustainable body size of an animal depends on the costs of maintaining the biochemical machinery and the food available to meet those costs. Large body size cannot be maintained by a suspension-feeder under a regime of low phytoplankton productivity unless the animal in question either has the means to reduce costs—by lowering or shutting down its metabolism, for example—or has an alternative food supply that does not depend directly on photosynthesis. Accordingly, a decline in planktonic productivity should be accompanied by a decrease in the growth rates and body sizes of large suspension-feeding animals such as bivalves, brachiopods, barnacles, and certain gastropods. Sponges, which are capable of feeding on bacteria-sized plankton, might be much less affected than the suspension-feeders just listed, because the latter typically take larger phytoplankton and are unable to capture most of the microbial fraction. On land, and in those marine environments with a vegetation of seagrasses and attached seaweeds, I would look for similar responses among herbivores. All else being equal, large-bodied herbivores should live in places where plant productivity is high enough to yield a reliable surplus available to consumers.

The geological record of extinction reveals that all the mass extinctions, as well as many of the lesser events, are associated with significant disruptions in primary productivity and with the disappearance of high-energy photosynthesizers. Marine phytoplankters seem to be particularly vulnerable, having suffered huge losses during each of the six great crises. This includes not only those organisms classically thought of as the planktonic algae—dinoflagellates, coccolithophores, and diatoms, for example—but also animal-like protists such as many radiolarians and foraminifers, which contain symbiotic algae. The phytoplankton crashes coincide with sudden changes in the carbon cycle as revealed by the ratio of the two isotopes of carbon in fossil organic matter.[21]

Marine animals with algal partners have also been prone to extinction during the great crises. None of the symbiont-bearing brachiopods or bivalves survived the end-Permian extinction, nor did any alga-bearing rudist bivalve at the end of the Cretaceous. Brian Rosen calculated that about 70 percent of coral genera with algal symbionts disappeared in the end-Cretaceous extinction, as compared to only 30 percent of genera without symbionts. Large bottom-dwelling fusulinid foraminifers may have evolved symbiotic relationships with photosynthesizers during the Late Carboniferous, but they did not survive beyond the end of the Permian.[22]

There was a time when land plants were thought to have been relatively resistant to extinction during the great crises, but new evidence has shown very clearly that they suffered catastrophically, along with land vertebrates and marine animals, during all the mass extinction events from the Late Devonian onward. At the boundary between the Middle and Late Pennsylvanian (Westphalian to Stephanian epochs in the Late Carboniferous), for example, canopy-tree lycopods became extinct in Europe and America, to be replaced by *Psaronius* treeferns, whose ancestors lived in the understory. The end-Permian crisis witnessed the disappearance of the cool-adapted southern-hemisphere members of the so-called glossopterid flora. In Greenland, at least two episodes of extinction among plants, both particularly affecting the competitively dominant gymnosperms, took place. The end-Triassic extinctions eliminated something like 90 percent of leaf species in the North Atlantic region. Plant extinctions ravaged the terrestrial world again at the end of the Cretaceous, when the dominants of the contemporary forests, especially gymnosperms, were annihilated over wide swaths of the northern hemisphere. According to Peter Wilf and his colleagues, about 30 percent of pollen species became extinct at the end of the Cretaceous in North Dakota, one of the few places where a continuous terrestrial record of plant life is available. As in the earlier extinctions of the end-Permian and end-Triassic, ferns temporarily dominated vegetations on a large scale irnmediately following the devastation.[23]

All the extinction events of the past 65 my are considered minor in the sense that they were either regional in extent or less disruptive than the mass calamities, but they were still dramatic in their effects, and like their more severe counterparts appear always to be associated with a decline in primary productivity. Regional collapse in the marine ecosystem, for example, is strongly indicated for large parts of the temperate and tropical Atlantic and for the northeastern Pacific during several episodes of extinction beginning about 3.6 Ma (Middle Pliocene) and continuing into the Early Pleistocene, 0.9 Ma. Throughout the tropical and temperate North Atlantic, large-bodied bivalve and barnacle clades became extinct, and those large-bodied clades that did survive underwent a decrease in body size. In the tropical western Atlantic, seagrass diversity decreased, with the probable result that many animals dependent on seagrasses for food, especially seacows, also became extinct. Suspension-feeding bivalves comprised 92 percent of bivalve molluscan specimens in Miocene samples from the Caribbean coast of Panama and Costa Rica, but only 72 percent of those after the Late Pliocene, implying a drop in abundance of animals directly dependent on phytoplankton for food. There is a similar drop in suspension-feeding snails. These declines are not well reflected in the species lists from successive assemblages, and are for all intents and purposes invisible in the global record. This state of affairs pointedly reminds us that economically important changes in abundance, size, and productivity can occur without leaving a signature that would be revealed by analyzing only lists of species names. We remain profoundly and frustratingly ignorant about how the extinctions and the reductions in marine productivity took place during the past 3.6 my.[24]

Extinction in the Midst of Plenty?

The idea that ecological collapse and extinction result from large-scale interference with primary production conflicts with the widely held view that the great and small extinctions chronicled in the fossil record are caused by an oversupply of nutrients, which causes instability and subsequent collapse. This latter view is founded on three claims: (1) reefs built by extinction-prone photosynthesizing corals thrive in low-nutrient waters and are destroyed by the input of plentiful dissolved nutrients, which enable fleshy algae and suspension-feeding animals to outcompete the corals for space; (2) organic-rich, oxygen-deficient sediments often occur in or near the sedimentary layers that record mass extinction events, and are interpreted as indicating a brief, large rise in primary productivity in the overlying plankton; and (3) nutrient enrichment in living ecosystems often increases the number of cohabiting species, because it enables a few species rapidly to exclude others through competition for food or space. These three claims are either not quite correct or are more consistent with a decrease in primary productivity as the primary cause for extinctions associated with a bottom-up ecosystem collapse.

Coral reefs are most luxuriantly developed in clear blue waters, in which concentrations of nutrients and phytoplankton are low. In such classically beautiful settings, corals are subject to intense grazing by fish and various invertebrates, which also keep seaweeds from settling and growing on the colonies. The nutrient-poor waters are unfavorable to the settlement and growth of such seaweeds and to suspension-feeding animals without algal symbionts, and thus free the corals from severe competition by these organisms. The weak link in the maintenance of corals and the reefs they build are the symbionts (zooxanthellae) that enable corals and several other reef-forming animals to function effectively as oxygen-producing plants. The symbionts are susceptible to interruptions in food supply and especially sunlight. Excessive heat or cold, prolonged darkness, smothering by sediments, and stress from immersion in fresh water arriving as rain or runoff from the land can cause symbionts to die or to be expelled by their hosts. Without symbionts, colonies lose some of their competitive edge, and may die back. Fleshy algae can thus settle on dead coral and quickly monopolize surfaces. Their smothering of coral colonies would therefore be a consequence rather than a cause of the demise of reef-builders. In reefs today, which worldwide have been heavily influenced by fishing and other forms of exploitation by people, the reduced presence of herbivores gives fast-growing algae an additional advantage when corals lose their symbionts and die back. As I noted earlier, this chain of events overcame Caribbean reefs, which had already been overfished for decades or centuries, when the last abundant herbivore (the sea urchin *Diadema antillarum*) was largely eliminated by a fast-acting disease in the 1980s.[25]

The idea that reefs are killed by excessive nutrients is deeply rooted among reef biologists and paleontologists. In the geological record, for example, reefs are often capped by large quantities of land-derived sediment, an observation that has prompted many scientists to conclude that sediment-laden food-rich runoff from the land was responsible for the demise of reefs.[26]

Proximity to sources of prolific nutrients, however, does not by itself doom corals, nor does it necessarily prevent reefs from being built. Reefs along the nutrient-rich coasts of the continents and large islands around Indonesia, the Philippines, New Guinea, and northern Australia are vastly richer in species and more prolific in individuals of all categories of species than the reefs in the planktonically less productive waters around the atolls of eastern Micronesia and Polynesia in the western and central Pacific. There are, for example, over three hundred species of coral on reefs in the Indonesian region, as compared to only thirty in the Hawaiian Islands. Reefs in the upwelled Bay of Panama have fewer coral species (about ten) than do those in the less seasonal and somewhat less planktonically productive Gulf of Chiriqui on the Pacific coast of Panama (about twenty species), but they nonetheless hold their own, evidently being maintained by intense grazing, especially by scraping and gnawing fishes. Consumers destroy an estimated one-third of the coral production each year at Señora Island, one of the Pearl Islands, in the Bay of Panama. An even higher proportion may be removed in the reefs in the Gulf of Chiriqui, where several additional consumers (including the seastar *Acanthaster planci*) are present. In Panama and elsewhere, many symbiont-bearing corals live in turbid waters and do not build reefs. For example, I have seen heads of the coral *Porites panamensis* in deep tidepools on the heavily polluted shore at Paitilla, in Panama City.[27]

Nearshore ecosystems, in fact, play an important role in reef economics because they subsidize the coral-rich outer parts of reefs nutritionally. These ecosystems include lagoons, mangrove swamps, and seagrass meadows. Some of the nutrients emanating from these inshore ecosystems are transported offshore in tidal currents, but others arrive in the form of fish, which typically hatch inshore and occupy the outer parts of reefs as adults. Highly productive seagrass meadows and mangrove swamps are most luxuriantly developed on the shores of continents and topographically complex large islands. The destruction of nearshore primary producers eliminates the inshore subsidy and in addition causes sediments from the land, which would otherwise have been effectively trapped and exploited near shore, to spread over the reef at the expense of corals and other sediment-sensitive reef-builders. Again, the destruction of reefs by nutrients and sediments is a consequence of inshore destruction of the productive machinery, not the consequence of an increase in reef primary productivity.[28]

Anoxia—the absence of oxygen in water or in sediments—has long figured prominently as a potential cause of mass extinction. Anthony Hallam, Paul Wignall, and others have pointed out that many horizons in the geological record marking great crises are chronicled by sedimentary rocks rich in organic matter and poor in oxygen. Sediments of this type are deposited today either in stagnant basins, such as those of the Black Sea, or beneath zones of upwelling along the western coasts of North and South America and southwestern Africa. The association between these organic-rich sediments and planktonically productive waters of upwelling zones has led most students of the mass extinctions to conclude that a catastrophic oversupply of nutrients was responsible for the anoxia, and thus for the extinction of many oxygen-dependent species—a noxious death indeed.[29]

As a direct cause of extinction, the elimination of environmental free oxygen is likely to be regionally but not globally effective. Kazutaka Amano and his colleagues, for example, have shown that anoxia in the bottom waters of the Japan Sea, an enclosed basin cut off from the rest of the ocean when sea levels are lower than today's, caused the extinction of some deep-water molluscs during the glacial epochs of the Pleistocene, when sea level was perhaps as much as 100 m lower than today. In the Paleozoic era, a similar fate overcame species that occupied the shallow seas that spread over large parts of North America and Europe. As sea level fell, these seas became isolated and turned anoxic, dooming to extinction all those species that occurred only in these so-called epicontinental seas. In these cases, the anoxia arose through isolation, and may not have been related to a rise in surface-water productivity.[30]

A more global mechanism for creating anoxic conditions in the deep ocean is sudden dramatic warming. As temperature rises, the concentration of free oxygen in water falls and the metabolic rate of oxygen-using organisms increases. Several of the mass extinctions—notably those at the end of the Permian, Triassic, and Cretaceous periods—are marked by sudden global warming, as indicated by the spread of warm-adapted plants, vertebrates, and marine animals toward the poles, and the disappearance of cold-adapted species. The most likely cause for the sudden, brief warming episodes is the mass release of methane, a potent greenhouse gas, into the atmosphere from methane hydrate deposits beneath the seafloor and near the poles. These releases, in turn, may have been triggered by extensive volcanism. Each of the three mass extinctions mentioned is very close in timing to the beginning of intensive flood-basalt volcanism, although there is still debate about the precise order of events. I return to this issue later in this chapter.

Another weakness in the argument linking anoxia with nutrient enrichment is that other, more compelling explanations for anoxia are available that at the same time cast light on other facts about extinction. Carbon-rich, oxygen-deficient sediments may accumulate on the seafloor not because of a sudden increase in productivity in the waters above, but instead because of mass mortality in the plankton. When plankton dies en masse, carcasses of dead organisms rain down on the seafloor at a rate high enough to overwhelm the oxygen-holding capacity of bottom waters and sediments as oxygen-using microbes consume the dead creatures. Sediments rich in organic matter do, of course, accumulate beneath sites of prodigious plankton production, but their deposition may take place mainly during episodes of mass death. After all, sediments contain dead, not living, members of the plankton. Moreover, evidence from organic-rich sediments in the Mediterranean Sea and elsewhere has revealed that periodic accumulations of mats of planktonic diatoms form beneath relatively unproductive waters. In other words, carbon-rich sediments are not necessarily indicators of high planktonic productivity, but they do offer convincing evidence of catastrophic mortality in the plankton. More importantly, anoxia would be the consequence rather than the primary cause of extinction, though of course it might be one of the mechanisms by which extinction is propelled through the ecosystem.[31]

An increase in productivity in living ecosystems typically leads to a short-term decrease in the number of species. This decrease occurs because a few competitively dominant, fast-growing species monopolize most of the resources, and locally exclude other species. Scaled up to the regional or global economy, the addition of nutrients could thus drive competitively inferior species to extinction if those species have nowhere to hide from their more vigorous rivals. In the real world—a heterogeneous place full of refuges—this scenario seems implausible to me. Competitors rarely if ever overlap completely, and rarely exclude each other to the point of extinction of one of the parties. Moreover, at large spatial scales, biotas and regions in which conditions of high nutrient input and high productivity are widespread are richer in species than climatically comparable regions in which unproductive environments predominate. Every major group of plants and animals is richer in species in the parts of the western Pacific and Indian Oceans adjacent to continents and large islands, where productivity is high thanks to abundant runoff, than in the nutritionally starved, planktonically less productive islands of Micronesia and Polynesia. In any case, I know of no examples of competition leading to extinction as a consequence of nutrient enrichment.[32]

The short-term reductions in the number of co-occurring species observed when nutrients are added to ecosystems all occur in human-dominated ecosystems, which are atypical in that they are plagued by the overexploitation of consumer species. The activities of such consumers—predators, herbivores, and suspension-feeders—at normal abundances would rapidly convert extra nutrients into organisms, and thus damp the bottom-up fluctuations that the injection of raw materials would bring about. Moreover, eutrophication—the addition of abundant nutrients—in lakes and coastal marine environments typically takes place at the same time as, and often as a consequence of, the destruction of wetlands, forests, and other productive ecosystems that store large quantities of potential nutrients in soil or in living things. In other words, observed disruptions thought to be the result of eutrophication alone are in fact perhaps due more to prior overexploitation of consumers or to the destruction of nearby ecosystems.

Support for this line of argument comes from recent experiments by Boris Worm and his colleagues. Working at Maasholm, a productive site on the Baltic Sea coast of Germany, and Bald Rock off Nova Scotia, Worm and coworkers tracked the number of species in plots in which they variously added or withdrew consumers and nutrients. Among other results, they found that the presence of abundant consumers substantially damped the deleterious influence of nutrient enrichment on diversity. As in the human economy, high demand prevents overproduction and turns the destructive addition of raw materials into an overall economic stimulus. In every economy, supply and demand are linked in complex feedbacks. As we eliminate enemies from ecosystems, we take away a negative-feedback regulatory mechanism, and magnify any destabilizing positive feedback that may result from oversupply—a *foe pas*, one might say.[33]

In short, evidence from living and fossil ecosystems, supported by arguments about how species respond to increases and decreases in nutrient supplies, strongly indicates that extinction and ecosystem collapse are brought about by a fall in primary productivity, and not by a rise as many paleontologists have supposed. An increase in the availability of resources, in fact, creates opportunities for proliferation and adaptation, especially if it occurs in the presence of a strong consumer demand (see chapter 8).

The Causes of Collapse

We are gaining some theoretical understanding of how the elimination of place-holding organisms such as trees and reef-forming animals causes disruptions for many species dependent on topographic complexity, and of how a decline in primary productivity ripples through the contingent of consumers in ecosystems, but the underlying physical causes of these collapses are far from well understood. It takes a substantial jolt to place populations of phytoplankton comprising trillions of individuals in jeopardy, or to eliminate forests and seagrass meadows over large areas. All the insults that humans have hurled at coastal marine environments have thus far resulted in very few extinctions, although our cumulative effect could easily and very quickly change this picture. Physical mechanisms powerful enough to bring about significant extinction must operate on a very large spatial scale, cause most individuals of many species to die or fail to reproduce, and for other populations to keep the rate of mortality higher than the rate of recruitment.

It has proved difficult to establish what is cause and what is effect during episodes of ecosystem collapse in the geological past. Part of the problem is that the time resolution in the record of sediments is too coarse to allow a detailed chronology of rapid events to be reconstructed. Sediments accumulate in an orderly fashion, one layer on top of another, but the rate at which they accumulate varies, and accumulation is often interrupted by periods of erosion. Moreover, once the layers are deposited, burrowing animals and the underground parts of plants disturb the original structure and thus mix up the layers, so that the original faithful chronology of sedimentary events is erased. Collapse of the phytoplankton-based marine ecosystems in the upwelling waters off South America occurs over a matter of weeks to months during El Niño years, far too fast to be followed in detail in fossil sediments. Sediments that have been little disturbed reveal that the changes precipitating the crises at the end of the Permian, Triassic, Cretaceous, and Paleocene were very rapid, from a maximum of 50 ky to as little as 8 ky according to some calculations. The effects of the crises are typically of much longer duration, a point to which I shall return later in this chapter. The point is that cause and effect are difficult to disentangle in the geological record of extinction. The trigger may be an essentially instantaneous event.[34]

Economic feedbacks, moreover, can either magnify or suppress physical triggers. A lake may be able to absorb a certain extra quantity of nutrients with little obvious consequence, but once a threshold has been passed, change can be sudden and dramatic. An ecosystem of attached submerged plants growing in clear water is transformed into one dominated by phytoplankton, floating plants, and fish, all living in turbid waters. If we observed such a switch, we would be tempted to ascribe it to a sudden increase in nutrient input, when in reality the increase could have been taking place gradually and for a long time without obvious effect until some threshold was exceeded. In the Sahel region of Africa, perennial vegetation established during a long run of wet years can persist for a time even when rainfall is below normal for several years, but eventually the vegetation switches to a sparser cover that holds and recycles water less effectively. The desert vegetation, in turn, lasts well into a prolonged series of rainy years because the available plants simply cannot exploit the greater supply of water, much of which runs off. These delays in response, and the sudden nature of switches among stable states in ecosystems, are widespread phenomena reflecting feedbacks among plants, consumers, and climate. The trigger that proverbially breaks the camel's back might itself be minor, even if the background of larger-scale, slower change is plain to see. The same event against a different background might be utterly inconsequential.

This phenomenon of threshold triggers is also well known in human history. The American Civil War might have begun with the bombardment of Fort Sumter, but the underlying tensions related to slavery and states' rights had been evident since the Constitutional Convention in 1787. Martin Luther's Wittenberg proclamation in 1517 may have precipitated the Protestant Reformation, but the excesses of the Roman Catholic Church to which Luther was reacting had festered for years. John Wycliffe, who already railed against the Church's wealth, the indulgences it extended for payments, and its idolatry in the fourteenth century, was succeeded in the fifteenth century by the Czech heretic Jan Hus. The fall of Rome in 476 was neither sudden nor unexpected; the city had been sacked in 411, and barbarian armies had threatened the Roman empire for centuries. The war of 1914–1918, during which airplanes and chemical weapons were employed for the first time on a large scale, was perhaps the direct outgrowth of rapid weapons development by Germany and Britain, powered by unprecedented expansion of colonial empires—Britain, Germany, France, the United States, Russia, and Japan—since 1870. Turning points in history can have seemingly trivial immediate causes—the assassination of Archduke Franz Ferdinand in Sarajevo precipitated World War I, for example—but they occur against a background of long-term circumstances whose nature makes the sudden dramatic shift more likely.[35]

If the correspondence between triggering mechanisms and economic response is imprecise on the short time scales on which we observe ecosystems today, surely the same lack of precise correspondence applies to extinctions of the past. A decline in productivity in the steppe tundra system—in this case brought about by the disappearance of large grazing mammals—evidently did not precipitate collateral extinction as the Alaskan ecosystem switched to moss tundra, but a further decrease might well have such an effect. A major challenge to biologists and paleontologists is to establish when, how, and under what circumstances a collapse in production would initiate a cascade of extinction. We need to know by how much photosynthesis must decrease, whether a certain production threshold must be passed before effects become evident, how long the interruption must last, and how much lag time there is between the physical change that triggers the disruption and the biological feedbacks that propel species to extinction. The answers to these questions are likely to be time- and place-specific, and thus to depend on the particular circumstance in which an ecosystem finds itself. Unfortunately, these answers remain elusive.

Organisms are apt to provide us with the most reliable clues about the causes of extinction. It is not those that die to which we must turn for such clues—after all, it is often difficult to infer what made these extinct creatures tick—but those that survived. Furthermore, we need to characterize the environments or regions in which survivors persisted.

Asphyxiation from carbon dioxide is the mechanism of extinction that Andrew Knoll and colleagues favor for the end-Permian disaster. At high concentrations, carbon dioxide poisons many animals, and it is only those animals with a tolerance for high levels of this gas that survived the crisis. Vertebrates, Knoll and coworkers argue, generally resist carbon dioxide better than such organisms as brachiopods and echinoderms. How or why plants would be affected is unclear. One scenario is that high concentrations of carbon dioxide cause unacceptable warming, which would have eliminated many plants with large heat-absorbing surfaces.[36]

The alternative interpretation I favor is that an ability to shut down metabolism was critical to survival during the end-Permian and other great crises. Despite their low metabolic rates, brachiopods and echinoderms appear to be very poor at entering a state of suspended animation. These groups suffered disproportionately at the end of the Permian. Some normally active animals, notably some land vertebrates, can hibernate or aestivate for months of inclement conditions by greatly reducing their metabolic rate, an ability accompanied by high tolerance of carbon dioxide in the blood. Tolerance of carbon dioxide is also connected with the short-term ability to engage in anaerobic metabolism, which can be crucial during unsustainable and intense muscular activity during escape or in bouts of aggression. I interpret such tolerance not as the result of having survived a sharp rise in carbon dioxide during the end-Permian and other crises, but as a frequent manifestation of the ability to suspend metabolic activity and to engage in unsustainably intense activity.[37]

The most likely immediate cause of a collapse in primary production is, I think, the absence of usable energy from the sun, that is, prolonged darkness. Experiments with various species of phytoplankton—diatoms, dinoflagellates, and coccolithophorid prymnesiophytes—have demonstrated that individual cells die after two to eight weeks of continuous darkness, the time depending on the species. With their stored nutrients in inert resting stages, many species—including land plants and even many diatoms and dinoflagellates—could presumably withstand months to perhaps years of darkness in some cases. In other words, to be effective as a cause of extinction, darkness should be essentially global and should last for a minimum of two months.[38]

Other causes of collapse—widespread acid rain, poisoning from gases such as methane or hydrogen sulfide, burning in gigantic fires, extreme cold or extreme heat—may well have contributed to extinction locally or even regionally, but for the most part such conditions are unlikely to have been global. Wildfires, for example, may indeed have destroyed terrestrial ecosystems on a large scale at the end of the Cretaceous and perhaps during other crises, but they alone could not have been responsible for these extinction events, for they would have left organisms in the sea and in freshwater unaffected. We now know that all the great extinctions from the Late Devonian onward involved the demise of both land and aquatic life forms. Interference with sunlight by clouds, dust, smoke, or something else in the atmosphere strikes me as the only plausible mechanism with a potentially global reach.

Despite the difficulties that beset the interpretation of extinctions in the geological past, there is overwhelming evidence that all of the global mass extinctions and many of the so-called minor ones are associated with disruptions in primary production triggered by unusual tectonic and celestial events. The triggers fall into three categories, which very likely are causally linked: (1) enormous volcanic eruptions, leading to the formation of huge volumes of so-called flood basalt over land areas of hundreds of thousands to millions of square kilometers; (2) collision of Earth with asteroids, comets, and meteorites; and (3) sudden eruptions of large volumes of methane from deposits of crystallized methane hydrates beneath permafrost and on continental shelves and slopes beneath the oceans. All have the capacity to interfere with photosynthesis over very large areas, and one or more of these triggers is associated with each of the mass extinctions as well as with many of the lesser events.

A link between mass volcanism and mass extinction has long been suspected. Indeed, Norman MacLeod holds that flood-basalt eruptions are more consistently tied with mass extinction events than any other proposed causes. Nevertheless, I shall argue that, although volcanism has contributed to many of the great crises, it and several other major disruptions may ultimately be triggered by the most important cause of all: collision with celestial objects.[39]

Evidence that violent volcanic eruptions interfere with biological production is tenuous at best. Volcanoes inject dust—much of it rich in sulfur, chlorine, and fluorine—into the stratosphere, where sulfur dioxide and water vapor combine to form droplets of sulfuric acid. The removal of water vapor as sulfuric acid causes the atmosphere to cool and dry. These effects have been well documented in the wake of the eruption of Mount Pinatubo in June, 1991; and eruptions of this kind at equatorial latitudes seem to be particularly effective at depressing air temperatures by scrubbing water vapor, a potent greenhouse gas, from the atmosphere. Mount Toba's eruption on Sumatra about 73.5 ka released 2,200 to 4,400 tons of sulfate, cooling the atmosphere by at least 3.5°C. This sounds dramatic, but there is no evidence of extinction following this enormous eruption, which exceeded the magnitude of Mount Pinatubo's eruption by a factor of at least ten thousand. An eruption of comparable magnitude, spewing some 4,000 km^3 of ash, occurred 454 Ma during the middle Caradocian stage of the Ordovician period, and deposited a thick ash layer in large parts of what are now Europe and North America. This eruption, too, was not accompanied by extinction. As I noted in chapter 7, eruptions like this may actually have stimulated rather than depressed biological production. Only if sunlight became so diminished by dust that fully sunlit leaves would receive light below the luminosity at which photosynthesis reaches a maximum would volcanic eruptions become a real threat.[40]

Nonexplosive eruptions into the atmosphere may be more destructive than explosive ones in part because the magma from them is richer in sulfate. The nonviolent eruption of the Laki fissure in Iceland in 1783 produced 14.8 km^3 of lava, 1.22×10^5 kg of sulfur dioxide, 1.5×10^4 kg of hydrogen fluoride, and 7×10^3 kg of hydrogen chloride, and had devastating effects in Iceland. Some one-quarter of the human population, and three-quarters of the population of livestock, died; and there was a persistent blue haze that blotted out the summer sun. Crops and pasture grasses were unable to photosynthesize, leading ultimately to famine. These effects were evidently more or less confined to Iceland, for in continental Europe the summer was exceptionally warm, possibly because of the greenhouse effect following the release of carbon dioxide into the atmosphere.[41]

Iceland's Laki eruption represents a small-scale example of flood-basalt volcanism. The geological record reveals striking evidence of huge eruptions lasting hundreds of thousands to perhaps a million years and covering hundreds of thousands to millions of square kilometers with a thick layer of basalt. The erupted volume of this flood basalt amounts to millions of cubic kilometers. All the great and many of the lesser extinction events of the last 250 my are very close in timing to at least the beginnings of flood-basalt eruption. Although such coincidences do not imply that the volcanism caused the extinctions, the conclusion that such upheavals contributed to global disruptions in the atmosphere and in the economy of production seems inescapable.

The trigger that has captured the public imagination more than any other as a possible cause of extinction is the collision of Earth with an asteroid or comet. The most convincing link between an impact and a mass extinction is that for the end-Cretaceous event. The Chicxulub crater on the Yucatan Peninsula of Mexico has been identified as the site of collision between some celestial body, perhaps 10 km in diameter, and Earth, at a time (66 Ma) indistinguishable from the mass die-off marking the end of the Mesozoic era. Material hurled into the atmosphere by this collision formed a 3 mm thick globally distributed layer, representing about 10^{16} g of droplets and dust. Although this amount is similar to the 10^{16} g of material erupted by Mount Toba in the late

Pleistocene, the volcano's erupted material included a very large fraction of dust particles that quickly fell to Earth. The material entering the atmosphere from the end-Cretaceous collision consisted of about 99% droplets, including vaporized sulfides and carbonates, and only 1% dust. Mount Toba's quantity of material in this category (at most 4.4×10^9 g) thus fell far short of that released by the end-Cretaceous impact. Possible consequences of the collision include clouds dense enough to shield Earth from the sun from months to years, vaporization of limestone at the site of impact, global wildfires, and gigantic tidal waves.

Vast volcanic eruptions and devastating impacts may both trigger a third potential cause of global extinction: the catastrophic release of methane gas into the atmosphere. Huge amounts of methane lie buried as crystalline methane hydrates beneath permafrost in the Arctic and in sediments on the continental shelves and slopes of the oceans. Additional dissolved methane resides in oxygen-starved waters of stagnant ocean and lake basins. These reserves of methane remain locked up as long as temperatures remain low, pressures remain high, and stagnant basins remain stratified, with oxygen-rich waters at the surface and methane-rich waters beneath. When these conditions change, methane can bubble catastrophically into the atmosphere. This could happen, for example, when heat from undersea volcanism warms part of the ocean floor, or when changes in ocean circulation cause deep oceanic waters to warm. A drop in pressure could result from a drop in sea level, which in turn could be brought about as sediments shrink while releasing methane. Tidal waves generated by volcanically triggered earthquakes or by celestial impacts can bring about mass slumping of sediments in the ocean, destabilizing the hydrates and causing the gas bubbles to rise. All these changes can also undo the stratification of stagnant basins. When this happens, rising methane bubbles expand, entraining oxygen-depleted water. The frothing water boils, inundating nearby lowlands while releasing vast amounts of the poisoning gas to the atmosphere. Organisms face mass mortality as oxygen-depleted water overwhelms them and as lightning-ignited fires on a grand scale reduce terrestrial vegetation to ashes. The atmospheric warming brought about by the release of methane (a greenhouse gas) and by its oxidation to carbon dioxide (another greenhouse gas) may initiate a positive feedback by stimulating the release of still more methane.

The quantities of methane involved in these catastrophic releases are huge. At the end of the Paleocene (55.5 Ma), a time of minor deep-sea extinction and major worldwide warming, an estimated 1.5×10^{18} g carbon was released in three closely spaced episodes. During the first episode, lasting some one thousand years, about 6×10^{17} g carbon was liberated, yielding a release of about 6×10^{14} g carbon per year. This estimate is similar to an annual output of about 10^{14} g carbon from human activity. The atmosphere and oceans together contain about 5×10^{19} g carbon. According to Robert Berner's calculations, the sharp reduction of light as compared to heavy carbon at the end of the Permian would require the release of some 4×10^{18} g carbon, almost three times that at the end of the Paleocene.[42]

Geologists have long speculated about a possible causal link between impacts and volcanism. Although Ivanov and Melosh (2003) argue on the basis of their theoretical calculations that volcanism is unlikely to result from a celestial collision at the site of impact, and Wignall (2001) remains skeptical of this link as well, my hunch is that a causal association among major volcanism, impacts, methane releases, and global bottom-up extinction will hold up as more precise dates for the various events become available. On the basis of his modeling of the Permian and Triassic carbon cycle, Berner (2002) is also inclined to the view that impacts caused the end-Permian extinction and triggered volcanism and the eruption of methane.[43]

The Geological Evidence

I now turn to the geological evidence for the causes of the various great and small extinction events that have been critically examined to date. The evidence is far more complete for recent events, not only because there are more localities and more rock units from which to glean clues, but also because some of the older extinctions are not associated with any of the triggers I discussed earlier.

The earliest recognized mass extinction, at 580 Ma, is associated with both an impact of an asteroid, which formed a crater in Australia's Lake Acraman, and a negative shift in the carbon-isotopic ratio. The extinction marking the end of the Neoproterozoic era and the beginning of the Cambrian period, at about 542 Ma, is likewise marked by a strong negative carbon-isotopic shift, perhaps indicating a pulse of methane release, but no crater indicating a collision with a body from outer space has yet been recognized. One of the two very closely spaced mass-extinction events of the Early Cambrian, the so-called Sinsk Event of the middle Botomian epoch, seems to have occurred at the same time as a flood-basalt eruption of unknown magnitude on the Antrim Plateau of Australia, dated at 513 Ma. This event was accompanied by the widespread appearance of oxygen-poor sediments, possibly indicating warming or the release of methane or carbon dioxide or both. Surprisingly, no further flood-basalt eruptions are known from the Paleozoic until the Middle Permian. Paul Wignall attributes this absence to the fact that continents tended to collide rather than to rift apart during much of the Paleozoic.[44]

One of the more puzzling extinction events of the last 600 my is the double pulse of major disturbance at the end of the Ordovician period, about 440 Ma. The first pulse occurred at the beginning of a brief (0.5 my) episode of glaciation in the southern hemisphere, affecting much of present-day South

America and Africa, whereas the second pulse came at the close of this cold spell. The first pulse appears to have coincided with a major decline in phytoplankton, as chronicled by the record of acritarchs. There is no evidence for volcanism or of collision with a celestial body, nor does the record indicate an isotopic shift in carbon consistent with a methane release; in fact, there is a pronounced shift toward heavier carbon. Lee Kump and Peter Sheehan favor the hypothesis that continental collision leading to the uplift of the Appalachian and Caledonian Mountains (in the eastern United States and Scotland, respectively) caused a rapid increase in the chemical weathering of silica-based minerals and therefore to a sharp reduction in the atmosphere's content of carbon dioxide, which in turn led to the advance of glaciers. Others have suggested that, as glaciation proceeded, currents in the deep ocean became stronger, ultimately leading to the upwelling of oxygen-starved water to the ocean surface. Presumably the phytoplankton would have succumbed to the lack of oxygen or to such accompanying poisons as hydrogen sulfide on a global scale. I am skeptical about all of these scenarios. The upwelling scenario might bring about regional collapse, but it is difficult to envisage this happening worldwide. My guess is that some as yet unknown extraterrestrial cause—perhaps a comet or asteroid falling into the ocean—will be discovered to have triggered the extinctions as well as the observed glaciation and anoxia.[45]

Equally mysterious are the extinction events of the Devonian. At 380 Ma, near the boundary between the Eifelian and Givetian epochs of the Middle Devonian, a substantial worldwide extinction involving many planktonic and other marine groups coincides with an impact by an extraterrestrial body. As many as three significant extinction events took place within a span of about 10 my: in the Middle Givetian, at the end of the Frasnian, and at the end of the Famennian (the last) epoch of the Devonian. All three of these events devastated the marine plankton, and all concide with episodes of carbon-enriched sedimentation. Unlike the event of 380 Ma, which witnessed a negative shift in the carbon-isotope ratio, the three Late Devonian events show no evidence of an isotopic disruption that would be consistent with a massive methane release. Indicators of impacts of extraterrestrial bodies exist, but there are unresolved controversies about the timing of these impacts relative to extinctions. At least one of two proposed impacts came between the last two of the Devonian extinctions. It is possible that the increase in the heavy carbon isotope during these extinction events is the consequence of vaporization of limestones exposed on the land. This, in turn, could have occurred through intense heating following the impact of a celestial body, as has also been proposed for the end of the Cretaceous.[46]

The first recognized flood-basalt event after the Cambrian came during the Middle Permian period near the end of the Paleozoic era, when most of the continents were assembled in a giant single block known as Pangaea. This small eruption, producing a rock volume of probably less than 10^6 km^3, created the Panjal volcanics of northwestern India and the Emesian flood basalts in South China. It appears to coincide with a minor extinction event at the end of the Guadalupian stage of the Middle Permian, notable for the demise of what Robert Bakker calls the first dynasty of large land vertebrates—herbivorous "pelycosaurs" and early "mammal-like" reptiles.[47]

To many paleontologists, the calamity marking the end of the Permian period and bringing the Paleozoic era to a close stands out as the watershed event of the Phanerozoic eon. This is probably somewhat of an exaggeration, given that the major renewals within marine and terrestrial ecosystems were not ushered in until 20 to 30 my after the great crisis, but many major animal and plant groups did become extinct at or near the end of the Permian, including trilobites, many groups of sedentary filter-feeding echinoderms, most cephalopods, cordaitalean conifers, and glossopterid seedferns, among others. An abundance of spores of fungi indicates catastrophic mortality and cessation of production among land plants. The end-Permian extinction coincides with at least the beginning, and perhaps with the peak, of flood-basalt eruption in Siberia, which produced the Siberian Traps. The volume of this pile of volcanic rock, including a good deal of material ejected explosively, is difficult to estimate because the geographic extent of volcanism remains poorly known, but current calculations imply a volume of not less than 2.3×10^6 km^3. The record of carbon-isotopic ratios across the boundary between the Permian and Triassic periods features a dramatic increase in the proportion of ^{12}C, the light isotope. This spike is best accounted for by the catastrophic release of methane, which is enriched with ^{12}C. If only 10 percent of the current 10^{19} g of methane beneath the oceans were liberated into the atmosphere, it would be enough to account for the observed isotopic pattern. The loss of high-latitude floras and faunas and the poleward expansion of warm-adapted seaweeds all imply that the extinction was accompanied by significant global warming, which would be expected from methane release and from volcanically produced carbon dioxide.[48]

Pieces of a meteorite recovered from Antarctica and elsewhere imply that a large celestial object slammed into the ocean at the close of the Permian. Such a collision could have triggered the end-Permian crisis.[49]

The end-Triassic extinction, which eliminated many planktonic species as well as large numbers of land plants, coincided with the beginning of one of the largest flood-basalt eruptions known. The eruption in the so-called Central Atlantic Magmatic Province (or CAMP) covers some 7×10^6 km^2 in eastern North America, Africa, and South America, and is estimated to have had an original volume of some 2×10^6 km^3. In the Newark Basin of eastern North America, where a continuous fossil record has made the detailed study of events possible, Paul Olsen and colleagues have also identified a peak in abundance of the element iridium at the horizon marking the extinction. The iridium, Olsen and coworkers argue, came from a celestial body rather than from the volcanic eruptions. The terrestrial extinctions studied by Olsen's group are dated at about 201 Ma, a little earlier than the 199.6 Ma and Pálfy and colleagues have calculated for the marine extinctions as chronicled in a rock sequence in the Queen Charlotte Islands of British Columbia. There is also evidence for a sharp decline in productivity and for a brief, large injection of methane into the atmosphere, triggering an episode of global warming. From this evidence I conclude that the end-Triassic extinction is associated with a reduction in primary productivity, and that, separately or in combination, this reduction was brought about or exacerbated by volcanism, methane, warming, and a collision with an extraterrestrial object. Celestial bodies may have slammed into Earth earlier during the Late Triassic (about 214 Ma) as well, coinciding with an as yet poorly documented extinction at the end of the Carnian stage of the Late Triassic.[50]

A minor extinction event during the Early Toarcian stage of the Early Jurassic period, about 183 Ma, coincides with the eruption of the Karoo Traps in South Africa and the Ferrar Traps in Antarctica, which together have a volume of 2.5×10^6 km^3 of lava. Once again, these events were accompanied by widespread deposition of oxygen-starved sediments in the ocean, by warming, and by methane release.[51]

The Cretaceous was a volcanically very active period. In the Valanginian and Hauterivian stages of the Early Cretaceous, about 133 Ma, a flood-basalt eruption with an estimated volume of 2.35×10^6 km^3 formed the Paraná Traps of South America and the Etendeka Traps of Namibia. No recognized extinction event coincides with this eruption. Andrew Gale suggests that the flood-basalt event caused carbon cycling to speed up globally and that it initiated separation between South America and Africa as the South Atlantic Ocean opened. From 124 to 118 Ma, during the Late Barremian and Early Aptian stages, massive eruptions beneath the Pacific and Indian Oceans ushered in a 30 my interval of at least seven episodes of minor extinction, each of which coincides with volcanic activity and with brief periods of widespread accumulation of fine, organic-rich sediments on large parts of the ocean floor. The first of these came in the Early Aptian, about 120.5 Ma, when many nanoconids—small (2 to 20 m) phytoplankters with a heavy skeleton of calcium carbonate—disappeared. Other minor extinctions in the sea occurred in the Late Aptian (116 Ma) with the onset of volcanism on the Kerguelen Plateau, the latest Aptian (113 Ma) during the Kerguelen eruptions, the Early Albian (110 to 109 Ma) with the formation of the Rajmahal Traps of India and further eruptions in the southern Indian Ocean, the Middle Late Albian (102 Ma) and latest Albian (99.2 Ma) perhaps accompanying the formation of Hess Rise in the Pacific, the Middle Cenomanian (96 Ma) coinciding with the resumption of volcanism on the air-exposed central Kerguelen Plateau, and the Cenomanian-Turonian boundary (93.5 Ma) with widespread volcanism beginning in the Caribbean and continuing in the Indian Ocean. Of these, the events at 120.5, 113, and 93.5 Ma were the most significant in the plankton. Methane release may have contributed to the extinctions of the Early Aptian (120.5 Ma) and perhaps other events as well. Finally, the Turonian to Coniacian separation of Madagascar from India (87.6 to 84 Ma) is not clearly associated with an extinction event.[52]

The end-Cretaceous extinction event, dated at about 66 Ma, is associated with extensive flood-basalt volcanism in northwestern India and nearby parts of the Indian Ocean. In India, the Deccan Traps cover as much as 2.5×10^6 km^2 and have an estimated volume of at least 2×10^6 km^3. The eruption seems to have begun about 66.5 Ma, and continued well into the Paleocene epoch of the Cenozoic era, 62.8 Ma. It coincided with a brief warming in the midst of a general cooling trend begun about 70 Ma. There was also a collision between Earth and an extraterrestrial body some 10 km in diameter, whose impact formed a crater about 300 km in diameter on the Yucatan Peninsula of Mexico. Slumping of sediments on the eastern American continental shelf resulting from this impact may have triggered the release of methane, but the global warming associated with the Deccan volcanism may also have contributed to this release. According to calculations performed by David Beerling and colleagues, the amount of carbon dioxide liberated in the atmosphere by Deccan volcanism fell far short of the observed spike of carbon dioxide just after the end-Cretaceous boundary. Based on the density of stomates on fern leaves from the Raton Basin of New Mexico, carbon dioxide concentrations could have reached 2,300 parts per thousand by volume (ppmv), six to seven times present-day values. The mechanism Beerling and colleagues favor is the vaporization of limestone on land owing to enormous heat generated beneath the point of impact. According to the estimates given by the Beerling group, at least 6,400 gigatons of carbon dioxide would have been injected into the atmosphere, enough to account for the sudden observed increase in the ratio of heavy to light carbon isotopes in sediments marking the end-Cretaceous crisis.[53]

In the North Atlantic region, flood-basalt volcanism began in the Early Paleocene, about 61 Ma, and a second larger episode commenced during the Late Paleocene, 56 Ma. About one-third of the estimated volume of 6×10^6 km^3 of the volcanic rock erupted during this interval formed as flood basalts on land, especially in southeastern Greenland and northwestern Britain. Around 55 Ma, there was explosive volcanism for about 100 ky in the Caribbean region. At 54.9 Ma, marking the end of the Paleocene epoch, a sudden warming coinciding with and likely caused by or resulting from a major release of methane ensued. Temperatures in the deep sea and on land at high latitudes increased by as much as 7°C. As a result of these climatic and physical events, mammals and land plants underwent significant changes in distribution (see chapter 8), but there was little extinction save for bottom-dwelling deep-water foraminifers. In the most careful study of extinction among shallow-water marine animals, David Dockery has shown that 69 percent of Late Paleocene molluscan species present in the fauna of the coastal plain of the Gulf of Mexico did not survive into the Eocene epoch. Very few of these extinctions represent terminations of lineages; most of the extinct species were replaced by close relatives or possibly descendants within the same genus. The magnitude of extinction at the Paleocene-Eocene boundary was smaller than the 97 percent loss sustained by the molluscan fauna earlier in the Paleocene, and smaller than the 94 percent loss incurred after deposition of the Early Eocene Bashi Formation.[54]

There have been only two relatively minor episodes of flood-basalt volcanism since the end-Paleocene. In northeastern Africa, the Ethiopian flood basalts were laid down in two closely spaced pulses between 31 and 28 Ma during the Early Oligocene epoch. By earlier standards, this eruption was small, producing an estimated volume of 0.75×10^6 km^3 of volcanic rock. The Ethiopian basalts come well after two notable extinctions, one at the Middle to Late Eocene boundary (about 36.5 Ma), and a second pulse at the end of the Eocene (33.7 Ma). They may, however, be related to two other extinctions, one during the earliest Oligocene (about 32.5 Ma), and the other during the Middle Oligocene (about 28 Ma). Interestingly, at least three impact craters—including the Chesapeake Bay and Toms Canyon in the eastern United States and the Popigai crater in northern Siberia—record collisions with extraterrestrial bodies during the Late Eocene at 35.7 Ma. Although the present consensus among geologists is that these impacts are not associated with either the volcanic eruptions or the Late Eocene and Early Oligocene extinctions, there are enough uncertainties in the dating of all these events that a causal link may yet be uncovered.[55]

An even smaller flood-basalt eruption occurred in western North America during the Late Early and Early Middle Miocene, from 17.2 to 15.8 Ma with a peak rate of accumulation at 16.1 Ma. The total volume of volcanic rocks produced by this so-called Columbia Plateau flood-basalt event may not amount to more than 0.17×10^6 km^3. There is mounting evidence for significant extinction among marine molluscs on both sides of the North Atlantic as well as on the Pacific coast of North America. Although cooling is usually considered the cause of these still poorly documented losses, the eruptions and perhaps as yet other as yet unstudied factors will be found to be responsible.[56]

Although they took place only within the last 3.5 my, the extinctions of the Pliocene are poorly understood. Important changes in world geography had dramatic consequences for the pattern of circulation in the ocean, and may have cut off mechanisms by which surface waters were enriched with nutrients from below in large parts of the world. The connection between Siberia and Alaska was broken by the opening of Bering Strait about 5.32 Ma. Events in Central America that had narrowed and shoaled sea passages between islands in a volcanic island arc culminated perhaps 3.5 Ma in the formation of a continuous isthmus, temporarily or perhaps permanently cutting the Atlantic Ocean off from the adjacent Pacific. Seaways may have formed once or twice later, but by 1.8 Ma the isthmus seems to have become a more or less permanent geographic feature. With some barriers eradicated and others newly in place, currents in the Atlantic Ocean may have quickened, and upwelling may have been curtailed. Whether geography is the only story is, however, quite unclear. At least two relatively small impacts are known from the Pliocene, one in the far-southern Pacific about 2.15 Ma, another in Patagonia at 3.3 Ma. The later event is thought to have produced an explosive force of 10^5 to 10^7 tons of TNT, at the low end of the range for a global effect. No flood-basalt eruptions are known from the Pliocene, and there is no record of exceptional methane emission. Just what caused productivity to decline, how rapidly the decline occurred, and whether the decline was a brief aberration or a more permanent characteristic of marine ecosystems remain unanswered questions.[57]

From this quick survey of the great and small extinction events of the geological past, I take away three lessons. First, the extinctions are all associated with a drop in primary productivity, which must come about through interference with the supply of energy and raw materials, and which destabilizes ecosystems from the bottom. Organisms directly dependent on living resources are particularly at risk during the production crises. Second, any of several circumstances and triggers, acting alone or in concert, potentially interfere with photosynthesis and diminish the supplies of necessities for primary producers. Although we remain in the dark about the details, a reasonable hypothesis is that the greatest crises are precipitated by the impact of large celestial objects on Earth's surface. Third, whether and how a given trigger causes a collapse in production depends not only on the magnitude and extent of the disturbance and on the adapted state of the producers, but also on the particular background context of climate and geography and on the extent to which positive feedbacks amplify the initial trigger. A celestial impact occurring at a time when ecosystems have already been stressed by some other disturbance, such as falling sea level, may be more devastating than an impact under more economically favorable circumstances. These possibilities will keep scientists busy for some time to come.

Disturbance and Civilization

No event in human history even approaches in magnitude the crises that marked the major and minor extinctions of the geological past, but the archaeological and written record of human affairs clearly documents the destructive consequences of climate-controlled and human-caused decreases in the supply of food and other sources of energy on society.

We ultimately depend on photosynthesis for food, and are therefore like other animals in being vulnerable to the same kinds of climate-related disruptions to primary production. For example, a 200-year drought between 5,200 and 5,000 years ago may have brought about the collapse of late Uruk society in Mesopotamia after some 800 years of vigorous agriculture and urbanization. About 4,200 years ago, a major episode of cooling and drought, associated with a 30 percent drop in rainfall, caused regional abandonment of civilizations from the Indus Valley to Egypt and Crete. Severe flooding, probably associated with an intense El Niño, conspired with severe drought during a thirty-year interval of the late sixth century A.D., leading to famine and to the destruction of the capital city of Peru's Moche civilization. Prolonged and severe drought also accounts for the demise of the classic Maya civilization in the ninth century, the collapse of agriculture in the Tiwanaku civilization of Peru in the central Andes, and the collapse of the Anasazi civilization of the southwestern United States at the end of the thirteenth century. In Europe, famine and the spread of the plague in the early and middle fourteenth century coincided with the beginning of the Little Ice Age, a period of cool summers and cold winters following a 500-year interval of warm climate and human population expansion. Human-caused deforestation and soil erosion doubtless exacerbated the effects of drought in all these cases.[58]

With an increase in long-distance trade, greatly helped by cheaper and faster transport of a greater quantity of commodities, the effects of regional climatic variation on agricultural output have been substantially blunted. As long as climate-related or weather-induced crop failures remain local or regional, modern civilization's relative immunity to this kind of bottom-up disruption should continue. But we run great risks by becoming too dependent on a very small number of highly productive regions, many of which are being degraded by soil erosion, unsustainable water use, increasing salt content of soil, and urbanization, to say nothing of the ever-increasing human population. The lesson that the extinctions of the past teach above all is that tampering with highly productive ecosystems through pollution, overexploitation, fragmentation, or replacement with built-up areas invites disaster.

Unlike other animals, humans have come to depend on other sources of energy—fossil and nuclear fuels, and perhaps in the future hydrogen—for much of our economic activity. A new type of "primary production"—extracting, converting, and distributing the new sources of energy and building and maintaining the industry that accomplishes these tasks—has thus been added to the traditional methods of gathering and growing food. Leaving aside the point that fossil fuels will eventually have to be replaced by other sources of energy, the supply of fuels from living as well as nonliving sources is substantially under human control, meaning that economic disruption is increasingly likely to be human caused. War, sabotage, trade embargoes, and political indifference and oppression today loom large as causes of famine, destruction, cutoffs in the fuel supply, and suppression of demand.

In human history as well as in the history of life as a whole, we can discern a general shift from external, bottom-up disruptions to crises created within economies themselves. Extinctions driven by climates, volcanic eruptions, and above all by celestial impacts will surely occur again, but relentless and cumulative selection has perhaps increasingly protected survivors from previous catastrophes from subsequent bottom-up disruptions. In a parallel way, climate-related famines have been largely absent in the last 150 years of human history. By contrast, extinctions due to other, mostly powerful, species may have become increasingly common through geological time. In the same way, economic disruptions stemming from human activities may have become more frequent and more destructive. The cause for this shift is the same in the human and nonhuman realm: intense competition, fed by an increasingly prolific and reliable supply system, has produced more powerful agents and larger, more productive economies, which have correspondingly acquired greater abilities to disrupt and destroy as well as to spread wealth.

Before leaving the subject of disruption and civilization, I want to say a few words about which parts of the human realm are most sensitive to disturbances from the outside and to crises of our own making. As I noted earlier in this chapter, species with few energy reserves or an inability to curtail metabolism are most at risk of extinction in the chain reaction set in motion by circumstances interfering with photosynthesis. This appears to hold in human societies as well. Brief intervals of widespread starvation, initiated by crop failure and often exacerbated by war or callous government policy, are most precarious for the peasants who eke out a subsistence living on the land. It is these peasants who died out disproportionately during the European famines of the early fourteenth century, the 1590s, 1740–1741, and 1815–1817. They were also the primary victims of the Irish potato famine (1845–1848) and the Ethiopian disasters of the 1980s. Similarly, the politically driven oil crises of 1973 and 1979 were especially devastating to those poor countries where the shock could not easily be absorbed by trade or by alternative sources of energy. With their reserves of money and energy, powerful entities are better able to survive and to recover from short-term interruptions in the supply of food and fuel. Longer-term disruptions, however, will imperil the powerful sectors more than smaller, less energy-intensive components. So far, we have not had to face this situation on a large scale, but I believe that bottom-up crises lasting decades or longer would have a devastating and disproportionate effect on those sectors that currently exercise the most powerful controls on modern economies.

In nature, disruptions emanating from the top—through the addition or elimination of powerful competitors—place the previous ruling elites at greatest risk of extinction while leaving the base of the ecosystem changed but intact, at least in terms of species composition. A similar case can be made for disturbances initiated by human disruptive competition—war, in other words. In countries that have lost wars during the last four centuries, it is the most powerful sectors of society—civil government, the military establishment, and industry—that have disintegrated. With a few exceptions—the Austro-Hungarian and Ottoman empires after World War I, for example—the losing nations and their populations have survived, often with their boundaries more or less unchanged. New, often more effective governments and industries eventually replaced those that succumbed in war.

Recovery and Reassembly

With all the gloom and doom of extinction, it is worth remembering that life has existed continuously on Earth for perhaps as long as 3.5 billion years despite numerous, often horrific assaults. Life forms are incredibly tenacious and resilient even if their more complex manifestations lie in tatters after a crisis. The process of recovery and repair following major disruption raises important questions. How does the reassembly of ecosystems proceed? When and how does evolutionary innovation play a role? What long or short shadow does extinction cast on the postcrisis biosphere? What effects do extinction and the less draconian versions of disruption have on economic systems?

Postextinction recovery begins with survivors, some of which were able to hang on in otherwise devastated ecosystems during the crisis, and some of which came from refuges, where ecosystems were less affected. Understanding recovery thus requires that we understand survivors as well as the environments in which they persevered. It also requires that we understand how supply and demand interact in the aftermath of economic collapse.

In a view widely held by paleontologists, extinctions offer opportunities for renewal. With so many incumbents eliminated, survivors repopulate devastated regions and environments, and vacant economic roles are opportunistically filled, first by generalists jacks-of-all-trades, later by more specialized entities arising through rapid and repeated formation of species. The result is a new world order.

This view, however, is overly simplistic. Nothing very innovative or revolutionary will happen until two key components of the economy are restored to some semblance of normalcy: (1) the productive machinery, the system whereby raw materials are assimilated by primary producers to supply the primary producers themselves and their consumers; and (2) consumer demand, the aggregate competition and consumption that exercises strong top-down control on the adaptive characteristics and distribution of all members of the emerging economy. Even when economic collapse emanates from the bottom, it ultimately destroys demand, especially the demand by high-powered consumers. It takes time for that demand to become reestablished through evolution. Meanwhile, the economy is in a mad scramble, with weedy opportunists vying for a place in the sun.

Ecologists understand quite well what kinds of species have an advantage in environments that recently suffered disruption. These species fall into two widely overlapping categories: (1) those with wide powers of dispersal, and (2) those with very rapid rates of growth of individuals or populations. Species in the first category—the wide dispersers—include small members of the plankton, organisms attached to floating wood or pumice, tiny wind-dispersed insects and spiders, ferns with wind-dispersed spores, flowering plants whose seeds are carried far by wind and water, and—following the evolution of birds migrating over long distances—plants and animals whose dispersal stages cling to the feet or feathers of birds. These species often have nearly cosmopolitan distributions and are well known as early colonists of temporary pools, lava fields after volcanic eruptions, islands after hurricanes, newly exposed surfaces on pilings or on the hulls of ships, large clearings in forests, and sunken ships. At least some of these organisms also figure prominently in ecosystems during and immediately after the great extinction crises. Fern spores, for example, are major components of the spore and pollen assemblages at the end of the Permian, Triassic, and Cretaceous periods. Ferns are well known for their ability to cross ocean barriers to distant islands, where widely distributed fern species form a prominent part of the vegetation.[59]

From an economic and evolutionary point of view, the second attribute of colonists—rapid growth of individuals and populations coupled with the ability to withstand or escape from unfavorable conditions in inert resting stages—is particularly important. This weedy syndrome is associated in the active phase of the life cycle with high productivity and rapid metabolism, and enables individuals, populations, and species to capitalize rapidly on suddenly available resources. To make hay while the sun shines is what weediness and rapid, unsustainable exploitation are all about. Organisms engage in it; so do some individual humans and corporations. The entities involved are typically small, poorly defended, and often unsuccessful in competition against larger, longer-lived entities. Paradoxically, the wanton exploiters often prepare the way for the successful invasion or evolution of better competitors, and, given their own high rates of activity, are themselves prime candidates to spawn descendants of high competitive and defensive power and greater persistence as economically active agents. Weeds, in other words, facilitate the evolution of a new dynasty of competitive dominants, and often themselves give rise to members of that new privileged class.

Although high powers of dispersal and high growth rates often characterize the same species, these two attributes of effective early colonists are not always associated with each other. Jared Diamond, for example, points out that some extremely widely dispersing birds in the South Pacific have a very small clutch size of one egg per brood. That same clutch size, however, also characterizes bird species that are loath to cross water bodies more than 5 km wide. Many of the wide dispersers have high per capita birth rates, often achieved by multiple broods per year, but six of fourteen good colonizers (43 percent) have only one brood per year. Good dispersers, in other words, do not necessarily capitalize quickly on a windfall of resources, and not all fast exploiters are effective at crossing barriers of inhospitable environment.

Recovery following great crises is complicated by a profound inequality in the response of producers and larger consumers. In surface waters of the ocean, the small cells comprising the phytoplankton can divide rapidly, doubling their numbers on the scale of hours to days. This means that surface productivity in the ocean can rise to precrisis levels within a very short time, perhaps not more than a few years, provided the sun shines and dissolved nutrients are available. Small protistan consumers could in principle also recover quickly, but larger consumers could not. The restored microbial loop of tiny producers and consumers in the plankton would export few resources to the seafloor except perhaps at the shallowest ocean depths, and therefore leave the ecosystems of many parts of the seafloor starved for nutrients until larger planktonic and bottom-dwelling consumers could become reestablished. Postcrisis ecosystems, in other words, suffer the conditions of recession and depression, in which there is insufficient demand to take up and regulate supply. The same scenario is likely to apply to ecosystems on land, except that restoration of precrisis production levels would take longer owing to the lower fecundities of land-plant weeds than of small phytoplankters.[60]

During the stage in which postcrisis ecosystems remain in recession, top-down selection from consumers will be relatively weak. It is for this reason that crises do not in themselves offer opportunities for innovation and growth. Instead, the growth in numbers that postcrisis conditions allow must be accompanied by effective consumer-driven selection, which will then strongly favor many attributes of power that underlie the emergence of the new economic order. In short, if innovations are to spread, their bearers must be under intense competition from powerful agents and must have the supply-side flexibility to be able to do something about it.

Events following the great end-Permian die-off illustrate these points well. Earliest Triassic marine ecosystems, though little known, seem to have been extraordinarily impoverished assemblages composed chiefly of species that had survived the rigors of the end-Permian crisis. Prominent among surviving bivalves were *Claraia* and *Monotis*, peculiar flat-shelled forms that led sedentary lives lying free on the seafloor. It is entirely possible that these bivalves depended nutritionally not on food in the plankton, as most other bivalves do, but on bacteria growing in their gills or other organs. Early Triassic snails are tiny micrograzers and detritus-feeders. Most of the Early Triassic new species of snails, bivalves, bryozoans, and brachiopods belong to essentially Paleozoic clades, and show none of the innovations that were to characterize groups diversifying later. Middle Triassic assemblages of Anisian age still have a Paleozoic character, even 10 my after the crisis. On land, there was a 7 my gap in the production of coal by swamp vegetation between the end of the Permian and the Middle Triassic. All the largest herbivorous and predatory vertebrates were eliminated by the crisis, leaving relatively small herbivorous dicynodonts, carnivorous theriocephalians, and predaceous reptiles related to the later dinosaurs. The depression following the end-Permian collapse thus lasted some 7 my, and came to an end only as both demand from consumers and supply mediated by producers rose.[61]

In the aftermath of the end-Cretaceous crisis, a period of 2 to 3 my of instability ensued in seafloor communities, during which groups that had become competitively subordinate in the Cretaceous briefly achieved numerical supremacy. These groups include reef-forming glass sponges, which in the Cretaceous and today normally live in deep and often cold water, and cyclostome bryozoans, which in warm waters are normally outcompeted by cheilostomes. Although eleven coral genera with algal symbionts survived the crisis, coral-dominated reefs did not appear during the first 2 to 3 my, or perhaps even the first 5 my, of the Paleocene epoch. In the plankton, photosynthesizing foraminifers were likewise absent 3.5 my following the mass extinction.[62]

Recovery may have been even slower on land. Although a verdant rainforest vegetation existed in Colorado 1.4 my after the end-Cretaceous event, terrestrial ecosystems elsewhere in North America felt the effects of the crisis for much of the Paleocene. In the northern Rocky Mountains, where the crisis extinguished about 80 percent of plant species as recorded by leaves, precrisis diversity was not reestablished until the Early Eocene, 10 my after the catastrophe, and herbivory, as measured by the frequencies of both generalized and specialized types of insect-induced damage to leaves, was low during the whole Paleocene as compared with earlier and later items. The Eocene increase in plant species and in herbivory may have been abetted by warming. Large herbivorous and carnivorous mammals did not evolve in North America until the Late Paleocene, some 8 to 10 my after the appearance of dinosaurs with comparable roles. In short, crises like those at the end of the Permian and Cretaceous periods were brought on suddenly, but had effects lasting millions of years during which agents of top-down selection seem to have been weaker than either before or since.[63]

Well-adapted incumbents that become extinct during bottom-up crises are not immediately replaced by comparably or better adapted newcomers, but disruption of the economic status quo occasioned by the temporary drop in top-down selection accompanying the disappearance of powerful consumers and competitors is nonetheless an important long-term amplifier of the very long-range trends dictated by the action of economic principles. The disruptions accelerate the trends because, once top-down control and sufficient resources have been restored, newcomers have the opportunity to forge a new order, often characterized by greater power among economically dominant entities. It is to the general long-range trends that I shall turn in chapter 10.

Chapter 10

PATTERNS IN HISTORY: TOWARD GREATER REACH AND POWER

Everything in the universe has a history. This would not be so if the universe and everything in it were at equilibrium; but things do change as time proceeds because the universe and everything in it are not at equilibrium. By history I mean a time-ordered sequence of events, entities, environments, interactions, and trends, cascades of causes and effects, things appearing and disappearing, growing and declining or staying the same. If we give the entities life, we get economic history, a sequence in which entities acting autonomously on their own behalf collectively create larger structures and environments that influence subsequent events. History is a sequence—or, more precisely, a branching and interconnected series—that we can reconstruct from the past, from an incomplete record of activities and players and conditions existing during times before our own.

In earlier chapters, I sketched my views on how economies work. Some concepts are universal: competition, inequality, the benefits of trade to the larger whole, the effects of scale on opportunity and constraint, the predominance of powerful entities as agents of economic change, and the positive feedbacks between production and consumption. Others are more specific to the particulars of our own biosphere: sources and sinks of energy and raw materials, the timing and location of opportunities and disruptions, and the mechanisms by which economically relevant information is transmitted, received, and applied.

I shall argue that these concepts, when applied to the history of life and to the much shorter narrative of humanity, reveal fundamental trends, which I would go so far as to characterize as economic laws of history. Economic conditions are determined disproportionately by those entities that gain more, or lose less, during competitive or cooperative interactions over resources. These dominant entities have predictable properties—high rates of energy use and productivity, great reach, and a large capacity to modify the environment—and also impose a basic directionality on economic systems. Within limits imposed by external conditions and by existing technology, economic systems tend to increase in productivity, diversity, and opportunity, powered by positive feedbacks to yield increasingly powerful top competitors, which collectively restrict less powerful entities to parts of the economy where power, productivity, and the intensity (or stakes) of competition are lower. They also increase the rate of supply and the predictability of resources, with the result that the realm of regulation expands while that of uncertainty and external dependence recedes. These trends are neither constant nor irreversible, but the conditions in which they prevail are vastly more common and longer lasting than are the disruptions that halt or temporarily reverse the trends. Competition-driven feedbacks ratchet economic systems toward greater productivity, power, and reach, and replace vulnerability to external disruptions with regulation and disruption generated from within.

This view of history combines the knowable consequences of economic principles with the accidents of particular circumstances. My aim is therefore to document the trends, and to arrive at an understanding of humanity and of why human history recapitulates the much more protracted history of life as a whole.

Chance and Necessity

Historians are confronted with a fundamental tension between two phenomena—chance and necessity—that they know must both be important in history but whose relative contributions to the inferred sequence is devilishly difficult to pinpoint. Chance—or contingency, as Stephen Jay Gould and others would call it—comprises the realm of accident, of unpredictability, of initial conditions that themselves cannot be known in advance but that have knowable consequences, of interactions and activities that propagate in unanticipated ways. The asteroid or comet that struck Earth at the end of the Cretaceous fell on what is today the Yucatan Peninsula of Mexico rather than in Africa or the Pacific Ocean; and many of the consequences of this event flow from this geographically and temporally unpredictable instant. The intelligence and economic power of humanity is an outcome of evolution among placental mammals, of a process that in its broad principles is comprehensible and knowable; but it could just as well have arisen in a lineage of large-headed dinosaurs if suitable conditions—geographical, selectional, and nutritional—and enough time had presented themselves, and had the ground-dwelling dinosaurs not been extinguished by the end-Cretaceous extinction event. Our mammalian heritage is an accident, an initial condition imposed long ago, an instantiation among many possible points of departure that cannot be predicted from first principles. Our time of first appearance is also, to an extent at least, a matter of chance. This point in time may well be characterized by certain conditions that stimulate novelty in other clades at other times, but even if many of the same conditions prevailed at some earlier time, animals capable of our level of sophistication would simply have been too far removed from the range of actual animals, with too many steps separating the likes of us from the entities that were then available.

The Spaniards rather than the English, Dutch, or Polynesians defeated the Aztecs and the Inca empires. This fact had important long-range consequences for the language, culture, and economic development of what we still call Latin America, but this historical event belongs firmly in the realm of the unpredictable, even if it "made sense" in the larger themes of history or in the particular political situation of the time. Similarly, Jared Diamond argues that the peoples of the world eleven thousand years ago had all reached comparable levels of economic sophistication, and that it was an accident of geography—the availability of abundant domesticable plants and animals in a large, fertile region of southwestern Asia—that propelled the tribes of the Fertile Crescent to take up agriculture and to initiate a long episode of expansion and control. The general conditions can be predicted; the specifics cannot. All of history is pervaded by such accidents, coincidences, unique events, singular players, and unprecedented circumstances.

Given the implausibility of discovering universal laws, many observers have concluded that biology and the social sciences, including history, can never reveal principles as powerful as the reductionist laws of physics. In the words of the noted philosopher of science Karl Popper, "There can be no scientific theory of historical development serving as a basis for historical prediction."[1]

On the other side is necessity. Parts and wholes interact according to the principles of physics, chemistry, and economics. As far as we know, these principles are universal and invariant; they apply regardless of the particulars of place, time, and participants. Most of the time, players capable of exerting greater power exercise control over the characteristics and distribution of players with less power. We can predict the attributes that make entities powerful, and we can characterize the conditions that create opportunities for the evolution of such entities. To some extent we can even identify when and where favorable conditions will occur; but the details of place and time, and the particulars of genealogy of participants, are matters of contingency.

In his thoughtful book *Chance and Necessity*, Jacques Monod confronted the problem of constructing a complete theory of the biosphere, including its evolution. "In a general manner," he argues, "the theory would anticipate the existence, the properties, the interrelations of certain *classes* of objects or events, but would obviously not be able to foresee the existence or distinctive characteristics of any *particular* object or event." Monod elaborates this point by noting that "the biosphere does not contain a predictable class of objects or of events but constitutes a particular occurrence, compatible indeed with first principles, but not *deducible* from those principles and therefore essentially unpredictable." In his book *Investigations*, Stewart Kauffman makes the same point. It is possible to imagine biospheres with a molecular architecture utterly different from our own, an instantiation of a class of biospheres. We are unable to predict what that instantiation and what the particulars of the system may be, but once we have a biosphere, or whatever else we might wish to call an economic system, we know that it will have certain attributes and, importantly from the perspective of this chapter, knowable general patterns of change through time.[2]

There are, of course, schools of historical thought that would assign essentially no role at all to necessity. Gould sees the history of life as "a story of massive removal followed by differentiation within a few surviving stocks, not the conventional tale of steadily increasing excellence, complexity, and diversity." To him, natural selection and adaptation are local processes, and the conditions to which organisms adapt do not last. Instances of improvement—of increasing control—may be far outweighed by the number of cases of diminishment, in which lineages evolve toward smaller body sizes and specialized habits. To him, the fact that bacteria and other microbes still constitute the bulk of life's biomass whereas the complex mammals and insects are only a recent veneer on the biosphere, testifies to the absence of long-term historical trends or the application of principles that would yield such trends. Trends, he and others would argue, arise as figments of the human imagination, inspired by expectations of progress rooted in eighteenth- and nineteenth-century thought, and supported only by a highly selective reading of the historical record in which human achievements stand as supreme. Humans have as long a history as does the bacterium *Escherichia coli*, and had we dwelled on the latter instead of on the former, we would never embrace the view that history reveals any tendencies toward general improvement. In short, Gould holds that there is nothing innate in the entities of life that could yield a universally applicable trend. Instead, there are variation and replacement—the directionless coming and going of lineages as events, large and small, swamp the short-term effects of selection.[3]

How can we resolve the tension between chance and necessity? An answer to this question is important, because it affects whether we can learn anything from history. If chance overwhelms the laws of physics and economics, then history is little more than a unique sequence, effectively unpredictable and without lessons for the future. If the particulars of time and place can be broadly predicted by application of general principles, then the long sweep of history cannot be ignored, for it would then constrain the possibilities that the future holds. Seen globally, history is indeed unique. There is but one history of life on Earth, only one physical past of the Earth's crust and oceans and atmosphere, just one branching sequence of events from human ancestry to our current place of dominance on Earth. Yet, although there are historical phenomena that play out on the global scale, and are influenced by events of worldwide scope, there appears to be a degree of independence among events, among participants, and among places where history takes place. Conditions that should, according to the principles of economics, be favorable to the evolution of powerful entities occur in many different places and at many different times. The same is true of the circumstances that limit or diminish power. Whether and to what extent these conditions have the predicted consequences can be known retrospectively by approaching history not as a single unique sequence, but as a multitude of semi-independent sequences. Comparative history provides a way of dissecting the roles of chance and necessity, of assessing how and to what extent the time-invariant and space-invariant laws intersect with the unique attributes of individuals and their surroundings. By examining history on the time scales and spatial scales appropriate to interactions, we can expect to make generalizations about the contributions of contingency and necessity.

Economic perspectives should apply to any biosphere. When we discover biospheres other than our own, we shall be in a position to assess which aspects of economic history are predictable and which are not. My position throughout this book is that biospheres will show parallels in economic development, with the timing and extent to which the trends are played out depending on the physical circumstances and chemical make-up of the biosphere in question. This chapter, then, explores those aspects of our own biosphere's history that I hold to be predictable from first principles; and it probes Earth's specific contingencies of when and to what extent major economic trends and innovations do take place.

The Establishment of Historical Pattern

We and other living things inhabit a genealogical world, one that, until we invented language and all its sequelae, was dominated by transmission of adaptive hypotheses along lines of descent from one generation to the next. With the elaboration of computer-aided methods to infer the order of evolutionary branching in the tree of life, evolutionists have sought to trace adaptation through time and to identify or evaluate adaptive trends by superimposing adaptations or inferred levels of performance on the evolutionary tree. For these practitioners, historical patterns must be studied within clades, because trends—insofar as they exist at all—arise through patterns of descent with modification, where descent is taken to imply a genetic continuity. These patterns reflect branching and differentiation, where each branch, once formed, is a genetically independent unit that may economically interact with other branches but which has become isolated from them genetically.

Evolutionists have long known that many organisms represent integral unions between two or more historically separate branches. As discussed in chapters 2 and 6, these unions—the eukaryotic cell, plant-fungal mutualisms, and many others—often succeed because they offer the cooperating participants collective advantages that would have been unattainable by the partners in isolation. Unions of this kind can be recognized through careful analysis of the genomes of their partners. Once formed as units on which selection acts, they evolve in the more traditional mode of descent with modification. For this reason, history as approached by evolutionary biologists has been predominantly seen in terms of branching genealogy.

Historians of human affairs have not labored under this constraint. To be sure, they concern themselves with genealogies of rulers, nation states, and other institutions. They study intellectual history by tracing ideas as they are transmitted from mentors and modified by students. But historians also place a heavy emphasis on revolutions, take-overs, and other phenomena that bear no close analogy to genetic transmission. The history of rulers in England, China, Japan, and nearly every other political entity is one of inherited power frequently punctuated by episodes of replacement. Religious institutions evolve and split and compete as many other types of entities do, but there are also many cases of conversion from one doctrine to another. Technology and information spread not just from one generation to the next, as in traditional biological evolution, but often in the opposite direction, as when children teach their elders. In short, human-economic activities, entities, and attributes have become increasingly liberated from the genealogical constraint.[4]

The perspective of replacement is, of course, familiar to paleontologists. It is common knowledge that "mammals replaced dinosaurs," that survivors of the Permian crisis came to fill ecological roles vacated through the extinction of many Paleozoic clades, and that flowering plants pushed more ancient conifers and other plants to higher latitudes and altitudes as they spread and diversified during the Cretaceous period. What these patterns tell us is that the history of performance in a clade depends very much on the competitive environment in which members of that clade find themselves. This competitive environment changes from place to place and over time, sometimes easing, sometimes intensifying. The Cenozoic descendants of Paleozoic brachiopods are living in environments—cold, dark, and oxygen poor—that are restrictive compared to the range occupied by members of this clade during the Paleozoic, when competitively more vigorous bivalved molluscs were less abundant and perhaps metabolically less active. Dominant ancestors can give rise to subordinate descendants, or weedy opportunists can spawn members of the economic ruling class. Clades can undergo ecological expansion, invading what for them are novel environments and new ways of life and adaptation, or they can suffer restriction, persisting in refuges on the economic margins.

In short, historical patterns arise in three complementary ways in economic systems: (1) through evolution—descent with modification—fundamentally a branching process of differentiation; (2) union or coalition—the coming together of separate entities to form integrated larger wholes, which can then evolve; and (3) replacement—one genealogical lineage or coalition partly or wholly giving way to another one which is not on the direct line of descent.

If we are tracking historical performance through time, or identifying any other historical pattern, we must follow events within lineages and within clades, identify instances of union, and determine instances of replacement. Performance, insofar as it is affected by the universal process of adaptation, is context dependent, where the context comprises both the prior history of the entity whose performance we are evaluating and the economic power and distribution of all the other entities—related and unrelated—with which our entity is interacting. This is as true for the explanation of the history of humanity and its institutions as it is for unraveling the history of life. Tracking people and their things through time is as much a matter of working out family trees as it is of documenting the births of collective entities—nations, tribes, religions, labor unions, scientific societies, social clubs, and the like—and the changing context in which life is led.

History, then, is not random change. Among competitive dominants, there is a trend toward increased power through time. Among economic systems, there is a general trend toward increased diversity of membership and increased productivity. Dominants competitively push subordinates into economically marginal positions, and therefore power an overall expansion of economic activity. These directions of change are superimposed on a great deal of "noise." If we think of an economy of entities as being analogous to a cloud of molecules in a gas, the more powerful ones form neighborhoods of instability that act as sources of convection, currents that spread heat and power and that entrain other molecules in active ways. In other words, the system as a whole is not one dominated by passive diffusion of adaptations, but one in which change initiating with external triggers and with internal powerful entities spreads through active transport of momentum and information. Subordinate entities are influenced by their competitively superior neighbors and move accordingly. In essence, economic change can be understood best as a herd response, in which individual entities follow those in power or respond to those in power rather than engaging in a collective random walk. Economic forces impart directions—an arrow of time—to history.

Diversity and Productivity

If there is one arrow of time in the biosphere on which a majority of historians would agree, it would be expansion. From an initially small range of actors and interactions, an ever larger number develops with time, resulting in an ever increasing number of ways of making a living or performing tasks. The number of interactions increases even faster than does the number of entities, ensuring a rise in complexity. By complexity I mean one or more of the following: (1) the number of steps needed to construct a given entity; (2) the number of levels of organization in a hierarchically structured entity, that is, in an entity organized so that units at one level are included in a unit at the next higher level; (3) the number of processes occurring within an entity at any given time; and (4) the number of different parts within an entity at any given time. An organism consisting of one hundred cell types organized hierarchically into tissues, organs, and organ systems would thus be more complex than one of just five cell types organized into only tissues. Complexity thus embodies the notion of diversity, measured as the number of kinds of things. With proliferation and increasing complexity comes increasing diversity, and on longer time scales the emergence of new levels of inclusion. In short, economic history is a general, though neither constant nor wholly consistent, trend of expansion. Diversity begets diversity, subject to the imposition of disruptions. Interactions that create or strengthen positive feedbacks may be favored over those that impose constraint; hence diversity and productivity will increase together in a biosphere whose expansion is driven by internally driven, but perhaps externally triggered, self-reinforcing processes.[5]

Elements of these ideas have formed part of the intellectual landscape of evolutionary biology for three decades. Stephen Stanley, for example, noted in a classic paper in 1973 that all clades show an increase in the range of body sizes of their members as the clades expand through repeated branching from a single species to many daughter species. Highly inclusive clades, which taxonomists like me would call phyla or classes in the hierarchical system of classification first rigorously adopted by Linnaeus, often began with a very small ancestor. In a process that could be described as passive evolutionary diffusion, repeated branching of the original lineage and its immediate descendants would yield many descendant species, most of which were larger than the founding member. A similar increase in range applied to the complexity of individuals, because most ancestors of ultimately diverse clades had an organization with few gene duplications, little differentiation in function, little redundancy, and a limited division of labor. In most episodes of expansion, the clade produces a greater number of descendants with a more complex organization than it produces descendants that are simpler in structure. Limitations imposed by organization and environment place upper (and probably also lower) limits on the size and complexity of descendant individuals, but these constraints are pushed back from time to time by rare organizational innovations or by environmental circumstances that are unusually forgiving of errors and obvious adaptive shortcomings.[6]

Evolutionary biologists have identified several key innovations in organization that, once invented, led to vast increases in the range of body size and in diversity. One of these was the symbiosis among various bacterial and archaeal cells to form the eukaryote. Whereas bacteria have a maximum diameter of about 0.74 mm, single eukaryotic cells range from about this size to a few centimeters (large fossil foraminifers, for example). With their new ability to engulf particles rather than to take up nutrients molecule by molecule through the cell wall as bacteria do, eukaryotes were able to become predators of whole organisms. Multicellularity in eukaryotic lineages evolved at least ten times, leading to plants, animals, and fungi. Once again, the size range expanded enormously to include organisms with linear dimensions of hundreds of meters. Diversity came to be expressed in behavior and morphology, rather than in biochemical metabolic pathways as in bacteria and Archaea.[7]

The evolution of sex, and later of internal fertilization, greatly increased the potential for life to diversify along dimensions of shape and behavior. Dimensions along which traits and performance could vary that simply did not exist previously now became relevant when first gametes, and later—with internal fertilization—whole organisms, began to choose or to be chosen as mates. In other words, the dimensionality along which performance is measured and competition occurs increased.

The emergence of complex social systems, in which genetically similar or identical individuals achieve high levels of coordination, again greatly extended the reach of entities. In concert with the evolution of language, writing, and other cultural means of transmitting knowledge and experience, social organization in humans spawned enormous diversity in occupations, inanimate products, and social institutions.

As each of these innovations came to be deployed, the old entities were restricted but for the most part not replaced. Instead, they continued to perform crucial roles in the new order, and overall diversity rose as each organizational advance was added to the overall economic hierarchy.

As I pointed out in the discussion of organization in chapter 6, two fundamental processes—differentiation and recombination—yield diversity. In differentiation, an initially homogeneous unit produces daughter units or simply expands, with new parts coming to differ slightly in both form and function from the old by virtue of the accumulating and pruning of errors. In essence, this is what happens in the branching process of evolution as traditionally understood and as first outlined by Charles Darwin. In recombination (a word used here in its general sense, not in the restricted sense of genetic recombination in biology), a small number of distinct types of units—nucleotides, elements, letters of the alphabet, amino acids, zooids, cell types, and so on—come together in a vast array of configurations to form larger structures. This kind of recombination may have characterized early prokaryotic life before the branching mode of evolution became dominant. It also characterizes languages, proteins, immune systems, neural networks, compounds, communities of species, human institutions, buildings, and the genome in sexual organisms (genetic recombination). Often, perhaps usually, recombination and branching go hand in hand, for most systems in which recombination occurs have an evolutionary history in which "lineages" can be traced by following them back through branching events to some initial state. The important point is that both processes, separately and together, create diversity. Every entity thus created has the potential to spawn yet more diversity through branching or through recombination in an expanding universe of possibilities.

At this point, the skeptical reader may well ask why, if variety has increased more or less consistently through time, we are living in a time in history when variety is decreasing. Species and languages are becoming extinct in the face of increasingly rapid and cheap transport and communication. Cells making up the bodies of animals, plants, and fungi have tended to lose organelles relative to their autonomous ancestors as these cells take on specific functions as part of an integrated larger whole. Even the union between the ancestor of the mitochondrion and the archaebacterium that was to be its host involved the loss of functional components in each of the two parties. Counteracting this loss of diversity among cooperating parts of the whole, however, is the vast increase in diversity of the wholes: books and occupations in the human realm, cell types and species in the biological realm. The level at which diversity is expressed becomes more inclusive as economies increase in size and in productivity; and as this happens, diversity at those more inclusive levels rises.[8]

The most compelling evidence for accumulation—for increased diversity—comes from analyses, first undertaken by Richard Bambach a quarter century ago, of the number of co-occurring species of fossil animals in local assemblages from physically similar habitats through time. The median number of species in North American assemblages from mud and sand bottoms rose from 17.5 during the Cambrian period (543 to 505 Ma) to 25 during the succeeding Ordovician period and 34.5 in the Silurian. It leveled off, with some extinction-related fluctuations, until the Cenozoic era (after 65 Ma), with the median numbers during the Miocene and Pleistocene epochs being 60 and 62 species, respectively. Similar patterns of increase—a rapid rise, followed by a plateau, followed by another rise—characterize assemblages of land plants and rock-boring marine organisms. All the studies point to a low diversity until the Early Cambrian, a rapid increase through the middle of the Ordovician period, a general plateau during the rest of the Paleozoic era and the early and middle Mesozoic to the mid-Cretaceous, and then a sustained rise until perhaps early Miocene or early Pliocene time, followed by a slight fall. In detail, as seen through the lens of a finer division of time, this pattern will undoubtedly prove to be more complex, with frequent short-term rises being punctuated by declines or plateaus of short duration; but the overall trend is toward an increase in diversity.[9]

By itself, diversity is a rather abstract notion. In its species-level manifestation, somewhat ostentatiously labeled biodiversity in recent years, it has become the object of intense study and fierce political debate about whether it is valuable, desirable, or necessary. As usually understood in this narrow sense, however, diversity is little more than a collection of names, a list of things, an expression of variety that, divorced from its context, tells us little about the system that produced it and maintains it or about the variants its comprises. Fortunately, diversity need not remain an abstract concept.

Diversity reflects and permits economic activity. Generally speaking, a positive feedback exists among diversity, competition, and productivity (or collective power). There is therefore a historical link between the general trend in economic systems toward increased diversity and a general trend toward increases in the power of consumers and the rate of production.[10]

Potentially at least, every new type of unit created in a diversifying economy itself becomes a resource which another entity could exploit as a habitat, source of energy, or ally. This potential becomes reality when the resource achieves sufficient stability and predictability, either through its own activity or with the help of others. Any economic activity that enhances an entity's own resource supply and therefore its own growth and persistence will therefore make it an increasingly profitable target for another entity to exploit or cooperate with. Increasing productivity through the harnessing and regulation of supply therefore spreads through an increasingly diverse economy, creating opportunity for old and new entities as it does so until a limit imposed by extrinsic size or some other circumstance is reached. To economists, this kind of feedback translates into the more efficient allocation of resources among competing entities, an allocation not achievable in a monopoly or in a system of inflexible top-down control in which economic units lose all autonomy. It is this kind of feedback that creates forests, in which consumers and producers enhance one another's prospects in spite of competition among them, and in which trees of one species often improve the soil not just for themselves but for their neighbors. Productivity and diversity beget themselves and each other, together creating a stable, relatively flexible, somewhat disturbance resistant, diversified economy with diffuse top-down control by high-energy consumers and producers.[11]

Critics will counter that an increase in productivity—or, at least, an increase in nutrient availability—almost invariably leads to a decrease in diversity, a trend exactly opposite to the one I just proposed. One or a few fast-growing species take over and exclude or marginalize all the others. But if the high-nutrient conditions persist, and members of the affected ecosystem adapt to the changed circumstances, diversity is apt to rise. This happens for several reasons. Small isolated populations, which under a low-productivity regime might disappear for want of offspring sufficiently numerous to keep the population going, are more likely to persist if nutrient supplies are high and consistent. If they persist long enough, these isolates can form new species, genetically independent of the parent population. Moreover, the fast-growing numerical dominants create new environments and new resources, which can become targets of species arriving from elsewhere or evolving in place. In short, the temporary decrease in diversity observed as nutrients become more available gives way to a long-term increase, orchestrated through evolution and adaptation to altered economic circumstances.[12]

The trend toward higher economic metabolism and productivity has been evident from the very beginning of the biosphere. From the earliest signs of life about 3.5 Ga to the present high-energy civilization presided over by humanity, a succession of increasingly powerful forms of metabolism has characterized economically dominant entities and has increased the rate of resource use in the economy as a whole. Each new, more powerful type of metabolism originates in the most productive parts of the biosphere, where it is superimposed on the already existing, slower forms of metabolism. The overall economic system thus builds up by adding increasingly rapid means of transforming energy and raw materials to an existing architecture of more ancient means, which remain functional even if their reach and influence contract as the newer metabolic pathways take overall economic and evolutionary control.

Among the earliest forms of metabolism to arise were anaerobic pathways in which microbes exploited the chemical energy in bonds between atoms of molecular hydrogen, carbon dioxide, organic substances, and various sulfur-containing compounds. Probably these pathways arose in warm conditions near volcanic and hydrothermal sources of raw materials. By 3.47 Ga or earlier in the Archean eon, bacteria operating in the absence of free oxygen were combining hydrogen or organic matter with sulfate (SO_4^{2-}) to yield energy and such sulfide minerals as hydrogen sulfide and pyrite. In the microbial mats that comprised Archean ecosystems, these sulfate-reducers occupied a layer on top of a layer composed of Archaea, which recycled nutrients present in the dead cells from above to produce hydrogen and methane. Still other microbes were able to process methane. When bacteria invented the ability to capture energy from light in a photosynthetic pathway that yields organic matter but not oxygen, they formed a layer in the microbial mat on top of the sulfate-reducers. Finally, still another layer was superimposed by Cyanobacteria, which evolved the more familiar oxygen-liberating form of photosynthesis no later than 2.7 Ga. Each new metabolic pathway led to higher productivity by the dominant groups in which it evolved, and each enabled life to expand to places increasingly remote from volcanic and hydrothermal sources of energy and nutrients. By the end of the Archean, 2.5 to 2.6 Ga, life occupied the land as well as the sea, and had therefore achieved a global reach.

Metabolic rates during the Archean and Paleoproterozoic era (2.5 to 1.8 Ga) were severely constrained by the very low concentrations of free oxygen and by low rates of supply of other essential resources. In the nearly oxygen-free atmosphere of the time, sulfur-bearing minerals in rocks had low solubility and therefore weathered out very slowly, with the result that the supply of sulfate was limited. Moreover, phosphorus was also in short supply, because any dissolved phosphorus in the ocean became unavailable to organisms as it combined with abundant iron oxide minerals in the ocean to yield insoluble minerals in the sediment. Organically produced nitrogen may also have been buried in sediment.

An increase in the concentration of oxygen in the atmosphere and near the ocean's surface from 2.4 to 2.0 Ma, during the Paleoproterozoic era, was made possible by the spread of microbes that liberated oxygen during photosynthesis and by the burial of organic matter in sediments. With the higher concentrations of free oxygen, sulfur-containing minerals weathered more quickly from rocks exposed on land, and therefore increased the supply of sulfate to the ocean. This enhanced supply, in turn, raised the concentration of sulfides in the deep sea. Under conditions of abundant hydrogen sulfide, phosphorus is released from insoluble minerals in the sediment, and therefore becomes more available to organisms. These conditions prevailed during much of Mesoproterozoic time, from 1.8 to 1.25 Ga.

Although phosphate was now more abundant, productivity in the biosphere as a whole may have become severely constrained by limits on the availability of nitrogen. In sulfide-rich waters, iron and molybdenum—elements that are crucial components of the enzymes that catalyze the fixation of nitrogen (transforming molecular nitrogen, N_2, to ammonia, NH_3), nitrification (combining ammonia with oxygen to yield nitrate, NO_3^-), and denitrification (transforming nitrate back to molecular nitrogen)—bind to sulfides to form insoluble minerals. These elements are therefore unavailable to organisms taking part in the critical nitrogen cycle, meaning that the rate at which nitrogen-containing proteins are synthesized is low.

An increase in free oxygen beginning 1.25 Ga and accelerating 400 my later ended the sulfide age in the world's deep ocean, and may have returned the biosphere to long-term limitation of productivity by phosphorus. This switch coincided with, and was likely related to, the emergence of abundant seaweeds—multicellular eukaryotic primary producers on shores and shallow seafloors—and eukaryotic phytoplankton. These new photosynthesizers, which seem to have arisen in productive nearshore ecosystems, were able competitively to dominate Cyanobacteria in environments where nutrient supplies were high. By the Late Neoproterozoic era, 600 Ma, productivity was high enough to support multicellular consumers.

Oxygen has continued to increase unevenly and episodically through the Phanerozoic. By 370 Ma in the Late Middle Devonian, at a time when forests of land plants had risen, oxygen was abundant enough—perhaps 15 percent of air by volume—to sustain wildfires, as chronicled by the oldest known charcoal. Oxygen levels in the atmosphere during the Carboniferous may have equaled or exceeded the concentration of 20.9 percent observed today. There is a general consensus that oxygen levels dropped somewhat during the Permian, but by the Jurassic and especially the Cretaceous, oxygen may again have made great increases in metabolism—the evolution of warm-bloodedness in the late Permian or Early Triassic, the evolution of insect flight during the Carboniferous and of vertebrate flight in the Late Triassic—possible.[13]

The rate of accumulation of biomass available for consumption continued to increase episodically over the course of the Phanerozoic eon. Perhaps the most compelling evidence comes from fossil shell beds. These are layers of sedimentary rock in which the skeletal remains of animals comprise the bulk of the volume, with grains of sand and silt filling the spaces between shells or other skeletal fragments. The oldest shell beds, dating from the earliest stage (Manykay or Nemakit-Daldynian) of the Early Cambrian, are generally less than 10 cm in thickness, and contain "small shelly fossils" of unknown evolutionary affinity. The abundance and thickness of shell beds slowly increase through the rest of the Cambrian. In the Tommotian stage, shell beds remain thin, but are now dominated by trilobites. Inarticulate (lingulate) brachiopods and echinoderms become more abundant from the Late Cambrian onward, but trilobite dominance continues into the Early Ordovician. From the Middle Ordovician to the Jurassic, brachiopods were the most important contributors to shell beds, although other groups—bryozoans, echinoderms, and molluscs—were locally abundant as well. Bivalve and gastropod molluscs form the bulk of later Mesozoic and Cenozoic shell beds, with barnacles locally joining them from the Oligocene onward in the Cenozoic. With these replacements, the thickness of shell beds generally increased. In the Ordovician and Silurian, 87 percent of shell beds are less than 20 cm thick, and the thickest concentrations have a thickness of 65 cm. By the Jurassic, 64 percent of the beds were still thin (less than 20 cm), but the maximum observed thickness is 2 m. Only 33 percent of Neogene (Late Cenozoic) shell beds can be classified as thin, and beds of this age can be up to 6 m thick. The most straightforward interpretation of this pattern is that productivity of skeleton-bearing animals has increased through time, and that more productive groups successively replaced each other as the numerically dominant members of their marine communities on level sea bottoms.[14]

An intriguing manifestation of a likely increase in productivity through time is Richard Bambach's observation that early animals are "flat, not fat." Compared to their modern counterparts, most marine animals of the Cambrian in particular and the Paleozoic in general consisted of little more than thin films of tissue covering or enclosed by a skeleton. For example, the predominant echinoderms of the Paleozoic were sessile suspension-feeders with food-gathering arms often elaborately branched into brachioles and pinnules, servicing a small central mass containing the reproductive organs. Such echinoderms still exist as sea lilies (crinoids) and brittle stars (ophiuroids), but the predominant echinoderms today are thick-bodied seastars (Asteroidea), sea urchins (Echinoidea), and sea cucumbers (Holothuroidea). Paleozoic arthropods were numerically dominated by trilobites, whose internal organs were small in one or two dimensions, much as they are still among living low-energy sea spiders (pycnogonids) and horseshoe crabs (limuloids). Decapod crustaceans, a major clade with Devonian origins, are bulky animals by comparison, which is why humans find crabs, shrimps, lobsters, and even large barnacles (Cirripedia) worth eating. Brachiopods—sessile bivalved suspension-feeders—predominated on seafloors from Ordovician to Jurassic time. They persist today in cool, deep, or cryptic environments as slowly growing, slowly metabolizing consumers. Much of the shell's internal volume is taken up not with tissue, but with the calcified supports for the filtering organ. In addition, up to 50 percent of the animal's biomass is actually situated in the shell wall, as tiny extensions of the mantle, the organ that builds and lies adjacent to the shell. Bivalved molluscs, on the other hand, tend to fill their shells with massive soft parts. Although bivalves and brachiopods have comparably ancient origins in the Early Cambrian, bivalves effectively eclipsed brachiopods in abundance after the Jurassic.[15]

What these patterns appear to document is an increase in the productivity of consumers, that is, of animals that nutritionally depend on primary producers. Whether the patterns also imply a rise in primary productivity is less clear, but I shall argue that they do. Feedbacks between consumers and primary producers, discussed in chapter 6, have had the effect of making a larger proportion of primary production available to animals as compared to decomposers (microbes and fungi) and of facilitating the recycling of nutrients back to the primary producers instead of being lost to the biosphere. To recapitulate briefly, primary producers make organic matter and release oxygen. The rate of production can be limited by any one of several elements, but phosphorus appears to be the most universal of these. With a larger supply of phosphorus, productivity can increase, and some of the organic matter produced settles to the seafloor, where it is buried in sediment out of reach of organisms that would combine it with oxygen for fuel. Burial of excess organic matter therefore leads to an increased concentration of oxygen, which in turn can support large, aerobically metabolizing consumers. Some of these consumers can burrow into the sediments, where they retrieve nutrients, which are then recycled back into the ecosystem for primary producers to use. The result of these and other feedbacks should be an increase in net primary productivity, that portion of photosynthetic production potentially available to consumers. The aim of the next few sections of this chapter is to document when these feedbacks were able to operate, what prompted the initiation of the feedbacks, and how the levels of performance of producers and consumers changed over the course of time.

Plankton and the Consumer Age

In the past as today, the overwhelming bulk of production in the sea was carried out in the plankton, notably by Cyanobacteria and other single-celled phytoplankters. Most of these organisms are small and have no mineralized parts, and therefore do not fall to the seafloor where they can leave fossil remains. Chemical markers perhaps diagnostic of Cyanobacteria have been recognized in rocks as old as 2.75 Ga in Australia, and this date may be taken as the best current estimate for the time when the first oxygen-producing photosynthesizers first evolved. At about 1.8 Ma, in the latest Paleoproterozoic era, these primary producers were joined by simple, spherical leiosphaerid acritarchs. For 1.3 billion years, planktonic acritarchs evolved little and remained at low apparent diversity until small, spiny, planktonic acritarchs make their appearance at the beginning of the Tommotian stage of the Early Cambrian.[16]

According to Nicholas Butterfield's penetrating analysis, informed by studies of modern phytoplankton in nutrient-poor waters of the open ocean, this pre-Cambrian phytoplankton may have been consumed almost entirely by single-celled zooplankters such as flagellates, ciliates, and amoebae. Like their tiny food organisms, these unicellular consumers would not have sunk to the sea bottom after death. Not only did they fail to leave a fossil record, but more importantly they recycled organic matter within the plankton in the so-called microbial loop. Dead cells or their disintegrated remains that did eventually reach the seafloor would have been consumed by bacteria and, beginning no later than 750 Ma during the Late Neoproterozoic era, by a diverse assemblage of vase-shaped bottom-dwelling amoebae and other single-celled consumers. It is possible that there were also very small multicellular animals present, forms resembling acoel flatworms and perhaps other minute metazoans that feed on tiny individual particles or on dissolved organic matter. No fossils of such animals have yet been recovered, but controversial inferences from molecular sequences of living animals have led some evolutionary biologists to suggest that multicellular animals could have arisen as early as 1.2 Ga. An alternative idea I now favor is that, although the lineage leading to multicellular animals may have been distinct from that of other single-celled protistans for hundreds of millions of years, multicellularity in animals may not have arisen until 800 Ma or as late as 600 Ma. In any case, early animals were small, and could have consumed not only individual particles, but also large (0.13 to 2.7 mm) spiny bottom-dwelling acritarchs, which evolved around 1.1 Ga and diversified later in the Neoproterozoic.[17]

This pre-Cambrian universal microbial loop of the plankton, which for the most part failed to produce a surplus usable by larger consumers, was replaced near the beginning of the Cambrian by an ecosystem in which animals living in the plankton as well as on the seafloor converted excess production into larger bodies. There was, in other words, a revolutionary transformation from a subsistence economy with little top-down control and modestly developed anticonsumer defenses among the primary producers, to a more complex economy productive enough to support large populations of larger, actively metabolizing consumers, which began to exercise strong evolutionary control on their food organisms. Thanks to this top-down influence, planktonic acritarchs beginning in Tommotian time during the Early Cambrian developed spines as defenses against consumers. Moreover, consumers initiated or greatly amplified positive feedbacks between consumption and the nutrient supply for primary producers. With the advent of these animals, therefore, the biosphere entered what we might call the consumer age. The economic regime of the Proterozoic eon gave way over an interval of perhaps tens of millions of years to the new order of the Phanerozoic eon.

The first multicellular fossils that are generally thought to have been animals belong to the so-called Ediacara fauna, and arose some 600 Ma during the Late Neoproterozoic era. These nonmineralized fossils have long puzzled paleontologists, not least because they appear to lack feeding structures and a digestive system. Perhaps they absorbed dissolved organic matter through the general body surface from the surrounding water. Minor elements in the worldwide Ediacara fauna could have been small bottom-dwelling suspension-feeders. There is, for example, evidence of millimeter-scale sponges as old as 560 Ma, and large mineralized sponges or cnidarians (*Namapoikia*) up to a meter in diameter and 25 cm high formed reefs in Namibia during the latest Neoproterozoic, 548 Ma. These and other nonmineralized forms could therefore have ensnared planktonic organisms and their remains.[18]

Undoubted suspension-feeders were present by earliest Cambrian time, but they were not similar to later types. They fall into two distinct categories: (1) relatively large sponges and cnidarians, and (2) very small bilaterians (animals with more or less equal right and left sides). Exceptionally preserved assemblages from the Early Cambrian of China have revealed the presence of large sponges, growing 20 to 50 cm above the seafloor. There was at the same time a substantial diversity of rigidly mineralized archaeocyathan sponges, which built loose reeflike structures. Today, sponges are renowned as active suspension-feeders capable of extracting bacteria-sized particles and dissolved organic matter from the surrounding water. Very likely they already had this capability by Early Cambrian time. Other suspension-feeders in the plankton and on the seafloor were for the most part strikingly small. They include Early Cambrian branchiopod crustaceans, the peculiar arthropod *Isoxys*, probably some agnostid trilobites, and an assortment of tiny bivalves, lingulide and other brachiopods, hyoliths, and various tube-dwelling organisms. It is far from certain whether all these animals were in fact suspension-feeders. The tiny bivalves, for example, more probably picked individual particles from the sediment surface. Although suspension-feeding larvae (trochophores) probably existed for short durations at or just above the seafloor, evolutionary studies strongly imply that plankton-feeding larvae such as molluscan veligers, echinoderm plutei, and others had not yet evolved in the Early Cambrian. I strongly suspect that the bacterial fraction of primary producers was far larger than the fraction composed of larger phytoplankters such as acritarchs, and that suspension-feeders capable of harvesting these larger forms were severely limited in size by a low food supply.[19]

Comparisons of suspension-feeders from successive intervals of time imply that the productivity of plankton, or at least the proportion available to consumers, increased over the course of the Phanerozoic eon. Consider, for example, the skeleton-bearing suspension-feeders that live attached to other shell-bearing animals. Such encrusters—hydroids, sponges, bryozoans, brachiopods, mussels, oysters, some gastropods, and barnacles, among others—occur prolifically on the shells of living surface-dwelling molluscs as well as on empty shells in coastal regions where planktonic productivity is high. By contrast, encrustation in regions where plankton is less abundant, as on many coral reefs, is a habit mainly for coralline algae and less commonly for small corals. Patterns of encrustation therefore reflect, at least in a qualitative way, the planktonic productivity of the water to which the encrusted shells are exposed.

My former student Halard Lescinsky has found that shell-encrusting suspension-feeders do not become common until the Middle Ordovician, at the same time that trilobite-dominated shell beds give way to brachiopod-dominated ones. The Cambrian and Early Ordovician (Arenigian) shells that Lescinsky examined—mainly brachiopods—are free of encrusters. The oldest encrusted shells in his survey come from the Middle Ordovician (Llandeilian) Bromide Formation of Oklahoma and the Benbolt Formation of Virginia. Shells in these formations are encrusted by chainlike and matlike bryozoans; some 1 cm long, wormlike *Cornulites*; a few cemented craniid brachiopods; and the holdfasts of crinoid and edrioasteroid echinoderms. The diversity and sizes of encrusting suspension-feeders increased in the Late Ordovician (Caradocian and Ashgillian), and remained locally high for the remainder of the Paleozoic era.[20]

The economic microcosm of shells as habitats for encrusting suspension-feeders reflects a more general trend from small, passive suspension-feeders to larger, more active types through time. Passively suspension-feeding echinoderms comprised an important element of the suspension-feeder guild during much of the Paleozoic, along with sedentary brachiopods, which produce weak food-gathering currents. Suspension-feeding bivalved molluscs averaged 10 mm in largest dimension during the Early Ordovician, and attained 25 mm or more only in the Llanvirnian stage of the Middle Ordovician. By the Late Devonian, bivalves had reached 100 mm in longest linear dimension. Mesozoic and especially Cenozoic suspension-feeders in sandy and muddy environments are overwhelmingly active feeders such as bivalves and barnacles, which produce currents to draw water and food particles into the digestive system. Particularly large active suspension-feeders are known from intervals of the Jurassic, Cretaceous, and the Miocene to Recent. Large passive suspension-feeders, especially corals and crinoids, persist mainly on reefs and in other situations where water movement is strong enough to replenish the supply of food quickly.[21]

The emergence of consumption and of surplus in the sea was manifested in two more or less distinct processes. First, bottom-dwelling organisms—burrowing animals and eventually rooted plants—began to mine nutrients that had accumulated in sediments beneath the seafloor and would previously had been out of reach of the primary producers in the waters above. Second, the products of photosynthesis, which had been packaged and recycled in dispersed unicellular organisms, came to be concentrated in larger packages—as living multicellular animals or their fecal pellets—that lived or settled on the seafloor. Because of this process, predators and herbivores shortcircuited the pathway of the products of photosynthesis, which before the consumer age would have passed mainly through microbes and single-celled protists including many decomposers. The result of these two interrelated processes working together was that more nutrients were being kept in circulation longer in the ecosystem. Each link in the transformation of raw materials to phytoplankton to consumer and back to the primary producers was represented by a population of organisms that, compared to those in the microbial loop, were far better defended against attack by the consumers that this more productive ecosystem evolutionarily created and supported. And all this, I would argue, happened because of an increase in nutrient supply, which led to an increase concentration of oxygen, which enabled larger-bodied life forms to arise.

The History of Mining

Sediments on the ocean floor offer two potential rewards to marine organisms. First, they provide safety from enemies. Animals that occupy permanent burrows in sand and mud, or that plow through the sediment, gain a certain degree of protection from predators that hunt on or above the seafloor. Second, sediments are a rich source of nutrients that accumulate from the waters above. Concentrations of nutrients in sediments are ten to one hundred times higher than in the overlying water. Animals burrowing into sediments retrieve these nutrients and the organisms that this source of food supports. In doing so, they impose two opposing effects, one enhancing and the other depressing production. On the positive side, burrowers increase the water content of the sediment and the oxygen concentration of the water between sediment grains, and therefore stimulate bacterial production. This stimulus makes the environment more hospitable both for the burrowers themselves and for other animals, because it increases the food supply. Animals partially or wholly buried in sand and mud become potential resources for larger consumers, which disrupt and aerate the sediment still further as they search for prey. Burrowing thus exemplifies an evolutionary positive feedback, or escalation, between sediment-dwellers and their consumers, possibly triggered by species seeking refuge from enemies.

On the negative side, burrowing animals interfere with primary producers anchored in the sediment. In experiments carried out by Susan Williams and her colleagues in the U.S. Virgin Islands, artificial plants were either buried or uprooted by the activities of burrowers. Whether this effect is important in nature is debatable, because the experimenters did not take into account the potentially helpful role of burrowers in fertilizing sediments with feces. On the whole, I agree with Charles Thayer that burrowers have a net positive effect on the productivity of sediments and the waters above them.[22]

Increased productivity probably caused, and was caused by, two concurrent temporal trends in the history of marine burrowing. The first of these trends is a marked increase in the intensity of movement of sediment by animals, measured as the volume of sediment swallowed or displaced per day by an individual burrower or by a population of burrowers. The second trend is an increase in the depth of disturbance.

From about 600 Ma, during the latest Neoproterozoic, the only known animal trails preserved in rocks representing sandy and muddy environments are narrow (1 mm diameter) meandering horizontal tracks probably made on or just beneath microbial mats occupying the top millimeter of sediment. Studies of tracks made by living animals suggest that these trails were made by small worms and bottom-dwelling cnidarians, which creep on the seafloor by means of cilia on ribbons of mucus. Sediment is carried in the mucus as the animals creep. Some deeper permanent burrows appear during the latest Neoproterozoic (545 Ma) in Namibia and the earliest Cambrian (543 Ma) in Newfoundland. These burrows were likely made by animals using forces applied by circular and longitudinal body-wall muscles acting antagonistically. At the beginning of the Cambrian, these burrows rarely penetrated more than 1 to 2 cm into the sediment, and the animals responsible for them remained uncommon except in shallow-water sandy environments. In the Tommotian stage (530 to 525 Ma), burrows extending to a depth of 6 cm are rather common, and some burrows in sandy habitats have been recorded to have a length of a meter. Most burrows, however, were permanent dwellings. Mary Droser and colleagues have shown that traces of animals plowing through sediments in search of food arose during the Tommotian, but remain uncommon and shallow (within the top 2 cm of sediment). Beneath this disturbed layer, sediments quickly compacted through the loss of water, and were therefore firm.[23]

To me, this suggests that most of the Early Cambrian burrowers entered the sediment in search of safety rather than food. The chief Cambrian burrowers—trilobites, minute bivalves, lingulide brachiopods, and probably various annelid and sipunculan worms—were low-intensity sediment-disturbers, likely displacing much less than 10 cm^3 of sediment per day, and perhaps rarely disturbing the sediment at all for most of their lives once a permanent burrow had been built.

During the Proterozoic and perhaps also the Cambrian, most of the sediment-dwelling animals may have been associated with microbial mats occupying the top surface of the sediment. Burrowers might have been micrograzers and detritus-feeders, perhaps picking up and ingesting individual particles rather than processing larger volumes of sediment and extracting nutrients from them in the gut. Nick Butterfield's detailed examination of sediment-filled guts of Cambrian arthropods such as *Leanchoilia* from the Burgess Shale of British Columbia reveals that the sediments entered the animals after death, and that these arthropods are better interpreted as predators. Butterfield, in fact, finds no compelling evidence for deposit-feeding during the Cambrian, and has suggested that this method of feeding evolved during Early Ordovician time. His views are consistent with the fact that major groups of deposit feeders—protobranch bivalves, many groups of annelid worms, crustaceans, heart urchins, sea cucumbers, and cephalochordates—are evolutionarily derived members within the major clades to which they belong. For example, if the tiny (1 to 5 mm) bivalves of the Cambrian are at all like modern animals of this size, they would have been feeding particle by particle rather than by ingesting packets of sediment as protobranch and tellinid bivalves do. No major clade of animals is primitively deposit-feeding. Mary Droser and her colleagues suggest that mat-dominated, firm sediments were not widely replaced by softer, less stable sediments until the Middle Ordovician.[24]

Larger burrowers emerged in the Early Ordovician period, and increases in the depth and intensity of burrowing occurred between middle and late phases of that period. Some burrows extended as much as 1 m below the sediment surface. Further intensification occurred by the Early Devonian, when some of the culprits may have been bottom-dwelling fish in search of buried prey. Still, the bulk of Paleozoic burrowers remained in Thayer's low-intensity category.

Many animal groups that previously had lived only on hard surfaces or on top of the sediment spawned lineages of burrowers after the great Permian crisis. Most became what Thayer classifies as intense and deep burrowers, those displacing 10 cm^3 of sediment per day or more and digging to a depth of 5 cm or more. Particularly important were bivalved and gastropod molluscs, sea urchins, crustaceans, fish and, from Early Eocene time onward, marine mammals. Some of these animals were deposit-feeders, others are predators, still others use the sediment as a hiding place. The most intense sediment-disturbers today are gray whales (*Eschrichtius robustus*), which displace more than 100,000 cm^3 of sediment per day as they hunt burrowing amphiopods in the Bering Sea; and dugongs and manatees, which uproot seagrasses in tropical seas. Large rays also disturb sediments on a large scale in many parts of the world as they exhume clams. The deepest burrowers are ghost shrimps—axiid crustaceans that construct elaborate burrows a meter to as much as 3 m below the sediment surface—which evidently culture bacteria on organic debris. All these intense and very deep sediment-disturbers belong to Cenozoic clades. In short, the intensity and depth of mining nutrients in sediments have both dramatically increased during the last 550 my, with the largest increases occurring in the earliest Cambrian, the Early to Middle Ordovician, the Late Silurian to Early Devonian, the Late Triassic to Early Jurassic, and the Cenozoic.[25]

Marine plants have done their part in mining nutrients from sediments and potentially making the retrieved matter available to other members of their ecosystems. The siphonaceous green algae—multinucleate, single-celled plants up to a meter long, belonging to the green-algal class Ulvophyceae—may have been the first primary producers to take up and transport ammonium, phosphate, and other nutrients from the sediment. They tend to have rootlike underground rhizoids that serve not only to anchor the plants in sand or sometimes in mud, but also to acquire nutrients from the water surrounding grains of sediment. It is not known how widespread this habit is or when or how many times it evolved. The earliest undoubted members of this group are dasycladaleans of Middle Ordovician age, but at least two Cambrian genera have also tentatively been attributed to the Dasycladales.[26]

Seed plants, with true roots, are probably far more effective at extracting nutrients than are seaweeds, but their history in the sea—or at least in coastal marshes and wetlands—is not well known. Some conifers may have formed a mangrovelike vegetation in coastal swamps during Carboniferous time, and there is evidence for such vegetation in the Early Cretaceous. "True" mangroves—members of fifteen families of flowering plants—are mainly Late Cretaceous and Cenozoic in origin. Perhaps more important are seagrasses, members of three major clades of monocots that invaded the sea during latest Cretaceous (Maastrichtian) time. Perhaps for the first time in the history of marine sediments, nearshore sediments were now extensively mined by primary producers, which today (and presumably ever since their appearance) for huge, highly productive meadows that support large herbivorous mammals, turtles, fish, and birds.[27]

Extraction of nutrients from sediments by organisms on land undoubtedly goes back to very ancient times, but the early phases of that history remain shrouded in mystery. Chemical weathering by lichens—cooperative associations between trebouxiophyte green algae and various fungi—and perhaps by early bryophytes (liverworts and mosses) must have affected the surfaces of rocks and sediments from at least the Proterozoic eon onward. Burrows possibly excavated by euthycarcinoid arthropods are known from coastal settings, as preserved in rocks of latest Cambrian or earliest Ordovician age in Ontario. These and other Ordovician burrowers indicate that animals were beginning to disturb sediments on land very early. The culprits likely ate detritus or searched for microorganisms living in the soil. Nutrient mining was thus well underway before vascular plants evolved in the Late Ordovician or later.[28]

On land, the most important miners by far have been plants with roots. Before the Devonian, chemical weathering by plants—lichens, bryophytes, and vascular plants without roots—was superficial, affecting only the top few millimeters of rock and soil. In the Early Devonian (Emsian stage), however, the roots of vegetation in well-drained soils preserved in Antarctica penetrated to a depth of 1 m. Depth of rooting had increased to 1.5 m by the Middle Devonian (Givetian stage) in a low-forest vegetation in New Brunswick, Canada. Today, roots in some tropical forests extend to a depth of 18 m. These depths indicate not only that plants are able to mine nutrients from layers that during Early Paleozoic times were largely out of reach of primary producers, but also that plants create and modify the subterranean environment to a degree no less profound than above ground.[29]

Plant roots, however, do not act alone. They take up nutrients in partnership with fungi, which live on or inside the root. Fungi belonging to the order Glomales, the clade containing many root-associated fungi, have been reported from Middle Ordovician rocks (460 Ma). With their appearance and with that of land plants, the rate of chemical weathering and soil formation on land may have risen by a factor of ten, contributing not only to a huge increase in productivity on land but also to an enormous enrichment of the nutrient base in the oceans. Just when increases began is very unclear. Very likely they occurred in steps, first with the emergence of rootless vascular plants in the Late Silurian (420 Ma), followed by the origins of roots and deeper penetration of the soil by the Early Devonian, followed by the first appearance of trees in the Middle Devonian, in turn followed by the appearance of seedplants during the latest Devonian (365 Ma). With the seed, which enabled plants to reproduce in the absence of liquid water, colonization of relatively drier environments was made possible, expanding the areal coverage of productive vegetation in the Carboniferous. Still another stepwise increase in productivity probably accompanied the diversification and spread of flowering plants, beginning about 120 Ma in the Late Early Cretaceous.[30]

Compared to marine and terrestrial environments, the bottoms of rivers and lakes are, and evidently always have been, relatively little affected by the mining activities of animals. Although potentially burrowing crayfish have existed since at least Permian time, along with mayfly nymphs and other larval insects, examination of preserved freshwater sediments by Molly Miller and her colleagues has revealed that the intensity of sediment disturbance was low until Jurassic time, when many insects specialized for burrowing—chironomid midges, for example—arose. Freshwater burrowing clams have existed since at least the Devonian, but no freshwater bivalve has evidently adopted the mobile deposit-feeding habits of many marine nuculanid and tellinid bivalves. Similarly, very few freshwater snails actively burrow.[31]

A History of Herbivory

The biosphere we inhabit is full of herbivores, ranging from the familiar rodents, hoofed mammals, and insects of the land to a wide assortment of snails, sea urchins, crustaceans, segmented worms, fish, turtles, and seacows in the ocean. This state of affairs is so familiar that it seems utterly natural, a condition hardly requiring comment or inspection. Yet, the evidence from the branching order of animal clades as well as the fossil remains of extinct forms indicates that the consumption of living green tissues is not just a fairly recent mode of feeding, but also an evolutionarily derived one.

Undoubted multicellular seaweeds—filamentous members of the still living red-algal family Bangiaceae—are known from as far back as 1.2 Ga during the Mesoproterozoic era, and had become rather diverse well before the latest Neoproterozoic and Cambrian appearance of skeleton-bearing animals. If these plants were consumed in living form, the culprits would have been unicellular, at most puncturing individual plant cells for their contents. Most of the biomass produced by these seaweeds would have entered the ecosystem by way of decomposers, which consume plants after death. With the evolution of herbivores, this link between producers and the rest of the ecosystem through decomposers was partially short-circuited and speeded up, as production was channeled directly into the bodies of animal consumers.[32]

There is surprisingly little concrete evidence for marine consumers of marine multicellular or multinucleate algae growing on the seafloor during the Paleozoic era. The most evolutionarily basal branches of the animal kingdom—sponges, cnidarians, and acoel flatworms—count suspension-feeders, detritus-feeders, and predators among their members, but no herbivores. Herbivore-containing clades—Mollusca, Annelida (segmented worms), Arthropoda, Echinodermata, and Vertebrata—all occupy more derived positions in the evolutionary tree, and likely arose somewhat later than the basal clades, though all were present no later than Early Cambrian time.

Whether Paleozoic members of these major derived clades in fact ate plants is uncertain. Among gastropod molluscs, for example, two large basal clades—the patellogastropods and vetigastropods, including limpets, topshells, abalone, and turban snails—were distinct by Late Cambrian or Early Ordovician times, but modern herbivorous members within these clades belong to highly derived groups with origins in the Triassic or later. Similarly among chitons

(Polyplacophora), comprising a clade of molluscs that was distinct by the Early Cambrian, all herbivorous members belong to the derived order Chitonida, which differentiated during the Jurassic period. Living members of basal gastropod and chiton clades feed on sponges, detritus, single-celled organisms, and other non-algal foods, and invariably belong to subclades traceable to the Paleozoic era.

The same conclusion applies to other major clades of marine animals. Sea urchins (echinoid echinoderms) are recorded as fossils from the Middle Ordovician onward, but the structure of the five-part jaw (the so-called Aristotle's Lantern) of Paleozoic and most Early Mesozoic forms is consistent with a diet of fine particles, including detritus. Jaws capable of tearing, scraping, or grinding algal food appear in the Late Triassic and do not become common until the Late Cretaceous. Herbivorous crustaceans—various isopods, amphipods, and at least four different clades of crabs—belong to clades with a Cenozoic origin. Herbivorous fish, which have evolved in more than ten clades of bony fishes, are also not known before the Cenozoic era, even though they are major algal grazers on modern reefs. Herbivorous marine turtles are no older than Late Cretaceous, and potentially marine herbivorous mammals—seacows or sirenians and fossil Desmostylia—appeared no earlier than the Early Eocene, about 50 Ma. In short, all living marine herbivores belong to evolutionarily highly derived clades, most of which cannot be traced further back than the Mesozoic era.

Another surprise, not noted by previous observers, is that conspicuously large plant-eaters arrived on the marine scene only during the Early Eocene, some 55 Ma. The largest modern marine herbivores are seacows (including the ten-ton, recently extinct Steller's seacow of the North Pacific, the dugong of the Indian and western Pacific Oceans, and tropical Atlantic manatees), the tropical green sea turtle (*Chelonia mydas*), and the 1.2-m-long tropical Indo-West Pacific parrotfish *Bulbometopon muricatum*. There is also a profusion of large gastropods, especially conchs of the genera *Strombus* and *Lambis*, which reach or exceed a length of 30 cm. Herbivorous snails 10 cm or more in shell length or diameter are represented among top shells (Trochidae and Turbinidae) and limpets (Lottiidae and Patellidae), mainly in tropical and warm-temperate waters. Although detailed evolutionary studies remain to be done, these herbivorous snails of modern seas mainly emerged after the Eocene. During the Eocene, large plant-eating snails were dominated by cowries (*Gisortia* and *Megalocypraea*, both at least 30 cm in length), campanilids (a group of turreted gastropods today represented by a single southwest Australian species, but known throughout the Eocene tropics by several species (including the 1-m-long *Campanile giganteum* of Europe, ampullospirid "moon snails," and perhaps the 10-cm-diameter neritid *Velates*. Seacows and parrotfishes likewise appeared during the Eocene.

The history of herbivory on land is more abbreviated but follows a similar course to that seen in the sea. The only types of herbivory that have been recognized among the early land plants of the Late Silurian and Devonian periods (410 to 360 Ma) are small penetrations associated with reactive (that is, responding) plant tissues, indicating the work of arthropods that pierce living plant tissues and suck out the fluid contents. The earliest evidence of attacks on leaf margins comes from the fronds of *Macroneuropteris* and other seedferns in the Mazon Creek beds of Illinois, which are of Middle Late Carboniferous age (307 Ma). In the latest Carboniferous, species-specific galls made by one of the earliest known insects with complete metamorphosis (in which the larva and adult have different forms) infested the base of petioles of the canopy treefern *Psaronius chasei* in Illinois. Beginning in the succeeding Early Permian period, consumption of leaves becomes common. Beck and Labandeira estimate that about 2.6 percent of 11,562 cm^2 of leaf surface area, mainly belonging to gigantopterid foliage, was removed by leaf-eating insects in a leaf assemblage of that age from Texas. Of the 1,346 leaves they examined, 31.8 percent were attacked in some way. The most frequently eaten leaf types sustained damage comparable in magnitude to that of living woody flowering plants. Consumption of wood in living stems began in the Triassic, and undoubted leaf mining—a form of consumption in which the insects and its feeding traces remain hidden inside the leaf—dates from the Late Triassic. With the ascendancy of flowering plants beginning in the Early Cretaceous, terrestrial herbivory possibly again intensified.[33]

The herbivores responsible for this damage reflect the trends observed in the leaves. Evidence from fossils and from phylogeny shows that herbivory is a derived condition among insects. Among the earliest recognized herbivores to arise were fluid-feeding bugs (hemipteroids), apparently from detritus-feeding ancestors, in the Early Permian. Other major insect clades in which herbivory evolved—Blattodea (including grasshoppers and other orthopterans), Holometabola (including beetles and lepidopteran moths and butterflies), among others—also began as feeders of dead plant material. Charles Mitter and his colleagues estimate that herbivory among insects arose more than fifty times independently, mainly during the Mesozoic era. Additional cases occur among mites—members of the arachnid order Acari—and land snails.[34]

By my conservative estimate, four-footed vertebrates (tetrapods) evolved the habit of eating the green parts of plants more than fifty times, apparently always from carnivorous (including insectivorous) ancestors. The first of these herbivores to evolve were *Desmatodon* and related diadectomorphs, comprising a clade of ancestral tetrapods interpreted to be the sister clade of amniotes ("reptiles," birds, and mammals). These slow animals are first recognized in rocks of latest Carboniferous (Stephanian) age, 303 to 295 Ma, one stage after the first appearance of amniotes. These herbivores were joined in the Early Permian by bolosaurid reptiles and by at least two groups in the large clade

Synapsida, *Edaphosaurus* and caseids, the latter including *Cotylorhynchus*, which at a length of 4 m and an estimated weight of 400 kg seems to have been the largest animal of its time. In what Robert Bakker dubbed the Kazanian revolution of the Middle Permian, a new cast of herbivorous tetrapods evolved from more generalized or carnivorous ancestors. They include two clades of Synapsida commonly referred to as "mammal-like reptiles," the Dinocephalia and the Anomodontia. Early anomodonts were evidently the first tetrapods capable of physically chewing (or breaking down into small fragments) their food in the mouth as some later dinosaurs and most mammals do. Other members of the Late Permian cadre of plant-eaters are the parareptilian pareiasaurs, which attained a maximum length of 3.5 m, and captorhinids.[35]

Another change in the complement of herbivorous vertebrates took place after the end-Permian extinctions. Among "mammal-like reptiles" the gomphodont cynodonts were among the largest Triassic herbivorous tetrapods. At a length of 2 m, the cynodont *Traversodon* from the Late Triassic of Argentina and India may have had cheeks, indicating that food was processed in the mouth before being swallowed. Dicynodont anomodonts also proliferated in the Triassic, as did the reptilian clade *Procolophonia*. In the Diapsida, the major clade later giving rise to dinosaurs and birds, herbivory evolved in the Middle and Late Triassic rhynchosaurs in the southern hemisphere, the Late Triassic Trilophosauridae in North America, and the Late Triassic prosauropod dinosaurs, first known from the southern hemisphere. The dinosaurs, which originated about 230 Ma, outlived the rhynchosaurs (extinct at the end of the Carnian stage of the Late Triassic, 215 Ma) and the other herbivorous groups (extinct about 200 Ma at the end of the Triassic) and then slowly proliferated and rapidly increased in size from a maximum of 8 m in length in the Late Triassic (the prosauropod *Plateosaurus* from Europe) to the Early and Middle Jurassic, when outrageous gigantism evolved. By the Late Jurassic, some sauropod dinosaurs such as titanosaurids and brachiosaurids may have reached a length of 30 m and a weight of 78,000 kg. Although dinosaurs continued to dominate the ranks of large herbivores to the end of the Cretaceous, they did not prevent a few other clades from adopting the plant-eating habit. There is a Mid-Cretaceous (Albian) herbivorous member of the mainly predaceous clade Sphenodontia (today represented by the New Zealand predatory tuatara, *Sphenodon*), as well as a Cretaceous plant-eating crocodile.[36]

It took some 10 my before the large dinosaurs of the Latest Cretaceous were replaced by relatively large plant-eating mammals such as the ungulate-like *Arctostylops* and the barylambdid pantodonts near the end of the Paleocene epoch. Large herbivorous mammals—amynodonts, brontotheres, dinoceratans, and true rhinoceroses—appeared in the Middle Eocene. Mammals with high-crowned, continuously growing cheek teeth suitable for chewing grass first evolved by the end of the Middle Eocene, but became very widespread among ungulates, elephants, and larger rodents during and after the Middle

Miocene. In South America and on many larger oceanic islands such as Madagascar, New Zealand, the Mascarine islands of Mauritius and Reunion, and the Hawaiian Islands the large-herbivore role was filled by various mainly flightless birds. Plant-eating evolved at least a dozen times among birds, as well as among tortoises and at least ten clades of lizards after the Cretaceous.[37]

A telling sign of how important herbivory has become is the grazing-resistant architecture of some of today's most abundant plants. In most seaweeds and land plants, new growth occurs at the tips (or apices) of branches. Removal of the apex by herbivores typically does not doom the plant, but instead interrupts growth, which resumes elsewhere through renewed branching away from the affected area. In the tree lycopods (club mosses) that characterize the canopy vegetation of forests during most of the Carboniferous period, however, there was a single zone of growth at the top of the tree. Damage to this vulnerable region would have resulted in the plant's death. Today's palms also grow in this fashion, but in these Cretaceous to Recent trees the apical growing region of dividing cells is hidden deeply below the expanding leaves and is therefore not especially vulnerable to damage from attackers. In Late Carboniferous time, lycopods were replaced in equatorial forest canopies in Europe and America by such treeferns as *Psaronius*, and in the Permian by conifers, which have multiple apices or at least the ability to grow even after the chief growing region is destroyed. This change in architecture in the canopy represents an example of plants formerly restricted to the forest understory becoming members of the canopy community after the demise of the old guard there. In this instance, it coincides with the appearance of significant herbivory by both arthropods and vertebrates.[38]

Grasses and related monocots have invented a novel means of growing continuously under constant assault from vertebrate consumers that remove the tips of the blades. Their zone of growth is situated near the base of the blade. As herbivores—or mowing machines wielded by humans in hayfields and on suburban lawns—remove the older parts of the blades, new tissue forms at the base, and is gradually moved in conveyor-belt fashion to the zone of consumption. Although the earliest monocot fossils date from the Turonian stage of the Late Cretaceous (90 Ma), molecular sequences imply that grasses (Poales, including also rushes, bromeliads, and cattails) diverged as early as the Aptian stage of the Early Cretaceous (115 Ma). Grasses did not become major elements of the vegetation until the Oligocene (30 Ma) in South America and the Miocene (15 Ma) in most other parts of the world. Some dicots—dandelions among Asteraceae, and the genus *Plantago* in Scrophulariaceae—have converged on the grass habit of growth, which also occurs among many other monocot groups. The only herbivores that appear to overwhelm the grass defense of growing at the base are sheep.[39]

Why the grass habit of growth did not evolve earlier among land plants remains an intriguing mystery. Herbivorous vertebrates, including anomodont mammal-like reptiles capable of chewing their food in the mouth, appeared as early as the Late Permian, and have existed ever since. The giant dinosaurs of the Jurassic and Early Cretaceous must have had ravenous appetites, especially those later dinosaurs with relatively large heads and the ability to chew food; and Robert Bakker has made the plausible suggestion that the regime of intense herbivory favored the evolution and diversification of initially weedy, fast-growing flowering plants in the Early Cretaceous. Perhaps the huge herbivores of the Mesozoic ate whole small plants, a type of consumption against which hiding growth zones would have been ineffective.[40]

Seaweeds, too, have found ways of sustaining herbivory without compromising growth. Uniquely among algae, kelps—members of the brown-algal order Laminariales—evolved basal growth of the blade. The blade is thus a bit like a conveyor belt, in which a cell travels from the basal growth zone to the end of the blade, where it is consumed by a grazer. This style of growth likely evolved in concert with the evolution of large grazing marine mammals—first desmostylians, later seacows culminating in the recently extinct North Pacific Steller's seacow *Hydrodamalis gigas*—during Late Oligocene time. Although grazing marine mammals originated in the Early Eocene and occupy both tropical and temperate seas, kelps are, and apparently always have been, restricted to temperate and polar latitudes. They originated in the North Pacific, and later spread (probably by long-distance dispersal) to the cold southern hemisphere and to the Atlantic. Seagrasses, which also grow from the base, have coexisted with marine grazing mammals since the latter's origin, but no tropical seaweeds have achieved the basal growth habit characteristic of the temperate kelps.[41]

Coralline seaweeds, which originated in their modern form during the Jurassic, have their reproductive organs and growth zones protected by an outer layer of cells. Grazers—chitons, sea urchins, snails, and fish—remove this outer layer while leaving at least some of the plant's reproductive structures and zones of growth intact. These remarkable seaweeds, whose bodies are heavily impregnated with calcium carbonate, form resistant crusts on the wave-washed seaward crests of tropical reefs as well as in many cold-water environments. Robert Steneck's careful studies have shown that they evolved in concert with grazers powerful enough to gouge hard surfaces.[42]

Still another architectural indication that herbivory increased greatly through time comes from the venation pattern of leaves. In all Devonian and most Carboniferous leaves, the veins that conduct water and nutrients to the leaf and the products of photosynthesis away from the leaf to the rest of the plant form an open branching system, in which repeated branching of conduits reaches all parts of the leaf without any of the branches coming together again. There is, in other words, only one pathway from the junction between the leaf and the stem to any point on the blade. By contrast, the overwhelming majority of plants today are characterized by a network pattern of venation, in which branches frequently join at two or more points. Multiple pathways exist between the base of the leaf and any position on the leaf. A network of this kind provides critical redundancy. Damage to veins by herbivorous insects, for example, need not compromise the water-conducting or food-conducting functions of the leaf, because the network ensures that pathways of transport around the damaged region are preserved.

One explanation for the near absence of network venation patterns in the early history of land plants is that appropriate mechanisms of leaf growth did not exist that would have permitted anything other than an open branching pattern of venation to develop. In order to achieve a network, it is necessary to have growth take place not only at the leaf margins, base, and apex, but also in areas away from the margin. Devonian and most Carboniferous plants evidently had only marginal leaf growth. That such a restriction is rather easily overcome is demonstrated by the fact that more than fifty lineages of ferns, as well as various fossil seedfern and gymnosperm clades and the major angiosperm (flowering-plant) clade, have evolved network venation independently. Although a few Late Carboniferous seedferns had already evolved a degree of connectedness in their venation, the vast majority of cases of evolution of network venation occurred later, especially in the Mesozoic and Cenozoic eras. Theory suggests that open branching is the most efficient means of distributing fluids from a central point to a large area or volume; in other words, a closed network makes no improvement in this function. The main advantage of the network, it seems to me, is in providing a measure of protection of leaves against herbivores. I therefore interpret the frequent post-Carboniferous evolution of network venation as an adaptive response to intensified herbivory, an interpretation that fits well with the increases in leaf damage in the Early Permian and Cretaceous as observed by Conrad Labandeira and his colleagues.[43]

In all these examples, primary producers have evolved growth habits that allow production and growth to continue under heavy assault by grazers. Indeed, as I discussed in chapter 5, some of these plants have become dependent on their herbivores. They tell a tale of increased rates of production in the face of increasingly intense consumption by powerful animals with high metabolic demands.

Low levels of herbivory are only part of a larger syndrome of slow ecosystem-level metabolism during the early history of forest vegetation. In the great lowland coal forests of the Carboniferous period of the Paleozoic, fallen leaves and other plant parts accumulated because the rates of decay were evidently every slow. In part, this slow decomposition reflects the extraordinary resistance of lignins, resins, and other chemical components of the forest trees to bacterial degradation. Indeed, it is likely that these compounds played a significant role in the evolution of land plants as agents against bacterial and fungal attack of living plant tissues. Perhaps not coincidentally, these substances are also highly effective against herbivores, but herbivory is unlikely to have been an important selective agency in the early biochemical history of land plants. Besides the abundance of resistant compounds, Carboniferous forests also were characterized by the absence of large decomposers. There were no wood-chewing termites yet, and no large basidiomycete fungi—the kind with gills or pores in their reproductive structures, the mushrooms. Instead, there were oribatid mites and early roaches that tunneled into and ate the contents of dead stems.[44]

All these observations point to the surprising conclusion that the coal forests of the Carboniferous, growing at equatorial latitudes in most lowlands, were relatively unproductive compared to the lush rain forests of today. The low rates of water transport through the trunks of the dominant lycopod trees, the slow growth rates and presence of growth-inhibiting, energetically expensive, decay-resistant compounds, the low levels of herbivory, and the limited recycling of nutrients all diagnose a low-energy forest, in which production may have been too slow to sustain a significant cadre of specialized vegetation-feeders. Things changed in the latest Carboniferous, when lycopods were replaced by more productive ferns in the canopy, and when herbivorous animals came on the scene; and they changed again during the Cretaceous, when still faster-growing angiosperms began to take over the world in the presence of ever more energetic and numerous plant consumers. Finally, with the development of closed angiosperm-dominated forests in the Paleocene or Eocene, and the spreading of lush grasslands beginning in the Oligocene, vegetations achieved modern high levels of productivity in seasonally or perennially warm regions with moderate to high rainfall. Production by plants, herbivory by animals, and decomposition by a diverse array of microbes and fungi and animals went hand in hand, all contributing to land vegetations in which nutrients move rapidly and through many organisms in a self-made microclimate that is particularly amenable to plant growth and consumer sustenance.[45]

Time and the Power of Predators

Herbivory is not the only form of consumption in which the Phanerozoic eon—the last 543 my of Earth history—is distinguished from earlier eons. Predation, too, emerged as a method of consumption, which affected both the characteristics and distribution of every species in the biosphere and which institutionalized escalation as a major economic and evolutionary process. The history of predation, like that of herbivory, is one of increasing power of predators and of adaptation along many different pathways by prey.

The fossil record shows unambiguously that predation had become an important evolutionary agency by the latest Neoproterozoic. An as yet unknown attacker commonly made holes through the wall of the tubelike fossil *Cloudina*, one of the earliest mineralized skeletons known, in the Nama Group of Namibia. Some of the holes penetrated the tube's wall completely, but others did not, indicating that this earliest record of predation already involved a prey species that had adapted to hole-drilling enemies. By the Early Cambrian, repaired injuries are known from trilobites, likely made by anomalocarid arthropods, the first known predators capable of breaking skeletons. The origin and evolution of skeletons, which affected at least eight phylum-level clades independently during the latest Neoproterozoic and Early Cambrian, therefore can be ascribed in large part to the emergence of predators whose modus operandi includes breaking and entering.[46]

Unlike the consumption of living multicellular plants by herbivores, predation—one animal eating part or all of another—appears very early in the evolutionary tree of animals. All phyla—even the Porifera (sponges), the earliest group to branch from the tree—contain predators; and although many forms of predation are highly derived modes of feeding, others are primitive and have led to herbivory and suspension-feeding. In general, the forms of predation emerging in the basal branches of the tree, including the basal branches of all the component phyla, are more passive and involve less force than the forms that evolved later in the upper branches. Swallowing small prey whole, and grazing parts of colonial victims, are ancient methods, whereas such methods as shell breakage, bone crushing, envenomation, and the use of force in extracting edible parts from the openings of shells are evolutionarily derived. Ancient predators moved slowly and probably detected victims at short distances or upon contact; many more derived ones were fast and could recognize prey from far away.[47]

Consider, for example, the predators that eat animals encased in shells. Predators employ a variety of methods to subdue and consume shell-bearing victims, including (1) swallowing the prey whole without damage to the shell; (2) inserting part or all of the body into the shell, so extracting the edible parts, but typically leaving the shell intact; (3) drilling a hole through the shell wall, followed by insertion of the feeding organ to extract the edible tissues within; and (4) breaking the shell in order to expose and consume the flesh. Breakage offers the fastest way of killing the victim; drilling is one of the slowest. Swallowing the prey whole limits the predators to relatively small food items compared to the other methods.[48]

Predators swallowing their victims whole belong to every animal phylum, and had already appeared by the Cambrian, when sipunculan worms and probable cnidarians evidently took prey in this way. When other groups acquired the prey-swallowing habit is difficult to say, because fossil predators with this habit reveal few diagnostic cues. Today, the method of swallowing whole prey is especially widespread among predators inhabiting sandy and muddy marine environments. Modern groups with such habits can be traced back to the Jurassic (many seastars), Early Cretaceous (various snails), and Middle to Late Cretaceous (many fish). Some snails, which envelop prey in the greatly expandable foot, are able to subdue quite large prey, and have therefore partially overcome the size limitation inherent in whole-prey ingestion. They include volutes, olive snails, and moon snails, most of which acquired the large foot during the Late Cretaceous and Cenozoic.[49]

Worms, snails, crabs, seastars, beetles, and snakes are among the predators that extract edible flesh by entering the shell opening of their prey. The earliest recorded instance of this habit dates from the Late Ordovician, when the seastar *Promolaeaster* forcibly inserted its stomach between the valves of clams. Whether this ability persisted in other Paleozoic seastars is unknown, but it reappeared during the Early Jurassic with the evolution of the family Asteriidae. Today, most seastars that insert the stomach into prey, with or without the use of force, occur at high latitudes. Without the use of force, the process can take days. With the ability to pull valves open using suckered tube feet, seastars shorten the subjugation phase, but predation typically still requires a day or two to complete. Snails using various insertion methods appeared in the Early Cretaceous. Often, they use the whole outer lip of the shell as a wedge to pry open the bivalved shells of clams, and snails with protruding lips capable of wedging appear independently in several clades. When beetles and snakes, which insert feeding organs into the shells of land snails, appeared is not known.[50]

Drilling is typically a slow means of predation, but given that subjugation occurs largely outside the body of the attacker, it does allow predators to kill and consume large prey. This is especially so from the Late Cretaceous onward. During the early phases of the history of drilling, from the latest Neoproterozoic to the Devonian, all drill holes were small (1 mm or less in diameter), and drilled individuals of affected species typically comprised less than 5 percent of populations. Larger drill holes made by naticid and muricid snails in molluscs and barnacles and by cassid snails in echinoderms appear only in the Late Cretaceous. Both naticids and muricids evolved during the Campanian stage of that period, and cassids are known first in the succeeding Maastrichtian stage. Modern high levels of drilling predation were established only during the Paleocene or Eocene epochs of the Cenozoic. Octopods, which generally drill molluscan and crustacean prey, take minutes rather than hours or days to complete the drilling process. They are known from the Late Cretaceous onward, but when modern levels of drilling were attained is unclear. Edge-drilling, a special form of drilling in which the predator excavates a hole at the junction between the two valves of a clam or brachiopod, is faster than other types of drilling, and today characterizes several derived groups of naticids and muricids. The earliest examples of edge-drilled prey are brachiopods from the Early Carboniferous of the United States, but edge-drilling has become the predominant mode of drilling only in Recent populations of a few tropical Pacific bivalves.[51]

Rapid dismemberment is a major theme in the history of predation. Jaws and appendages that bite, grasp, hold, cut, smash, or crush victims arose among predators in the Early Cambrian and have evolved many times since. Anomalocarid arthropods from the Early and Middle Cambrian may have been the earliest predators to evolve skeleton-damaging claws. From their first appearance in the Early Cambrian, trilobites too included predators among their ranks, grasping and shredding soft-bodied prey with spines and basal parts of their appendages and then moving the food forward to the mouth, but their feeding organs were unspecialized. Stomatopod crustaceans appeared in the Late Carboniferous and decapod crustaceans—lobsters, crabs, and their relatives—predominate from the Late Triassic onwards, evolving claws or enlarged mandibles to pinch, crush, cut, and smash prey.[52]

Cephalopods, which arose as small swimming predators during the Late Cambrian, probably became the top predators in Ordovician seas, some endocerid nautiloids reaching a shell length of 9 m. Possibly from the beginning, they had adhesive armlike appendages of the head, with which they caught prey. The grasping function of the arms was improved by the presence of hooks in the Early Devonian to Late Cretaceous belemnites, cephalopods in which the primitively external shell was replaced by a mineralized, cylindrical internal rod. Suckered arms, typical of today's squid and octopus, evolved in the Late Triassic in cephalopods in which the internal shell was little more than an internal flexible organic support. Delivery of a venomous bite could have evolved as early as the Late Cretaceous, the time of appearance of the earliest known octopod. Jaws capable of crushing small prey, and perhaps also functioning as protective doors when the soft parts were retracted into the shell's living chamber, had appeared by the Early Devonian. From the Early Jurassic to the Late Cretaceous, many ammonoids developed mineralized jaws, which may have cut prey flesh or been used to scoop up small items from the sediments on the seafloor.[53]

Among the earliest known members of the Deuterostomia—the great branch of animals that includes echinoderms and vertebrates—are 25-mm-long Early Cambrian fossils from China tentatively regarded as the earliest chaetognaths (arrow worms). Today, chaetognaths are planktonic marine predators that use grasping spines to catch prey. Jaw-bearing vertebrates are thought to have appeared no later than the Caradocian stage of the Middle Ordovician (450 Ma) on the basis of diagnostic elements of body armor of acanthodian and chondrichthyan (sharklike cartilaginous) fishes in the Harding Sandstone of Colorado. Perhaps by the Silurian and certainly by the Early Devonian—no later than 400 Ma—jaw-bearing vertebrates had replaced cephalopods as the top predators in the sea. Acanthodians preying on other fish appeared in the Late Silurian, joined in the Early Devonian by the earliest bony fishes (palaeoniscoid Osteichthyes) and placoderms. The 6-m-long Late Devonian placoderm *Dunkleosteus*, whose gill-arch bones were beset with toothlike structures that grasped and cut large prey including the slow predatory shark *Cladoselache*, was the largest Devonian predator.[54]

Top predators in the sea as well as on land show long-term, though interrupted, trends toward a stronger bite and greater flexibility of jaw movement. The early fishes of the Paleozoic had the upper and lower jaws quite rigidly bound to the rest of the skull, so that the jaws could do little more than open and close. The teeth lining the jaws therefore functioned largely to grasp and puncture prey. Reorganization of bones and muscles in the head of many Mesozoic lineages of fish provided greater flexibility, permitting the fish to expand the mouth cavity as the mouth opened, and so allowing water and prey to be sucked in. In advanced bony fishes—the teleosts, originating in the Middle Triassic and greatly diversifying during the Early Cretaceous, Late Cretaceous, and Early Cenozoic—the upper jaw can be made to protrude well beyond the lower, so that hidden prey can be plucked from crevices, and shell-encased edible tissues can be levered out of their shells. Greater jaw mobility enabled Early Jurassic and later sharks to become meat-cutters, with teeth modified for sawing and slicing rather than for holding and piercing as had been the norm in Paleozoic antecedents. Beginning in the Triassic, fishes were joined or in some cases surpassed as top predators by reptiles in the Mesozoic and whales and seals in the Cenozoic. Although no information is presently available on the comparative feeding mechanics of these tetrapods, per capita power requirements probably increased through time. The Mesozoic marine reptiles—ichthyosaurs, nothosaurs, plesiosaurs, and mosasaur lizards, among others—were effectively cold blooded, whereas the Cenozoic mammals, as well as some tunas and sharks among predatory fish, produce copious internal body heat.

The earliest predators of vertebrates on land—Early Permian sphenacodont "pelycosaurs" such as *Dimetrodon*—likely depended on the inertia of the rapidly closing jaws, so that the slicing teeth could hold and penetrate prey. In the later Early to Middle Permian, 275 to 260 Ma, sphenacodonts were succeeded as top predators by larger, more powerfully biting dinocephalians, such as the 5-m-long, 400 kg *Anteosaurus*. Like most other Permian to Middle Triassic top predators on land, dinocephalians belong to the great synapsid clade of tetrapods, which is represented today by mammals. The last 10 my of the Permian, from about 260 to 250 Ma, witnessed the replacement of dinocephalians by gorgonopsids and therocephalians, which included the first known saber-toothed predators among their ranks. The enlarged saberlike canine teeth enabled some predators to eviscerate very large prey rapidly.

Although therocephalians extend as small-bodied relicts into the Triassic, large representatives along with gorgonopsids disappeared during the end-Permian crisis, to be replaced by the Early Triassic small cynodonts weighing less than 100 kg. In the Middle Early Triassic, the cynodonts were in turn eclipsed as top predators by erythrosuchids and rauisuchids, animals weighing up to 600 kg. These latter animals belong to the Diapsida, the great clade of tetrapods that includes dinosaurs and living birds and that would supply the top predators for the remainder of the Mesozoic era. In the Late Triassic, top predators were represented by running ornithosuchian and saltoposuchian crocodiles, and by the first theropod dinosaurs including *Eoraptor* and *Herrerasaurus*. Theropods included a succession of top predators during the Jurassic and Cretaceous, joined later by such gigantic crocodiles as the Early Cretaceous *Sarcosuchus* of Africa (up to 12 m long and weighing as much as 8,000 kg) and the only slightly smaller Late Cretaceous North American *Dinosuchus*. The Jurassic theropods, including *Allosaurus*, had relatively lightly built skulls and a jaw mechanism that relied on inertia and speed to deliver the killing bite. Later theropods, such as the Late Cretaceous *Tyrannosaurus*, executed a slower, more powerful bite. Interestingly, saber-tooth dinosaurs seem not to have evolved, the raptorial functions of the saber canines probably having been taken over by claws on the forelimbs, which were able to specialize to predation because only the hindlimbs are involved in the characteristically bipedal locomotion of theropods.

The role of top predator reverted to the synapsid clade after the end-Cretaceous demise of the dinosaurs. Compared to earlier predators, Cenozoic mammals showed a much greater differentiation into grasping incisors, stabbing canines, and grinding or crushing molars. This dental differentiation was associated with the ability of the lower jaw to move in a fore-aft direction relative to the upper jaw, as well as from side to side, an ability already evident in less specialized form in Triassic cynodonts. The first more or less predatory mammal was the coyote-sized mesonychid *Dissacus* of the Mid-Paleocene (62 Ma), but it was not until the latest Paleocene or earliest Eocene, about 55 Ma, that large creodont mammals more or less specialized to eat other vertebrates reappeared in North America. During the Early to Middle Eocene, the sabertooth habit reappeared among predators (the creodont *Machaeroides*), and for the first time some animals with bulbous premolar teeth specialized for cracking bone evolved, first in the creodonts *Patriofelis* and *Palaeonictis*. Later bone-crackers evolved among Middle to Late Miocene borophagine dogs in North America, in Old World hyenas beginning in the Miocene, and perhaps even in Late Oligocene to Early Miocene boarlike entelodont ungulates. Similarly, saber-tooth predators appeared in several successive groups, including Oligocene nimravid carnivorans, in several groups of Miocene to Late Pleistocene cats, and in the extinct South American marsupial *Thylacosmilus*. Although all Cenozoic top predators are much smaller than their Mesozoic antecedents, their high metabolic rates indicate a much greater food requirement than was the case among the diapsid predators of the Mesozoic and the synapsids of the Permian and Triassic.[55]

Skeletal breakage by animals with claws or mouthparts specialized to deliver large forces and bearing thick, broad, flattened teeth or plates, is traceable to the Early Devonian. Among early predators with such specializations were various fish—placoderms, lungfish, and blunt-toothed sharklike forms—and pterygotid eurypterids, huge arthropods distantly related to scorpions and spiders. The succeeding Carboniferous period witnessed diversification among sharks, early bony fishes, and ancestral mantis shrimps (stomatopods). The great push toward shell breakage, however, came during Mesozoic and Cenozoic times. Hybodont sharks may have been the only animals with crushing specializations during the Early Triassic, but later in the period they were joined by bony fishes (neopterygians in the Norian stage) and placodont reptiles (Middle Triassic). Advanced teleost bony fishes and true brachyuran crabs diversified explosively during the Late Cretaceous, especially the last two stages (Campanian and Maastrichtian, 80 to 66 Ma), but distinctive weapons specialized for crushing—molar-toothed claws in crabs, molarlike teeth and fused thickly mineralized tooth plates in bony fish and rays—arose only in the Early Cenozoic. The Eocene to Recent eagle rays (Myliobatidae) and spiny puffers (Diodontidae) surpass any earlier vertebrates or any living shell-crushers (except perhaps the loggerhead turtle) in the specialization and crushing force of their jaws. The worldwide work of these and other shell-destroying predators is well illustrated by the observation by Tatsuo Oji and his colleagues that broken shells become common constituents of fossil shell beds in Japan only during the Eocene. In short, the last 80 my, and especially the last 50 to 60 my, witnessed the evolution in many clades of fishes, crustaceans, and tetrapod vertebrates of predators with shell-damaging feeding organs capable of generating enormous forces.[56]

Armored animals emphasizing passive resistance defense show general trends toward a greater expression and increasing incidence of features that deter or slow predatory attacks. These trends are evident within clades as well as among successive clades in physically similar environments.

For example, gastropods with a small or inaccessible shell opening—a narrowly elongate aperture or one whose rim is beset with inwardly directed thickenings or folds—are essentially unknown in the Paleozoic. They appear first in the Jurassic, increase in frequency during the last 20 my of the Cretaceous, and increase against during the Early Eocene and the Late Oligocene to Early Miocene intervals of the Cenozoic in warm seas. Antipredatory spines first appeared or became frequent among brachiopods, sedentary gastropods, and crinoids during the Wenlockian stage of the Silurian, and increased in frequency and length during the succeeding Devonian period. Cementation, which secures one valve firmly to the substrate and which in bivalved animals affords some measure of protection from predators that dislodge prey from their moorings, appeared in brachiopods during the Silurian, and in bivalved molluscs during the Late Carboniferous. The habit subsequently appeared in many Mesozoic and Cenozoic bivalves and in Eocene barnacles. Shell-reinforcing pleats occur in 15 percent of Late Middle Ordovician (Caradocian) North American brachiopods and in double that percentage of Late Devonian species, and continued to increase during the Late Paleozoic. Bivalves frequently evolved these features in the Mesozoic and especially the Cenozoic. Other post-Paleozoic antipredatory features that repeatedly evolved among bivalves include overlapping valve margins, where one valve protrudes slightly beyond the other, and drilling-resistant internal organic shell layers. Among shell-bearing cephalopods, species with breakage-resistant, spirally coiled shells and forms with reinforcing ribs appeared in the Ordovician. Coiled shells constituted the majority of shell-bearing cephalopods from the Devonian onward. Ammonoid cephalopods show trends from their Early Devonian appearance to the Turonian stage of the Late Cretaceous toward increases in external sculptural relief and toward increasingly complex wave patterns in the junctions between internal septa and the outer shell wall. Both trends enabled ammonoids to increase resistance to loads imposed by crushing predators without burdening the swimming ammonoid with a heavy shell. The post-Turonian decline in ammonoid defensive sculpture may reflect the diversification of predators that overwhelmed even the most sophisticated defenses in these thin-shelled creatures. Among bottom-dwelling crustaceans, the crab form, in which the egg-bearing abdominal segments are tucked beneath a broad one-piece carapace, evolved in at least five clades of decapods, first in the Sinemurian stage of the Early Jurassic with the appearance of the true crab *Eocarcinus*. Finally, whereas Paleozoic sea lilies (crinoids) show no evidence of the ability to autotomize arms, those evolving in the Mesozoic and Cenozoic have specialized joints between some arm segments that enable the sea lilies to sever parts of the body when under attack.[57]

These trends in architecture are generally reflected by the incidence in repaired injury. In the oldest known examples of shell repair among gastropods, dating from the Late Middle Ordovician of Sweden, fewer than 10 percent of gastropod individuals had repaired injuries. The incidences were higher among well-preserved gastropods from warm-water assemblages in the Late Carboniferous of Texas and the Late Triassic of Italy, but modern high warm-water incidences were not achieved until the Campanian stage of the Late Cretaceous. Increases in frequencies of repair have been noted from the Early to the Middle Jurassic and again during the Late Cretaceous among ammonoid cephalopods, from the Turonian to the Campanian and Maastrichtian stages of the Late Cretaceous in inoceramid bivalves, and in various groups of oysters during the Late Cretaceous. Regeneration of nipped arms was rare in Paleozoic crinoids, but became common in the Mesozoic and Cenozoic as crinoids evolved specialized means to autotomize these feeding structures.[58]

Locomotion

There is, of course, much more to predation than shell breakage and adaptation to it. Predators are also pursuers, and prey adapt by taking quick evasive action. We should and do discern increases in speed and maneuverability of top predators through time, as well as locomotion-related adaptive responses in those prey species that have evolutionarily chosen to emphasize escape over other forms of antipredatory adaptation.

Consider, for example, the jet-propelled cephalopods. In the earliest forms from the Late Cambrian, jet propulsion was probably first achieved by expelling water from the living chamber of the shell as the head retracted into the shell, a maneuver of escape or protection. Contraction of the funnel, a tube-like modification of the shell-secreting mantle, enlarged the water-propulsive capacity, and therefore the animal's power, as water was expelled in discrete pulses from the cavity enclosed by the mantle. The conical to cylindrical, straight to gently curved, horizontally oriented, external shell contains chambers filled with gas, which provides buoyancy for the animal during swimming. The weight of the shell and the internal deposits that ballasted and stabilized the animal made for slow acceleration, slow top speeds, and a limited ability to change direction. The evolution of tightly coiled shells in several Early Ordovician lineages eliminated the ballasting requirements inherent in a horizontally disposed straight shell, and also lightened the shell and decreased resistance to turning. The shell's living chamber and mantle cavity became larger, increasing the volume of water available for expulsion and therefore providing more power to the swimmer. By the Late Devonian, cephalopods with coiled shells had surpassed the straight and curved ones in diversity, indicating a general rise in locomotor performance through time. Beginning in the Early Devonian, one or more lineages evolved internal shells in which the buoyancy function was eliminated, so reducing the constraint on the volume of expelled water still further and allowing propulsive musculature to increase in mass. Although many lineages of shell-bearing cephalopods evolved disk-shaped, smooth shells in which drag—the friction of water against a moving object and the wake of the water behind the swimmer—was reduced, David Jacobs' studies of ammonoid cephalopods indicate that an increase in power was more important as a means of increasing locomotor performance. Despite these improvements, shell-bearing cephalopods other than nautilids disappeared after the Cretaceous. The squid, cuttlefish, and octopus of the Cenozoic are powerful, often very fleet animals compared to their Paleozoic and Mesozoic shell-bearing counterparts, and have locomotor performance levels on a par with those of many fish.[59]

An increasing emphasis on locomotion is evident in many groups, even when armor remains an important means of defense. Streamlined gastropods and bivalves capable of active burrowing in sediments are essentially unknown in the Paleozoic. Species with shell ribs oriented so as to promote forward motion and resist backward slippage did not appear until the Early Eocene. Large-footed, fast-burrowing moon snails and olive snails appeared in the Late Cretaceous, but early members had smaller shell openings and a less expandable foot that did those in post-Eocene time. Even groups that led sedentary lives in the Paleozoic gave rise to clades of motile animals in the Mesozoic and Cenozoic. Crinoids, for example, were permanently attached suspension-feeders during the Paleozoic, and deep-sea stalked crinoids today still live this way; but comatulid crinoids, which are diverse on reefs in shallow water today, evolved the ability to shift position after the Paleozoic. There are even motile Cenozoic bryozoans.[60]

Vertebrates, which have been the top predators in the sea from Silurian time onward, have likewise explored several pathways toward greater locomotor power. Most, including ancestral forms, employ short-burst swimming that enables them to attack and overtake moving prey from a short distance. The ancestral form of locomotion in vertebrates is eel-like swimming, in which the flexible body pushes water to the side and backward as it is contorted into waves traveling backward from the head. Most Paleozoic fish had relatively small tails and fins with which they produced forward thrust, and therefore accelerated and swam slowly. Rapid, more sustained swimming is made possible when propulsion is limited to either the forelimbs or hindlimbs, as in penguins (Eocene to Recent), plesiosaurs (Mesozoic), and mosasaurs (Cretaceous), or to the large tail, as in whales (Early Eocene to Recent), tuna and mackerel (whose time of origin is not known), or ichthyosaurs (Late Triassic to mid-Cretaceous). The earliest swimmer with this locomotor mode was the Late Devonian shark *Cladoselache*, which may nonetheless not have been particularly fast. By the Toarcian stage of the Early Jurassic, ichthyosaurs such as *Stenopterygius* could have attained average sustained swimming speeds of 1.2 to 1.6 m/s, comparable to that of tuna, but whether this or any other Mesozoic reptile could have achieved such high top speeds as tuna attains (21 m/s) is doubtful. Richard Cowen speculates that ichthyosaurs, like penguins and many cetaceans and some lamnid sharks, were able to increase their speed by jumping out of the water, which would eliminate the strong backward forces acting on the animal as it swims at the water surface. Ichthyosaurs declined in diversity during the Early Cretaceous, their place as fast sustained swimmers perhaps being taken over by fish. Fast-swimming lamnid and carcharhinid sharks are first known from the Tithonian, the last stage of the Jurassic. They were joined in the Late Cretaceous by a host of teleost bony fish, and in the Eocene by cetaceans. Other teleosts, meanwhile, became adept at maneuvering in cluttered places by relying on fins for fast turning, often in conjunction with the tail as propulsive organ.[61]

When vertebrates first ventured as adults on dry land, no later than 340 Ma during the Early Carboniferous, their limbs extended out to the side, as in amphibians and most lizards today. In this position, forward motion of the limbs compresses the lungs and therefore interferes with breathing. Rapid, sustained locomotion is therefore constrained by its incompatibility with continued delivery of oxygen to the body. This constraint was removed when limbs acquired a more erect orientation beneath the body. Erect limbs first evolved in the bipedal bolosaurid reptile *Eudibamus* about 290 Ma, during the Early Permian, and evolved again during the Late Triassic in at least three clades of archosaur-like reptiles (including the dinosaurs) and in mammals. With them came the potential for faster running and greater endurance. According to Philip Currie, predatory theropod dinosaurs display a general trend toward longer limbs and faster locomotion from the Late Triassic to the Late Cretaceous, especially among smaller, light-bodied forms. Interestingly, long-distance pursuit of prey, involving sustained fast running over distances of 300 m or more, did not arise among mammals until the Middle Miocene in the Old World, when some hyenas acquired it, and the Pliocene (5 Ma) elsewhere with the evolution of wolves (genus *Canis*).[62]

Flight in air is perhaps the most effective escape mechanism, and also opens up the possibility of catching highly dispersed, often aerial prey. The earliest known fliers are insects, which had evolved wings and powered flight no later than the middle Carboniferous. The current consensus is that powered flight evolved just once among insects. Vertebrates took to the air later. The earliest known glider—an animal gaining lift in air by passive means—is the Late Permian reptile *Coelurosauravus*. Gliding has evolved repeatedly since among various clades of amphibians, snakes, lizards (already in the Late Triassic), and mammals; it has even appeared in several groups of fish (earliest in the Late Triassic) and in squids. True powered flight arose thrice among vertebrates: pterosaurs (Late Triassic to Late Cretaceous), birds (Late Jurassic to Recent), and bats (during or before the Eocene). Gliding is known in some nonpredatory mammals (flying squirrels, flying lemurs, and phalangers, for example), but powered flight has apparently always evolved in clades of predators.[63]

As is the case with rapid and powerful methods of predation, the derived, rapid, sustained locomotion of top predators and the increased emphasis on locomotion in many victim species were superimposed on older, less energy-intensive means retained by animals with smaller energy budgets. To some, these trends exemplify only an increase in the range of functional possibilities, practically a statistical necessity if overall diversity is increasing. To me, however, the generally increasing performance of the most powerful members of successive ecosystems represents a general raising of the bar, not just for the top predators that lead the way, but for many of the species with which they interact.

The History of Regulation

In many ways, the supply of resources is everything in an economy. As I discussed in chapter 5, there can be little consumption unless and until the rate of supply of essentials is both high enough and predictable enough for entities relying on these essentials to be sustained. It should therefore come as no surprise that, through the collective activities of living things, organisms have played an increasingly dominant role in regulating and raising the rate of supply of raw materials over geological time. Externally driven variations in the sources, supply rates, and losses of materials have largely been replaced by internally regulated processes controlled, and often enhanced, by the collective efforts of organisms.

Earlier in this chapter, as well as in chapter 6, I discussed these ideas in connection with the regulation of nitrogen, phosphorus, iron, and oxygen. The regulation and enhancement of these critical elements occurs not through top-down decree, as in an authoritarian regime, but by self-organization of the participants, nudged and prodded particularly by the more powerful competitors among them. These points are underscored by two further examples, drawn from the regulation of minerals used for the construction of plant and animal skeletons.

The most widespread element in skeletons is calcium. It is a key component of calcium carbonate (the skeletal mineral in most foraminifers, cnidarians, annelids, molluscs, brachiopods, bryozoans, and echinoderms, and in many sponges and crustaceans), hydroxyapatite or calcium phosphate (typical of vertebrate skeletons), and calcium sulfate (in some protists). Before 2.5 Ga, during the Archean eon, waters in the ocean were highly supersaturated with calcium carbonate, and deposits of this mineral formed as crystals directly on the seafloor. Seafloor crystallization, which evidently proceeded with little or no intervention by organisms, was gradually superceded during the Proterozoic eon (2.5 to 0.54 Ga) by a different style of accumulation, in which calcium carbonate crystals formed in seawater and then settled as sediment to the seafloor, where mats of Cyanobacteria trapped and cemented the crystals. This mode of calcium carbonate precipitation dominated after 850 Ma during the Neoproterozoic era. Organisms had begun to intervene in the process, at least in well-lit, shallow-water environments.

The role of organisms greatly expanded when, beginning about 560 Ma during the latest Neoproterozoic, calcium carbonate came to be used in the formation of protective skeletons. During the next 45 my, extending into the Early Cambrian, many groups of plants, animals, and protists evolved calcium carbonate skeletons in ecosystems on the seafloor. With their evolution, the regulation of calcium carbonate extended to environments below the reach of sunlight. Biological control expanded again when, beginning in the Late Triassic, calcified single-celled plankton evolved. For the first time in Earth's history, calcium carbonate began to precipitate as skeletons in sediments beneath the deep open ocean far from land.[64]

Organisms have also taken an increasingly important hand in controlling calcium carbonate by dissolving it from biologically precipitated limestones. In the sea, the erosion and chemical dissolution of limestones by organisms has a history strikingly parallel to that of burrowers in sandy and muddy sediment. Cyanobacteria were boring into and dissolving limestone during the Late Neoproterozoic, but they did not penetrate more than a few millimeters into it. Animals making deeper burrows appeared in the Early Cambrian and the Arenigian stage of the Early Ordovician, but these were isolated occurrences, evidently representing animals that did not persist geologically. Borings in rocks more than a few millimeters deep have a continuous record from the Middle Ordovician onward, but those extending deeper than 5 cm are known only from Late Mesozoic time to the present. Rock-eroding animals that graze algae are, as I pointed out earlier in connection with the history of herbivory, also known only from the Mesozoic onward. Precipitation and dissolution of calcium carbonate therefore both became processes largely orchestrated by life.[65]

The history of control of silicon is remarkably similar. Before the latest Neoproterozoic, the precipitation of silica (silicon dioxide) was unregulated by life. Sponges became perhaps the earliest organisms to employ silica in their skeletons. As bottom-dwelling animals, they became the chief biological regulators of silica precipitation during the Paleozoic and Early Mesozoic. Siliceous single-celled plankton existed (as radiolarians) from the Early Ordovician onward, but major diversification of diatoms and silicoflagellates during the Cenozoic broadened the control of silica by organisms to the ocean as a whole. On land, silica may have become important for the first time as a component of plant tissues when Cenozoic grasses incorporated it in their stems and leaves.[66]

Trends versus Random Walks

The trends I have discussed in this chapter emerge over the long term, over millions to tens of millions of years in the case of nonhuman life, and decades to millennia in the case of human history. As such, they conflict with evidence from short-term studies for directionless evolution, characterized by aimless fluctuations, or "random-walk" change. The Middle Jurassic ammonoid cephalopod genus *Kosmoceras* in the Oxford Clay of England, for example, changed in many shell characters over a 1.2 my interval, but these characters fluctuated frequently and aimlessly over time. In a remarkable thirty-year study of evolution in two species of groundfinch at Daphne Major, one of the Galápagos Islands, Peter and Barbara Grant observed numerous increases and decreases in body size, as well as changes in bill size and proportions. Some trends lasted two or three years before reversing, but none of the characteristics the Grants examined showed a consistent evolutionary direction over the period from 1972 to 2001. The changes in the finches appear to have a genetic basis, and occurred in response to fluctuations in local climate and food supply.[67]

I believe that the apparent conflict between short-term aimless fluctuation and long-term trends toward greater power and productivity can be resolved by considering the role of opportunity. As I noted in chapter 8, opportunities on which many economic entities can capitalize simultaneously arise as possibly global triggers, which then set in motion positive feedbacks of escalation and adaptation. These triggers and their aftermath are not everyday events. In order to have evolutionary effects, they must increase the supply of resources on large spatial scales, and be accompanied by intense competition; and they must not be immediately reversed by interruptions in supply. The long-term trends therefore arise because many entities in many economies at many spatial scales capitalize on opportunities that last long enough for feedback mechanisms to yield adaptive evolution. Most of the time, opportunities are either too local or too fleeting in time to have a lasting evolutionary effect. The directions of economic change and of adaptive evolution are dictated by the realization of universal economic principles of inequality and competition, but the timing and extent of such change depend on the timing, extent, and duration of opportunity.

To me, one of the most fascinating parallels to emerge from the comparative study of history is between trends through time and trends in space. Moving forward in time, toward greater power and productivity, is like moving in space from the poles to the equator, from high to low altitudes, from deep to shallow water, from islands to continents, and from countryside to city in the human arena. Power-enhancing innovations arise preferentially in the most productive economies and spread outward in space and forward in time. Neither the temporal nor the spatial patterns are wholly consistent, and there is much about modern low-productivity settings that would strike a time-warped visitor from the distant past as strange, but the parallel does suggest an intriguing path for studying ancient conditions. Broadly speaking, the characteristics of participants and of economies in unproductive regions should resemble those in ancient, more productive settings. We may therefore gain insight into the deep past by understanding how marginal economies of today work in comparisons to the more productive ones.

The Human-Historical Parallel

Our current position as the most powerful species on Earth is the product of a remarkably brief history that recapitulates in less than 2 million years the chronicle of life as a whole, which played out over a span of some 3.5 billion years. To do justice to our economic history would require a book or two and more expertise than I possess. Instead of the full-length treatment this history deserves, I reiterate some of the major trends, which transformed our species from just another high-energy consumer to today's superspecies.

As noted in chapter 3, the power used by individuals, and by the human species as a whole, as measured in units of energy consumed annually, has risen by a factor of five over the past century or so, and has shown a general upward trend from the time some 1.8 million years ago when humans first learned how to use fire. Over the millennia, and today over decades, we have added energy sources to our arsenal, from sun and wind to domesticated animals, fossil fuels, and nuclear energy. We have expanded the food supply through a combination of more intensive farming and a larger variety of crops. None of these trends would have been possible without a consistent and perhaps increasing stream of technological innovations. The energy sources were always out there, but they remained unavailable until we acquired the technology to tame them. In concert with these changes, the human population has vastly and perhaps episodically increased, expanding our collective take of the world's living and nonliving resources. The emergence of banking in the Middle Ages, the establishment of modern business practices in the seventeenth century, and the elaboration of patent and property law during the past three centuries have importantly contributed to the stabilization and regulation of capital, with the result that, in advanced capitalist economies at least, investment can proceed in a more predictable, more secure economic environment. Increased trade reduced local limitations on land and labor while cushioning other local populations from externally driven ups and downs in food production.

As the human population rose, societies became larger, and with the emergence of dense, urbanized societies came new social adaptations. A hierarchical power structure replaced a more egalitarian arrangement of small bands and tribes. Religious institutions hierarchically organized to expound and enforce monotheistic ideologies sprang up in southwestern Asia and largely replaced a multitude of polytheistic systems, which were perhaps better suited to smaller societies. The unification and cohesion of large groups made possible by these and other, ostensibly nonreligious belief systems, were further enhanced by increasingly sophisticated military structures, which progressively developed and used weapons of ever-increasing power.

These gains have not been equally enjoyed by all human populations, but they have spread out from centers of power and innovation. The Industrial Revolution began in Britain, but spread worldwide in short order. In all but a few of the very poorest countries of Africa, life expectancy has increased during the last century thanks to the advances in medicine and public health that began in Europe and North America. The difference between rich and poor may be greater today than ever before, but the least affluent are in an absolute sense richer today than they were in the past. Individually and collectively, therefore, the human species has become more powerful, more productive, and more wide ranging. For better or worse, our transformation has been achieved almost entirely through cultural rather than genetic evolution, by the emergence of hierarchical, top-down social systems coupled with the collective power of multiple, bottom-up components.

Comparative economic history is in its infancy, but the patterns that it reveals imply the action of simple principles as they intersect with unique entities and events. In the last chapter I consider what these patterns tell us about ourselves and our future.

Chapter 11

THE FUTURE OF GROWTH AND POWER

IF THE THESIS OF THIS BOOK is correct, economic life on our planet has exhibited a long-term, though occasionally interrupted, trend toward increased power and independence, a trend ultimately caused by competition among economic units for resources and by the disproportionate influence of competitive dominants. Human history recapitulates these same trends over a greatly compressed time span, and the human species falls in line with the larger trend as the historically most powerful species yet to have evolved. The trend continues not only because prolific resources have made it possible, but importantly because powerful economic players have helped to increase the supply of resources and thus to stimulate production. The trend is not merely an empirical generalization; it is the inevitable consequence of first principles, of the ways in which economic units compete with and influence each other in an environment of opportunities and constraints, rewards and risks, success and failure, adaptation and restriction, predictable change and unpredictable disruption.

Can or should this trend continue? History and the principles governing it strongly imply that the future will be like the past, that power and independence will increase at least among economic dominants, that technological innovation will continue to ensure further growth. As the current unchallenged economic dominants, we humans expect and assume that that this scenario will apply to us and that we shall be able to maintain our position of global power.

Most economists embrace some version of the doctrine—hope might be a better word—that an increase in living standards for humans will continue indefinitely. Resources, they argue, will not limit us globally, either because we can find more of them or because we use them more efficiently. The Australian economist H. W. Arndt, for example, maintains: "It is most unlikely that the world will ever 'run out' of any natural resource, in the sense that suddenly there will be none left. Rather, if and to the extent that, with population and economic growth, demand for any resource, such as petroleum, runs ahead of supply, its price will tend to rise. This, if it happens to essential materials for which no substitutes are available, will in itself tend to slow down economic growth." Others who are more aware of the possibility that resources may eventually limit human economic growth advocate sustainable development—development without growth—as a way to sustain our species in a finite world. In Herman Daly's words, "It is a subtle and complex economics of maintenance, qualitative improvement, sharing, frugality, and adaptation to natural limits. It is an economics of better, not bigger." But even this view implies growth as resources are used more efficiently. Is this a rational expectation, or is sustainable development a myth, as Douthwaite maintains?[1]

Still others think that material growth is reaching its limits even in the human domain, at least in the most highly developed economies in a globalized, market-driven world. "Affluence has, in practical if not in absolute terms, reached saturation." That is the view of C. Owen Paepke, who foresees a world in which genetic engineering and neural modification will produce an economy dominated by minds rather than matter. Writing in 1993, he predicted that "as one form of progress ends, another begins that may more than compensate for the stagnation in living standards. Beginning early in the next century, people will enjoy vastly expanded intelligence and other abilities, finally escaping the natural limitations under which the species has labored since it first appeared." But it is hard to see how such expansions, if they come, would not further increase our use and control over resources. A perceptive observer 5 million years ago would likely also have claimed that life on Earth—and especially its advanced manifestations, the dominant animals of land and sea—had approached the limits of resource extraction given the technology that had managed to evolve in 3.5 billion years. But then humans evolved and, even without the interventions of technology envisaged by Paepke, proved the observers wrong.[2]

Those who hold that economic growth cannot continue without limit can point to a large body of evidence from history, biology, and economics showing that every economic system—from the cell to the biota, from the household to the empire—reaches a state of dynamic equilibrium, in which growth effectively ceases as costs and risks mount to curb expansion. Individual organisms develop and grow rapidly when young, reach a peak in vigor early in life, and then grow slowly or not at all as they mature and begin to reproduce. In the history of individual species, an interval of rapid evolution at and shortly after the time of origin is followed by a long period of stability in the features that distinguish species from one another. Were it not for this well-known pattern, dubbed punctuated equilibrium by Niles Eldredge and Stephen Jay Gould, geologists confronted with sequences of sedimentary rocks and their enclosed fossils in different parts of the world would be unable to use widely distributed species as time markers. Vegetations on land display a succession beginning with a disorganized collection of small weedy plants and ending with a mature climax forest ruled over by light-intercepting large trees and a cadre of powerful consumers. The number of species in a clade or in a newly constituted biota first rises exponentially, but it often reaches a plateau of sorts, a dynamic state in which addition through species formation or invasion is more or less balanced by loss through extinction. Even in the history of humanity, there are long periods of relative stability or even decline—the "Dark Ages" in Europe, for example—rather than the rapid growth that for the past few centuries has come to be taken as normal.[3]

On very long time scales, too, there are intervals when the overall trends toward greater power and productivity falter. Compared to the Paleoproterozoic era before and the Neoproterozoic era after, the Mesoproterozoic era seems to have been a time of rather low, constant productivity. Stability was lifted only when, about 1.25 Ga, tectonic activity made critical nutrients and oxygen more available to life. A shorter interval of generally static productivity, at least in the ocean, may have prevailed during the Late Carboniferous to Early Triassic periods, 310 to 230 Ma, and perhaps on land during much of Jurassic time. Communities of shell-bearing animals on level seafloors show an increase in the number of species in the Early Paleozoic, followed by a plateau that, with some extinction-related interruptions, lasted until mid-Cretaceous time.[4]

We must be careful of these claims that all systems eventually stabilize around a dynamic equilibrium. Successful evolutionary lineages are characterized by stable individuals replacing each other. On a longer time scale, a period of stable diversity within a clade may be followed by dramatic increases. This is what happened to mammals, which after their origin some 225 Ma during the Late Triassic remained a small, unobtrusive clade of mostly nocturnal land animals for 160 my until the demise of their economically dominant cohabitants, the dinosaurs. Once the dinosaurs and the other major groups of Mesozoic vertebrates had disappeared, the various clades of mammals diversified rapidly and extensively, ultimately producing a richness of species that both locally and globally far exceeded the numbers of Mesozoic dinosaurs. The longer stability of the Mesoproterozoic and the marine Late Paleozoic was ended by renewed growth, probably triggered by tectonic activity.

Perhaps more importantly, stability and dynamic equilibrium in the number of entities in an economy may be phenomena that characterize only the more energetic components. For example, mammals and birds seem to obey the rule that the number of co-occurring species increases exponentially with area among islands in an archipelago. Insects, plants, parasites, and marine invertebrates are much more lawless. Numbers of their species vary greatly even among islands or habitats of the same area. There are no well-defined upper limits to diversity, meaning that many potential ways of making a living are not realized. Dov Sax and his colleagues (2002), for example, find that when islands are invaded by human-introduced plants, diversity on the islands always rises—sometimes by a factor of two—and never falls. The introduction of birds leaves diversity about the same or a little higher, but again the number of island species does not decrease.[5]

We thus have the makings of a paradox. On the one hand, many systems, including their dominant elements, tend toward a limit imposed by supply and the available technology. On the other hand, there is long-term growth, triggered by disruptions, propelled by positive feedbacks between competitors and resources, and usually engineered by replacement of one cadre of controlling agents by another, more energy-intensive one. This long-term growth is made possible by technological innovation, which is favored in conditions of abundance and is spread by conditions that upset the established order. Given our exceptionally rapid growth and our very heavy demand on Earth's resources, can the human economy continue to grow—whether by adding to the resource base or by using existing resources more efficiently—through further innovation? Can we or should we continue to transcend economic constraints in order to raise still further our level of control over the biosphere? Alternatively, can we or should we voluntarily limit our individual and collective economic power?

Problems

In order to tackle these questions, we must clearly identify and evaluate the obstacles and objections to further growth. These crystallize around seven points: (1) finite resources; (2) pollution; (3) degradation of ecosystems on which we depend for production and many other services; (4) military annihilation; (5) inequality of wealth and the related problem of overproduction; (6) the positive feedbacks underlying speculation (socially enhanced confidence in the future) and downturns (reduced trust and confidence) comprising the boom-and-bust business cycle; and, encompassing all the rest; (7) monopoly, at the level of our species as a whole and within our species.

First consider the question of resources. Skeptics argue that the rapid growth of the world economy during the last 250 years is unsustainable because resources cannot be exploited, created, or replenished fast enough over the long run to maintain the current rate of expansion. According to Joel Cohen, the human population since the year 1600 A.D. has increased tenfold. Even since 1955, the human population has more than doubled, from 2.4 billion to 5 billion. World output of goods and services has more than tripled, and per capita income increased by more than a factor of two. In 1990, per capita use of oil amounted to 1,567 kg per year, or about 19 megawatt hours per person. This use of the single fossil fuel oil compares to an estimated 0.9 megawatt hours per person for all fossil fuels in 1860, and 8.2 megawatt hours per person for all fuels in 1950. Given that there is a tenfold difference in per capita energy consumption among nations, an increase in living standards in the developing countries would mean a continued dramatic rise in the use of fossil fuels even if population levels stabilize. Fossil fuels are certain to run out in the next few decades (oil) or centuries (coal). In principle, there are substitutes for them, and market forces will likely facilitate the switch to them; but previous substitutions often followed stagnation or decline as the replaced fuel became depleted. John Perlin, for example, makes a persuasive case that most early civilizations in southwestern Asia and around the Mediterranean declined after local and regional forests had been cut down for the wood that served as fuel and building material. The increased use of coal for industry and transport, together with the importation of wood from North America, enabled Great Britain to industrialize in the eighteenth and nineteenth centuries.[6]

If humanity is already in control of 40 percent of terrestrial biomass and 25 percent of living marine resources, how much higher can we drive the rate and magnitude of exploitation before we overwhelm the productive and replenishment capacity of the world? Part of the answer evidently depends on how the extraction of resources competes with other economic pursuits. Traditionally, the production of food, fuel, fiber, and building materials has required land; but land is also needed for housing, roads, industry, and thousands of social institutions. Water for drinking and irrigation is also drawn from sources on land. The use of artificial fertilizer and the introduction of such crops as potatoes and new, more productive varieties of cereals can dramatically increase yields. Such benefits are potentially threatened by intensive agriculture. The application of herbicides, pesticides, and artificial fertilizers temporarily drives up yields, but continued use depletes the soil of its native nutrients, destroys regulation of resources and pathogens by the soil's microfauna, and makes the soil completely dependent on energetically expensive chemical subsidies. Overly generous irrigation allows growth-inhibiting salts to build up in soils. Currently, about 15 percent of cultivated land worldwide is irrigated, accounting for some 30 percent of agricultural production. Erosion of unvegetated topsoil permanently reduces productivity and is a worldwide problem.[7]

In a penetrating comparative analysis of economic conditions throughout the world during the past few centuries, Kenneth Pomeranz notes that land limitations were beginning to constrain economic development in all densely populated parts of Europe and Asia during the eighteenth and nineteenth centuries. By importing land-intensive resources such as food, timber, and textile fibers from other regions, notably the New World, Europe was able temporarily to escape this limitation. The land and labor of Europe, dependent on the land and forced labor elsewhere, thus became available for industry, which itself produced such labor-saving and production-enhancing innovations as steam-powered machinery, chemical fertilizers, and institutionalized scientific inquiry. Without a relatively underpopulated New World with which to trade, the densely peopled regions of China and India were unable to overcome the land limitation until well into the twentieth century. In the words of Paul Hawken, "Our means of forestalling the feedback from our environments is to take over other environments (changing tropical forests into farms as an example) as a way to increase our drawdown of resources." Subsidies of this kind will not be available when the human population has become dense everywhere.[8]

According to a recent analysis, the world may already be mining biological resources, that is, using land-intensive resources unsustainably, since the 1980s. The rate at which the biosphere replenishes resources on land and in the sea is, in other words, lower than the rate at which humanity is harvesting those resources. This imbalance arises both because the world's population continues to increase and because per capita consumption continues to rise.[9]

Second, increased per capita wealth and global prosperity are causing horrendous pollution. In Herman Daly's pessimistic words, "Throughput begins with depletion and ends with pollution." Under the term pollution I include chemical alterations to the environment that compromise human health and welfare or that diminish production. Effects range from the local and temporary to the global and persistent. Perhaps the best known global pollution is the increasing concentration of carbon dioxide in the atmosphere. This concentration has risen from 280 parts per million by volume (ppmv) before 1750 to about 360 ppmv today. Carbon dioxide released by humans now exceeds that erupted from volcanoes by a factor of one hundred (10^{13} as compared to 10^{11} kg per year). Together with water vapor and methane, carbon dioxide is a greenhouse gas, which prevents the radiation of Earth's heat out to space. Global warming—seemingly a reality as I write—is the probable consequence of increasing concentrations of carbon dioxide and methane under human governance.[10]

Air, water, soil, and sediment are polluted by a thousand other substances released into the environment by human activity. Without belaboring this obvious point, I would simply point to ground-level ozone, nitrous oxide, sulfur dioxide, toxic metals, pesticides, antibiotics, hydrocarbons, crude oil, dust, radioactive wastes, organophosphates, plastics, sewage, refrigerants, road salt, bleach, paint thinners, explosives, formaldehyde, carbon monoxide, acid-laced rain, hormonelike substances, dyes, and industrial solvents. Some of these—carbon dioxide, methane, organic poisons, and salts—are not really novel, but are being introduced at relatively high rates, although the concentration of carbon dioxide remains far below that during much of Earth's history. Other substances, however, are products of the industrial and postindustrial age, and are wholly new additions to the biosphere. The same is true of highly distilled alcohol and drugs. Primates, birds, and fruit flies commonly eat fermenting fruits that contain alcohol, and are able to withstand the relatively low concentrations of methanol that naturally occur in such food sources; but they are unprepared for the alcohol concentrations made possible when alcohol distillation was invented about 200 B.C. and when it became widespread after 1100 A.D. Life, including humanity, is not adapted to these new substances, nor perhaps to the very rapid rates of change humans are inflicting on the biospere.[11]

Third, we are fundamentally altering ecosystems on a global scale. Hunting with weapons of increasing range, precision, and power has eliminated top consumers worldwide or depressed their numbers to the point where these animals no longer fulfill their previous ecological roles. Fishing has had the same effects in the sea. Logging and burning have transformed forests into grasslands, farm fields, or cities. Agriculture has resulted in the erosion of topsoil and the transfer of nitrates and phosphates to rivers, lakes, and coastal marine waters, with the consequence that the nutrient content has doubled in the North Sea, the waters off the eastern United States, and elsewhere. Sediments in the Gulf of Mexico have become so enriched with nutrients and with oxygen-using life that they have become dead zones, devoid of oxygen and of animal life. Most ecosystems have been fragmented into small remnants through the construction of roads or the transformation of habitats, in effect turning formerly continuous habitats into archipelagoes of habitat islands where species with large per capita area requirements and low population densities cannot survive. Commerce, aided by rapid worldwide transportation, has allowed the deliberate and accidental spread of thousands of species, some of which dramatically alter the recipient environments they invade. Dams and other water diversions cut off sources of nutrients and sediments to such places as the Gulf of California, the Black Sea, and the eastern Mediterranean. Much land has been paved over with roads and urban sprawl, reducing the ability of soils and rivers to soak up and transport water and therefore increasing the intensity of devastating floods. Industry, agriculture, and urbanization have simply eliminated ecosystems. It is especially the most productive ecosystems—in estuaries, on continental shelves, in continental lowlands, on plains, and in river valleys—that humanity has exploited most. Life of no direct economic importance to people hangs on in ecosystems of low productivity.

Our activities have had unexpected important ecosystem-level consequences. By eliminating top consumers, we have created the ecological equivalent of recession in that production and supply outstrip consumption. This has created various perturbations. The effect may be to reduce redundancy of tasks, and therefore to place ecosystems at greater risk of fundamental restructuring. Perhaps the most dramatic example comes from the transformation of luxuriant coral reefs in the Caribbean to seaweed-choked ecosystems. Chronic overfishing had long ago eliminated most of the top predators in Jamaican and other coral reefs before two events in the 1980s brought about a fundamental change. The first event was a devastating hurricane, which particularly damaged corals. Then, in the mid 1980s, an infectious disease spread like wildfire around Caribbean populations of the sea urchin *Diadema antillarum*, a major herbivore in the system. More than 99 percent of the population was eliminated by this disease. With many herbivorous fish already having been exploited, this left large parts of the Caribbean without effective herbivores. Aided by intense runoff from the adjacent land, much of which was being developed for agricultural and urban use, seaweeds settled and grew with few controls, and overwhelmed the coral recruits that had managed to settle. With the death and lack of replacement of corals, much of the three-dimensional complexity afforded many habitats for the enormous diversity of small animals and fish on the reefs disappeared.[12]

Fourth, greater wealth and more powerful technology have for the first time in the history of life made it possible for a single species to destroy itself and the biosphere as a whole. There is nothing new about weapons escalation among enemies—organisms have, after all, been at it since life's beginning—but the rate and extent of weapons development and deployment have far surpassed anything that happened before humanity asserted its economic control over the world.

Like biological escalation, military escalation appears to depend on growth and wealth. We can therefore expect weapons improvement to be with us as long as the human economy expands and humans resolve their conflicts with force. Under conditions of surplus, resources—money, materials, and people—are available for tasks beyond basic maintenance, and can thus be deployed in the development and use of weaponry.

The historical record confirms that increased power of weapons and conflicts inspired by expansionism often follow the attainment of a certain affluence and at least the capacity, if not always the realization, of a formidable organization to design, produce, and execute wars with powerful weapons and a professionalized soldiery. Organized warfare may have begun about 5,500 years ago, in Mesopotamia, after the establishment of the first urban centers, with the development of bronze weapons. Greek advances in military organization—the use of the phalanx, or rows of soldiers fighting in head-on combat with opponents in the open—and in weapons during the eighth century B.C. were made possible by trade and by an increasingly reliable system of food production. The same conditions were favorable to literature, philosophy, mathematics, and empirical inquiry. In China, steel-making and other industrial innovations during the eleventh and twelfth centuries A.D. set the stage for naval expeditions to Java and the Indian Ocean. From the second half of the fifteenth century onward, however, this kind of adventurism and the accompanying development of weapons and warships abruptly ceased as the state, under the despotic rule of the emperor, turned away from foreign trade and conquest to concentrate instead on internal security. Many of the Chinese military inventions were brought to the Arab world, but deteriorating climates in North Africa and southwestern Asia led to economic stagnation, a circumstance not conducive to military investment. Only in Europe, where the later fourteenth and fifteenth centuries witnessed a rise in population and in economic activity, were the weapons systematically improved. Mobile large guns that hurled metal balls existed during the last third of the fifteenth century, and defenses—the *trace Italienne*, a system of vertical earthen walls and cannon—suitable for withstanding bombardment by these new weapons were in place by 1500. The Industrial Revolution and its attendant economic expansion of the nineteenth century made the mass production of cheap, effective, long-distance weapons possible. The arms buildup by the superpowers after World War II would have been unthinkable without economic growth in the United States and Europe.[13]

The development of weapons of global destruction has been defended as a mechanism of deterrence, a way of avoiding war. Indeed, this style of deterrence worked during the Cold War as the United States, Great Britain, the Soviet Union, and later several other countries developed and improved nuclear weapons. Deterrence is effective as long as rational people and governments unwilling to risk the destruction of their own countries are in control. It is sobering to remember, however, that most weapons that humans have invented have, at one time or another, been used. As in the domain of nonhuman escalation, the expenses of developing and maintaining weapons cannot be economically (or evolutionarily) "justified" unless the weapons effectively fulfill their offensive or defensive function. The continued presence of powerful weapons, and the existence of the expansionist economic conditions that make possible their further development, thus poses a very real threat, especially if one of the parties either believes it can win or perceives that it has nothing to lose. This threat is vastly increasing now that extremely potent chemical, microbial, and nuclear weapons have become cheaper to produce and more readily available readymade.

An additional problem with large investments in potent weaponry is the false claim that the weapons provide absolute security in the event they are used. When President Ronald Reagan proposed the missile defense system in 1983, he argued that such a defensive shield would protect the United States from all incoming enemy rockets. It seems very doubtful that any well-informed military leader would sanction such a claim, for as long as there are enemies capable of responding to improvements in weaponry, there exists the great likelihood that even the most potent offensive and defensive weapons can be partially compromised. Certainly no weapons or defenses that have evolved in nonhuman organisms can be described as insurmountable, even if many of them have achieved extraordinary sophistication and effectiveness. Claims of the insuperability of proposed weapons strike me not only as misleading and false, but as dangerous. They lull people into unwarranted confidence and dependence, and promote the perception that risks of annihilation are declining when in fact they are rising. Nothing in evolution and economics is perfect, and that includes weapons and the people who use them.

Fifth, the modern human economy is characterized by a pervasive imbalance between production and demand. Technology and an abundance of investment capital have enormously increased the production capacity of industry and agriculture, whereas the demand for products has not kept pace. In capitalist economies, moreover, the proportion of wealthy individuals who are in a position to buy the products is diminishing. Thus the low demand relative to supply is in part the result of highly inequitable distributions of wealth.[14]

The sixth problem, which has become especially prominent in the last four hundred years of economic history in the developed world, is the socially driven instability of the business cycle. A mood of optimism and confidence created by risk-taking entrepreneurs and investors spreads to others willing to invest capital in ventures that sometimes do, but often don't, turn out to yield reliable, long-term profits. The classic early case was the tulip craze of the 1630s in the Netherlands, in which bizarre cultivated varieties of tulips fetched enormous prices until the bottom fell out of the market and thousands of investors and buyers were left with nothing. With an abundance of capital in the modern globalized human economy, such speculative schemes have become common; and, although they often lead to disaster, they are also essential in funding offbeat projects that on occasion lead to astonishing technical breakthroughs, such as computers. Both the boom and the later bust part of this cycle are driven by a social phenomenon, in which the actions of one individual are influenced are those of others. If leaders or influential commentators have confidence, that confidence will spread; if, on the other hand, they decide that conditions will turn bad, they lose confidence, and their prediction will become a self-fulfilling prophecy by virtue of the social spread and acceptance of their prognoses. This socially driven boom-and-bust business cycle seems to characterize all modern human economies, and creates significant instability that can on occasion result in deep depressions, as during the 1930s. The magnitude and rapidity of response are amplified by instant communication and by the limited controls on the movement of capital.[15]

Finally, the human species as a whole has evolved to the status of an economic monopoly in the biosphere. Moreover, the emergence of the United States as the current single superpower among nations underlies the reality that monopolies can arise as governments and corporations within our species. Monopolies exercise power without restraint or modification by other competitors, and therefore do not act for the larger common good of those around them. The selfish, unconstrained, and often destructive actions of a monopoly may therefore damage the web of interactions in the economy that sustains the monopoly, and lead to economic instability and to a greater vulnerability to disruption. Ultimately, then, the bullying arrogance of the monopoly may bring down not only the monopoly itself, but also the entities under its control.

In his global survey of genera of animals, Richard Bambach found that the rates at which predators appear and disappear in the fossil record are on average 1.3 times higher than the rates at which other kinds of animals—herbivores, suspension-feeders, and detritus-feeders—arise and become extinct. This contrast would presumably be even greater if the category of predator were restricted to the most energy-intensive consumers. Blaire Van Valkenburgh, in fact, has uncovered this pattern in a comparison of large carnivorous mammals, which come and go rapidly, and the longer-lasting, less evolutionarily volatile small-bodied predatory mammals. The greater vulnerability of monopolistic or other very powerful entities seems to rest on the exploitation rather than the enhancement of their resource base, that is, on the erosion of feedbacks that promote the larger common good. Local monopolies, such as trees poisoning their neighbors and ants annihilating most other insects in their vicinity are rare and short-lived in nature.[16]

Solutions

It would be folly to think that the problems I have just sketched have solutions readily at hand. A good solution for one party, or for one point of view, might be judged a disastrous solution by another. Answers, after all, involve judgments about what is "good" and "bad," and for whom under what circumstances. My aim is not to come up with some magic bullet that has so far eluded a thousand other observers. Rather, I want to explore whether and how the various potential solutions to our problems would be consistent with fundamental economic principles and with the long sweep of economic history. Some solutions are likely to be more to be more consistent with "human nature"—economic nature, to be more precise—than others. Some may call for actions that are unprecedented; others may conflict less with biological and economic imperatives. I want to look to the economic structures in nature—structures tried and tested for billions of years under every imaginable circumstance—for insights into what our economic future might look like under various proposed solutions.

What, then, do the economies of nature tell us? Do growing organisms or ecosystems exhibit problems similar to those confronting human economies? Do economic units that have achieved dynamic stability in nature suffer from the dramatic inequities and the chronic problem of oversupply that now characterize so many human economic systems? What would the solutions proposed for the human economic future look like in nature?

Before proceeding to such an investigation, it is worth recapitulating seven realities about economic systems, realities that arise from first principles and are supported by overwhelming empirical evidence. These are: (1) there is always competition for resources; (2) there is always inequality for the parties involved in competitive interactions; (3) there is always adaptation by economic units; (4) there are always disturbances, some common and mild enough to be incorporated into adaptive hypotheses, others so rare and intense that they disrupt economic systems; (5) no adaptation, and no adapted system, is perfect; (6) adaptive response is less disruptive when resources are abundant and when the economic system is growing than when resources are scarce and the economic system is static or in decline; (7) successive economic dominants show a pattern of increasing per capita energy use and power through time and create positive feedbacks with resources, so that they and the economy they control become more independent of their environment even as they increasingly modify internal conditions.

These realities set limits to the kinds of solutions that are available to us. If we decide that we can or should violate these principles, we must be made keenly aware that violations would be without precedent. In my view, this probably means that solutions embodying a departure from economic realities are bound not to work. They are doomed to failure not because of some unshakable economic determinism or the complete abdication of moral and esthetic responsibility in favor of the amoral material marketplace, but because competition and all the other principles are to economics what gravity is to the construction of buildings and bridges. It is simply not possible to fashion a workable economy without taking competition—conflicting interests among parties—into account or without accepting self-interest as a point of departure.

To most human observers, the notion of solving a problem implies the existence of economic agents who examine the evidence, diagnose the problem, weigh the options, make a decision, and implement a solution. In nature and in the idealized free marketplace of human commerce, however, this adaptive process occurs "by itself," in a self-organized, diffuse way involving many agents. Some agents have more control over the process than others, but there is a minimum of centralized planning. This ideology is the modern-day version of Adam Smith's competitive marketplace, his "invisible hand," unencumbered by centralized constraints on economic activity yet acting for the common good. The use of resources is determined by price, which fairly reflects supply—the amount for which an agent is willing to sell something—and demand—the amount people are willing to pay for it—with competition among sellers and among buyers providing the mechanism of stability and fairness. Adam Smith believed that the common good of the group is best served when the market preserves individual initiative and self-interest as reflected in the ownership of land and other assets. Diffuse control importantly also limits the damage that errors made by particular agents cause.

It is important to examine the underlying claims of this diffuse, free, and democratic style of control. When Adam Smith espoused his system as the best suited to benefit the common good, he did not consider which segments of human society would be the chief beneficiaries. The European economic system about which Smith wrote in the eighteenth century depended in large measure on the toil of slaves in the New World and of forced colonial labor in other parts of the world. Although Smith abhorred slavery and condemned it as unjust and uneconomical, he nonetheless judged the common good largely from the perspective of a member of a privileged class in a powerful nation. The common good, it seems, applied mainly to those in power, to the Europeans and the colonists who held the slaves and who ruled the lands overseas. Even the United States, a nation uniquely founded on principles of equality and freedom, extended these privileges mainly to white males, and denied them entirely to slaves. Notions of the common good broad enough to include nonhuman species did not surface until the last third of the twentieth century. In short, the breadth of inclusion is critical to the discussion of whether and how economic systems operate for the benefit of participants. Up until now, they have benefited largely the high-powered consumers and producers who have been the chief architects of our system and of nonhuman systems in the past. To be sure, selection and the actions engendered by high-powered participants have created opportunities and occupations for many other entities, and through a complex self-organized regulation have even ensured the survival of the less powerful, but these effects have been secondary to the self-interest of the major players.

As the point about slavery already makes plain, Adam Smith's and others' free-market economy is free mainly for those in power. Moreover, in today's world of big government and powerful corporations, the democratic, inclusive ideal envisaged by Adam Smith is not closely approached even in those societies that think of themselves as democratic and free. Representative, democratic government with checks and balances ensuring that no segment of government can wield unrestricted power may characterize some technologically advanced societies today, but it coexists with corporations whose power rivals that of governments and whose structure is anything but representative. Although corporations of modern style did not exist in Adam Smith's day, their equivalents did. The East India Company in Great Britain and the Verenigde Oost-Indische Compagnie in the Dutch Republic wielded monopolistic power in the colonies. In principle, these early companies and modern corporations work for their investors, but these investors are small in number even in today's market, and have increasingly become large powerful institutions themselves. The board of directors and the officers who run the corporation comprise an oligarchy which exercises strong top-down control different in principle from the ideal of representative government, which through frequent evaluation by election exercises diffuse, if often flawed, control. Even democratic representative government is not immune from taking reckless action, as events in Germany in the 1930s demonstrate all too plainly.

At the other end of the spectrum of control is strong centralized authority exercised in the human domain by governments, corporations, and religious institutions. Self-interest is suppressed through various mechanisms in favor of the common good, or for the aggrandizement of those in power. Control is concentrated in an oligarchy or even in a single individual. Control of this kind takes many forms and has been used for many ends, but it often explicitly counters the free market by insulating groups, prices, and trade from prevailing competitive pressures.

One modern version of control that seeks to regulate traditional concepts of the free-market economy is that advocated by the so-called "green" economists. Through the implementation of policies restraining power and trade of corporations, the "greens" emphasize regional or local self-sufficiency and long-term sustainability over globalized interdependence, community values and needs over individual wealth and wants, the efficiency of resource use over environmentally costly and often wasteful exploitation, and a pricing system that incorporates not just supply and demand, but also the hidden long-term costs that no one involved in buying and selling bears directly. Hidden costs include disposal, pollution, and other external costs that markets generally ignore, as well as high wages for workers. Although this economic view, like the free-market approach, rests on economic growth, it is motivated by the belief that the economy will stabilize around the sustainable exploitation of resources, with a more equable distribution of wealth than most free-market economies have been able to achieve. If, within a system of centralized restraint and regulation, most people would be able to invest capital, and if they therefore had a stake in the enterprises for which they work, aggregate as well as individual economic behavior would aim toward long-term economic improvement and not just short-term gains. The more equable distribution of wealth would increase demand for goods and services, and would thus ease the chronic oversupply that marks modern capitalist economies and fuels economic recessions.

Just as the free-market economy remains an ideal that in practice has not been closely approached, so the sustainable economy is likely not to be easily achieved. Inequalities in performance remain, and no system of centralized restraint can be so effective that everything is perfectly regulated. In fact, a major shortcoming of centralized control is that the system becomes highly sensitive to error and to cheating. An unintended or deliberate error can be corrected or rendered harmless by redundant diffuse control, but tends to spread and do damage if it is made by an unchallenged central authority. The sad history of societies under twentieth-century Communism offers many examples; so do the expulsion of the Jews from Spain in 1492 and the invasion of the Soviet Union by Hitler's Germany in 1941.

Economists and sociologists have been surprisingly unwilling to consider the long-term consequences of the economic choices that may be forced upon us. Twenty-five years ago, Arndt complained that "There was surprisingly little careful analysis of the likely political and social problems of the 'ecological transition' from a growing to a stationary economy." This remains as true today as it was then. Growth-addicted economists and politicians of every stripe have been unwilling to consider what a nongrowing economy could look like except by pointing to depressions and recessions. Those advocating limits to growth have often glossed over the social consequences of the rigidities that would likely accompany a transition from growth and expansion to maintenance and stability.[17]

Biologists, on the other hand, have been studying economies in all states of maturity, from rapidly growing ones to those in dynamic equilibrium to those in decline. In an influential paper in *Science* more than twenty-five years ago, Rapport and Turner quite reasonably suggested that "a study of how natural communities come to grips with resource limitations and achieve a nongrowth economy may provide guidance for the management of human communities faced with the challenge of making the transition to a steady state economy." That this expectation has not been fulfilled testifies to the great rift that still exists between the biological and the human-oriented sciences.[18]

Consider first those solutions that are consistent with a diffuse "free" market. The most general of these is more knowledge, more technology, more innovation. Adam Smith wrote at a time just before the Industrial Revolution, when scientific inquiry was less institutionalized than today. Since then, we have come to understand that technological innovation is a primary driver of economic growth. Empirical inquiry and the development of scientific theories had been a European trademark since the late Middle Ages, and attained more formal recognition with the founding of scientific societies in Rome (1603), Paris (1635), and London (1660); but before the second half of the nineteenth century, innovation and invention were largely in the hands of gifted individuals who, with or without formal education, devised ingenious techniques and machines. Scientific exploration on expeditions began in the late eighteenth century with James Cook's voyages, but systematic research centered on experimentation and theorizing began to flourish in the late 1800s. The benefits of this organized effort were enormous. Medicine vastly improved individual and public health and lengthened individual life expectancy by decades. Improvements in crops, food conservation, soil management, fertilizers, and transport increased agricultural production and stabilized the food supply so that many parts of the world came to be shielded from famines and chronic shortages. Institutionalized science, though very much dependent on increasing affluence, became a powerful engine that created more growth and more wealth.[19]

Even in today's establishment of organized research, great breakthroughs depend on the talents of a few individuals. The pool of highly talented scientists, inventors, and entrepreneurs clever enough to solve the most intractable problems and with enough savvy to bring the solutions into the marketplace is likely to be proportional to the overall size of the population. This is the crux of the argument by Julian Simon and other economists in favor of an increase in the human population as the best way to innovate ourselves out of our collective predicaments. In order for population growth to have this effect, however, everyone must be afforded the educational and employment opportunities to realize their intellectual or entrepreneurial potential. This means that population growth must be closely linked either to an abundance of resources and to a permissive attitude toward innovation in society, or to an increase in resources and social mobility. Growth in the number of individuals by itself is neither sufficient nor even necessary; it must be accompanied or replaced by per capita economic growth, by individual liberty, and by inclusive participation in the discovery and implementation of solutions and social goals. It is no accident that Greece, with perhaps the earliest version of democratic government in some of its city-states, produced a disproportionately large number of extraordinary thinkers despite a small population. Similarly the Netherlands in the seventeenth century, and Scotland in the eighteenth, produced a remarkable crop of scientists, merchants, philosophers, and inventors, who took advantage of their relatively tolerant and permissive societies to make great intellectual, scientific, and sometimes artistic strides even though these countries had tiny populations compared to those of more regimented ones like India and China.[20]

A reliance on talented people to solve problems through application of the scientific methods is contingent on three other circumstances, which all require a permissive environment in which curiosity and empirical investigation are unencumbered. These are (1) the existence of institutions where research is carried out; (2) a diminished role for systems of thought based on proclaimed untested doctrine; and (3) the conscious decision to subsidize inquiry, including seemingly useless inquiry. The aim is to create an environment favorable to the development of ideas and prototypes, to generate variation in the same way that organisms have evolved systems for generating genetic and phenotypic variation. Moreover, the new ideas and prototypes must for a time be sheltered from lethal competition with the products and the patterns of thought that comprise the status quo. The system must be tolerant of error; it must preserve the flexibility needed to generate and test novelties.

Early protection from market forces is important. As discussed in chapter 3, many organizational mechanisms have evolved in the nonhuman domain that both generate and preserve variants. Prototypes are rarely as good as the incumbent products or adaptations with which they must compete eventually. Thus even potentially very useful innovations are eliminated if their poorly adapted early versions are exposed to the full complement of economic forces from the beginning. This is a cogent argument for sheltering the research side of innovation from the profit-making function of enterprises, and for sheltering the creation of knowledge for a time from results-oriented administrators who wish to turn universities, museums, and libraries into business enterprises. It is also an argument for keeping the research and development functions of private enterprise separate from the more directly market-oriented side of the business. Without such organizational safeguards, no long-term solutions are likely to be sought or nurtured; only ideas and products with short-term economic benefits will reach the marketplace. Systematic inquiry by scholars and tinkering by inventors are best practiced, and yield the greatest and most unexpected rewards, when these activities are carried out in the absence of an economic imperative. Human curiosity is a far more potent weapon of progress than is inquiry motivated only by economic gain, even if that curiosity arises from the desire to celebrate and glorify the work of an imagined Creator. An adaptable society must find and maintain the means to explore new ideas and new methods, no matter how outlandish these new approaches may initially appear to be.

Talent and institutional support cannot engage in empirical exploration and innovation in the absence of a cultural environment that accepts observation, experiment, and deduction as the most reliable methods for gaining and organizing knowledge. As I emphasized in chapter 2, these methods—collectively dubbed the scientific method—are in essence identical to the economic processes that underlie evolutionary adaptation. The hypotheses, products, theories, and policies resulting from or informed by the scientific process are not necessarily the best that could be devised, but they are workable approaches that have the great virtue of being improvable as more is learned. Most students of comparative history have rightly assigned a leading role to cultural freedom as an essential precondition for invention, curiosity-driven science, and enterprise.

Whenever organized religion, built on a faith-based unshakable dogma, achieves social control through the powers of the state as a theocracy, enterprise and empirical inquiry tend to be suppressed, and the civilian economy is tightly controlled to prevent the unequal distribution and undue accumulation of material wealth. As David Sloan Wilson rightly points out in his book *Darwin's Cathedral*, strong religions achieve adaptive social cohesion by motivating people, as no secular law can, to "do the right thing," to adhere strictly to a moral code benefiting the group and serving to separate that group from potential competitors. The benefits of allegiance and group cohesion, however, come at the expense of intellectual exploration, and often emphasize internal security and sometimes militarism. Theocracies and functionally similar authoritarian regimes led by a charismatic dictator—many governments in today's Middle East, inquisition-ravaged Portugal and Spain during the sixteenth and seventeenth centuries, Mao's China and Stalin's Russia during the twentieth century, and Calvin's Geneva in the 1500s, come to mind as examples of pervasive top-down control of thought and action.[21]

The cultural systems that are most receptive to the scientific enterprise and that are most adaptable are those in which controlling bodies—the state, religious institutions, and corporations—intrude least on freedom of thought. People everywhere, the rich as well as the less affluent, must have the freedom and the confidence to satisfy their curiosity and to explore solutions to perceived problems; and there must be some connection between effort, or performance, and reward. The fates of individuals should, to an important degree, reside with the individuals themselves, and not be in the hands of a capricious deity or despot or a remote oligarchy. Hope and well-being must be retained through just rewards in the face of frequent calamities, which so often breed a culture of fatalism in parts of society at the edge of subsistence.

Some scholars credit religion, a system of symbolic beliefs underpinning a code of conduct of individuals in a group, with providing just this kind of motivation to people who find themselves in a world of harsh realities and incomprehensible evil. Arguing that ancient emotion evolved as a means to motivate individuals into adaptive behavior, David Sloan Wilson explains how adaptive religions achieve cohesion and compliance: "We might . . . expect moral systems to be designed to trigger powerful emotional impulses, linking joy with good, fear with wrong, anger with transgression. We might expect stories, music, and rituals to be at least as important as logical arguments in orchestrating the behavior of groups." Comparing religion with the factual world of science, Wilson notes: "At times a symbolic belief system that departs from factual reality fares better." As I pointed out in chapter 2, the measure of effectiveness of an adaptation is whether it works, not whether the adaptive hypothesis is, in a scientific sense, true.[22]

Moreover, the Protestant Reformation, and especially the rise of Calvinism in the late 1530s, could have triggered the cultural transformation that eventually spawned scientific societies and the Industrial Revolution. Writing of Calvinism, Tawney asserted that "it is perhaps the first systematic body of religious teaching which can be said to recognize and applaud the economic virtues."[23]

Yet, it is possible to look beyond such connections and to argue instead that Calvinism and the later drift toward a more scientific world view are both consequences of economic growth and increased permissiveness that became well before the Protestant Reformation. Europe entered an age of geographic exploration and an expansion of intercontinental trade during the fifteenth century. This state-sponsored reconnaissance, like the empirically grounded philosophies of Plato and Aristotle in ancient Greece and the development of science in post-Reformation Europe, is the product of urban society, rooted in the surplus economies that enabled cities to arise, gain independence, and grow through trade. I would not credit Calvinism with establishing a permissive climate in which unbridled inquiry could proceed. Instead, I suggest that growth beginning in the late fourteenth century was conducive to departures from conventional wisdom, and that these departures came both as religious heresy and as increased emphasis on practical and theoretical science. In turn, the new philosophies—even perhaps Calvinism—not only condoned growth and individual enterprise, but actively encouraged them. In Tawney's words: "Such teaching, whatever its theological merits or defects was admirably designed to liberate economic energies, and to wield into a disciplined social force the rising *bourgeoisie*."[24]

In this view, the new teachings are a cultural adaptation to the economic circumstances of growth, as well as a catalyst for still more change. As I have stressed repeatedly, cause and effect are inextricably intertwined in economic systems because of the prevalence of positive feedbacks. The important point is that mechanisms generating new ideas, which in human society are expressed culturally, are as important as access to abundant resources for economic growth and economic adaptation. Moreover, social permissiveness and resource-based opportunity for adaptation are linked. When unification of Scotland and England was ratified in 1707, Scotland gained access to the foreign trade that in previous decades had been largely in the hands of the English. The wealth that flowed from this treaty transformed Scotland from a subsistence economy and a rigid theocracy based on the teachings of the late sixteenth-century theologian John Knox to a prosperous province in which educational institutions, enlightenment thinkers such as David Hume and Adam Smith, and a culture of thought and action founded on experience took root and flourished. Clearly, the cultural climate catalyzes growth and free inquiry only as long as dogma is prevented from becoming entrenched and social adaptation remains possible. When dogmatic rigidity sets in, partly because economic growth subsides, the social-adaptive status quo is re-imposed. This happened, for example, when, in Wilson's apt words, "Calvin's catechism turn[ed] faith from a belief designed to be modified by experience into a fortress designed to protect the belief system from experience."[25]

The redistribution of wealth is another widely desired goal that may be facilitated by conditions of economic growth. If a society's collective wealth remains constant, a narrowing of the gap between rich and poor is possible only when the incomes of the destitute are raised or when those of the wealthy are reduced. In other words, those with power and influence lose. Those in power, whether they be people or nonhuman entities, do not willingly give up their privileged status. Resistance from those quarters would likely doom any enforced redistribution or incomes, and lead to violent confrontations. It is the wealthy who, in times of economic decline, stability, or equalization have the most to lose, and it is they who are in a position to retain their advantage, violently or otherwise. In a growing economy, by contrast, some redistribution can proceed without anyone suffering absolute losses. Instead, the top brackets of income will gain less than the bottom.

Now, growing capitalist economies tend to be characterized by a widening rather than a narrowing of incomes as the rich get richer and the poor continue to get by with near-subsistence wages. This means that a narrowing of the gap will require a policy that is antagonistic to the usual course of events in a growing economy, but I would argue that such a conscious policy is far more likely to succeed under conditions of expansion that when economies are in stable or contracting mode. Such a policy would dictate increased overall taxation during times of growth, with the heaviest proportional burden falling on the richest segment of the population. The noted American economist Joseph Stiglitz has pointed out that this kind of redistribution of wealth accompanying growth was achieved at least for a time during the spectacular period of growth in East Asia beginning in the 1970s. He notes further that the unregulated market, left to itself, would not have achieved a narrowing of incomes. Instead, just as in twentieth-century Europe, top-down government intervention promoting employment and imposing high taxes that are graduated according to income was necessary to accomplish the broad social good of greater equality of wealth as the economy expanded.

In many parts of the world, such as Ethiopia and North Korea in recent years, chronic shortages of food and water persist despite a global glut in agricultural production. In principles, these regional shortages can be overcome by improving the transportation network required to distribute essential commodities. As is true for some many other solutions to economic, problems, however, such improvements cost money, and such outlays are more willingly borne by governments overseeing growing economies than by those in charge of static systems. In the long run, moreover, the alleviation of famine creates more demand. This would spur further economic growth by soaking up the overcapacity of production, but such growth will only add to the problems that those who find growth unsustainable are trying to solve in the first place.[26]

It is clear to all that inequalities in income will remain. Daly and Cobb propose that an ideal and, in their view, achievable goal is a tenfold difference in income between the wealthiest and poorest individuals. In the United States today, the gap is more like a factor of a million. The difference between the richest and poorest nations—Switzerland and Mozambique, for example—is about a factor of four hundred today, but it would have been only a factor of about five at the beginning of the nineteenth century. Differences among species in nature, as measured by per capita biomass or per capita rates of metabolism, are much more than those observed among individual people in advanced countries, and much higher than the tenfold difference advocated by Daly and Cobb. It thus remains to be seen how far the current inequalities can be reduced without some potentially draconian policies enforced by a strong centralized authority.[27]

Not only can growth fail to continue indefinitely, meaning that the advantages of growth must at some point cease to apply, but growth also engenders many risks and problems. It is therefore important to consider what an end to economic expansion would mean for humanity and the rest of the biosphere.

In order to arrive at some sort of dynamic equilibrium, two separate goals would have to be met: (1) a stabilization of the size of the human population, so that on average every person is replaced in the next generation by not more than one new person; and (2) stabilization of the per capita use of resources or, if these resources are being mined on an unsustainable basis under the current system, a reduction in per capita use. If the second goal is accomplished by increasing the efficiency of resource use, some economic growth might still be expected, at least until a limit in efficiency is reached. Ultimately, however, stabilization of individual and collective demand will usher in an age of what growth-oriented economists would call stagnation and more sympathetic observers would label sustainable stability.

The fundamental problem with cessation of growth is a reduced ability to adapt. The costs of improvement in one task are exacted by lower performance in another. Advancement of one party means retreat for others. Competition is brutally intense, but the ability to respond to it by means other than maintaining the conservative status quo is severely restricted. Perhaps most importantly, the environment that permits inquiry and innovation is replaced by one favoring allegiance to proclaimed dogma. In Tawney's words, "the object of statesmen is not to foster individual initiative, but to prevent social dislocation."[28]

Historians have pointed out that the prophets who founded the great monotheistic religions come from the ranks of herdsmen and farmers, thousands of years after the origins of pastoralism and agriculture. My hypothesis is that these religions are cultural adaptations to an economy that operated close to the carrying capacity given the level of technology of the time. By assigning a dominant role to a supernatural being whose actions are largely beyond the control of mortal humans, they reinforce the perception that little can be done to improve the lot of people. Importantly, they create a system of ethical rules that reduce conflict within the society. They therefore preserve and enhance the status quo, in effect reconciling society and its members to their likely fates given the improbability of betterment.[29]

The success of some religions, notably Christianity and Islam, in helping to motivate societies to engage in expansion and conquest implies that some religions may also be social adaptations to growth. Ideologies proclaimed by charismatic leaders similarly often create expansionist sentiments and are thus like deity-based religions in forging social cohesion sufficient to propel societies to grow. Two major unanswered questions emerge that beg for answers. First, what characteristics do ideologies leading to growth and expansion have in comparison to ideologies that retain their function of stabilizing society? Second, do ideologies based on dogma and myth work better in human society than does a scientific system based on facts and alterable theory? In Wilson's view, competition between faith-based and rational belief systems will reveal the former to be the better adaptation in human society: "If there is a trade-off between the two forms of realism, such that our beliefs can become more adaptive only by being factually less true, then factual realism will be the loser every time." Given the general trend toward secularization in Europe and Japan and in some parts of the United States, I am more hopeful about the ultimate success of factual realism in forging a moral code to which a majority in society will adhere.[30]

There may be debate about the long-term role of religion in growing societies, but there can be little doubt about the nature of social adaptation or accommodation to the cessation of growth. We need look no further than to human societies which have undergone economic decline or stagnation to glimpse the cultural consequences of stability. In his evocatively written book *Villages*, Richard Critchfield offers a sympathetic picture of life in economically stable agricultural village societies as he experienced them in Egypt, China, India, Indonesia, Brazil, and Mexico. He notes that the people in such societies are strongly religious, fatalistic, skeptical or even hostile to new ideas or new methods of production, socially stratified, and relatively harmonious. Speaking of villagers in stable agricultural communities, Critchfield writes: "For them, the spheres of reason, order and justice are terribly limited and no progress in our science and technology has yet convinced them otherwise. . . . There is no use asking for rational explanations. Things are what they are, unrelenting and absurd."[31]

For humanity, increased reliance on rigid dogma as the cohesive glue of society, and the increased entrenchment of incumbent ideas and social structures, has profound implications. We can expect that the hostility to evolutionary thinking, which today resides largely among religiously conservative people who have chosen or been forced to accept an economically lower position or who have culturally inherited a religious orientation, will spread from its current base to other parts of society as communities can no longer afford either the investments or the permissive attitudes required for a scientific, fact-based style of adaptation. Social rigidities will ensue, in which opportunities for individuals will shrink. Security may replace freedom as the more predominant social want, and perhaps even the style of government—the nature and extent of top-down control—will shift from a more diffuse, more democratic form to an increasingly centralized regime run either by some version of a nation-state or by a few dominant corporate entities supported by a religion-like ideology.

An Adaptive Design for the Future

The entire question of what to do about our predicament hinges on adaptation. We must adapt to ourselves and to our circumstances before the ultimate positive feedback destroys us. There are two extreme models of how to accomplish such adaptation. One is a centralized government, a master controller, ideally composed of the best, most benevolent experts and the most thoughtful leaders who, through a system of merit-based advance, successively come to occupy positions of economic and political power. The second is a system closer to the free market envisaged by Adam Smith, in which ideas and technologies from all quarters are tried and tested, with ideally the most workable and the most beneficial for the largest number becoming adopted. In this more diffuse system, solutions to problems would come from a wide variety of sources rather than from one central authority.

As earlier sections of this chapter have already made clear, I favor the more diffuse form of control. In most of nature, control resides in the bodies of several players, dominants to be sure but not monopolies. The rich and powerful of nature for the most part benefit a large number of species with which they interact, or the genes they control, by creating a more resource-rich, more reliable environment. The styles of biological organization that have stood the test of time best are those whose flexibility, tolerance of error, and compartmentalization into semiautonomous parts enable entities to respond quickly to many circumstances. The power of life is limited by resource supply and by the size of the larger economies of which individual entities are a part, and by competition among entities. These elements have given life its long-term persistence as well as its adaptability in the face of disruptions ranging from the forces of every day to the rare catastrophes of mass extinction.

Humanity has so far been able to design representative governments that work reasonably well because of specifically mandated internal controls on power of the various branches. Governments of this type are still in the minority, but they have arisen and persisted with varying degrees of success in many nations. We have been rather less successful in designing religious institutions and corporations in which power is diffused by checks and balances. I would argue that these institutions, too, should be organized along democratic lines. We must, I believe, look to the organization and adaptability of nature to uncover the rules of how best to design economic, educational, and social institutions that act always in the common interest of humanity and of the biosphere as a whole. Institutions must preserve the self-interest of individuals as well as serve the long-term common good at scales of inclusion as large as possible, and do so in ways that maintain adaptability and opportunity.

Even under conditions of stability, we have a choice where to allocate resources. No cultural characteristic, it seems to me, is more important to preserve than the one system of thought that closely mirrors the economically based process of evolution. That system comprises two essential elements: the scientific way of knowing and a democratic, representative organization. The scientific way of knowing is a world view that allows explanations and solutions to be tested and improved according to observational and experimental evidence organized into an emerging, modifiable body of theory. This approach allows us to understand and to craft codes of conduct that regulate the selfish or destructive behavior of individuals, the self-interest of corporations, and the power of leaders. Myth and sentiment, of course, occupy central positions in society, because they are inextricably bound up with status and therefore ultimately with our own economic and reproductive success. But, as in other living things, it is the honest signals—those manifestations of status that in some measure reflect underlying performance that are most successful in the long run. I embrace the scientific world view not as unshakable dogma—in fact, it allows us to revise our conceptions of the world—but as the best means yet devised to discover, to solve, and to innovate. How the method is applied, and to what ends, must be informed by society's values. These values flow from our evolving conceptions of the common good, about ourselves and our surroundings.

The hypotheses formulated under the scientific method are not perfect. In everyday life, people grossly overestimate their chances of getting rich in the lottery and of being attacked by a terrorist, and they tend to underestimate the likelihood of automobile accidents. Even when they understand the statistics, many people cling to the idea that they can beat the odds, or that their particular circumstances are not adequately captured by the statistics that describe the whole. Their hypotheses are affected by distortions in the media and exacerbated by the social spread of rumor, a herd phenomenon akin to speculative investment. But there is also a conflict between intuition, itself the product of previous experience as well as incorrect information, and the often counterintuitive realities embodied in the statistical and probabilistic nature of most things. Accuracy and education help, but they will never wholly undo these misapprehensions and conflicts.

Moreover, it goes without saying that application of the scientific world view creates its own problems. Science-based technology may be responsible for healthier lives and greater awareness of the natural and irreplaceable riches of the ecosystems with which we must coexist, but it also has enabled us to destroy that richness and to annihilate civilization. If the framing and testing of hypotheses are to serve the common good, they must be accompanied by an enforceable body of ethical rules that limit behaviors by individuals and groups where selfish gains conflict with the larger public purpose. No individual or group must be allowed to proceed without some version of oversight or accountability, so that actions and hypotheses can be evaluated with a minimum of distortion. Diffuse control must, in other words, be tempered and regulated to some degree by top-down, centralized intervention, which accomplishes ends that the free market by itself cannot.

As a means of learning about the world and adapting to it, the scientific method cannot stand alone. It must, I believe, be practiced in a context of representative, democratic organization in which individuals have opportunity and freedom. In the words of philosopher Hilary Putnam: "Democracy is not just one form of social life among other workable forms of social life; it is the precondition for the full application of intelligence to the solution of social problems." Liberty and inclusiveness do not guarantee adaptation and innovation, but they provide the best possible opportunity for the process of adaptation to proceed, and limit the destructive tendencies toward alienation and fatalism.[32]

The history of life and its economic context offers us a wealth of solutions that have proved successful in the long run and that are consistent with unbreakable economic rules and realities. Evolution and economics—different expressions of the same principles that govern life and the emergence of life's organization—show us what works and what does not, which adaptive hypotheses preserve adaptability and opportunity and which lead to rigidity and constraint. Our future depends on them.

Appendix 1

ABBREVIATIONS

C	Celsius: 0° C = 273K
g	grams
Ga	billion (10^9) years ago
K	degrees Kelvin
kg	kilograms (10^3 g)
km	kilometer (10^3 m)
J	joules: 1 joule = 1 kg × m^2/s^2
m	meters
Ma	million (10^6) years ago
mg	milligrams (10^{-3} g)
ml	milliliters (10^{-3} liter)
mm	millimeters (10^{-3} m)
my	million years (10^6 years)
N	newtons: 1 N = 1 kg × m/s^2
s	seconds
W	watts: 1 W = 1 J/s = 1 kg × m^2/s^3

Appendix 2

THE GEOLOGICAL TIME SCALE

Archean Eon—4500 to 2500 Ma

Possible earliest life on Earth at 3460 Ma; oldest sulfate reduction 3470 Ma; oldest stromatolites (calcareous growths possibly associated with bacteria) 3000 Ma; possible earliest eukaryotes 2700 Ma; oldest Cyanobacteria 2700 Ma; oldest known land life 2600 Ma; first assembly of large stable continents 3000 to 2400 Ma; first extensive carbonate platforms 2600 to 2300 Ma

Proterozoic Eon—2500 to 544 Ma

Paleoproterozoic Era—2500 to 2000 Ma

Rise of oxygen levels in atmosphere 2400 to 2200 Ma; oldest redbeds (oxidized iron-bearing sediments) 2200 Ma; tropical-zone glaciation 2200 Ma

Mesoproterozoic Era—2000 to 1000 Ma

Widespread precipitation of inorganic calcium carbonate on seafloor; first acritarchs 1500 Ma; first bangiophyte red alga 1200 Ma; supercontinent in existence 1700 to 1200 Ma; continental breakup 1200 to 900 Ma

Neoproterozoic Era—1000 to 544 Ma

Rifting and breakup of supercontinent Rodinia 800 to 700 Ma; tropical-zone glaciations 760 to 700 Ma; first vase-shaped amoebae and increasing ecosystem complexity about 750 Ma; further equatorial-zone glaciations 620 to 600 Ma; multicellular animals appear 600 Ma; skeleton-bearing animals appear 565 Ma

Phanerozoic Eon: 543 Ma to Present

Paleozoic Era—543 to 250 Ma

Cambrian Period—543 to 505 Ma: Rifting of Siberia and Laurentia 543 to 530 Ma; Antrim Plateau volcanism 513 Ma; major diversification of life 543 to 520 Ma; extinctions in Early Cambrian; possible earliest multicellular land life at end of Cambrian

Ordovician Period—505 to 440 Ma: Great diversification of animals in first half of period; mass extinction at end of period

Silurian Period—440 to 405 Ma: Increasing evidence for land plants and animals; increased defensive attributes in marine animals

Devonian Period—405 to 360 Ma: Diversification of land-plant life; earliest forests 370 Ma; several extinction events in latter half of period; first tetrapods in last stage (Famennian)

Carboniferous Period—360 to 290 Ma: (often divided into the earlier Mississippian and later Pennsylvanian periods): coal forests; first significant terrestrial herbivory 300 Ma; first land vertebrates 340 Ma

Permian Period—290 to 250 Ma: Mass extinction and associated volcanism, methane release, and perhaps celestial impact at end of period

Mesozoic Era—250 to 65 Ma

Triassic Period—250 to 200 Ma: First flying vertebrates, first dinosaurs, and first mammals 230 to 225 Ma; coalescence of Pangaea 230 Ma; extinctions toward end and at end of period; major Central Atlantic flood basalt volcanism beginning near end of period

Jurassic Period—200 to 145 Ma: Karoo flood basalt and major extinction 182 to 185 Ma, accompanied by methane releases; onset of continental drift 175 Ma; diversification of sediment-burrowing animals early in period; formation of Shatsky Rise late in period

Cretaceous Period—145 to 65 Ma: Origin of flowering plants 140 Ma; origin of social insects in early Cretaceous; formation of Paraná-Etendeka flood basalts 127 to 137 Ma; formation of Ontong-Java and other submarine volcanic Plateaus beginning about 124 Ma and continuing intermittently to 80 Ma; warm interval 120 to 90 Ma; major diversification of plankton beginning 120 Ma; Deccan Traps volcanism, asteroid impact, and mass extinction at end of period

Cenozoic Era—65 Ma to Present

Paleogene Period—65 to 24 Ma

Paleocene Epoch—65 to 55 Ma: North Atlantic flood basalt volcanism 62 to 55 Ma; major methane release and deep-sea extinction 55 Ma

Eocene Epoch—55 Ma to 34 Ma: Major diversification of life throughout early and middle parts of epoch; extinction events at end of middle and late Eocene

Oligocene Epoch—33 to 24 Ma: Ethiopian flood basalts in first half of epoch; major diversification in late Oligocene

Neogene Period—24 Ma to Recent

Miocene epoch—24 to 5.3 Ma: Major diversification in early Miocene; extinction, Columbia flood basalts, and Antarctic glaciation in middle Miocene

Pliocene Epoch—5.3 to 1.8 Ma: Diversification in early part of epoch; hominids appearing by earliest Pliocene; first northern-hemisphere glaciation 2.73 Ma

Pleistocene Epoch—1.8 Ma to 0.01 Ma: Classic ice ages, origins of modern humans about 0.1 Ma; first major human-caused extinction at end of epoch

Recent (Holocene)—0.01 Ma to present: Human civilization emerges

NOTES

CHAPTER 2
The Evolving Economy

1. See E. O. Wilson (1998). Oddly enough, Wilson paid very little attention to economics in his conception of consilience or holism.

2. G. S. Becker (1976:3).

3. See the excellent discussion by D. S. Wilson (1997). Leigh (1994, 1999) has also widely explored the theme of shared fate in the particular context of rain forests.

4. The two quotes are from S. A. Kauffman (2000), pages 4 and 35, respectively. Chaisson (2001) offers an alternative characterization of life: an open, organized, carbon-based system operating in a medium of liquid water, maintained far from thermodynamic equilibrium by a flow of energy through it. Chaisson argues that life is a special case of a general class of systems based on matter, in which order (or organization) and complexity (division into different components) are generated by energy flows. Still another characterization of a living system is due to William Martin and Michael J. Russell (2003:63): "compartments separated from their surroundings that spontaneously multiply with energy gleaned through self-contained thermodynamically favorable redox reactions." This definition emphasizes the point that life takes place in compartments, or cells, which in free-living organisms are surrounded by membranes composed of lipids.

5. The general importance of energy in the evolution of life has long been recognized. Chaisson (2001) provides a good historical account. Van Valen (1976) explicitly described evolution in terms of energy. J. H. Brown (1995) extended Van Valen's views and advocated the use of energy as the currency by which economic systems and evolutionary processes could be compared.

6. Morowitz (1968) fully appreciated the nonequilibrial condition of life. Others who have discussed this condition and its implications for the evolution of information and order include Riedl (1978), Weber et al. (1989), Eigen (1992), and Maynard Smith and Szathmáry (1995, 1999). The most lucid discussion of thermodynamics, information, order and disorder, and related concepts is Eric Chaisson's (2001) superb book *Cosmic Evolution*.

7. The origin of life and the acquisition of such essential characteristics as metabolism and replication are lucidly laid out by Eigen (1992), Maynard Smith and Szathmáry (1995, 1999), and Kauffman (2000). The general consensus is that life originated only once on Earth, or possibly on Mars or another nearby planet well before 3.8 Ga (see Nisbet and Sleep 2001). The molecular unity of life may reflect a single origin, but may also reflect a pattern of selection among early life forms. In the latter scenario, life could have arisen many times in many places, and competition would have culled much of the early variation at the molecular level. Although I lean toward the idea that life arose more than once, the issue of whether life is unique or a common outcome of chemical evolution has no significant bearing on my treatment of the properties and history of economic systems, nor does it in any way invalidate the concept of evolution (descent with modification).

8. Information theory has been a popular recent theoretical framework for evolution. Riedl (1978) was one of the first to set out clearly the relationship among information, complexity, and evolution; but many others have contributed as well, including Wiley and Brooks (1982), D. R. Brooks et al. (1989), Weber et al. (1989), Margalef (1991), Sterrer (1992), G. C. Williams (1992), and Dennett (1995), among others. Information as applied to theoretical issues in economics has been explored by N. Clark and Juma (1987), Ayres (1994), and Hodgson (1999), among others. Although information as hypothesis is useful when applied to adaptation, as Sterrer (1992)

emphasized and as I show later in the chapter, I find information too abstract a concept for most applications to economic systems. Information is a means to an end, that end being performance and function related to the acquisition and retention of resources. In the restricted and informal use I advocate here, the word information more or less conveys the sense in which most laymen use the term, as conveying meaning. Dennett's quote (1995:203) about meaning as an emergent property is elaborated by him on page 427: "Real meaning, the sort of meaning our words and ideas have, is an emergent product of originally meaningless processes—the algorithmic processes that have created the entire biosphere, ourselves included."

9. Hodgson (1999:108).

10. Competition among the parasites of fish was discussed by Rohde (1991).

11. For the relationship between competition and catastrophe, see Gould (1985). I countered Gould's argument in my chapter (Vermeij 1996a) and later paper (Vermeij 1999). In a short commentary accompanying a paper on long-term competition among bryozoans by Sepkoski et al. (2000), Gould (2000) seems to have softened his position. He now accepts that competition in the broad sense has influenced the course of evolution and history in major clades of animals. Ecological statements against competition as an agency of population regulation have been made by many authors including Strong et al. (1979). P. R. Grant and Abbott (1980) effectively rebutted these arguments. Strong and colleagues confused the ecological importance of competition, observable as population regulation, and the evolutionary importance, expressed at the individual level. They assumed that if competition does affect population size, the phenomenon plays no important role in the species or at the individual level either.

12. Corallines and grazers: Steneck (1983, 1985).

13. Scarlet gillia and its herbivores: Paige and Whitham (1987). African grasses and their grazers: S. J. McNaughton (1984), McNaughton et al. (1997). Conditions for dependence between grazers and grazed: de Mazancourt et al. (2001).

14. Maynard Smith and Szathmáry (1995, 1999) and Eigen (1992) provide readable accounts of scenarios for the origin of life. They tend to favor the scenario in which RNA replicated itself, perhaps using proteins as catalysts in the reactions in which components of RNA broke and formed longer chains. Kauffman (2000) may well be right that this scenario is incorrect, that RNA was never the sole replicator. Instead, he favors the idea that the first replicator was a community, perhaps a proto-cell, in which cooperation between molecules produced self-reproduction as an emergent property. For organisms as survival machines, see Dawkins (1976).

15. Nisbet and Sleep (2001) proposed the hypothesis that Cyanobacteria are composite organisms. One of the best accounts of the origin of the eukaryotic cell as a symbiosis is that of Margulis (1981). W. Martin and Müller (1998) proposed a specific hypothesis for the origin of mitochondria. They suggested that hydrogen, liberated by a free-living bacterium, could be used by a methane-producing (methanogenic) member of the Archaea that served as the energy-poor host. See also Blackstone (1995, 1998) and Maynard Smith and Szathmáry (1999) for discussions of highly integrated symbioses becoming units of evolution and competition. Buss (1987) and Michod (1997) highlighted the importance of suppressing the competition among the eukaryotic cell's organelles for the establishment of the more inclusive unit. Margulis (1991) argued that symbioses and the cooperation associated with them are more important in evolution than are competition and natural selection. Maynard Smith (1991), however, effectively countered her thesis by noting that symbioses become units of evolution, subject to both competition and natural selection.

16. Douglas (1994) offers an excellent summary of symbioses in nature. The critical role of symbioses in the ecology of rain forests is thoroughly treated by Leigh and Rowell (1995) and Leigh (1999). For the role of mycorrhizae and the origin of land plants see Atsatt (1991), D. H. Lewis (1991), and L. E. Graham et al. (2000). Cowen (1983) provides a highly readable summary of symbioses involving dinoflagellates or other photosynthesizers with marine animals. See also Coates and Jackson (1987) and Rosen (2000).

17. See Buss (1981). Buss argued in his paper that competition led to cooperation, but the reverse—cooperation making for a competitive, larger whole—seems more likely to me. For a general discussion of the advantages of colonial over solitary life in marine invertebrates, see J.B.C. Jackson (1977a, 1979).

18. Predation by hunting dogs: R. Estes and Goddard (1967). Competitive power of cooperation and organization in human society: A. Smith (1776), M. Olson (1982), Galbraith (1983, 1984). The quote is from S. A. Levin (1999:36).

19. It is the sensitivity of RNA to external conditions that enables RNA to change its shape and to function as an enzyme as well as a replicator (Maynard Smith and Szathmáry 1995). For a superb discussion of degradation of information as information is being transmitted, see Chaisson (2001).

20. Van Valen (1976), J. H. Brown (1995), and Chaisson (2001), among many others, favor energy as the universal currency of science in general and life in particular. Van Valen and Brown, however, admit that power may be more directly applicable to systems based on living things.

21. In the so-called free market, price is determined by supply and demand. Slawson (1981), however, has drawn attention to the increasingly important role that advertising, mainly by producers of goods and services, has played in the determination of price. For additional insightful discussions of prices and advertisement, see Galbraith (1973) and D. H. Fischer (1996).

22. The sensitivity of rates to the time interval over which the rates are measured has been investigated in two quite different realms. Gingerich (1983) pointed out that rates of evolutionary change, when measured over long time intervals, tend to be slower than those measured over brief durations. The same conclusion applies to the measurement of rates of sedimentation. The longer the time interval, the more likely a period of erosion or lack of sedimentation will be included (see Sadler 1981; Dingus and Sadler 1982; Dingus 1984; Schindel 1982a,b; Sheets and Mitchell 2001). These outcomes are exacerbated by the tendency of researchers to make measurements during times of evolutionary change or times of sedimentation. Bias thus works in conjunction with temporal variation to yield a strong tendency for rates of processes to be lower when the interval of measurement is longer.

23. Application of the concept of selection to nongenetic contexts has been explored by Dawkins (1976), Sober (1984), G. C. Williams (1992), Maynard Smith and Szathmáry (1995), D. S. Wilson (1997), Sober and Wilson (1998), and Chaisson (2001), among others. Lenski and colleagues (2003) have demonstrated that, given some simple rules of competition and replication, "digital organisms" can evolve very complex functions by building on and modifying older, less complex functions.

24. As I shall emphasize again later in this chapter, it is important that we generalize and extend concepts that were first clearly articulated in one branch of scientific inquiry, in this case the study of evolution as pioneered by Darwin. The evolutionary perspective is extremely powerful both inside and outside biology; it powerfully informs all the historical sciences, ranging from astronomy to anthropology, a point also forcefully made by Chaisson (2001). Kauffman (1993, 2000) holds that life and the biosphere are on the "edge of chaos," close to the transition between an ordered and a chaotic state. In this position, error is introduced as novelty at a rate sufficiently low for selection to operate but high enough to allow the system to incorporate innovation as it changes and adapts.

25. The definition and criteria of adaptation adopted in this book are those I developed in *Evolution and Escalation* (1987) and in my 1996a chapter in Lauder and Rose's *Adaptation*. Reeve and Sherman (1993) independently arrived at a similar conception. In this view, adaptation refers to current, not past, performance and is silent on the origin and history of the traits and processes involved. Another school of thought, championed by Coddington (1988) and Larson and Losos (1996), among others, holds that adaptation can be recognized only when the pattern of descent is known. According to this view, the demonstration of current utility of a trait is insufficient. I find this conception unnecessarily restrictive, because it bars the

examination of many instances of function (including that in fossil organisms) where neither the history nor the genetic basis of traits and benefits can be known. See my 1996a discussion on this controversy.

26. For discussions of error thresholds, cooperation during repair, and mutation rate in relation to genome size see Eigen (1992), Maynard Smith and Szathmáry (1995), and S. A. Frank (1996). Kauffman (2000) notes that error thresholds are akin to phase transitions, such as those between a solid and a liquid or a liquid and a gas. Above the threshold, error overwhelms selection and the system descends into chaos. Systems without error cannot respond to change and will rigidify and be outcompeted. The trick is not to eliminate error, but to channel it in potentially useful ways. I return to this matter in chapter 6.

27. Sterrer (1992:106). A. Hoffman (1989) has also expressed the view that adaptation is the expected universal condition of life.

28. Role of nongenetic means of adaptation and of transmitting information: Dawkins (1976, 1995, 1996), Dennett (1995), Maynard Smith and Szathmáry (1995, 1999), and Frank (1996). It is simply not possible to create enough genetic code to specify all the information necessary to deal with all the environmental stimuli to which an active, complex living thing is exposed and with which it must deal. See also the superbly clear discussion by Kirschner and Gerhart (1998) and Newman and Müller (2001).

29. For a discussion of armor and locomotion, see Vermeij (1987, 1993), Bertness (1981a,b,c). Plant growth and defenses: J. M. Robinson (1990), Coley et al. (1985). Role of ant defenses in plants: Davidson and McKey (1993).

30. Reproductive tradeoffs received a theoretical underpinning in Charnov's (1997) paper, accompanied by a highly readable commentary by Godfrey (1997). The work of my student Mark D. Bertness on growth and reproduction in hermit crabs in Panama (Bertness, 1981a,b) stands out as one of the finest empirical evaluations of tradeoffs.

31. Imperfection of Rubisco: Nisbet and Sleep (2001); Geider and MacIntyre (2002).

32. For further discussion see Vermeij (1987, 1996a); Reeve and Sherman (2001). The best defense of optimality theory I have seen is by Seger and Stubblefield (1996).

33. Adam Smith (1776) was one of the first to recognize the importance of division of labor, or specialization. For ecological discussions, see MacArthur (1965, 1972), Levins (1968), and Ghiselin (1974).

34. For discussions of tree height and the role of shade and canopies, see Horn (1971), Terborgh (1985), King (1990), and Leigh (1999).

35. Data on weapons yields: Sagan (1997).

36. Dawkins (1995, 1996) has eloquently made the point that evolution is the story of unbroken successes. His, A. Hoffman's (1989), and my conception of adaptation as the universal attribute of organisms and as the most fundamental manifestation of evolution contrasts with that of Gould (1985, 1989, 1996), who maintains that adaptation is a short-term process whose scope is severely limited by the imposition of occasional extinction-causing catastrophes. I shall argue in chapter 9 that these crises in no way limit the power of adaptation in defining the pathways of evolution and of economic history in general.

37. Scholars beginning with Adam Smith (1776) have long recognized inequality as an integral if unwelcome characteristic of the human condition. I explicitly discussed inequality and its implications for economic structure and change in my 1999 paper in the *American Naturalist*. Obviously, I am not claiming inequality to be a newly discovered principle. My aim is simply to explore its many manifestations and implications.

38. The idea that catalyzed reactions are favored over uncatalyzed ones is eloquently laid out by Kauffman (2000), who also notes that if there is a sufficient number of such reactions, some of them will come to depend on one another and form the kind of reaction (or autocatalytic) loop that constitutes the work cycle of life.

39. Gould (1999) takes a quite different position from mine. He sees moral codes as entirely separated from anything that could be construed as adaptation. The domain of religion, Gould holds, should be strictly and wholly separate from the domain of science. The scientific method can inform us about how things work and how they evolved, but in Gould's view it cannot impart values. With Sober and Wilson (1998), de Waal (1996), and D. S. Wilson (2002), I hold that moral codes and the institutions that enforce them are the outcomes of adaptation at the level of societies that generally also align with the interests of individual members. I return to the question of religion in chapter 11.

Chapter 3
Human and Nonhuman Economies Compared

1. It is the zeitgeist of our time that has made principles of emergence a popular topic for those writers who have sought to achieve a certain unity between biology and human affairs. Leslie White (1943) laid the foundations in his "Energy and the Evolution of Culture," and expanded his views in his 1959 book, *The Evolution of Culture*. Jane Jacobs applied evolutionary principles in several books, including her short work in 2000, *The Nature of Economies*. Robert Wright's (2000) *Nonzero: The Logic of Human Destiny*, and Steven Johnson's (2001) *Emergence: The Connected Lives of Ants, Brains, Cities, and Software*, are other examples.

2. I have long criticized biology and paleontology for not taking the economic context of life into account (see Vermeij 1987; Vermeij and Leighton 2003). Nomothetic paleontology took its inspiration from theoretical ecology, which had its origins in the 1920s. For the foundational papers in this field, see Raup et al. (1973) and Raup and Gould (1974). My principal complaint about this approach is not only that the context of life varies dramatically from place to place and over time, making any global attribute of taxa or clades an amalgam incorporating inconsistent and heterogeneous signals, but also that the global scale of analysis is irrelevant to the lives and evolutionary context of organisms, species, and ecosystems (Vermeij 1987; Vermeij and Leighton 2003).

3. Hodgson (2001) presents an excellent account of the history of economics, including the rise of the historical perspective in that field.

4. For good summaries of the approaches mentioned, see MacArthur (1972), Schoener (1969, 1974a,b), Ghiselin (1974), and H. T. Odum (1983).

5. See Raup and Seilacher (1969), Cowen (1981).

6. This list of differences between human and nonhuman economies is my synthesis of many discussions I have had with economists and biologists. Sterrer (1993) drew up a somewhat different list, but I have incorporated his main points into mine. The discussion to follow expands upon that of Vermeij (1999) and Leigh and Vermeij (2002).

7. For evolutionary perspectives on economics, see Mokyr (1990, 1991), and especially N. Clark and Juma (1987) and P. T. Saunders (1999), who provide excellent historical accounts.

8. Paine (1980) highlighted the different roles that the single species *Pisaster ochraceus* plays in the northern and southern parts of the range of this common northeastern Pacific seastar.

9. Ofek (2001) held that humanity is unique in that division of labor occurs among genetically unrelated members of just one species, whereas in nature it occurs either among species or among very close relatives (such as castes in a colony of social insects, or differentiated polyps in a Portuguese man-of-war colony) of the same species. Accordingly, Ofek argues that mutually beneficial trade among individuals is a uniquely human economic process. Although such trade undoubtedly offers huge advantages, as Adam Smith (1776) knew long ago, those same advantages are evident when members of different species interact.

10. Woodpecker finches: P. R. Grant (1986). De Waal (1996) also discusses tool use in nonhuman primates.

11. Hodgson (1993:264). For the conservative role of organizations, see M. Olson (1982).

12. Vitousek et al. (1997a) provide an excellent general summary of human effects on the world's ecosystems. In the same issue of *Science* in which their paper appeared, there are important contributions by Chapin and colleagues on alterations in ecosystem function and on the intensification of agriculture (1997), and Botsford and colleagues (1997) on the consequences of fishing and other human activities on the marine biosphere. This last topic was also treated by J.B.C. Jackson and colleagues (2001), who stress the dramatic repercussions of the removal of large vertebrates from marine ecosystems; see also chapter 9 in this book. Estimates of human appropriation of annually produced biomass are from Rojstaczer et al. (2001); these range from 10 to 55 percent of global biomass, depending on how production and its use are calculated and to what extent heterogeneity (variation across regions and across ecosystems) is taken into account (see also the helpful commentary by Field 2001). Bird and Cali (1998) document the history of African fires as inferred from the abundance of elemental carbon in offshore sediments.

13. I shall treat the topics of human-caused species invasion and extinction in more detail in chapter 9. A good summary paper on the destabilizing effects of nonnative species on ecosystems is that of Vitousek et al. (1997b). Diamond's (1984) chapter is an outstanding summary of human-caused extinction.

14. Data on human energy use and population growth are from J. E. Cohen (1995). According to Hall and colleagues (2003), hydrocarbon fuels (coal, oil, and natural gas) have increased in global human use by a factor of 800 since 1750, and by a factor of 12 since 1950. The growth in per capita and global energy consumption continues despite the rise of the "information economy."

15. Coal-forest waste: J. M. Robinson (1990). Toxic leaves and bark dropped by selfish trees: Leigh (1994, 1999). The fate of mobile surface-dwelling marine animals: Thayer (1979, 1983), McKinney and Jaklin (1993).

16. Estimates of temperature with and without photosynthesis are given by Lenton (1998) and Schwartzman (1999). Glaciation resulting from the deployment of land plants during the Late Paleozoic has been suggested by Schwartzman.

17. Increased control of silica and calcium carbonate by organisms through time: Maliva et al. (1990), Knoll et al. (1993); see also chapter 10. Effects of diatoms on silica extraction by other organisms: Maldonado et al. (1999), Racki (1999), Racki and Cordey (2000).

18. For a gripping and learned account of globalization in the human economy, see Greider (1997).

19. Asian mammals spreading to North America: Beard (1998). Trans-Arctic marine invasions: Vermeij (1991a). Movements of species across the Central American isthmus: Stehli and Webb (1985). I revisit the topic of species invasions in chapter 8.

20. Role of migration in globalization: Vermeij (1999). Imbalances between production and consumption because of movements of species and biomass between ecosystems: Polis et al. (1997).

21. Dennett (1995) has ably summarized the distinction between the self-organized character of control in nature and the intentionality that characterizes human enterprise.

22. For a thorough account of how human social structures arise through altruism, see Sober and Wilson (1998). D. S. Wilson (2002) notes, however, that altruism is by no means the only mechanism by which social groups arise or are held together. Fehr and Fischbacher (2003) note that enhancement of reputation is another factor underpinning altruistic behavior in humans.

23. I give here only the bare bones of de Waal's (1996) arguments, elegantly and compellingly laid out in his important book, *Good Natured*. See also Sober and Wilson (1998) for a wide-ranging discussion of how social structures arise.

24. For the evolution of long-term benefits, see Vermeij (1996a) and Lenton (1998). Leigh (1994) and D. S. Wilson (1997) have developed the arguments about good-neighborliness.

25. S. Johnson (2001): 204.

CHAPTER 4
The Role of Enemies in Nature

1. Predation by bacteria: Guerrero et al. (1986). For a more detailed treatment of the logic of consumption in general (and predation in particular) as an agency of evolution, see Vermeij (2002b).

2. The importance of unsuccessful predation and the division of the predator-prey encounter into three phases were discussed by Vermeij (1982a). Stages in predatory attacks were clearly distinguished by Holling (1966) and Griffiths (1980).

3. Good introductions to cryptic coloration as antipredatory defense may be found in M. H. Robinson (1969), Edmunds (1974), Gilbert (1983), and Endler (1986). The egglike spots on passion-vine leaves were investigated by Gilbert (1971) and Benson et al. (1975). Cladoceran defenses and other adaptations of planktonic animals are documented by Kerfoot et al. (1980) and Morgan (1995). For transparent plankton see McFall-Ngai (1990). False-head butterflies: Robbins (1981). Clams shutting valves on prey: summary in Vermeij (1993). Encrusted snails: Feifarek (1987), Vermeij (1993). Debris-covered sea urchins and crabs: Kilar and Lou (1986), Stachowicz and Hay (2000).

4. Interaction between *Hyla cinerea* and insects: Freed (1980).

5. The crab spider example comes from Théry and Kasas (2002).

6. The evolution of warning coloration has been investigated experimentally and with observations by Wiklund and Jårvi (1982) and Wiklund and Sillén-Tullberg (1985) on butterflies. The cost of aposematism was studied by Jårvi et al. (1981) in larval *Papilio machaon* when great tits (*Parus major*) caught them. For the relation among warning coloration, distastefulness, and body toughness, see Carpenter (1942).

7. It used to be thought that warning members of one's own species or group is an unselfish act of altruism from which the sender of the message does not profit. As summarized by Sober and Wilson (1998), however, it is clear that warning calls benefit the caller as well as other individuals who hear and heed the call.

8. Speeds of ungulates are from Garland (1983a). Hyena predation: Kruuk (1972). Lions: Schaller (1972). Wolves: Mech (1966).

9. *Meyenaster* predation: Dayton et al. (1977).

10. Escape from individual predators, supplemented by defensive writhing and shaking upon contact with the enemy, has been documented in many marine molluscs (Margolin 1964a, 1975; Ansell 1967, 1969; Nielsen 1975; Garrity and Levings 1981; Schneider 1982; Savazzi and Reyment 1989) and in freshwater snails (Wilken and Appleton 1991). All-purpose avoidance of predatory snails, seastars, and crabs by marine prey snails is also widespread (Hadlock 1980; McKillup 1982; Geller 1982). Various neogastropods, such as nassariids and olivellids, are capable of detecting individual moon snails at a distance (Gonor 1965; R. N. Hughes 1985). The vast majority of predatory neogastropods living on hard surfaces are very slow crawlers and show no escape responses to any predator. In a landmark study of water flow through the body of snails, Lindberg and Ponder (2001) laid out the anatomical basis for the absence of long-distance chemical detection in patellogastropods, vetigastropods, and other primitive gastropod clades. According to their insightful analysis, long-distance chemical detection requires that water enter the body at just one point (the front end). This directional detection is often associated with the presence of a notch or channel at the leading (front) end of the rim of the snail's shell. In some olivids and stromboideans, there may also be a posterior notch or gutter, which accommodates an extension of the mantle that may likewise have a chemical-detection function.

11. Observations and experiments on the effectiveness of molluscan shells as armor for the molluscs that built the shells and for hermit crabs that occupy empty shells are set out in detail by Zipser and Vermeij (1978), Bertness and Cunningham (1981), Vermeij (1982b, 1987, 1993), Bertness (1981b,d, 1982), and A. R. Palmer (1979). The method of growth of shells—the addition

of skeletal material at the rim—produces a hollow structure that approaches, but rarely precisely matches, the form of the so-called logarithmic spiral. For detailed discussions of shell form and growth, see Vermeij (1993, 2002a) and D. W. Thompson (1942).

12. A. R. Palmer (1982) experimentally documented the effectiveness of reduced suture number in protecting barnacles against drilling predatory gastropods, which attack their victims preferentially at the junctions (sutures) between plates. Sidor (2001) argues that a strong long-term trend in all terrestrial clades of tetrapods toward fewer skull bones through time increases the ability of the skull to withstand stresses absorbed when jaws exert strong bite forces or when the animal is attacked by predators. For excellent accounts of turtle armor, see Bramble (1974) and Bramble et al. (1984).

13. For a general review of caddisfly cases, see H. H. Ross (1964). Helicopsychid caddisfly larvae build cases remarkably like snail shells in form; see Berger and Caster (1979). The ones I have seen (near Cincinnati, Ohio) are topshell shaped, with the apex pointing up, a shape that is rare in freshwater snails but common among marine species. Otto and Svensson (1980) have documented the antipredatory function of caddisfly cases in Sweden. J. D. Taylor et al. (1999) vividly described Australian venerid bivalves with a "concrete overcoat" of coagulated sand grains on their shells and reviewed other examples of augmenting the shell defense with an extra coating of sediment. The shell-attaching habits of xenophorid gastropods were discussed by Feinstein and Cairns (1998), but no conclusive experiments on antipredatory function have been carried out. The habit is also known for Miocene fossils of the turritellid gastropod genus *Springvalea* and for the Recent, peculiar, small snails of the unrelated genus *Scaliola*, still living in the Indian and Pacific Oceans (El-Nakhal and Bandel 1991).

14. For a summary of secondary shell-dwellers and their history, see Vermeij (1987, chapter 8). The use of cephalopod shells by fossil occupants has been inferred by Davis et al. (2001) for some Ordovician and Silurian trilobites and by Fraaye and Jäger (1995) for Jurassic erymid lobsters. In these cases, researchers suggest that the shells might have served mainly as safe sites for molting rather than as mobile housing, as is the case in sipunculans and living crustaceans with the shell-dwelling habit.

15. The data on loads required to break palm nuts are from Kiltie (1981, 1982), who measured loads in kilograms rather than in newtons. Other nut-cracking mammals, such as squirrels, exert smaller forces.

16. For a summary of molluscan shell-armor defenses, see Vermeij (1993). Shell strengths of *Drupa morum* and related tropical and temperate species are given by Vermeij and Currey (1980). Little is known about the food habits and mechanical performance of the loggerhead turtle; see Randall (1964) for comments on its possible feeding on the large West Indian conch *Strombus gigas*. Many authors have noted size refuges from drilling predation; see, for example, Kirby (2001) for thick-shelled oysters from the Miocene of California; and Vermeij et al. (1989) for the cold-temperate East Asian clam *Pseudocardium sachalinense*.

17. See Cowen (1973). Simkiss (1989) similarly noted that the sudden appearance of shell-encased animals in the Cambrian required a certain minimum oxygen concentration. The presence of collagen, a protein crucial to connective tissue, also requires oxygen (Towe 1970).

18. The literature on chemical defenses is vast. Good overviews exist for land plants (Swain 1978; McKey 1979) and seaweeds (M. E. Hay and Fenical 1988; M. E. Hay and Steinberg 1992).

19. For further discussion, see Davidson (1997) and Coley et al. (1985).

20. See Dumbacher et al. (1992) and Diamond (1992) for documentation of toxic birds. The speculation I offer for the rarity of toxicity in warm-blooded animals is my own.

21. Sea cucumber autotomy: Kropp (1983). Scales of moths and butterflies in spider webs: Olive (1980). Autotomy of siphons in bivalves: Stasek (1967), Peterson and Quammen (1982). Crinoid autotomy: Oji (2001). Crab-claw autotomy as antipredatory defense: M. H. Robinson et al. (1970).

22. Protection of plants by ants: Davidson and McKey (1993), Davidson (1997). Crabs protecting corals against predators: Glynn (1983). Protection of bryozoans by hydroids and corals:

Osman and Haugsness (1981). Defenders of hermit crabs against octopus and other predators: D. M. Ross (1971, 1974), W. R. Brooks (1988). Anemones also protect fish (Mariscal 1971).

23. For further discussion of the antipredatory origins of sociality, see E. O. Wilson (1971), de Waal (1996), and Ofek (2001).

24. For extensive discussion of soldier castes among social insects see E. O. Wilson (1971) and Hölldobler and Wilson (1990). For the role of avicularia in protecting bryozoan colonies, see Winston (1986).

25. *Vasula melones*: for shell characteristics see Vermeij and Currey (1980); behavior is discussed by Bertness and Cunningham (1981), L. West (1988), and S. L. Miller (1974).

26. See Marden and Chai (1991), Chai and Srygley (1990), and Srygley and Chai (1990).

27. I presented this argument in my 1987 book, inspired by West-Eberhard's (1983) point that the search for mates among internally fertilizing species requires elaborate long-distance signals.

28. Dawkins and Krebs (1979) discussed the life-dinner principle in their famous paper on coevolution. I discuss the evolutionary consequences of unsuccessful predation in several publications (Vermeij 1982a, 1987, 1994, 2002b). Although I initially thought of the effects of predation in evolution as a process of reciprocal adaptation (or coevolution) between predator and prey, I applied the term escalation to it in my 1977a paper. Differences in the number of opportunities for selection between predator and prey were noted in Vermeij (2002b).

29. See J. D. Taylor (1980) Huey et al. (2001), Fritz and Morse (1985), and B. Morton and Britton (1993).

30. The bombardier beetle case was described by Nowicki and Eisner (1983) and Conner and Eisner (1983). Byssal thread entrapment: Wayne (1987), R. W. Day et al. (1991). Predator injury between clam valves: for birds' toes see De Groot (1927); for predatory gastropods Radwin and Wells (1968), R. R. Alexander and Dietl (2001), Dietl (2003). Limpets clamping enemy snails: Branch (1979). Snails biting seastar tube-feet: Margolin (1964b), Harrold (1982), Branch (1979).

31. Defense against bombardier beetle: Nowicki and Eisner (1983), Conner and Eisner (1983). Grasshopper mouse handling scorpions: Langley (1981). Herbivorous insects preventing mobilization of chemical defenses by plants: Rhoades (1985). Spiders and noxious prey: Nentwig (1983).

32. For an excellent account of phenotypic plasticity, see Agrawal (2001). In my previous writings, I tended to downplay the importance of this phenomenon, and overlooked the possibility that plasticity allows predator and prey to respond to each other over very short time scales, well within the lifespans of the individuals concerned. As discussed later in this chapter, such reciprocity on short time scales may mean that long-term reciprocal evolution (coevolution) plays a more important role than I had surmised previously (see Vermeij 1982a, 1987, 1994, 2002b).

33. Crab-induced shell defenses in marine snails: Appleton and Palmer (1988), A. R. Palmer (1990), Trussell (1996, 1997, 2000), Trussell and Smith (2000), Behrens Yamada et al. (1998). Crab-induced defenses in mussels: Côté (1995), Leonard et al. (1999), L. D. Smith and Jennings (2000). Mussels evidently also respond nongenetically to the presence of shell-drilling dogwhelks (*Nucella*) by thickening the shell (Smith and Jennings 2000). Food-induced modification of crab claws: L. D. Smith and Palmer (1994).

34. Induction of cichlid fish jaw form by food: Kornfield et al. (1982), Smits et al. (1996). Modification of the digestive system by food in wading birds: Piersma et al. (1993).

35. Food-induced changes in insect heads: Bernays (1986). Food-induced changes in primate jaws and teeth: Corruccini and Beecher (1982). A similar phenomenon has been noted in sea urchin jaws (Levitan 1992).

36. Cheetahs competing with other carnivores: Schaller (1972). Bats competing with nighthawks: Shields and Bildstein (1979).

37. Competition between whelks was described by Kent (1983), who referred to the species as *Busycon spiratum* and *B. contrarium*. Weldon (1981) also described fighting behavior in whelks.

38. Van Valkenburgh and Hertel (1993) published a truly exceptional piece of comparative scholarship. Its authors provide evidence that the conditions prevailing at La Brea also existed elsewhere in Pleistocene America.

39. Bakker (1983) forcefully advocated the interpretation that the claws of carnivores evolved not because of selection imposed by the prey but because of competition with rival predators and scavengers. The jaws of the sea urchin *Echinometra* proved to be defensive weapons in fights over holes (Grünebaum et al. 1978). Prey stunning by sound is known in whales (K. S. Norris and Mohl 1983; M. A. Taylor 1986) and in alpheid snapping shrimps (Versluis et al. 2000).

40. The multiple uses of crab claws have been well documented by Schäfer (1954), Vermeij (1977a), and S. C. Brown et al. (1979). The role of claws in competition for mates has been revealed in fine experiments by L. D. Smith (1995), Levinton and Judge (1993), Levinton et al. (1995), and Seed and Hughes (1995).

41. I return to this matter later in the chapter when I take up coevolution. The idea that enemies play a leading role in evolutionarily directing specialization is not new (see Brower 1958; Schoener 1974b; Futuyma 1983; Bernays and Graham 1988; Rosenzweig 1995), but it has not been able to budge the apparently more appealing notion that reciprocal adaptation between species accounts for most observed patterns of specialization to food and habitat.

42. I discussed this hypothesis briefly in my 2002b paper.

43. For excellent discussions of the multiple origins of vision see Dawkins (1995), Raff (1996), and especially de Queiroz (1999).

44. See Lindberg and Ponder (2001). The so-called siphonal canal, which indicates the presence of a siphon and potentially of long-distance chemical sensation, is not a reliable indicator that the snail is a predator. It also occurs in such micrograzers as cerithiids and in herbivorous conchs (Strombidae), and some predators (naticid moon snails and several opisthobranchs) lack a siphonal canal.

45. For an excellent general review of chemoreceptors, see Laverack (1988). Long-distance chemical detection of prey by decapod crustaceans is well documented by Lawton (1989) and Hirtle and Mann (1978), among others.

46. Burrowing-snail speeds: Vermeij and Zipser (1986), E. C. Dudley and Vermeij (1989), Trueman and Brown (1989). Burrowing speeds in bivalves: Stanley (1970), Ansell and Trevallion (1969), Trueman and Brown (1989), Donn and Els (1990). Northwest American shore animals: Schmitt et al. (1983), Vermeij et al. (1987). Running speeds in mammals: Garland (1983a), Schaller (1972), Eaton (1974). For somewhat lower estimates of pronghorn speeds (66 km/h), see Lindstedt et al. (1991). Burst swimming speeds in fish: R. M. Alexander (1977), Beamish (1978), P. W. Webb (1984). Herbivorous birds and insects: E. S. Morton (1978), R. Dudley and Vermeij (1992). Freshwater plankton: Kerfoot et al. (1980).

47. Important contributions on locomotion in fossil and living marine vertebrates are those of P. W. Webb (1984), Braun and Reif (1985), and Cowen (1996). J. J. Day et al. (2002) estimated the running speed of a theropod dinosaur from the Middle Jurassic (Middle Bathonian stage, 163 Ma) of Oxfordshire, England, from a trackway of footprints that indicated a running rather than a walking gait. J. R. Hutchinson and Garcia (2002) argued that *Tyrannosaurus rex* could not have been a fast runner because its estimated proportion of body mass devoted to leg-extensor muscles (7 to 10 percent) is too small to allow the animal, which weighed up to 6000 kg, to run. Runners of that size would, according to Hutchinson and Garcia's calculations, require 86 percent of the body mass to be devoted to muscles that extend the leg. For estimates of dinosaur running speeds, see Farlow (1981).

48. For the history of mammalian locomotion, see Bakker (1983), Janis and Wilhelm (1993), and Van Valkenburgh (1999). Witmer and Rose (1991) investigated *Diatryma*.

49. Importance of acceleration to African carnivores: Elliott et al. (1977). Goshawk locomotion: Goslow (1971).

50. For bite forces in vertebrates, see the summary by Erickson et al. (1996). Snail-shell resistances and inferred maximum bite forces of fish and crabs are given by Vermeij and Currey

(1980) and Preston et al. (1996). Measurements of crab and lobster bite forces and analyses of the functional specialization of their claws can be found in Schäfer (1954), Warner and Jones (1976), Vermeij (1977a), S. C. Brown et al. (1979), Elner and Campbell (1981), Blundon (1988), and G. M. Taylor (2000, 2001). Biting and resistance to it can also be measured in terms of pressure (force per unit area of contact between load-bearing surfaces). This may in fact be biologically more relevant than force, but pressure is hard to measure because of great uncertainties in the estimation of contact surface area.

51. See Greene (1997) for a discussion of the factors leading to the evolution of envenomation. In ants, envenomation is associated with tough-bodied prey. The sting has either been lost or been modified to deliver contact surface poisons to soft-bodied prey and to other ants in several lineages of superabundant tropical tree-dwelling ants (Davidson 1997). Lichtenegger et al. (2002) described envenomation of soft-bodied prey by bloodworms (marine glycerid polychaete annelids). In muricid and other predatory snails, anaesthetizing secretions come from the hypobranchial gland and, in some cases, from salivary glands. In cone snails (family Conidae), an impressive diversity of highly specific proteins makes up a potent venom, which targets all function of a prey's nervous system and causes rapid paralysis in the prey, which, depending on the cone species, includes worms, snails, and fish (Olivera 2002).

52. Home ranges of predatory and herbivorous mammals: Garland (1983b), Kelt and Van Vuren (2001).

53. For primate (including human) food habits, see Kay et al. (1997), Wrangham et al. (1999), Wrangham (2001), and Kaplan and Robson (2002). The omnivory of tropical superabundant ants has been nicely documented by Davidson (1997).

54. Sizes of land herbivores and predators: E. Anderson (1984), Burness et al. (2001).

55. The literature on the adaptive significance of sex is vast and complex. The problem facing evolutionary biologists has always been that sex is wasteful, and that good genomes are broken up. Sterrer (2002) is to my mind the first author to propose a credible hypothesis for the origin of sex. A major consequence of sexual reproduction is that sex creates genetically unique individuals. With more variants among which to choose, evolution can proceed more rapidly, and the focus of natural selection responsible for that evolution shifts from parts within organisms to the organisms themselves. These consequences, important as they are, constitute indirect benefits of sex, and therefore cannot themselves be favored by natural selection.

56. Grosberg and Strathmann (1998) were the first authors even to ask the question why multicellular organisms always pass through a single-celled stage. Although their explanation is in many ways similar to the one Sterrer (2002) proposes for the origin of sex, Sterrer seems to have been unaware of their paper.

57. I have argued this point previously (Vermeij 1983, 1987).

58. For caterpillar distributions, see Heinrich and Collins (1983). Snail distributions: Bertness and Cunningham (1981). Armored catfish: Power (1984). The term enemy-free space was coined by Jeffries and Lawton (1984); the idea was further elaborated by Holt and Lawton (1994). Brittlestar distribution: Aronson (1987, 1988, 1991).

59. The case of the ochre seastar (*Pisaster ochraceus*) is documented in classic papers by Paine (1966, 1974) and Dayton (1971). Paine (1976) noted the local abundance and large size of California mussels below the low-tide line. The effects of sea otters on the distribution of sea urchins has been beautifully documented in a series of papers by James A. Estes and colleagues (Estes and Palmisano 1974; Estes et al. 1989, 1998). For the case of the armored catfish, see Power (1984).

60. See Farmer (1999).

61. In an important series of papers, theoretical biologist Robert Holt and his colleagues have laid out the population-genetic rationale for this argument and explored its implications. See Holt (1996), Holt and Gomulkiewicz (1997), and Holt and Hochberg (1997).

62. Coevolution between attine ants (including leaf-cutting ants), their food fungi, and the parasitic fungus *Escovopsis* has been documented on the basis of molecular evidence by Mueller

et al. (1998) and C. R. Currie et al. (2003). Leigh (1999) has discussed the ecological ramifications of this mutualism for Panamanian rain forests. Platypodine and scolytine beetle phylogeny and that of fungi cultivated by ambrosia beetles were documented by Farrell et al. (2001). Mueller and Gerardo (2002) provide an excellent overview of fungal agriculture by insects, and Diamond (1998a) has drawn attention to the parallels between agriculture as practiced by ants and that practiced by human farmers.

63. For an excellent summary of the fig work, see Herre (1999). The point that competition and cooperation often (always?) occur together in symbiotic relationships has been explored in a general review by Herre et al. (1999). Pellmyr and colleagues (1996) document a case strikingly similar to that of the fig story in the relationship between yuccas, which serve as food, and yucca moths, which feed on and pollinate the yucca plants.

64. The varying relationship between *Greya politella* and *Lithosperma parviflorum* is beautifully documented by J. N. Thompson and Cunningham (2002). In this paper and in earlier work, Thompson (1994, 1999a,b) notes that mutualistic coevolution arises only in some populations, while the same species may remain antagonists in other, often nearby, habitats.

65. Lice and pocket gophers: Hafner et al. (1994).

66. Insects switching to sugar cane: Strong et al. (1977). Butterfly switching to new hosts in Sierra Nevada: Singer et al. (1993). Host switching by fig wasps: Molbo et al. (2003).

67. For a good introduction to parasites, see Price (1980). Rarity of host-specific marine herbivores: M. E. Hay and Steinberg (1992), Vermeij (1992).

68. Role of predators in enforcing host-guest specialization in the sea: M. E. Hay and Fenical (1988), Hay et al. (1989, 1990a,b). Limpets and surfgrass: Fishlin and Phillips (1980). For examples on land, see Brower (1958), Bernays and Graham (1988), and Lill et al. (2002).

69. Advantages of symbiont-bearers: Cowen (1983), Wood (1993), Birkeland (1989). Fidelity of symbioses: Rowan and Powers (1991), Diekmann et al. (2002). For a more general review of fidelity, see Herre et al. (1999). Another possible benefit to the symbionts is the production of dissolved carbon dioxide in those hosts that produce skeletons of calcium carbonate. The process of calcification (combining calcium and carbonate ions to form calcite or aragonite) involves the release of carbon dioxide, which is then used by the symbionts for photosynthesis. See McConnaughey (1991, 1994) and A. T. Marshall (1996).

70. Schwartz and Hoeksema (1998) employed the mutualism between land plants and mycorrhizal fungi chiefly as an example of the relative advantages of trade, a theory worked out by Ricardo in the early nineteenth century. My point that the plant partner is the economically dominant one emerges from their analysis, but Schwartz and Hoeksema did not comment on it.

71. *Buchnera* and aphids: Moran and Telange (1998), M. A. Clark et al. (2000).

72. Sexual conflict is undoubtedly widespread; see T. Chapman and Partridge (1996). The example from *Drosophila* is from Rice (1996). Coevolution of genitalia in insects: Arnqvist (1998), Arnqvist and Rowe (2002), Rowe and Arnqvist (2002).

73. See Darwin (1859) and West-Eberhard (1983).

74. Skeptics are apt to dismiss G. F. Miller's (2000) arguments as being "just-so" stories. They should carefully examine the wide range of sociological, historical, and psychological evidence Miller has marshaled. Darwin (1859) had already speculated about the importance of sexual selection in the evolution of the human brain.

Chapter 5
Production and The Role of Resources

1. The paper by Hairston et al. (1960) began a vigorous debate among ecologists about which factors regulate populations. See also the important contributions by Schoener (1983), DeAngelis (1992), and Brett and Goldman (1997).

2. Landes (1998:40).

3. Chapelle and colleagues (2002) describe an Archaea-based ecosystem from a depth of 200 m beneath the Earth's surface in a hot spring in Idaho, where hydrogen-consuming Archaea produce methane. See also Margulis (1981) and Nisbet and Sleep (2001) for other early chemical sources of energy.

4. Calculations of total sunlight are from Chaisson (2001). Data on the use of sunlight in rain forests are from Leigh (1999) and Leigh and Vermeij (2002). Horn (1971) explored the inability of green plants to use more than 20 percent of sunlight in photosynthesis. Global estimates of solar uptake and of primary productivity are from Field et al. (1998). I have recalculated the estimate of global net primary productivity given by Field and colleagues (104.9×10^{15}g carbon per year) in terms of seconds instead of years in order to facilitate comparison with estimates of solar radiation absorbed by the Earth. Recall, however, that the interval over which a rate such as productivity is measured or expressed affects the observed value. The numbers I quote are in any case rough approximations, suitable mainly as indicators of the order of magnitude of the phenomena and processes in question. Although solar energy indeed drives respiration in land plants, Tanner and Beevers (2001) have experimentally demonstrated in sunflowers that this solar-powered transport is not alone responsible for the uptake and upward movement of minerals from the roots to the leaves. Metabolically driven processes are essential for mineral acquisition by plants.

5. Smetacek's (2001) argument applies mainly to single-celled phytoplankton, and does not consider competition among larger free-floating plants. It is quite unclear to me why free-floating multicellular photosynthesizers are so rare in the sea. Currents might eventually cause such passively floating organisms to become stranded on the world's shores, but partnerships between a passive alga and an animal capable of moving under its own power could circumvent this problem. My estimate of the number of times free-floating plants evolved is based on C.D.K. Cook's (1999) compilation of aquatic embryophytes, but Cook did not explicitly address himself to the evolution of floating plants. Plant groups involved include liverworts (*Ricciocarpus, Riccia*), ferns (Salviniaceae, Azollaceae), and flowering plants in the families Hydrocharitaceae (including *Stratiotes*), Lemnaceae (duckweeds), Ceratophyllaceae (hornworts), Euphorbiaceae (South American *Phyllanthus*), the insectivorous Lentibulariaceae (Utricularia) and Droseraceae (*Aldrovandra*), Scrophulariaceae (*Bacopa*), and Onagraceae (*Ludwigia*).

6. For discussions of light levels and shading, see Horn (1971) and Duarte (1995).

7. M. E. Hay (1981) compared photosynthesis in single and turf-associated plants of several tropical seaweeds in Panama. Sousa et al. (1981) discussed the advantages of vegetative propagation relative to nonvegetative space-holding in seaweeds. Similar studies of colonial as compared to solitary attached marine animals were undertaken by J.B.C. Jackson (1977a, 1979, 1983). Tiffney and Niklas (1985) discussed the merits and history of clonal land plants (those with vegetative propagation) and aclonal ones.

8. For detailed, quantitative accounts of diffusion see LaBarbera (1990) and Denny (1993). Briefly, the rate of transport of a substance, dS/dt, is proportional to the concentration gradient dC/Dx and a diffusion coefficient D of the substance: $dS/dt = D \times dC/dx$. The diffusion coefficient D has dimensions of length$^2 \times$ time^{-1}, or velocity times distance.

9. Koehl and Alberte (1988) explored the role of ruffled surfaces in promoting nutrient exchange in kelps. Vogel (1970) showed how the flow of air around a leaf is affected by leaf shape, and argued that lobes and other irregularities allow leaves in full sunlight to shed heat and to avoid overheating. It would be interesting to examine the ecological, geographical, and temporal distribution of turbulence-inducing leaf morphology and thallus form in algae. To my knowledge this has not been done.

10. The hypothesis that precipitation of calcium carbonate promotes photosynthesis was verified experimentally in the freshwater green alga *Chara* (McConnaughey 1991; McConnaughey and Falk 1991) and marine corals (A. T. Marshall 1996). Calcification is known in a few brown algae, such as *Padina*, and in many red algae (Corallinaceae and related groups), but in these groups it is not associated with the occupation of muddy and sandy bottoms. McConnaughey

(1994) has suggested that photosynthesis-enhancing calcification is responsible for the bulk of the world's carbonate (limestone) rock.

11. Nadkarni (1981) has shown that the uptake of nutrients in dust found in water condensed on or pooling in epiphytes may add significantly to the nutrient base of rain-forest ecosystems. The partial independence of transpiration and mineral uptake in land plants was demonstrated by Tanner and Beevers (2001).

12. Experiments with radioactively labeled nitrogen show that certain green algae can extract nutrients from pore waters in sediments (S. L. Williams 1984; Williams and Fisher 1985; Williams et al. 1985). It is likely that many seaweeds can absorb nutrients from sediments accumulating in the interstices of their holdfasts. If so, the absence of a wide diversity of seaweeds on muddy and sandy bottoms is all the more surprising.

13. C.D.K. Cook (1999) offers a detailed account of all embryophytes (land plants) that have invaded freshwater and marine habitats. The numbers of independent transitions from land to water that he infers are based only on living aquatic plants. If fossils were included, these numbers would likely rise, though perhaps not by much, for many freshwater clades with a terrestrial origin appear to have extremely ancient origins (hydropteridalean water ferns in the Late Cretaceous, nymphaealean water lilies in the Early Cretaceous, and the lycopod club-moss family Isoetaceae perhaps in the Permian, for example) (see also Friis et al. 2001). One of the most ancient flowering plants, the Early Cretaceous (125 Ma) *Archaeofructus* from China, appears to belong to an extinct freshwater clade (Sun et al. 2002). Some major groups of flowering plants are like gymnosperms in lacking freshwater members. Prominent among these are the pea family Leguminosae, the rose family Rosaceae, the tree order Fagales, and the heath family Ericaceae. My estimate of five invasions of the sea by freshwater monocots is based on Les and Cleland's (1997) phylogenetic studies; see also Vermeij and R. Dudley (2000).

14. Goldman and colleagues (1979) and Smetacek (2002) make the point that average concentrations of nutrients in water mean nothing to individual phytoplankters. Being near a zooplankter that happens to be expelling nitrogen or phosphorus means a nutritional bonanza at the microscale; remaining far away from such sources could mean starvation.

15. Mancinelli (2003) has given a masterful summary of the use of nitrogen by organisms. According to his interpretation, organisms first tapped ammonia (NH_3) and ammonium (NH_4^+) in the ocean, later the more abundant inorganically produced nitrate (NO_3^-) and nitrite (NO_2^-), and finally dinitrogen (N_2) as sources of nitrogen for both biomass and energy. Dinitrogen is returned to the atmosphere by denitrifying bacteria through the energy-enhancing process of denitrification in the absence of oxygen. For other helpful discussion of the nitrogen cycle in the ocean see Codispoti (1997) and especially Falkowski (1997). Data on the present-day abundance of various forms of nitrogen in seawater are from Tyrrell (1999). For identification of nitrogen-fixing microbes, see Kolber et al. (2001), Lilburn et al. (2001), and Mancinelli (2003).

16. Nutrient limitation (and especially nitrogen limitation) in terrestrial ecosystems is discussed by DeAngelis (1992) and Melillo et al. (1993). For an excellent discussion of nitrogen fixation by plants, see Douglas (1994). Klironomos and Hart (2001) document the strange case of the partnership between the pine and the fungus for the acquisition of nutrients from collembolans in the soil.

17. The importance of iron as a limiting nutrient for plankton in many parts of the world ocean was originally recognized by J. H. Martin and Fitzwater (1988) and experimentally confirmed through short-term additions of iron to surface-dwelling phytoplankton communities in the Pacific by subsequent investigators (A. J. Watson 1997; Takeda 1998; M. L. Wells et al. 1999, among others). Researchers at first thought that the amount of iron available to the plankton came primarily from dust and from river runoff entering the community directly, but later work has revealed that more remote points of entry may be involved. K. S. Johnson and colleagues (1999), for example, showed that the burrowing activities of animals on the seafloor beneath the plankton communities release iron that had accumulated in the sediment back to the water. Latimer and Filippelli (2001) have shown that the variations between high iron availability dur-

ing glacial episodes of the Pleistocene and low availability during the warmer interglacials are too large to be accounted for by differences in the amount of dust falling into the southern ocean. Instead, they argue, the iron is transported by bottom currents and becomes available largely through upwelling of cold bottom waters to the surface. The intensity of this upwelling, and therefore the iron concentration of surface water, does adequately explain the fluctuations in productivity between glacial and interglacial times in the plankton ecosystem of the southern ocean. Hydrothermal fluid as a source of iron may be relatively unimportant in today's ocean, but likely played a key role during the mid-Cretaceous period, when submarine volcanism was widespread and vigorous (Vermeij 1995; Leckie et al. 2002). The primary role of silicon has been documented in detail for diatoms (see Dugdale and Wilkerson 1998; Smetacek 1998; and especially K. G. Harrison 2000, and Tréguer and Pondaven 2000).

18. The estimates are taken from Vitousek et al. (1997a).

19. Maldonado and colleagues (1999) describe the skeletons of siliceous sponges raised under different silica concentrations and document the restriction of siliceous sponges to colder and deeper waters since the Mesozoic; see also Maliva et al. (1990). Racki and Cordey (2000) emphasize the role of diatoms in depressing silica levels in the ocean and blame trends toward lighter skeletal construction in radiolarians on the rise of these algal primary producers.

20. See B. N. Smith (1976) for an excellent discussion of C_4 photosynthesis and its evolution in land plants and for the estimate of how much of the world's carbon is available to organisms. DeLucia and colleagues (1999) and Osborne and Beerling (2002), among others, document the fertilizing effect of increased concentrations of carbon dioxide in the atmosphere. In an atmosphere rich in carbon dioxide, the reduced conflicts among carbon dioxide uptake, water loss, and respiration in the light enable trees to grow to enormous heights (commonly 40 to 60 m during the Late Jurassic and mid-Cretaceous, times of very high carbon dioxide levels) not just in regions relatively free of disruptive storms and fires, but across broad swaths of the tropical and temperate zones (Osborne and Beerling 2002). The fertilizing effect of carbon dioxide applies to C_3 plants such as the conifers studied by Osborne and Beerling, but evidently not to plants employing C_4 photosynthesis, nor to calcifying aquatic photosynthesizers (Kleypas et al. 1999; Riebessel et al. 2000; Gattuso and Buddemeier 2000; Raven 2002). Although the C_4 and CAM pathways are typically associated with a specialized leaf anatomy, which may protect enzymes from excess oxygen as well as protection against herbivores (Caswell et al. 1973), at least one plant (a Russian goosefoot, *Borsczowia aralocaspica* in the family Chenopodiaceae) has been described in which C_4 photosynthesis proceeds in the absence of diagnostic anatomy (Voznesenskaya et al. 2001).

21. These experiments were carried out by Hibberd and Qu (2002) and discussed by Raven (2002). As Raven points out, the ability of C_3 plants to take up carbon from the soil and from nonphotosynthesizing parts of the body may be, or may have been, very widespread, and likely accounts for the frequent evolution of the C_4 and CAM pathways.

22. For carbon limitation in diatoms, see Riebessel et al. (1993). The C_4 pathway in diatoms was identified by Reinfelder et al. (2000); see also Riebessel's (2000) clarifying comments accompanying that report.

23. For good reviews of water relations in land plants, see Bloom et al. (1985) and Orians and Solbrig (1977).

24. Tiffney (1981), among others, has nicely summarized ways in which early plants dealt with a shortage of water.

25. J. B. Graham and colleagues (1995) first proposed that increases in oxygen concentration during the Devonian and especially the Carboniferous were responsible for increased body sizes in animals (see also Gans et al. 1999; R. Dudley 1998), but they did not remark on the functional challenges of invasion of the dry land by animals.

26. History and extent of irrigation: Hillel (1991). Role of irrigation in stimulating the growth of cities: J. Jacobs (2000), Algaze (2001), Hillel (1991).

27. For excellent discussions of the energetics of aerobic metabolism, see Hochachka and Somero (1973), W. Martin and Müller (1998), and Pfeiffer et al. (2001).

28. Relationships among oxygen, external environmental temperature, and physiology in air and water have been thoroughly discussed by Denny (1993), J. B. Graham et al. (1995), R. Dudley (1998), and Gans et al. (1999).

29. This argument follows that of Tyrrell (1999), who assigns the role of ultimate limiting nutrient to phosphorus. For additional discussions, see Togweiler (1999). Citing a general deficit of nitrate in comparison to phosphate in waters in much of the surface ocean, Falkowski (1997) and Cullen (1999) maintain that nitrogen rather than phosphorus is the ultimate limiting nutrient. As discussed in chapters 7 and 10, Anbar and Knoll (2002) believe that the Mesoproterozoic era was characterized by global nitrogen limitation, whereas eras before and after were marked by phosphorus limitation.

30. Indirect benefits of waves and seaweeds and suspension feeders: Leigh et al. (1987). Excellent discussions of waves and currents and their effects on organisms are given by Denny et al. (1985) and especially Denny (1988, 1999).

31. K. M. Brown and Quinn (1988) document the reduced feeding times and body sizes of wave-exposed snails (*Nucella ostrina*, called by them *N. emarginata*) as compared to individuals at more sheltered sites. Tiny animals living in the boundary layer of relatively slow-moving water within a few millimeters of surfaces over which water flows will be less affected by severe turbulence and by lift forces, but their small sizes make these animals minor consumers. Denny (1988, 1999) has explored the physics of flow and discussed how flow limits animal performance. Leonard and colleagues (1998) have done the most comprehensive study to date on the ecosystem-level consequences of flow. They show that wave-swept shores have high densities of plants and suspension feeders in part because recruitment of these organisms is very high and rates of consumption are low, both because of turbulence.

32. Species that rely on rapid growth and high fecundity during short life cycles were dubbed *r-selected* by MacArthur and Wilson (1967). These are contrasted with *K-selected* species, which are characterized by slower individual and population growth rates, longer lifespans, and the tendency to maintain large populations for extended periods, purportedly at densities close to the maximum allowed by the environment. Van Valen (1971) and Grime (1977) subdivided the *K*-selected group into two categories, one consisting of species living under conditions where metabolism is chronically slow, the other comprising competitively superior long-lived species in environments conducive to rapid metabolism.

33. For detailed discussions of the three categories of marine primary producers, see Littler and Littler (1980), M. E. Hay (1981), Lubchenco (1978, 1983), Birkeland (1989), Steneck and Dethier (1994), and Duarte (1995). Worm's (2000) work in the Baltic Sea and in Nova Scotia represents the most ambitious experimental verification to date of the role of added nutrients in promoting short-lived seaweeds at the expense of perennial species. Smothering of corals by weedy algae, and prevention of coral recruitment by a cover of seaweeds, have been demonstrated as effects of removing grazers from reefs (T. P. Hughes 1994; Birkeland 1977; J.B.C. Jackson 1997).

34. For the characteristics of deciduous leaves and the conditions under which these leaves are expected to evolve, see Givnish (1978) and Reich et al. (1992).

35. For the relationship between mode of transmission and the virulence of parasites, see Bull (1995).

36. Storage and growth in kelps: Mann (1973), Mann et al. (1980). For a general discussion of storage from an economic perspective in plants, see Bloom et al. (1985).

37. Innovations in storing food: Mokyr (1990).

38. Hermit crabs waiting near sites of predation on gastropods: R. B. McLean (1974, 1983), Wilber and Herrnkind (1984), Rittschof et al. (1990). See also Vermeij (1987, chapter 8) for implications of stabilizing the shell resource for the evolution of the shell-dwelling habit in crustaceans.

39. For a general discussion of secondary shell-dwellers, see Vermeij (1987, chapter 8). Bertness (1981a,b) presents compelling evidence for intense competition among hermit crabs for shells. Symbiotic relationships between hermit crabs and encrusting bryozoans, hydroids, and sea anemones have been well described by Muirhead et al. (1986) and P. D. Taylor (1994), among others. Some of the shells made by these encrusters so closely resemble real snail shells that they were originally described as belonging to molluscs. These shells are golden or orange in color, have a chitinous consistency, and are marked by growth lines on the outside. These shells, named *Stylobates*, were always "empty." The mystery of their origin was solved by Daphne Dunn and her colleagues (1981), who showed that the builder is a deep-water sea anemone always associated with hermit crabs of the genus *Parapagurus*.

40. The statement that there are no freshwater hermit crabs is not quite true. McLaughlin and Murray (1990) described a new species, *Clibanarius fonticola*, from a small coastal freshwater pool in Vanuatu in the southwestern tropical Pacific. The extreme localization of this purported species only serves to underscore the extraordinary rarity of the freshwater secondary shell-dwelling habit.

41. See K. G. Porter (1973, 1976).

42. For discussions of mast fruiting, see Janzen (1974) and Leigh (1999).

43. See Clifton (1997).

44. Mass coral spawning has been described in many papers; see Wallace (1985) for an early account.

45. I had the rare and enormous pleasure of observing an outbreak of three species of seventeen-year cicadas in Maryland in 1987. For predation by red-winged blackbirds on periodical cicadas, see Steward et al. (1988).

46. Experimental evidence for the relationship between concentrations of suspended food particles and the growth rates and maximum body sizes of suspension-feeding bivalves is available for oysters (Lescinsky et al., 2002). The South African work by Bustamante et al. (1995b) highlights comparisons among three coastal zones, the highly productive west coast, the less nutrient-rich south coast, and the relatively planktonically unproductive east coast. Maximum suspension-feeder biomass is about the same in each region, because differences in wave action among sites within the same region override between-region differences in nutrient concentration. Central Pacific productivity and suspension feeders were discussed by Highsmith (1980) and Vermeij (1990b).

47. Data on planktonic and algal productivity come from Bunt (1975). Those for Nova Scotia are from Mann (1973) and Mann et al. (1980). My calculation of average seaweed productivity is derived from S. V. Smith's (1981) estimates of total world production of attached plants (1×10^{12} kg) and total area of seafloor occupied by these plants (2×10^6 km^2). Baltic data are from Worm (2000).

48. For data on herbivore biomass in relation to primary productivity in South Africa see Bustamante et al. (1995a,b). J.B.C. Jackson (1997) and Jackson and colleagues (2001) emphasize the intense herbivory exercised by large vertebrates in seagrass meadows. For the role of seaweeds in providing particulate food to suspension-feeding bivalves and barnacles, see Duggins et al. (1989), who took advantage of the difference in carbon-isotopic composition between seaweeds and phytoplankton to discover that attached plants contribute as much as 50 percent of the carbon to these consumers along the kelp-rich shores of the Aleutian Islands of Alaska.

49. The Florida mangrove example comes from the work of Onuf et al. (1977). For the relationship among fertility, productivity, and herbivory, see Coley et al. (1985), Coley and Barone (1996), and de Mazancourt et al. (2001).

50. Nitrogen limitation on herbivores: T.C.R. White (1978), Cebrián (1999). Mass survival of *Acanthaster planci* larvae when phytoplankton is abundant: Birkeland (1982).

51. Diversity, productivity, and spatial scale of comparison among ponds: Chase and Leibold (2002). For further discussions of diversity and productivity, see especially the landmark paper by Rosenzweig and Abramsky (1993).

52. Ghost shrimps culturing bacteria: Ziebis et al. (1996). Gardening by damselfish (Pomacentridae): Kaufman (1977), Russ (1987). Gardening by nereid worms: Woodin (1977). Territorial limpets and their algal gardens: Stimson (1973), Branch (1981).

53. The hypothesis that climatic instability during the Pleistocene inhibited the development of agriculture (Richerson et al. 2001) is weakened by the fact that the postglacial period (the last 11,000 years or so) is likewise characterized by great and rapid climatic fluctuations.

54. The role of agriculture in stabilizing and increasing food supplies and in providing conditions for the rise of urban elites has been thoroughly explored by many anthropologists and historians. I found the accounts by Diamond (1997), J. Jacobs (2000), Hillel (1991), and especially Algaze (2001) helpful.

55. Pomeranz (2000) argues that Europe and coastal parts of China had achieved comparable levels of trade, industry, and wealth by 1800 or so, but only Europe entered the Industrial Revolution, with Britain leading the way, to be followed by parts of mainly western Europe and the United States. Although I agree with Pomeranz that technological advancement in Europe was made possible by the partial transfer of land-hungry agriculture to the Americas and to some extent other colonies, the European predilection toward expansion and technological innovation is, as David Landes (1998) argues, much older. It is this cultural inclination to explore and to exploit that requires further explanation.

Chapter 6
Technology and Organization

1. Ecologists identify several categories of economically dominant species. Those that consume competitive dominants have been called keystone species (Paine 1969); species that serve as food for a wide variety of consumers are key-industry species (Birkeland 1974). Structural species are those that provide an ecosystem's three-dimensional structure. In the genome, genes that switch other genes on or off are known as regulatory genes, because they determine whether and how much genetically based traits are expressed in the phenotype. Whole biotas that "export" species to neighboring biotas are donor biotas (Vermeij 1991a,b).

2. The topic of power has been very widely discussed by scholars concerned with human affairs. I find Galbraith's (1983) treatise one of the most readable and incisive.

3. The idea that efficiency has not been important in the history and evolution of economic systems is not new. It was advanced by Bakker (1980), Vermeij (1987), and J. H. Brown (1995), among others. What is new is my claim that it is in the interests only of the powerless to become efficient in the use of resources. Data on the efficiency of oxygen uptake in *Nautilus* are from Wells and O'Dor (1991). Brachiopods as low-energy, highly efficient animals: LaBarbera (1977, 1984), Rhodes and Thayer (1991), Rhodes and Thompson (1993), Thayer (1986). Data on energy assimilation and production in cold-blooded and warm-blooded vertebrates are from Bakker (1973). Cost of locomotion in relation to body size and locomotor speed: Bakker (1973), Garland (1983b), Altmann (1987).

4. Temperature is the average kinetic energy of a molecule in a gas. Temperature T in degrees kelvin (K) is given by $T = m \times u^2/3k$, where m is the molecule's mass in grams, u is the average velocity of the molecule in centimeters per second, and k is Boltzmann's constant, 1.38054×10^{-23} JK^{-1}. Denny (1993) gives a superb, biologically informed discussion of temperature. For temperature-related physiology, see Hochachka and Somero (1973) and A. Clarke (1980, 1983, 1993). I treated the evolutionary implications of temperature in relation to the range of adaptive options (Vermeij 1978, 2003).

5. Thermal dependence was documented empirically by A. Clarke and Johnson (1999) and Peck and Conway (2000). Gillooly and coauthors (2001, 2002) derive a general formula for resting metabolic rate B as a function of body mass M, activation energy E_i (the energy required

to transform reactants to products), and temperature T (in degrees Kelvin): $B = M^{3/4}e^{-E_i/kT}$, where k is Boltzmann's constant, 1.38054×10^{-23} JK^{-1}). The three-fourth-power dependence of resting metabolic rate on body mass is thought to arise from the fractal (self-similar) nature of the transport networks that distribute energy and raw materials into, through, and out of the bodies of living things (G. B. West et al., 1997), but Darveau and colleagues (2002) question the assertion that the delivery of energy or oxygen supply is the limiting factor in resting metabolism.

6. For excellent summaries of molecular adaptations to different temperature regimes, see Hochachka and Somero (1973) and Rothschild and Mancinelli (2001).

7. For comparisons between brachiopods and bivalves, see Steele-Petrovič (1979), Thayer (1985, 1986), Vermeij (1987), Rhodes and Thayer (1991), and Rhodes and Thompson (1993).

8. Temperature and the scope of adaptation: Vermeij (1978, 2003). For thermal conditions of flight in insects, see Heinrich (1993).

9. Technically, the viscosity I discuss here is dynamic viscosity, measured in newtons times seconds per square meter. Kinetic viscosity is the ratio of the dynamic viscosity to the density of the fluid. Denny (1993) and Podolsky (1994) give particularly fine discussions of viscosity and its relation to temperature.

10. The best study of the cost of calcification is A. R. Palmer's paper (1992). Calcification as limit to soft-tissue growth: Palmer (1981). Sizes of Antarctic animals: Arnaud (1974). Graus (1974) suggested that molluscs in cold water are more efficient in their use of calcium carbonate than warm-water ones in that they enclose a larger volume for a given quantity of shell material. However, the narrowly turreted shells of many cold-water snails in Lake Baikal and the deep sea violate this efficiency principle, as do many bivalves in cold water with a compressed shell form. These shapes depart very far from the ideal spherical enclosure expected for a high-efficiency shell (Vermeij 1978, 1993). For data on thermal extremes, see Schwartzman (1999), Heinrich (1993), Markel (1971), Garrity (1984), and Rothschild and Mancinelli (2001).

11. For date on thermal extremes see Markel (1971), Garrity (1984), Heinrich (1993), Schwartzman (1999), Rothschild and Mancinelli (2001).

12. Data on Antarctic crustaceans come from A. Clarke (1983). In Clarke's (1980, 1983) view, the low metabolic rates of polar (and especially Antarctic) marine animals are attributable more to a seasonally low nutrient supply than to the direct physiological effects of low temperature. I am skeptical of this argument. Although phytoplankton is indeed scarce during the long dark winter at high latitudes, there are many regions of very high biomass, so much so that large warm-blooded mammals (whales and seals) and birds (penguins) exist in very large numbers. If such energy-intensive vertebrates can feed year-round, I see no reason why the metabolic rates of metabolically less demanding animals would be low on account of food scarcity.

13. The mechanisms of suspension-feeding are described in detail by Rubenstein and Koehl (1977) and LaBarbera (1984) for invertebrates, and Collin and Janis (1997) for vertebrates. Strathmann (1987) offers an excellent summary of feeding (including suspension-feeding) in larvae.

14. For data on feeding by bryozoans, see McKinney (1984, 1992, 1993). McKinney's Adriatic Sea data are from his 1992 paper. In their studies of cold-water bryozoans in Alaska, South Georgia, and the Falkland Islands, Barnes and Dick (2000) found that 74 percent of the interactions experienced by cheilostomes were with other cheilostomes, and that cyclostomes did not disproportionately lose bouts as they did in the Adriatic. This result would be consistent with my view that differences in performance are less well expressed in the cold, where all biological processes proceed slowly, than in warm water; and that the advantages of greater power are most evident when potential differences in performance are large, as is the case when external sources of energy are easily accessible and prolific.

15. The comparative histories of cyclostome and cheilostome bryozoans, together with functional comparisons between these groups, have received substantial attention from paleontologists; see especially Lidgard (1990), Lidgard et al. (1993), McKinney (1993, 1995), McKinney et al. (1998), Jablonski et al. (1997), and J.B.C. Jackson and McKinney (1990).

16. For mechanisms of temperature regulation and heat production, see Ruben (1995) for birds and mammals, Block et al. (1993) for tunas and related scombroid fish, and Heinrich (1993) for insects.

17. Production of heat in flowers: Knutson (1974), Seymour and Schultze-Motel (1996). Heat-producing flowers occur in Araceae (including *Philodendron*), Arecaceae (palms), Nymphaeaceae (water lilies), Cyclanthaceae, Annonaceae, and Magnoliaceae. Scarab beetles of the species-rich tropical American genus *Cyclocephala* gain energetic benefits from spending the night in the warm flowers of *Philodendron* (Seymour et al. 2003).

18. Reproductive advantages of endotherms: McNab (1980), Hennemann (1983). Data on speeds and endurance in relation to metabolic rate are from Hertz et al. (1988), Block et al. (1993), and Ruben (1995). Flight and temperature in insects are thoroughly documented in Heinrich's (1993) excellent book; see also Heinrich and Bartholomew (1979) for the work on dung beetles, and Heinrich and Mommsen (1985) for flying by winter moths near freezing. For additional discussions of the costs and benefits of endothermy, see Pough (1980), Bakker (1980), Vermeij (1987), and J. P. Hayes and Garland (1995).

19. Comparative data are taken from Pough (1980), Hertz et al. (1988), Denny (1993), and Ruben (1995). Lack of limb regeneration at high body temperatures: Goss (1965).

20. Maximal performance in orchid bees (R. Dudley 1995) exceeds that in hummingbirds and other organisms (Chai and Dudley 1995). Although the data were obtained in an experimental gas mixture not encountered in nature, they provide a good approximation to the upper limit of metabolic power attainable by animals. Marden and Chai (1991) have noted that, for butterflies, the highest power outputs provide advantages in escape from predators and in mate selection.

21. See Ruben (1995) for an excellent account of endothermy (the production of internal heat). In very large animals, such as dinosaurs weighing 5000 kg or more, high and stable body temperatures of 30°C or above can be maintained under most climatic conditions, because the body accumulates and stores heat and ensures that heat is lost slowly through the relatively small surface area. A high metabolic rate is therefore unnecessary for these large animals. The benefits of a high body temperature are thus available without the high cost of body-heat production. This circumstance has led J. P. Hayes and Garland (1995), Seebacher (2003), and others to suggest that the principal advantage of a high and stable body temperature is the greater efficiency of complex enzymes involved in temperature-sensitive biochemical reactions. I suggest instead that the large size of dinosaurs provides competitive advantages, and that endothermy provides additional major advantages in endurance and other power-related benefits.

22. *Rhea* speed and metabolism: Bundle et al. (1999). Comparisons among *Nautilus*, squid, and fish: M. J. Wells and O'Dor (1991), J. A. Chamberlain (1991). Relation between speed and power output in fish: P. W. Webb (1984).

23. Leaf-cutter ant metabolism and musculature: Roces and Lighton (1995).

24. Relative brain size, and relative size of the visual cortex within the brain, correlate well with sensory and other neural functions; see Jerison (1973) and Diamond (1996).

25. For additional discussion of the relationship between shell growth rate, shell shape, and metabolism, see Vermeij (1993, 2002a).

26. Maximum human power output from muscles was calculated by Hammond and Diamond (1997) for athletes in the Tour de France. These bicyclists expend about 7000 kilocalories per day (1 kCal = 4.2×10^3 J). For estimates of the increase in metabolic rate in humans thanks to outside sources, see L. A. White (1943, 1959) and especially J. E. Cohen (1995). The power consumption of modern people was calculated by Chaisson (2001).

27. For an excellent introduction to the physics and form of transportation systems in organisms, see LaBarbera (1990). Denny (1993) also discusses important aspects of transport, including a detailed account of diffusion. Transport systems in leaves have been thoroughly reviewed by Roth-Nebelsick et al. (2001). The evolutionary significance and taxonomic distribution of lactiferous canals in flowering plants were explored by Farrell et al. (1991).

28. G. B. West et al. (1997, 2001) have provided a theoretical foundation for the three-quarters power scaling of metabolic rate relative to mass, based on the fractal nature of the hierarchical transport system that supplies metabolizing cells with raw materials. Darveau et al. (2002) question the assumption that supply limits metabolic rate, and suggest instead that constraints are imposed by demand, that is, by the rate at which the machinery of metabolism draws energy from the incoming stream. They emphasize the fact that the three-quarters power rule applies mainly to basal, or resting, metabolism, and that the scaling factor for maximum metabolic rate lies close to unity. Averaged over all organisms, and therefore over a wide variety of adaptive types, the three-quarters rule may apply, and perhaps reflects space limitations for either supply or demand machinery. However, the three-quarters rule is violated in many instances, and it cannot be interpreted as a very powerful constraint on either size or organic form. Moreover, it is misleading in the sense that it directs attention away from the fact that adaptation can significantly alter the relationships among size, metabolic rate, and the proportion of tissues or organs with differing rates of metabolism.

29. Newman and Muller (2001) incisively discuss self-organization in the context of developing embryos. Maynard Smith and Szathmáry (1995, 1999) explain the molecular basis for the self-organization of two-layered biological membranes. The best general discussion of self-organization is by Kauffman (2000). According to Kauffman, organization is required to do work. The integration of tasks, measurements, records, and linkages by and among parts is for Kauffman a reasonable definition of organization. S. Johnson (2001) offers perhaps the most accessible account of self-organization, a concept he applies to ant colonies, the development of cities, the emergence of collective intelligence, and the development of software that learns.

30. Redundancy's advantages have been discussed at length with respect to social insects (E. O. Wilson 1971; Davidson 1997), marine colonial animals (J.B.C. Jackson 1977a, 1979), genetic systems (Kirschner and Gerhart 1998; Nowak et al. 1997), and developing embryos (Raff 1996).

31. McNeill (1982) and Diamond (1998a), among others, suggested that the decision by the Ming emperor of China to stop wide-ranging exploration by ocean-going fleets after 1433 represents an example of an error initiated by an overly centralized authoritarian regime. In Europe, similar action would not have stopped government-sponsored exploration, because there were multiple nations that could decide whether exploration was desirable. This interpretation, however, is complicated by the fact that Chinese exploration after 1433 fell largely into the hands of private interests. Moreover, much of the colonization and exploration undertaken by European powers was also carried out by private entrepreneurs, and it is they who largely established European control over such domains as the Dutch East Indies and British India.

32. Gatesy and Dial (1996) have given a detailed account of the locomotor system in birds and their apparent theropod dinosaur ancestors.

33. The incompatibility between air-breathing and running in early tetrapods was first described by Carrier (1987), and elaborated by Cowen (1996). The pursuit option made possible by a change from a sprawling to an erect posture would have been most accessible to warm-blooded animals. As Janis and Wilhelm (1993) and Van Valkenburgh (1999) point out, however, this pathway was rarely taken even by endotherms. Blob (2001) noted that the posture observable in Permian synapsids (the group from which mammals evolved in the Triassic) was initially facultatively erect. Only later, Blob argues, would the erect condition be genetically and anatomically fixed. See also chapter 10.

34. Cisne (1974) was the first to note that progressively younger clades of arthropods show increased function specialization both in limbs and in body-segment structure. More recent work on the genetic basis of arthropod development (Shubin et al. 1997; Nagy and Williams 2001) shows that regulation by genes begins globally and in the course of development becomes more regional, the degree of regionalization being higher for the functionally more differentiated, more recently evolving, arthropod clades.

35. Early one-piece molluscan shells are usually cap shaped, planispiral (coiling in a single plane), or conispiral (spirally coiled in three dimensions) with a round or broadly oval aperture. Later conispiral forms, characterizing sorbeoconch, opisthobranch, and at least some pulmonate gastropods, have a greater variety of apertural shapes, in which two or more sectors of the rim vary independently in sculpture, shape, and function. This increase in adaptive versatility from early to later major clades (Vermeij 1971, 1973b, 2002a) thus matches that seen in arthropods and other groups (Vermeij 1973b). D. K. Jacobs and colleagues (2000) identified a gene responsible for regulating the formation of skeletal elements in molluscs and other animals, and showed that the number of domains in which expression of this gene occurs has decreased from spicule-bearing Aplacophora to eight-valved chitons (Polyplacophora) to other molluscs, which have one-piece or two-piece shells.

36. Sidor (2001) presents an excellent discussion of the progressive reduction in skull bones among successive clades of tetrapods, and notes that similar reductions, accompanied by functional differentiation, apply to other skeletal elements of vertebrates as well. For reductions in terminal digits of limbs, see also Lande (1978), Raff (1996), and Wiens and Slingluff (2001). Work on dental evolution (Ziegler 1971) shows that the number of teeth has generally decreased, especially in the front of the jaw, and that front-to-back specialization has increased in successive major clades of tetrapods. There are a few exceptions, especially in groups in which additional molars appear at the back of the jaw.

37. Many authors have emphasized the importance of organizational structures that promote exploration of possibilities (Eigen 1992; Frank 1996; Kirschner and Gerhart 1998; Newman and Müller 2001; Kauffman 2000). Favorable combinations and interactions can then be brought under genetic control through a process of genetic assimilation (Waddington 1962; Pigliucci 2001).

38. I presented this argument in my 1996a chapter. I was led to it by Mulcahy's (1979) paper, in which the large role of randomness in fertilization in wind-pollinated or water-mediated unions between male and female gametes in plants was contrasted with the more overtly competition-related fertilization characteristic of animal-pollinated plants.

39. Most authors who have written about self-organization have justifiably emphasized the diffuse nature of learning (see Waldrop 1992; Krugman 1996; and S. Johnson 2001, among many others). This bottom-up view of learning, in which many simple components follow a few simple rules of pattern recognition, developed in reaction to the predominating top-down view of learning, in which information passes from a central authority to many naive individuals. Reading this literature on self-organization, one could come away with the impression that self-organization is sufficient to create central intelligence. However, I believe that a feedback between diffuse learning and the emergent central authority is necessary to evolve the kind of central intelligence that we associate with human beings. Until such feedbacks are built into systems of artificial intelligence, no humanlike intellectual abilities and actions can be achieved by smart machines.

Chapter 7
The Environment

1. A.G.B. Fischer (1935:3–4). The two quotes from W. A. Lewis (1955) are from pages 420 and 423, respectively. For an excellent account of the history of thought about economic growth, see Arndt (1978). Galbraith (1984) and Slawson (1981) emphasize the increasing importance of advertising in the twentieth-century human economy. I previously made the argument that competition must be accompanied by economic growth if competition is to be an agency of adaptation (Vermeij 1995, 2002c). In a brief essay (Vermeij 1998b), I suggested that social mobility and opportunities for individual improvement, spending on the arts and science, and many other social goods that the growth-oriented cultures of the West have come to expect would disappear under a regime of stasis. I return to this matter in chapter 11.

2. This ecological view of constraints (see also Vermeij 1978, 1987) differs from that held by most other biologists, who think of constraint as inflexibility in the pattern of development or in the genome. Tampering with development, especially at stages where the body plan appears to be very conservative, would create unworkable organisms. Raff (1996) offers an extremely clear and nuanced discussion of this kind of constraint.

3. For a lucid discussion of the relationship between population size and the precision and extent of adaptation, see M. Kimura (1983). For J. L. Simon's views, see his 1977 book.

4. I have long been fascinated by Fernando de Noronha, which I had the privilege of visiting in 1969 in pursuit of my Ph.D. research. For accounts of its marine biota, see Eston et al. (1986) and Leal (1991).

5. Darwin (1859:87–88). The only point with which I take issue is that the production of new species necessarily leads to extinction, or, as Darwin calls it, extermination. Darwin generally underestimated the importance of refuges, both geographic and ecological, for competitively subordinate species. Extermination is likely only when appropriate refuges do not exist, a common situation on today's oceanic islands, where humans and the continental species they have brought with them have radically changed habitats and routed many species to extinction.

6. For the relationship between diversity and area and the roles of immigration and extinction in setting species numbers on islands, see MacArthur and Wilson (1967). Heaney (1986, 2000), Rosenzweig (1995), and Losos and Schluter (2000) have highlighted the additional role that speciation plays in controlling diversity over the longer term. For a detailed treatment of Darwin's finches, see Grant's (1986) excellent book. For years it was thought that the Galápagos Islands were not more than 3 to 5 million years old, but recent geological work has demonstrated that precursor islands existed by Middle Miocene time (14 Ma) (Werner et al. 1999). For the origin and evolution of Hawaiian finches, see Bock (1970) and N. K. Johnson et al. (1989).

7. This is a summary of Burness et al. (2001). These authors quantified what Bakker (1980) had concluded qualitatively twenty years earlier. For herbivores, the relationship between mass M in kilograms and area A in square kilometers was given by Burness and colleagues as $M = 0.47A^{0.52}$. For carnivores this relationship is $M = 0.05A^{0.47}$.

8. Sizes of carnivores on islands and continents: E. Anderson (1984), Burness et al. (2001). Spiders as top predators on tiny islands: Schoener and Schoener (1983), Schoener and Toft (1983), Schoener and Spiller (1995). Top predators in New Zealand: Worthy and Holdaway (2002). Top predators in the Hawaiian Islands: S. L. Olson and James (1991).

9. The herbivorous diets of extinct birds on islands have been inferred from analyses of preserved coprolites (feces) and gizzard contents. Some large New Zealand moas, such as *Dinornis giganteus* and probably *Pachyornis elephantopus*, consumed twigs up to 6 mm in diameter, as well as the fibrous leaves of New Zealand flax (*Phormium tenax*), whereas smaller species ate mainly softer leaves and fruits (Worthy and Holdaway 2002). The Hawaiian moanalos may have eaten mainly or exclusively ferns (James and Burney 1997). For the structure and diet of the dodo and solitaire, see Livezey (1993). Morse (1975) gives an excellent summary of the feeding habits of birds. Estimated weights of birds are from Worthy and Holdaway (2002) and are generally lower than older estimates in the literature.

10. For the evolution of island carnivores and herbivores, see Bakker (1980). Island trees evolving from herbaceous mainland ancestors: Carlquist (1966), Francisco-Ortega et al. (1996), S.-C. Kim et al. (1996). Weedy opportunists growing larger on islands: Heaney (1978), Damuth (1993).

11. Channel Islands plants compared to mainland Californian species: Bowen and Van Vuren (1997). New Zealand plant architecture in relation to browsing by moa: summary and discussion in Diamond (1990) and Worthy and Holdaway (2002). Spines as defenses against vertebrates, especially large mammals: Cooper and Owen-Smith (1986). Vancouver newts and snakes: Brodie and Brodie (1991).

12. Jersey red deer: Lister (1989). Reduced energy budgets and flightlessness on islands: McNab (1994a,b). Galápagos cormorant: Livezey (1992).

13. I had been aware of the large ranges of top marine predators for a long time (see Vermeij 1978), but the evidence I have discovered since has considerably strengthened the case for it (Vermeij 2004).

14. This discussion expands that in Vermeij (2003). For additional helpful perspectives, see Raven (1998) and Kasting and Siefert (2002). The process of denitrification—the production of inorganic nitrogen (N_2) from organic sources—remains little understood. Recent experiments in the Golfo Dulce (on the Pacific coast of Costa Rica) and in the Black Sea show that the formation of N_2 through the anaerobic oxidation of ammonium with nitrite by so-called anammox bacteria significantly depletes biologically available forms of nitrogen in oxygen-starved sediments and deep ocean waters (Dalsgaard et al. 2003; Kuypers et al. 2003).

15. Estimates of erupted rock volume: Erwin and Vogel (1992), Sigurdsson (1990).

16. Chemical characteristics of tephras and volcanic soils: Shoji et al. (1993). Effects of volcanic aerosols on the distribution of light and on tree photosythesis: Gu et al. (2003); see also the helpful commentary by Farquar and Roderick (2003).

17. Coffin and Eldholm (1994) recognized that massive undersea volcanism would release nutrients into the ocean and therefore stimulate production there. I independently came to the same conclusion, and made the case for economic stimulus by submarine volcanism more general (Vermeij 1995).

18. Effects of mountain ranges and plateaus on climate: R. D. Norris et al. (2000b). Early onset of Asian monsoon: Guo et al. (2002).

19. The episodic nature of erosion was emphasized by Stallard (1988, 1992). An excellent review of dust and its sources is given by Rea (1994).

20. See Peizhen et al. (2001).

21. For bioerosion, see Vermeij (1987, chapter 5). I consider here only the mechanical bioeroders. Many other organisms—algae, fungi, sponges, phoronid worms, polychaete annelids, certain hipponicid gastropods, and various bivalves—rely predominantly on chemical means to dissolve rocks.

22. Chemical weathering: Stallard (1988), Moulton and Berner (1998), Lenton (2001).

23. Calculations of turnover times: S. V. Smith (1981), Field et al. (1998). Obviously, there is great variation in average turnover time. The numbers quoted are meant to give rough indications of the large differences among planktonic, aquatic attached, and land plants.

24. Accounts of the minerals precipitated by organisms: Lowenstam (1981), Simkiss (1989). The importance of sinking organisms and fecal pellets as sources of buried organic carbon was recognized by A. G. Fischer (1984) and especially by Logan et al. (1995).

25. To my knowledge, no serious efforts have been mounted to model the optimal pattern of increase in the supply of essential nutrients. My intuition is that the increase should be gradual and sustained, neither too slow nor too fast, but I have no quantitative feel for the ideal rates. The overall community-wide or ecosystem-wide metabolism—the aggregate demand for resources—is no doubt an important factor; the higher the metabolism, the higher the rate of increase in supply can be before the system is overwhelmed and materials are dumped in sediments. If aggregate metabolism has increased through time, as I shall suggest it did (chapter 10), then more episodes of enhanced supply, even those that are sudden and of short duration, could be turned into opportunities than would have been the case earlier in economic history, when metabolism was generally lower. This line of reasoning deserves quantitative attention from oceanographers, ecologists, and geologists brave enough to tackle the dynamics of a complex system.

26. Herbivory and predation as factors increasing the availability of nutrients: DeAngelis (1992), Vermeij and Lindberg (2000), de Mazancourt et al. (2001). Effects of sediment disturbance by animals on primary productivity: Thayer (1979, 1983), Bertness (1985), McIlroy and Logan (1999).

27. For a quantitative assessment of the sources of oxygen, see Lenton (2001). Duarte and Agustí (1998) pointed out that the central oceanic gyres are net oxygen-consuming areas even though primary production, mainly by Cyanobacteria, takes place there.

28. Although Schopf (1980) already understood the link between oxygen in the atmosphere and the burial of biomass in the sediment, Berner and Raiswell (1983) provided the first quantitative geochemical model for this relationship. The complex feedbacks among carbon, oxygen, and minerals are carefully outlined by Berner and colleagues (2000, 2003), Kasting and Siefert (2002), and especially Lenton (2001) and Berner (2003). The burial of pyrite similarly leads to the liberation of free oxygen (Canfield and Teske 1996). Kump and McKenzie (1996) and Lenton (2001) underscored the pivotal role of phosphorus. The claim that higher oxygen levels in the atmosphere depress the rate of photosynthesis is based on experiments in which living plants were exposed to an artificial atmosphere in which oxygen comprised 35 percent by volume, a value 1.5 times that in today's atmosphere (Beerling et al. 1998). It is unknown if this effect would persist once plants adapted evolutionarily to higher oxygen concentrations.

29. H. D. Holland (1984) suggested the link between iron oxides and phosphate for the Archean and Early Proterozoic, a link further investigated and confirmed by Bjerrum and Canfield (2002). See also the helpful comments on Bjerrum and Canfield's paper by J. M. Hayes (2002). Anbar and Knoll (2002) suggest nitrate instead of phosphate limitation for the Mesoproterozoic era, but like the strong phosphorus limitation of the Archean and Paleoproterozoic proposed by Bjerrum and Canfield, this limitation kept oxygen levels low and prevented free oxygen from becoming significant in the deep ocean.

30. Calculations are from Bunt's (1975) data on marine productivity and from data of Lieth (1975) and Melillo et al. (1993) for productivity on land. Field and colleagues (1998) give an excellent overview of global patterns in primary productivity.

31. For the hydrogen-escape hypothesis of oxygenation of Earth's atmosphere, see Catling et al. (2001). The role of nitrogen-fixing Cyanobacteria in releasing methane and the subsequent dissociation of methane into carbon and hydrogen atoms were discussed by Hoehler et al. (2001) and Barker Jorgensen (2001). The speculation to follow about the possible role of methane pulses as a mechanism to increase oxygen later in Earth's history is my own.

32. Subsidy of Hawaiian soils by Asian dust: Chadwick et al. (1999). For a general discussion of soil fertility, see the excellent book by Shoji et al. (1993).

33. For data on productivity in upwelling waters, see Bunt (1975) and Walsh (1988). Geographical distribution of upwelling: Riggs (1984), J. T. Parrish (1987), Walsh (1988), W. H. Berger et al. (1989).

34. See Butterfield (1997), Field et al. (1998), and Smetacek (2002) for discussions of the microbial loop. Butterfield (1997) and Smetacek (2002) pointed out the implications of larger size for nutrient cycling in the ocean, following the insight by Logan et al. (1995) that fecal pellets provided a new and powerful mechanism for transporting nutrients from the upper layer of the ocean to the seafloor.

35. Data on river inputs of phosphorus are given by Froehlich et al. (1982); those for silica are from Tréguer et al. (1995). Annual loads of dissolved and detrital material in rivers are from W. W. Hay et al. (1988) and McLennan (1993). The total water flux of rivers to the sea was calculated by Tardy et al. (1989).

Chapter 8
The Geography of Power and Innovation

1. The distribution of gliding and prehensile-tailed arborial mammals in rain forests was first pointed out by Emmons and Gentry (1983). R. Dudley and DeVries (1990) proposed the idea that different forest structure, notably the presence of tall emergent trees in Asia, is responsible. The presence of gigantic herbivores in African and Asian but not American forests has many potential consequences for plant defenses, types and abundances of flowers in the understory, and many other features of forests (Cristoffer and Peres 2003). Although some very large herbivorous

mammals occupied South America and became extinct there during the Late Pleistocene, potentially making South American forest ecosystems more like those of the Old World, Cristoffer and Peres argue that most of these animals were not forest-dwellers. Instead, these animals were savanna-adapted species, as were the Old World ancestors of the large African and Asian forest mammals. For the distribution and evolution of horned mammals, see Janis (1982). Interoceanic differences among shells: Vermeij (1978, 1993). Contrasts between marine polar regions: Dayton (1990). Explanations for many of these contrasts remain obscure.

2. Cosmopolitanism of free-living microbes: Finlay and Clarke (1999), Finlay (2002), Hutchinson (1960). For the prokaryotic perspective, see Torsvik et al. (2002).

3. For the biomechanical perspective on how organisms experience the world, see Denny (1993), Smetacek (2002), and Mitchell (2002).

4. Stanley (1973) also made the point that an increase in body size enforces specializations and habitat distinctions that small organisms do not make. Gould (1977) noted that a reduction in body size accompanying progenesis—the developmental process in which reproduction takes place in a small, juvenile-like, young individual—releases the clade in which it occurs from specialization, and allows the clade to follow a wholly novel evolutionary pathway.

5. Rosenzweig (1995), whose book on biological diversity I find one of the most readable and incisive on the subject, has also made much of the principle that specialization and the division of labor result from competition. Holt (1996) and Holt and Gomulkiewicz (1997) mathematically demonstrated that selection is most intense in, and populations are best adapted to, the most productive parts occupied by a population.

6. These molluscan examples are drawn from Vermeij (1973a, 1989a).

7. Kohn (1978b) has given details of cone-snail diets of the endemic Easter Island species *Conus pascuensis* (considered by him to be a subspecies of *C. miliaris*) in comparison to those of *C. miliaris* and other cones elsewhere in the tropical Pacific. Cones are by no means the only predatory snails, or the only predators, either at Easter Island or anywhere else in this region. Easter Island, however, is exceptionally impoverished in its diversity of predators, owing largely to its extreme isolation; see also Kohn 1978a. Interestingly, none of the more than 500 species of *Conus* (taken in the broad sense) preys on sipunculan worms, which form the dominant food items of several other gastropods, including species of *Vasum*; these worms are the exclusive diet of all members of the large family Mitridae (J. D. Taylor 1984, 1986, 1989).

8. See Losos et al. (1994).

9. Coley and Barone (1996) provide an excellent overview of herbivory in tropical and temperate forests. For data on plant alkaloids, see D. A. Levin (1976) and Levin and York (1978). Levin (1975) summarizes a wide range of evidence that "pest pressure" is higher in the tropics than in the temperate zones.

10. The study of Pennings et al. (2001) is the best to date on the geography of herbivory. One of several virtues of this study is that the authors compared plants from truly contrasting temperature regimes, a cool-temperate and a subtropical one.

11. The figures for predation on birds' eggs are quoted in Orians and Janzen (1974). Experimental work on predation by ants (Jeanne 1979) and spiders (Rypstra 1984) was done with local prey species, which are presumably adapted to the local collection of predators.

12. For details see Vermeij (1978, 1993).

13. See Bertness et al. (1981), Vermeij et al. (1989).

14. Shell-crushing crabs: Vermeij (1977a); fish: A. R. Palmer (1979). See also Vermeij (1978, 1993) for general discussions of predators and latitude.

15. Geographic distributions of highly toxic sea cucumbers and sponges: Bakus (1974), Bakus and Green (1974). There are various studies of chemical defense on seaweeds that contradict the observations in sponges and sea cucumbers (see Van Alstyne and Paul 1990; Targett et al. 1992), but the comparisons among seaweeds were made between tropical and warm-temperate plants. I predict that, when these warm-water algae are compared to species in cold-temperate waters, a difference in chemical defense will be detected, just as it was in sponges and sea cucumbers. In

addition, only some classes of chemical defense (polyphenols, similar to all-purpose tannins in land plants) have been examined in marine algae from a geographic point of view.

16. This partial summary is taken from Vermeij (1983).

17. See Moynihan (1971).

18. Meyer and Macurda (1977) suggested that stalked crinoids were restricted by the evolution of predatory fish in the Cretaceous, and that motile tropical comatulid crinoids can live on shallow-water reefs not only because these animals can hide from fish during the day, but also because their arms are beset with spines and other features preventing or slowing attacks by fish. See also Meyer (1985), Meyer and Ausich (1983), and Oji and Okamoto (1994). Oji (1996) confirmed these ideas. The Bahamian work by S. E. Walker et al. (2002) was done with initially empty shells, which presumably were subsequently broken by predators that either mistook the shells as being occupied by something edible, or found hermit-crab occupants inside. Vale and Rex (1988) conducted surveys of breakage in deep-sea gastropods along a transect from Gay Head, Massachusetts, to Bermuda in the northwestern Atlantic.

19. Predation and adaptation in freshwater molluscs: Vermeij and Covich (1978), Vermeij and E. C. Dudley (1985), Vermeij and R. Dudley (2000). Drill holes in brackish-water molluscs: Il'ina (1987).

20. Relationships between freshwater animals and sediments have received much less attention than those in the sea. Good accounts are given by C. K. Chamberlain (1975), McCall and Tevesz (1982), and M. F. Miller et al. (2002). Jones (1969) described *Caobangia* and its shell-boring habits.

21. Data on transfers between sea and land are from Polis and Hurd (1996) and Polis et al. (1997).

22. I have discussed the iron hypothesis, proposed by J. H. Martin and Fitzwater (1988), in chapter 6. Regional and temporal variations in the availability of iron and silica to open-ocean ecosystems were discussed by Tyrrell (1999), J. Wu et al. (2000), Harrison (2000), and especially Sigman and Boyle (2000).

23. I suspect that the divergence dates based on molecular sequences of plants, fungi, and animals given by Heckman et al. (2001) are too early. Oldest trackways made by animals: R. B. McNaughton et al. (2002). Oldest fungi: Redecker et al. (2000). General accounts of the early history of the eukaryotic land biota: Shear (1991), Vermeij and R. Dudley (2000). Tetrapod vertebrates, known from the latest Devonian (Famennian stage) as the clade Ichthyostegalia, had reached the land no later than the Tournaisian stage of the Early Carboniferous. Clack (2002) interprets the genus *Pederpes* from the Tournaisian of Scotland as the oldest terrestrial tetrapod. Evidence from mitochondrial gene sequences indicates that springtails (Collembola) adapted to land independently from the other six-legged arthropods, the insects (Nardi et al. 2003).

24. Data from Vermeij and R. Dudley (2000). We do not know when various groups of nematodes, flatworms, annelids (earthworms), and filamentous green algae invaded the land. Hibbett and Binder (2001) report that the "higher" basidiomycete fungi—the group containing mushrooms—have given rise to three or four aquatic and marine lineages, often associated with mangrove trees.

25. Comparative productivity: Field et al. (1998).

26. For a summary of times and places of invasion, see Vermeij and R. Dudley (2000). It will be interesting to see if birds follow the same patterns as mammals. According to S. L. Olson (1985), auks (Alcidae) became marine in the productive North Pacific, but times and places of a transition from terrestrial to marine or from freshwater to marine habits in birds are not well known. Greene's (1997) data on snakes show that these reptiles produced marine clades largely in southeast Asia, where marine productivity is high though perhaps not as high as in adjacent forests.

27. Details are given in Vermeij (2001a).

28. This very rough calculation is based on Addicott's (1970) monograph.

29. Earliest cold-temperate hominids in North China: Zhu et al. (2001).

30. Regional differences in host-guest specialization: Vermeij (1983, 1989a). *Sabia*: Vermeij (1998a).

31. Interoceanic differences in shell architecture: Vermeij (1974, 1978, 1989a, 1993), E. C. Dudley and Vermeij (1989).

32. Coralliophilids: Lozouet and Le Renard (1998). *Sabia*: Vermeij (1998a).

33. J. C. Briggs (1967a,b) suggested that the Indo-West Pacific biota was the main exporter of species among tropical marine biotas. He based his conclusions mainly on patterns of diversity among fishes, and did not consider the fossil record. We confirmed Briggs' suspicions in an analysis of the fossil record of molluscs (Vermeij and Rosenberg 1993). At the time, we thought there was a single exception to the rule that no species had colonized the Indo-West Pacific from elsewhere, but Kowalke (2001) has shown that the Indian Ocean records of the snail *Potamides conicus*, which occupies hypersaline lagoons, are probably all due to human introductions from the Mediterranean region. I have slightly modified estimates of the number of invading species from those in Vermeij and Rosenberg (1993) in the light of new evidence. The status of the Indo-West Pacific as the primary donor biota may be relatively recent, because pre-Oligocene faunas there were small compared to those in the Mediterranean region and tropical America (see Vermeij 2001a).

34. Data are modified from Vermeij and Rosenberg (1993). It is very likely that some habitats are more affected by invasion than others. Immigrants to West Africa, for example, seem to be concentrated among hard-bottom species. Future work on these interesting patterns must take the ecological context into account.

35. In my original paper on the trans-Arctic interchange of species between the North Pacific and North Atlantic (Vermeij 1991a), I listed 34 molluscs invading to the Pacific and 265 species (including descendant species) invading the Atlantic. Given the minute size, poorly known fossil record, and taxonomic uncertainties for several groups, ten of the invaders to the Pacific are suspect, so that I would now estimate that only 24 species (including descendants) took part in the Atlantic-to-Pacific expansion. Gladenkov and colleagues (2002) have shown on the basis of the earliest Atlantic bivalves (species of *Astarte*) and diatoms in the Pacific that the Bering Strait opened in the latest Miocene (5.5 to 5.4 Ma), earlier than I and others had thought. They noted that invasion to the Pacific began at least 0.7 my earlier than invasion to the North Atlantic (see also Marincovich and Gladenkov 1999; Marincovich 2000). Although Marincovich (2000) suggested that this pattern of invasion reflects a reversal in the direction of the prevailing current through Bering Strait at 3.5 Ma, a reversal resulting from the diversion of currents northward once the Isthmus of Panama closed, an additional factor may be the substantial Pliocene extinctions that affected the Atlantic but not the Pacific temperate biota (Vermeij 1991a,b). There would have been less resistance from incumbents in the postextinction North Atlantic than in the North Pacific.

36. These data are from Vermeij (1991b). All other cases of invasion, including those on land to be discussed shortly, conform to this pattern.

37. Beard (1998) made a strong case for Asia's being the place of origin of many of the major clades of modern mammals. Mounting evidence indicates that Asia's dominance applied even during the Cretaceous, when many groups of dinosaurs also extended from their place of origin in Asia to other continents, including North America.

38. Great American interchange: general papers by L. G. Marshall (1981) and S. D. Webb (1991); also see Stehli and Webb (1985). Geographic distribution of mammalian horns and antlers: Janis (1982). Bone-crushing carnivorous mammals: Van Valkenburgh (1999). Geography of running speed in mammals: data from Garland (1983a).

39. For a general account of island extinctions, see Diamond (1984). Extinctions of birds in Oceania: Steadman (1995). Role of rats in island extinction: Atkinson (1985). Extinction of plants on Ascension Island: Cronk (1980). Extinction of palm on Easter Island: Dransfield et al. (1984). Extinctions of reptiles on islands: Honegger (1981). Land snails on Moorea: B. Clarke et al. (1984). Role of disturbance in extinction: Diamond and Veitsch (1981).

40. See Diamond (1996). Cultural factors do, of course, influence patterns of expansion as well. European expansionism beginning in the late eleventh century was built on military traditions of annihilating the enemy with overwhelming, organized force, traditions that according to Hanson (2001) were inherited from the Greeks.

41. Africa as home to competitive species: Janzen (1976). Early technological advances by humans: Ambrose (2001). Probable first use of fire in cooking: Wrangham et al. (1999), Wrangham (2001).

42. In my synthesis paper on the evolution of the labral tooth in gastropods (Vermeij 2001b), I suggested that the tooth evolved fifty-eight times. New information has forced me to revise this number upward slightly to sixty. This revision strengthens the conclusions I reached in the 2001b paper.

43. I explored the contrasting pattern of first appearance and distribution of clades with a labral tooth and clades in which lefthandedness arose in my 2002c paper. This paper also presents a statement of the hypothesis of evolutionary opportunity and reports the first comparative test of that hypothesis.

44. The tropics as the source of innovation: Jablonski (1993). Environmental origins of cheilostome bryozoans: McKinney (1995), Lidgard (1990), Jablonski et al. (1997).

45. Des Marais and colleagues (1992) and Des Marais (1994) were early champions of the idea that Earth's early biotic history, as well as the history of atmospheric oxygen, were driven by tectonics. See also Knoll and Canfield (1998) for a general biotic history of the Archean and Proterozoic eons, and Anbar and Knoll (2002) for the link between nutrients and life in the Mesoproterozoic era. I discuss Anbar and Knoll's intriguing hypothesis further below.

46. Anbar and Knoll (2002) suggested that the Mesoproterozoic era, from 1.8 to 1.25 Ga, was a time of generally low primary productivity enforced by global limits on the supply of iron, molybdenum, and other metals crucial for the construction of enzymes required by prokaryotes to convert inorganic nitrogen into forms of nitrogen that organisms can use. Earlier, Brasier and Lindsay (1998) had also identified a long interval in the Mesoproterozoic as a time of remarkable stability, which they attributed to a relative tectonic quiescence. Moores (1993, 2002) argues that ocean crust during the Archean and to a lesser extent the Early and Middle Proterozoic was thicker than today, and that this thick oceanic crust could not be as readily subducted into the Earth's mantle as was the case when the crust beneath the oceans became thinner, beginning about 1.0 Ga. The reasons for the thinning ocean crust are unclear, but if thinning did indeed occur, it is possible that tectonic activity intensified during the Neoproterozoic and Phanerozoic eras, stimulating primary production and all the other economic activities that depend on the conversion of tectonically provided nutrients into biomass.

47. Rise of oxygen after 800 Ma: Kadko et al. (1995), Canfield (1998), Knoll (1996). Diversification of life at this time: Knoll (1994), Karlstrom et al. (2000).

48. Brasier and Lindsay (2001) offer an excellent account of the environmental and geographic history of the latest Neoproterozoic and Cambrian. For additional important perspectives, see Pelechaty (1996), Condie (1998), and Waggoner (1999). P. F. Hoffman and colleagues (1998) document several ice ages, which in their view were so extensive that ice covered much of the tropical belt. Although I remain skeptical about the equatorial extent of ice during the pre-Ediacarian Proterozoic (see also Pierrehumbert 2002), this period in Earth history does seem to have been one of wide fluctuations in temperature, including episodes of widespread glaciation.

49. Hallam (1994) gives a comprehensive account of geographic changes during the Paleozoic, as well as a history of sea level. In chapter 10 I return to ecological expansion during the Late Cambrian to Middle Ordovician interval.

50. Biological characterization of the Middle Paleozoic: Signor and Brett (1984). Hallam (1994) emphasized continental collisions for the Middle Paleozoic, but Racki (1998) noted important rifting during the Late Devonian and Early Carboniferous. The earliest charcoal deposits are documented by Cressler (2001).

51. Late Paleozoic geographic events: Hallam (1994). Cleal and colleagues (1999) documented the rise in carbon dioxide for the latest Carboniferous. For a further discussion of evolutionary events in the Late Paleozoic, see chapter 10. Saltzman (2003) proposed the hypothesis that interruption of circumequatorial ocean currents by a land barrier in the Middle Carboniferous led to the glacial episodes of the Late Paleozoic, which are best documented from the southern continent of Gondwana. He suggested that this geographic explanation for ice ages holds also for the glaciation of the past 2.5 my, which according to this hypothesis would be initiated by diversion of circumequatorial currents by the formation of the Central American isthmus (see also Weyl 1968). Ice would form at high latitudes as warm water and warm air rich in water vapor are forced away from the equator and cause an increase in precipitation near the cold poles.

52. Triassic geographic events: Veevers (1989, 1990), Hallam (1994). The uniformity of the world's tetrapod vertebrate fauna was documented by Shubin and Sues (1991). See also Sereno (1999) for the geography of Late Triassic dinosaurs.

53. Hallam (1992, 1994) has ably summarized changes in geography and sea level for the Jurassic. W. W. Hay and colleagues (1988) used the distribution of limestones to estimate the latitudinal extent of warm marine conditions. As was the case in the mid-Cretaceous, to be discussed below, the Middle Jurassic (Callovian to Oxfordian stages, 165 to 160 Ma) was marked by high rates of production of crust, underpinning the spread of warm conditions and the rise in sea level at this time (Sheridan 1997). Osborne and Beerling (2002) discuss the effects of high levels of carbon dioxide on the stature of conifers in the later Mesozoic. Whether the exceptionally large size of dinosaurian herbivores is directly related to the prevalence of exceptionally tall forests is unclear.

54. Coffin and Eldholm (1993, 1994) surveyed episodes of undersea volcanism. Further details are available for events of the Jurassic-Cretaceous boundary (Sager and Han 1993), Valanginian (Föllmi 1995; Gale 2000), Barremian to Early Aptian (R. L. Larson 1991a,b; Tarduno et al. 1991; Gale 2000), Albian (Bercovici and Mahoney 1994; Föllmi 1995; Gale 2000), and Turonian to Coniacian (Tarduno et al. 1998; Hoernle et al. 2002). For the history of sea level, see Haq et al. (1987) and Hallam (1992). Landis and colleagues (1996) and I (Vermeij 2003) suggest that oxygen concentrations increased in the atmosphere even as carbon dioxide levels also rose during the tectonically most active interval of the Cretaceous. Many authors have inferred that tropical sea surface temperatures for the Mid-Cretaceous were higher than equatorial temperatures today (Coffin and Eldholm 1994; Sellwood et al. 1994; R. D. Norris and Wilson 1998; P. A. Wilson and Norris 2001; Pierrehumbert 2002). The latitudinal extent of carbonate deposition was estimated by W. W. Hay et al. (1988). These estimates are in line with other indicators for relatively warm climates at high latitudes during much of the Cretaceous (D'Hondt and Arthur 1996; Tarduno et al. 1998).

55. For summaries see Condie (1998) and Gale (2000).

56. Summaries of Cretaceous biological events: Vermeij (1987, 2003, 2002b). Kemper (1982) described the great changes that took place in the Early Cretaceous plankton and connected these with the enlargement of warm, productive seas. Leckie and colleagues (2002) provide a highly detailed account of the relationship among volcanism, nutrients, and evolutionary events in the mid-Cretaceous plankton. The ancient and single origin of army ants was documented from molecular evidence by Brady (2003).

57. Dates of eruption and sizes of submarine volcanic deposits are from Wignall (2001), supplemented with data from Coffin and Eldholm (1993). Effects of mountain-building in Asia and South America during the Late Cenozoic: Filippelli and Delaney (1994), Filippelli (1997), Raymo and Ruddiman (1992).

58. Summary of Cenozoic climates: Zachos et al. (2001). Warm spells in the Cenozoic seem to have been associated with unexpectedly low concentrations of carbon dioxide (Pagani et al. 1999a,b; Pearson and Palmer 1999, 2000), though glaciations during the Pleistocene had still lower concentrations (Raymo et al. 1996; Petit et al. 1999). The only period of markedly high carbon dioxide concentration occurred at and perhaps just after the Paleocene-Eocene boundary

(McElwain 1998), probably resulting from a mass release of methane which quickly oxidized to carbon dioxide and water.

59. Cenozoic biotic events: Vermeij (1987, 2001b, 2002b).

60. See Kirchner (2002). For a further discussion of triggers, see chapter 9.

61. I proposed this idea of increasing control of innovation in my 1995 paper. Students of the human economy (Mokyr 1990, for example) are similarly inclined.

62. Sandoval and colleagues (2001) carried out a global analysis of evolution in ammonoids for the Jurassic, as well as a more detailed study of evolution in this group in relation to sea level on the Iberian peninsula. Cornette and colleagues (2002) established the relationship between genus-level diversification of marine animals and the concentration of carbon dioxide.

63. Archaeologists and historians who have linked events in human history to climate have almost always emphasized the calamities wrought by droughts, cold spells, rainy growing seasons, floods, and the like (see, for example, Lamb 1982; deMenocal 2001; Fagan 2001; Weiss and Bradley 2001). The likely explanation is that both the triggers and the short-term effects are readily identified, just as are the disruptions and their extinction effects in the history of life (see chapter 9).

64. See also Mokyr (1990). Historians such as Hanson (2001) have derided Diamond's (1997) view that geography played a large role in determining which peoples became dominant in human history, but I think his and others' doubts about Diamond's hypothesis can be reconciled by arguing that internal dynamics related to social permissiveness and the role of scientific research have become increasingly important as humanity transformed itself from a tribal species, in which hunting and gathering were central economic activities, to an agricultural species, and finally into one in which urbanization, industry, and global trade have become the norm.

Chapter 9
The Role of Disturbance

1. Disturbances following a power law: Raup (1992), W. Alvarez and Muller (1984), Kauffman (2000).

2. The first quantitative description of the five post-Cambrian mass extinctions was by Raup and Sepkoski (1982). Zhuravlev and Wood (1996) identified the Late Early Cambrian extinction as a sixth possible Phanerozoic event. Hallam and Wignall (1997) offer a fine descriptive account of these Phanerozoic mass extinctions and of some of the minor events. Genus-level quantitative descriptions of marine extinctions are given by Sepkoski (1990, 1993). Plant extinctions (Knoll 1984) and crises for vertebrates (Benton 1985, 1989) and insects (Labandeira and Sepkoski, 1993) round out the picture for the Phanerozoic eon. The gigantic size of some insect families—forty thousand living species of Curculionidae (weevils), for example—and the general arbitrariness of the notion of family and genus make direct comparisons among plants, vertebrates, and insects difficult to make. In particular, the insect record may be more volatile—and thus more like that of vertebrates—than the currently available family-level analysis would indicate. The recognition of pre-Cambrian mass extinctions has only very recently come about (Amthor et al. 2003; Gray et al. 2003). More events will doubtless be documented than are currently known.

3. Excellent accounts of Late Pleistocene extinction on islands and continents are given by P. S. Martin and Steadman (1999) for the world as a whole, P. S. Martin (1984) for North American mammals, Steadman and Martin (1984) for North American birds, Flannery and Roberts (1999) and Roberts et al. (2001) for Australia, and Burney (1999) for Madagascar. Extinctions since 1600 are reviewed by Diamond (1984) for species on land, Carlton et al. (1999) for marine species, and I. J. Harrison and Stiassny (1999) for freshwater fish. As these studies show, small mammals and birds suffered no extinctions on continents during and after the Late Pleistocene. In the sea, the human-caused extinctions are concentrated among large warm-blooded vertebrates.

Perhaps three molluscs, found in coastal seagrass meadows and salt marshes, have also died out, in part because of habitat destruction and, in the case of the eastern American seagrass limpet *Lottia alveus*, because of a devastating eelgrass disease whose microbial cause may be linked to invasion by a pathogen from elsewhere (Carlton et al. 1991, 1999). Humans are in any case implicated directly or indirectly in all these cases. Until recently it was thought that the Australian megafaunal extinction occurred some 9 my after the arrival of humans in that continent, but reevaluation of the oldest human-occupied site in Australia now shows that the extinction of species and the arrival of humans are statistically coincident in time (Bowler et al. 2003).

4. Zimov and colleagues (1995) described the vegetational changes wrought by the disappearance of Beringian Late Pleistocene mammals. Zazula and colleagues (2003) provide details of the species composition of the Late Pleistocene steppe vegetation in eastern Beringia.

5. This account follows the well-informed speculations of Janzen and Martin (1982). Extinction of the fruit-eating dodo, a large flightless pigeon in Mauritius, and of specialized pollinating birds in the Hawaiian Islands, has similarly endangered native plant species dependent on these animals for reproduction and dispersal (Temple 1977; Cox 1983).

6. Domning (1978) gave a detailed account of the evolution, anatomy, and ecology of Steller's seacow. J. A. Estes and colleagues (1989) explored some of the ecosystem-level changes wrought by the disappearance of this herbivore. Important papers on sea otters and their ecosystem-level effects include those by Estes and Palmisano (1974), Simenstad et al. (1978), and Estes et al. (1998).

7. J.B.C. Jackson (1997) gives a riveting and well-documented account of the history of Caribbean reefs and their exploitation since the voyages of Columbus. T. P. Hughes (1994) gave additional details about the transformation of coral-dominated to seaweed-dominated West Indian reefs. Lessios and colleagues (2001) showed that the sea urchin *Diadema antillarum* has been an abundant grazer on Caribbean reefs for at least the last one hundred thousand years despite the abundance of large predators and competitively superior larger grazers.

8. Silliman and Bertness (2002) carried out their experiments in salt marshes on Sapelo Island, Georgia. My own data show that the periwinkle, with a thick shell and the ability of the foot to withdraw deeply into the shell, is frequently broken and repaired, suggesting that the population control exercised by shell-crushing blue crabs falls largely on young, predation-vulnerable snails. The snails gain some protection from crabs as well as from predation by the snail *Melongena corona* by climbing grass stems (Warren 1985).

9. Changes in the Lago Guri islands are documented by Terborgh et al. (2001). S. J. Wright (1979), Karr (1982a,b), and Leigh and colleagues (1993) have documented aspects of change of the biota of Barro Colorado Island and nearby islands in Gatun Lake, Panama. Tiny forested islands near Barro Colorado, for example, host mainly lizards as insectivorous vertebrates, the insectivorous birds of larger forest tracts having been largely eliminated as deaths are not being replaced by colonists. Despite their ability to fly, many forest birds are loath to cross narrow water barriers (Wright 1979; Diamond 1973, 1974).

10. For extinction caused by competition in the broad sense, see Darwin (1859), Van Valen (1973), and Maynard Smith (1989). The Bahamian work on lizards was carried out by Schoener et al. (2001) as part of a comparative study of populations of plants, spiders, lizards, and other animals on islands of various sizes and biological make-up. Intriguing as these results are, I am skeptical that they can be scaled up to environments the size of continents or marine biogeographic units. The islands Schoener and colleagues studied are small and relatively uniform, unlike larger land masses and water bodies. Experimental ecology is, by its nature, more or less limited to comparisons among manipulated and control communities of very small size (microcosms and mesocosms), and for this reason is generally bedeviled by the uncertainties associated with scaling up the findings to larger, more complex, and biologically much more diverse ecosystems.

11. Fox (1981) and Fox and Morrow (1981) observed that most insect species that live locally on only one host species have several hosts when their entire distribution is taken into account.

J. N. Thompson similarly shows that host preference varies geographically and according to habitat type in many plant-eating and pollinating insects (Thompson 1994, 1998, 1999a,b). Novotny and colleagues (2002) showed that this lack of specialization even applies to forests in New Guinea, where an analysis of 900 species of herbivorous insects and 51 plant species revealed that herbivores tend to consume several related hosts, and that insect species specialized on a single host are in the minority. The interpretation that top-down changes in consumption bring about extinction only under unusual conditions is my own.

12. Plants on the Channel Islands off southern California have reduced antiherbivore defense over the course of the ten thousand years that the islands have been isolated from the mainland (Bowen and Van Vuren 1997). Sloths on the islands of Bocas del Toro on the Caribbean coast of Panama have also evolved rapidly (in this case to smaller body size) over a comparable or perhaps shorter interval (R. P. Anderson and Handley 2002), but the modifications have remained at or below the subspecies level.

13. Diamond (1984) has given a comprehensive overview of extinctions taking place during the past four hundred years. My conclusions are drawn largely from his careful analysis, supplemented by my own review of about fifty additional studies. Worthy and Holdaway (2002) give a persuasive account of the important role of introduced mammals, especially three European mustelids (weasels) and the Pacific and Norway rats, in dooming to extinction many small birds.

14. L. G. Marshall (1981) has made this circumstantial case, but, like me, he would probably be unwilling to submit it to a jury trial. Even so, it remains the strongest case for top-down extinction apart from the cases of the last ten thousand years in which humans are involved.

15. My summary is based on Van Valkenburgh's (1999) excellent documentation of the history of what she calls hypercarnivores, mammals that eat other, smaller carnivores as the main part of their diet.

16. Morgan (1995) has carefully reviewed the relationships among the food supply, survivorship, and settlement of planktonic larvae of marine animals. Valentine (1986) has particularly noted the vulnerability of species with such larvae to extinction during the catastrophe near the end of the Permian. Jablonski's (1986) data, however, show that snails with planktonically feeding larvae did not suffer a higher magnitude of extinction at the end of the Cretaceous than did snails whose larval stages were spent on the seafloor. Similarly, for sea urchins during this crisis, species with plankton-feeding larvae were not more prone to extinction than those without such larvae (A. B. Smith and Jeffery 1998; Jeffery 2001).

17. This analysis was done by Kitchell and colleagues (1986). See also Thierstein (1982).

18. For additional data in support of the hypothesis that bottom-up crises placed the most powerful entities at greatest risk, see Vermeij (1987, 1999). Several authors have expressed skepticism about this conclusion. In their analysis of extinction patterns during the end-Permian crisis, Knoll and colleagues (1996) argue that groups with a very high tolerance for carbon dioxide in the blood were less prone to extinction than those with a low tolerance. The high-tolerance group, including land vertebrates and fish, often have a high metabolic rate, and therefore have greater power than the low-tolerance groups, including brachiopods, echinoderms, and bryozoans. I would note, however, that within the high-tolerance group it is the largest and most powerful members that suffered most. Moreover, the same pattern can be explained, more satisfactorily I think, by the inability of vulnerable groups to shut down their metabolism during unfavorable times, whereas the less vulnerable groups are able to shut down or to pass periods of inclemency in an inert resting stage. Kelley and Hansen (1996) and Hansen and colleagues (1999) analyzed molluscan assemblages before and after the end-Cretaceous and end-Eocene extinction events, and found that antipredatory armor—thick shells, presence of shell-strengthening external sculpture, and narrow apertures—was more prevalent in the assemblages following the crises than in those before, or there was no difference. These analyses are flawed, however. A test of the hypothesis that more heavily defended species, or more powerful species, are more vulnerable to extinction during a crisis requires a comparison of assemblages before the crisis with the survivors from that assemblage after the crisis. Kelley, Hansen, and their colleagues instead included species in

their postcrisis assemblages that had either newly evolved or newly arrived. They therefore conflated extinction with rediversification and invasion. It may be true that passive armor, insofar as it protects against enemies as well as against hostile chemical and physical conditions, does confer an advantage during crises.

19. The volume edited by Glynn (1990) presents an excellent overview of the causes of the biological effects of the 1982–1983 El Niño event, one of the most intense of the last four hundred years in the eastern Pacific. My chapter in that volume (Vermeij 1990a) makes the case that few extinctions resulted from El Niño events in that part of the world during the last few million years.

20. The interpretation of carbon-isotopic ratios is subtle and fraught with problems. Good accounts are given by Brasier (1995), Knoll and Canfield (1998), and Beerling et al. (2002).

21. For a general summary of extinctions, see Hallam and Wignall (1997). Tappan and Loeblich's (1973) account of the history of phytoplankton still stands as the most accessible summary of extinction in this group despite later additions to our knowledge. Direct data from phytoplankton abundance, as well as inferences from carbon-isotopic shifts, strongly support catastrophic reductions in planktonic productivity during crises at the end of the Ordovician (Colbath 1986; Wang et al. 1993), Late Devonian (McGhee 1988; Racki 1998), end-Permian (Bowring et al. 1998), end-Triassic (Ward et al. 2001), and end-Cretaceous (Hsü et al. 1982; Kitchell et al. 1986; M. A. Arthur et al. 1987). The general case has been made by Hsü (1986) and Sheehan and Hansen (1986).

22. Patterns of survival among corals: Rosen (2000).

23. Plant extinctions have been documented for the Late Carboniferous (Labandeira and Phillips 2002), end-Permian (Retallack 1995; Looy et al. 2001; Ward et al. 2000; Rees 2002), end-Triassic (McElwain et al. 1999), and end-Cretaceous (Tschudy et al. 1984; Saito et al. 1986; Wilf et al. 2001). Based on stage-level regional to global analyses of Permian and Triassic floras, Rees (2002) expresses doubts about the reality of a large-scale end-Permian extinction event among plants. The decreases in plant diversity across the Permian-Triassic boundary are, in Rees's view, better explained by the northward movement of the supercontinent Pangaea into climatic zones that support fewer species, and by artifacts of sampling. To me, the end-Permian changes in local floras from widely separated parts of the world imply catastrophic reductions in diversity and productivity.

24. In his analysis of extinction among bivalves during and after the Pliocene in the eastern United States, Stanley (1986) found no difference in susceptibility between large and small species. More detailed studies within clades, however, have revealed that large-bodied species were more prone to extinction, and that surviving large-bodied species often became smaller after extinction events. Examples include Roopnarine's (1996) study of the venerid bivalve *Chione*, L. C. Anderson's (2001) work on corbulid bivalves, and Allmon's (1992b) study on suspension-feeding turritellid gastropods. Allmon and colleagues (1996) found evidence for decreased upwelling and productivity after the Pliocene on the west coast of Florida, and documented many extinctions among cormorants, seals, and other marine vertebrates typically associated with food-rich waters. Domning (2001a) argued convincingly that the disappearance of various herbivorous seacows in the Caribbean after the Pliocene is consistent with the extinction of several seagrasses. Decreases in the numerical abundance of suspension-feeding bivalves over the last 10 to 12 my on the Caribbean coast of Central America are most easily explained by a decrease in planktonic productivity, especially after the early Pliocene (Todd et al. 2002). There is also geographic evidence in support of a general decrease in Caribbean planktonic productivity after the early Pliocene. Clades with a broad distribution in Atlantic and Pacific tropical America became restricted either to the planktonically productive eastern Pacific and/or to productive Atlantic Central and northern South America. This restriction was recognized first by Woodring (1966) and later interpreted as a response to declining productivity in the insular Caribbean by Vermeij (1978, 2001a), Vermeij and Petuch (1986), Allmon (1992b), and Roopnarine (1996).

25. Effects of stress on the algal symbionts of corals: Rowan et al. (1997). The activities of grazers are important for the maintenance of corals by eliminating settlement and growth of algae; see, for example, Wonders (1977).

26. For versions of this argument see Hallock and Schlager (1986), Birkeland (1989), Caplan et al. (1996), and Allmon and Ross (2001).

27. The greater species richness of reefs along the shores of continents and high islands with prolific nutrient input into nearshore waters has long been known, and was particularly well documented by Birkeland (1989) and Allmon and Ross (2001). Rosen (2000) noted that many photosynthesizing corals live in turbid waters and often do not form reefs. For descriptions of reefs in Panama and estimates of annual removal of coral biomass, see the excellent paper by Glynn et al. (1972).

28. Allmon and Ross (2001) recognized the nutritional subsidy of outer reefs by ecosystems closer to shore.

29. For the causal relationships among oxygen, excessive nutrient supply, and extinction, see Magaritz (1989), R. E. Martin (1995, 1998), Hallam and Wignall (1997), Racki (1998, 1999), Crouch et al. (2001), Wignall and Hallam (1992), and Wignall and Twitchett (1996).

30. See Amano et al. (1996) for the extinctions in the Japan Sea. Sheehan and colleagues (1996) have argued the case for anoxia in epicontinental seas of the Paleozoic.

31. The mass sinking of diatom mats beneath relatively unproductive waters has been documented by Kemp et al. (1999), Villareal et al. (1999), and Sachs and Repeta (1999). Brongersma-Sanders (1957) argued long ago that sediments beneath productive upwelled waters record mortality more than faithfully indicating high productivity, but her point seems to have been ignored subsequently.

32. E. P. Odum (1969) forcefully argued that high productivity leads to low species diversity. See also Rosenzweig and Abramsky (1993), Huston (1993), and McGowan and Walker (1993). For the stimulatory effect of high nutrient supply and high productivity on the number of co-occurring species, see Vermeij (1978, 1990b).

33. Details are given by Worm et al. (2000, 2002) and Hillebrand et al. (2000). As these authors note, it is misleading to consider consumption, nutrient enrichment, and other ecosystem-wide phenomena in isolation.

34. Estimates of the duration of the initial phases of the extinction crises depend on calculations of sedimentation rate in a pile of sediment that has been deposited more or less continuously and at the same rate over the time interval encompassed by the event. Estimates run from a maximum of 50 ky to as short as 8 ky for the initial collapse of the marine and terrestrial ecosystems at the ends of the Permian (Jin et al. 2000; Rampino et al. 2000), Triassic (R.M.H. Smith and Ward 2001; Ward et al. 2001) and Cretaceous periods (D'Hondt et al. 1998; Mukhopadhyay et al. 2001). In fact, D'Hondt and colleagues note that minute phytoplankters such as Cyanobacteria have doubling times of as little as six days, with the implication that phytoplankton production could have recovered very quickly indeed.

35. Sutherland (1974) was one of the first ecologists to point out that ecosystems can switch between stable states. An excellent comprehensive review and synthesis of this interesting topic was recently published by Scheffer and colleagues (2001), who, besides providing details about the lakes and the Sahel ecosystem I discuss here, also discuss transitions between forest and grassland and between alga-dominated and coral-dominated tropical reefs, as well as theoretical models that emphasize the role of feedbacks in these switches. The switches themselves have often been described as "phase changes" analogous to the transitions between solids and liquids or between liquids and gases. For a wonderfully readable account of the long history of turmoil in the Church before the Reformation, see Bobrick (2001). For military history leading up to World War I, see McNeill (1982).

36. For the end-Permian scenario, see Knoll et al. (1996). For end-Triassic plants, McElwain and her colleagues (1999) suggest that warming resulting from release of copious carbon dioxide

into the atmosphere was responsible for extinction, which was most severe among large-leaved species. Methane release would have had similar effects.

37. See Vermeij (1987) and Dorritie and Vermeij in R. E. Martin et al. (1996).

38. Griffis and Chapman (1988) carried out experiments on a variety of species of phytoplankton in order to assess the susceptibility of actively metabolizing cells to prolonged darkness. Given that inclement seasons never last more than a year in all but a few desert environments, no primary producer on Earth can be expected to tolerate darkness as an active organism for longer than a year. The only way to survive longer periods is in an inert state such as a seed, cyst, underground tuber or bulb, or spore.

39. MacLeod (2003) analyzed the timing of volcanic eruptions in relation to the mass extinctions. In his comprehensive account of this topic, Wignall (2001) suggested that, although land-based eruptions coincide with many of the great crises, not all eruptions lead to extinction, and not all extinction can be blamed even in part on volcanic activity. The following account draws on his summary and other recent findings.

40. Soden and colleagues (2002) document the effects of Mount Pinatubo's eruption on world climate. Handler (1989) suggested that eruptions at low latitudes more effectively distribute aerosols around the world and therefore have more potent climatic effects than higher-latitude explosive eruptions. The Pleistocene eruption of Mount Toba has been characterized by Rampino and Self (1992), Erwin and Vogel (1992), and Wignall (2001). The comparably large Ordovician eruption was described by Huff et al. (1992). There is broad agreement among authors that these huge eruptions did not precipitate extinction events recognizable in the fossil record. Zielinski and colleagues (1994) documented forty-nine explosive volcanic eruptions in the last two thousand years, recorded as ash layers in ice cores in Greenland. The largest of these was the 1815 eruption of Mount Tambora on Sumatra. Links between these eruptions and climatic alterations large enough to perturb human civilization are tenuous at best.

41. Sigurdsson (1990) and Wignall (2001) ably summarized the characteristics and effects of the Laki Fissure eruptions in Iceland.

42. Estimates of quantity of droplets and dust released into the atmosphere by the end-Cretaceous impact: Pope (2002). In Pope's view, dust contributed only about 1 percent of this material and was not the primary shield preventing the sunlight from reaching Earth's surface. The idea that methane contributed to end-Paleocene extinction (Dickens et al. 1995, 1997) has been strongly supported by subsequent stratigraphic work and by quantitative assessments of the methane released (Bains et al. 1999, 2000; R. D. Norris and Röhl 1999). Sediment slumping associated with methane release has been proposed for the end-Cretaceous extinction (Max et al. 1999; R. D. Norris et al. 2000a). Estimates of annual volcanic and human-caused releases of carbon: Sigurdsson (1990), Wignall (2001). End-Permian release of carbon as methane: Berner (2002). Mechanisms and general consequences of methane release: Ryskin (2003).

43. Speculations about the link between impacts and large-scale volcanism (Oberbeck et al, 1993) are supported by Glikson's (1999) calculations of shock energies released by large-body impacts. Wignall (2001) remains skeptical of this link, but my guess is that a close causal and temporal association among major volcanism, impacts, methane releases, and global extinctions will hold up as more precise ages for the various events become available. On the basis of his modeling of the Permian and Triassic carbon cycle, Berner (2002) is also inclined to the view that impacts caused the end-Permian extinction and triggered volcanism and the eruption of methane into the ocean and atmosphere through dissolution of methane-hydrate crystals.

44. Amthor and colleagues (2003) characterized the mass extinction of 580 Ma and documented its temporal coincidence with the Acraman impact and with a worldwide negative shift in the carbon-isotopic ratio. The end-Neoproterozoic extinction has long been suspected (Hsü et al. 1985) but has become temporally better defined and documented only recently (Gray et al. 2003). The Sinsk event is one of two episodes of extinction near and at the end of the Early Cambrian (Zhuravlev and Wood 1996; Hallam and Wignall 1997). The age of the Antrim Plateau volcanism was quoted by H. Kimura and Watanabe (2001).

45. For summaries of end-Ordovician events, see Sheehan et al. (1996), Sheehan (2001), and Hallam and Wignall (1997). The upwelling hypothesis was proposed for extinctions in general by Berry and Wilde (1978) and Wilde and Berry (1984), and for the end-ordovician by Wang et al. (1993). Mountain building in Europe and North America began during the Late Middle Ordovician. If it led to increasing rates of weathering, reduced carbon dioxide concentration in the atmosphere, glaciation in the southern hemisphere, and global extinction, as argued by Kump et al. (1999) and Sheehan (2001), some critical threshold must have been crossed that unleashed the crisis. Colbath (1986) documented the crash of the phytoplankton at the end of the Ordovician.

46. The best general discussion of the Late Devonian extinctions is by Hallam and Wignall (1997). Joachimski and Buggisch (1993) and Racki (1998), among others, emphasized the role of oxygen-poor waters invading normally oxygen-rich regions of the ocean. Claeys and colleagues (1992), Wang (1992), and Wang and colleagues (1991) have linked Late Devonian microtectites and other evidence of debris from impacts to the extinctions, but Racki (1998) believes the horizons in which these sedimentary indicators are preserved do not coincide with the horizons chronicling extinction. The speculation about vaporized limestone is my own. Ellwood and colleagues (2003) established contemporaneity between an extinction event at 380 Ma, near the end of the Eifelian, and an impact in Morocco, as well as with a negative shift in the carbon-isotope ratio.

47. Wignall (2001) discussed the Panjal and Emesian volcanics and their relation to Middle Permian extinction; see also Bakker (1977) for a characterization of extinction among large vertebrates on land.

48. For recent accounts of the end-Permian extinction, see Knoll et al. (1996) and Bowring et al. (1998). Coincidence between extinction and eruption of the Siberian Traps was proposed by Campbell et al. (1992) and has generally been confirmed by later work, although Wignall (2001) believes that the bulk of the volcanism may have come slightly later than the extinction. Reichow and colleagues (2002) provide the most recent estimate of the volume of the Siberian Traps. Methane release at the end of the Permian may be connected to the extinction, but questions about timing and mechanism remain (Vermeij and Dorritie in R. E. Martin et al. 1996; Wignall 2001).

49. L. Becker and colleagues (2001) have presented circumstantial evidence for an impact by noting the presence of noble gases, presumably of extraterrestrial origin, in fullerenes found in end-Permian deposits. In their careful dissection of end-Permian events, Bowring and colleagues (1998), as well as other authors (Jin et al. 2000; Berner 2002), have stressed the likelihood that celestial-body impacts could be responsible for all the end-Permian calamities. The most compelling evidence for an impact comes from precisely dated meteorite fragments in the Antarctic (Basu et al. 2003).

50. Marzoli and colleagues (1999) documented the timing and extent of volcanism in the Central Atlantic Magmatic Province. Using Milankovitch cycles of sedimentation to calibrate sedimentary deposition in lakes, P. E. Olsen and colleagues (2002) determined that the end-Triassic extinction of land plants and dinosaurs preceded the earliest indication of volcanic eruption in the Newark Basin by twenty thousand years, but in England and Greenland, extinction and the onset of volcanism seem to have occurred simultaneously (Hesselbo et al. 2002; A. S. Cohen and Coe 2002). The Queen Charlotte Islands sequence yielded the estimate of the timing of the marine extinction (Pálfy et al. 2001). The suggestion that productivity at the end of the Triassic declined precipitously (Ward et al. 2001) was based on a sharp decrease in the heavy to light carbon-isotopic ratio. Large-scale methane release may also explain this isotopic pattern (Hesselbo et al. 2002) but is not at odds with reduced productivity. At the Triassic-Jurassic boundary, the density of stomates in tree leaves fell sharply, an observation that McElwain and colleagues (1999) interpret as an indication of suddenly higher levels of carbon dioxide in the atmosphere brought about by, or leading to, warming. Impacts of extraterrestrial bodies during and at the end of the Triassic have been discussed by Hodych and Dunning (1992) and Spray et al. (1998).

51. Extinction and volcanism in the Early Jurassic: Pálfy and Smith (2000). Methane release at this time: Hesselbo et al. (2000).

52. The Paraná (or Serra Geral) and Etendeka flood-basalts were characterized by Renne and colleagues (1992). Early Aptian volcanism: R. L. Larson (1991a,b), Tarduno et al. (1991); Turonian to Coniacian volcanism: Storey et al. (1995), Tarduno et al. (1998). Gale (2000) presents an excellent summary of Cretaceous climates and physical history. Leckie and colleagues (2002) provide a detailed account of extinction, volcanism, oceanographic conditions, and climate for the mid-Cretaceous period (Aptian through Turonian). My summary largely follows their synthesis. However, Leckie and colleagues attribute many of the minor extinctions to the periodic imposition of anoxia, increased productivity, and the upwelling-related elimination of distinct depth zones each inhabited by specialized plankton in the ocean, or to some combination of these interrelated factors. As I noted earlier in the chapter, the fossil record is insufficiently detailed to allow a chronology of climate and of episodes of species loss and gain. Leckie and coworkers treat extinction and species formation together as species turnover, and look to oceanographic conditions to explain these changes in the oceanic plankton. Theory suggests, however, that times of species formation are distinct from (and usually follow) times of extinction, and that the conditions providing evolutionary opportunity—high productivity, warm climate, elimination of incumbents, etc.—differ from the productivity-disrupting circumstances that mark extinction events. Paleoceanographers will, I think, have to look beyond conditions and stratification in the ocean's circulation to uncover causes of extinction in the plankton.

53. Deccan Traps volcanism: Courtillot et al. (1986), Coffin and Eldholm (1994), Rarizza and Peucker-Ehrenbrink (2003). Officer and Drake (1985) and Officer and colleagues (1987) proposed volcanism as the chief cause of the end-Cretaceous mass extinction; see also D. M. McLean (1985) and M. A. Arthur et al. (1985). A link between flood-basalt volcanism and extinction and extinction was emphasized by Rampino and Stothers (1988), Rampino and Caldeira (1993), and Olsen (1999), among others. Wignall (2001) favors the Yucatan impact as the direct trigger of the extinction. Sediment slumping associated with methane release has been proposed as part of the end-Cretaceous calamity (Max et al. 1999; R. D. Norris et al. 2000a). Beerling and colleagues (2002) inferred high levels of carbon dioxide just after the boundary, based on the analysis of stomates in the fern genus *Stenochlaena*, and plausibly suggested vaporized carbonates as the source of this gas.

54. North Atlantic flood-basalt volcanism: Coffin and Eldholm (1993), Wignall (2001). Dickens and colleagues (1995, 1997) proposed methane release as a cause of extinction among deep-sea foraminifers. The contribution of circum-Caribbean volcanism in the end-Paleocene crisis was emphasized by Bralower et al. (1997). The deep-sea extinction was documented by Kennett and Stott (1991). Warming and the biological response to it were chronicled by Kennett and Stott (1991), Rea et al. (1990), Wing (1998), Wing and Harrington (2001), Wilf and Labandeira (1999), Wilf et al. (2001), and Zachos et al. (1993, 2001). Gulf Coast molluscan extinctions were documented in detail by Dockery (1998).

55. Dating of the Ethiopian basalts: Hoffmann et al. (1997). For excellent accounts of later Eocene and Oligocene marine extinctions, see Hickman (2003) and Dockery and Lozouet (2003); for mammals in North America, see Prothero (1985) and Berggren and Prothero (1992). Late Eocene impacts: Poag et al. (2003). A good statement of the current consensus about the timing and causes of environmental and biological changes in the Late Eocene and Early Oligocene can be found in Ivany et al. (2003). Major Earliest Oligocene cooling brought on first by the opening of Drake Passage between South America and the Antarctic Peninsula about 37 Ma, and later by the separation of Tasmania from East Antarctica about 33.5 Ma, is generally held responsible for the extinctions and for the spread of cold-adapted marine species during this interval.

56. Columbia Plateau basalts: P. R. Hooper (1990), Coffin and Eldholm (1993), Wignall (2001). Petuch (1993) recognizes several Miocene extinction events in marine eastern North America. One occurred in mid-Serravallian time, more or less coincident with the Columbia

flood-basalts, but other regional events later in the Miocene are not known to correspond with any recognized physical triggers.

57. Opening of Bering Strait: Marincovich and Gladenkov (1999), Marincovich (2000), Gladenkov et al. (2002). The best documentation of the closing of the Central American seaway is by Cronin and Dowsett (1996). Gersonde et al. (1997) described the southern-ocean impact of a possible asteroid, and P. H. Schultz et al. (1998) documented an impact event in Patagonia during the Pliocene (3.3 Ma).

58. The examples are taken from DeMenocal (2001) and Weiss and Bradley (2001). For a detailed climatic record (from stalagmite growth in caves) in relation to Anasazi cultural change, see Polyak and Asmerom (2001). For the role of deforestation in exacerbating the effects of climate, see Perlin (1989). Lamb (1982) and Fagan (2000) provide accessible accounts of the Little Ice Age, with an emphasis on Europe.

59. Lewontin (1965) clearly identified the characteristics of initial colonists in newly created or greatly altered environments. Studies of colonization of real and artificial islands have identified various birds, corals, fast-growing colonial bryozoans, barnacles, ants, and land plants that are able to cross wide barriers of unsuitable habitat, either under their own power (birds) or by hitchhiking on floating wood, pumice, seaweeds, and ships, or on birds. Important accounts are by E. O. Wilson (1961) and Simberloff and Wilson (1969) for ants, Diamond (1974) for birds, J.B.C. Jackson (1977a,b) for marine colonial animals, Jokiel (1984, 1989) for corals attached to pumice, A. R. Smith (1972) for ferns, Heatwole and Levins (1972) for flowering plants, Wesselingh et al. (1999) for snails dispersing on birds, and Finlay and Clarke (1999) and Finlay (2002) for freshwater planktonic organisms. Abundance of ferns has been noted in sediments marking the Permian, Triassic, and Cretaceous mass extinctions (Tschudy et al. 1984; Looy et al. 2001). Polis and colleagues (1997) give an intriguing summary of the large quantity of insects carried around the world by high-altitude winds.

60. D'Hondt and colleagues (1998) point out that the faster response of phytoplankton than of larger consumers in the aftermath of the end-Cretaceous extinction created an ecosystem imbalance.

61. The impoverished marine and terrestrial communities of the Early Triassic have been characterized by Retallack et al. (1996) and Woods et al. (1999). All evidence points to a plankton-poor Early Triassic (Fensome et al. 1996). Bakker (1977, 1980) discussed impoverished Early Triassic vertebrate faunas.

62. Early Paleocene reef-building glass sponges were described from Alabama by Bryan and colleagues (1997). Brief numerical dominance of cyclostome over cheilostome bryozoans for an interval of about 2 my during the Early Paleocene was documented by McKinney et al. (1998). Rosen (2000) analyzed the pattern of survival and the history of reef-forming corals. R. D. Norris (1996) documented the first appearance of photosynthesizing foraminifers after comparable types disappeared at the end of the Cretaceous.

63. Depressed herbivory and plant diversity prevailed through much of the Paleocene (Labandeira et al. 2002) and in most places did not increase until the Early Eocene (Wilf et al. 2001; Jaramillo 2002). K. R. Johnson and Ellis (2002), however, described a diverse assemblage of rainforest species from Colorado fossilized just 0.4 my after the end-Cretaceous crisis. They did not comment on the intensity of herbivory in this community. For the history of mammal faunas after the Cretaceous in North America, see Van Valkenburgh (1999) and Janis (2000).

Chapter 10
Toward Greater Reach and Power

1. Popper (1964:vii). Popper did not seriously consider holistic principles, which predict and explain general consequences given specific initial conditions.

2. The two quotes are from pages 42 and 43, respectively, of Monod (1971). See also S. A. Kauffman (2000) for a similar perspective.

3. Gould (1989:25). W. B. Arthur (1989, 1999) has emphasized contingency in economics. He points out, for example, that many products and many routines, such as the layout of typewriter and computer keyboards, are determined very early in historically contingent ways.

4. Vrba (1983) is just one of many evolutionists who argue that trends must be understood within clades. I view this stance as an attempt to get paleontologists and others to think more explicitly in terms of evolutionary trees—"tree thinking," as some would call it. I first argued for the importance of thinking about replacement in addition to phylogeny in my 1994 paper.

5. See Kauffman (2000) and Chaisson (2001). Egbert Leigh and I have made the same argument on quite different grounds, as discussed further below. My characterization of complexity is based on McShea's (1998) incisive examination.

6. See Stanley (1973). Gould (1988) extended Stanley's insights by noting that much of evolutionary history can be interpreted as increases (and sometimes decreases) in variance. He elaborated this idea still further in his 1996 book, *Full House*. McShea (1994, 1996, 1998) has very carefully dissected trends, and with data from various sources has confirmed and extended Stanley's and Gould's assertions. Knoll and Bambach (2000) pointed out the role of rare innovations in extending the range of possibilities and therefore enlarging the observed range of body sizes among major clades of organisms.

7. Knoll and Bambach (2000) showed how the various great innovations increased the range of morphological and behavioral states available to life. A more genetically based version of their argument was put forth by Maynard Smith and Szathmáry (1995, 1999) and Carroll (2001).

8. W. Martin and Müller (1998) noted the reduction in complexity and functions of the ancestral mitochondrion after it entered its host. McShea (2002) showed empirically that diversity within the cells of multicellular organisms has decreased (or perhaps increased less rapidly) relative to the diversity of parts within unicells.

9. See Bambach (1977) for the first assemblage-level analysis of diversity through time. T. J. Palmer (1982) followed with his study of diversity among rock-boring and rock-encrusting organisms. The plant record has been ably documented by Knoll et al. (1979).

10. The last ten years have witnessed an explosion of studies in which experimenters have varied diversity (the number of species) in a community in order to track ecosystem-level responses in such properties as nutrient retention in the soil, productivity, and the predictability and stability of species composition. These studies have generally shown that an increase in the diversity of primary producers, mostly from a low of 2 to a high of 16 or more, brings about greater productivity, predictability, and stability (Naeem et al. 1994; Naeem and Li 1997; Tilman and Downing 1994; Tilman et al. 1996, 1997, 2001; D. U. Hooper and Vitousek 1997; McGrady-Steed et al. 1997; Hector et al. 1999). I agree with Mittelbach et al. (2001) and Paine (2002), however, that the studies are flawed in that diversity and community characteristics were measured only among primary producers—without considering the diversity or roles of consumers—and that nearly all the plants are herbaceous. In his experimental study on a wave-swept seashore in Washington State, Paine (2002) found that productivity rose dramatically, from 8.6 to 86 kg (wet weight) per square meter per year, in a regime of reduced grazing between the years 1996 and 2001. This rise in productivity, caused by the removal and continued near absence of grazers (sea urchins, chitons, and limpets), was accompanied by the overwhelming dominance of a single fast-growing primary producer, the kelp *Alaria*, which accounts for 80 percent of the production in the grazer-free plots. Consumers therefore play a critical role in this community in determining local diversity and productivity. Furthermore, as Loreau and colleagues (2001) note, most experiments (Paine's excepted) treat diversity as the independent variable, that is, as the variable which is experimentally manipulated. This procedure was adopted not just in the land-plant experiments to which I referred above, but also in experimental manipulations of phytoplankton and of bottom-dwelling animals in aquatic settings (Persson et al. 2001; Cardinale et al. 2002; Downing and Leibold 2002). I admit to considerable discomfort with this practice, because I tend to think of diversity as an outcome—a manifestation or result—of interactions; it is an epiphenomenon. Although diversity is linked with other individual and community properties in feedbacks,

I find it difficult to think of diversity as an instigating or triggering attribute in economic systems. To be sure, the presence of many species (or some other expression of high diversity) probably provides a certain redundancy of economic functions, a redundancy that likely protects the economy from disturbances that are unique to one or a few players or interactions; but given that most species are rare and seem to play extremely minor or peripheral roles in ecosystems, I am inclined to agree with Birkeland (1996) that diversity above some fairly low threshold is more ornamental than functionally vital. I hasten to add that this interpretation should not be taken as a rationale for eliminating those rare species or for stripping economies down to their bare bones of variety. Diversity is stunningly beautiful, interesting, and worth preserving.

11. Leigh (1994) introduced me to the importance of good-neighborliness in the establishment of forests and other ecosystems. He and I elaborated this and related ideas in our 2002 paper, in which we argue that diversity and productivity at the ecosystem level have both increased through time, with a few short-term interruptions corresponding to mass-extinction events. The idea that species that facilitate each other's well-being are more successful than selfish species was arrived at independently by Mark D. Bertness and his students in their work on New England salt marshes (see, for example, Bertness and Callaway 1994; Bertness and Leonard 1997). Connell and Slatyer (1977) recognized that, in the course of plant succession, plant species predominating early in the development of communities facilitate the establishment of later species.

12. Diversity decreasing after increase in nutrient concentration and productivity: E. P. Odum 1969), Huston (1993), Rosenzweig and Abramsky (1993), Rosenzweig (1995). It has been widely recognized that evolution reverses this short-term effect (Rosenzweig and Abramsky 1993; Rosenzweig 1995; Mittelbach et al. 2001). An explanation for this reversal, based on the likelihood that isolated populations persist and then diverge as new species, was proposed by Allmon (1992a), Allmon and Ross (2001), and Vermeij (1995).

13. The succession of metabolisms, inferred from analyses of the evolutionary tree of life, is documented by Xiong et al. (2000), Knoll and Bambach (2000), Shen et al. (2001), and Nisbet and Sleep (2001), among others. The hypothesis that successive forms of metabolism began in more productive parts of the biosphere was proposed by Nisbet and Sleep (2001). Anbar and Knoll (2002) provide evidence for the predominance of sulfides in the deep ocean during much of the Mesoproterozoic, and discuss the interrelationships among sulfides, oxygen, phosphorus, and nitrogen for this time interval. Earliest life on land, 2.6 GA: Watanabe et al. (2000). Models of the history of oxygen in the atmosphere in the Phanerozoic: Berner (1994), Landis et al. (1996), Beerling and Berner (2000), Lenton (2001). Oxygen as enabling factor in increasing metabolic rates and activities: J. B. Graham et al. (1995), R. Dudley (1998), Gans et al. (1999).

14. General increase in shellbed thickness: Kidwell and Brenchley (1994, 1996). Droser and Sheehan (1997) and Droser and Li (2001) provide details of Ordovician and Cambrian shellbeds, respectively, in the Basin and Range Province of the western United States.

15. Bambach (1993) noted that modern marine animals have much chunkier bodies than their Paleozoic predecessors. Curry and colleagues (1989) documented the distribution of biomass in individual brachiopods.

16. Earliest evidence of oxygen-producing Cyanobacteria: Brocks et al. (1999), Summons et al. (1999); see the commentary by Knoll (1999). Early history of plankton: Knoll (1992, 1994), Javaux et al. (2001).

17. For the reconstruction of pre-Cambrian marine plankton ecosystems, see Butterfield (1997, 2001). Although earlier authors had considered the large spiny acritarchs of the Neoproterozoic as belonging to the phytoplankton, their large size as well as observed attachment of some large forms to grains of sediment convince Butterfield and me that these were bottom-dwelling organisms. Vase-shaped bottom-dwelling amoebae were described from the Chuar Group of the Grand Canyon by S. M. Porter and Knoll (2000) and Karlstrom et al. (2000). The time of origin of animals remains highly controversial. Wray and colleagues (1996) suggested that some of the major groups of animals diverged as early as 1.2 Ga, but others (Vermeij 1996b; Erwin 1999;

Valentine et al. 1999; Lynch 1999) prefer later dates, from 800 to 650 Ma. These younger dates would more or less coincide with a rise in oxygen levels in the atmosphere. Ruiz-Trillo and colleagues (2002) showed that acoel and nemertodermatidan flatworms are the most evolutionarily basal living bilaterian animals. These flatworms, and presumably ancestral metazoans as well, are particle-feeders on the seafloor, and lack planktonic stages in their life cycles.

18. Small sponges have been recognized from the Late Neoproterozoic of South Australia and Mongolia by Gehling and Rigby (1996) and Brasier et al. (1997) respectively. Wood and her colleagues (2002) described *Namapoikia*, one of the largest known organisms of its time, from the Omkyk Member of the Nama Group (548 Ma) of Namibia. Buss and Seilacher (1994) interpreted many elements of the Ediacara fauna as belonging to a sister group to the Metazoa (multicellular animals). It is possible that many of these forms were cnidarians instead, or perhaps animals of the cnidarian grade of organization that lacked nematocysts with which particles of food could be captured. For further discussions of the Ediacara fauna, see Seilacher (1989).

19. Yuan and colleagues (2002) make a strong case on the basis of large sponges in exceptionally preserved Tommotian to Atdabanian assemblages that large sponges were important elements of Early Cambrian ecosystems. Debrenne and Reitner (2001) discuss the life habits of Cambrian sponges and cnidarians. The existence of suspension-feeding arthropods was demonstrated for Early Cambrian branchiopods from Canada (Butterfield 1994) and for the arthropod *Isoxys* from China (Vannier and Chen 2000). Based on these findings, Butterfield (1997, 2001) argued that the Early Cambrian witnessed the introduction of consumption of phytoplankton by small planktonic animals, and that these zooplankters in turn provided an economic base for larger consumers both in the plankton and on the seafloor. He disputed our earlier claim (Signor and Vermeij 1994) that suspension-feeding in the Early Cambrian was poorly developed except among sponges, which are especially effective at exploiting the smallest particles. Our case was bolstered by phylogenetic analyses showing that plankton-feeding larval stages are derived in molluscs, arthropods, annelids, and perhaps even in Cnidaria (Chaffee and Lindberg 1986; Haszprunar 1992; Haszprunar et al. 1995). Echinoderms and chordates, however, may, as Strathmann (1978a,b, 1993) maintains, have had plankton-feeding larvae since their origins in latest Neoproterozoic or earliest Cambrian time. Ancestral bilaterians are likely to have had larval forms that did not feed in the plankton (Ruiz-Trillo et al. 2002).

20. The early history of encrustation is told by Lescinsky (1996). There are excellent accounts of North American and European encrusting communities on shells from the Late Ordovician (Richards 1972, 1974), Silurian (Hurst 1974), Devonian (Ager 1961; Sparks et al. 1980; F. Alvarez and Taylor 1987), and Carboniferous (Lescinsky 1997).

21. The data documenting the trend from passive to active suspension-feeders are in Vermeij (1987, chapter 4). Sizes of Paleozoic bivalves: Cope (1996). Bambach and colleagues (2002) divide animals into active forms, capable of moving under their own power, and passive types, which lead sedentary postlarval lives. They found that genera of passive animals outnumbered genera of active ones during the Early Cambrian, equaled them in numbers from the Atdabanian stage of the Early Cambrian through the Llandeilian stage of the Middle Ordovician, then again achieved numerical dominance from the Late Ordovician to the end of the Paleozoic. Passive types were a large minority (about 44 percent of genera) from the Middle Triassic to the end-Cretaceous, but eventually dwindled to less than 30 percent for most of the Cenozoic. It is important to note that Bambach and coworkers' categories of passive and active animals do not coincide with passive and active suspension-feeders, because many active suspension-feeders such as sponges would be classified as passive in the classification used by Bambach's group.

22. Rhoads (1970) and Rhoads and Young (1970) showed that animals plowing through sediment and those processing sediment through their digestive systems cause the water content of sand and mud to increase. Higher productivity of sediment and overlying waters because of burrowing has been documented by Thayer (1983), A. G. Fischer (1984), Bertness (1985), and others; it was seen as an important feedback mechanism in Early Cambrian ecosystems by Logan et al. (1995) and McIlroy and Logan (1999). As discussed below, however, the enhancement of

productivity by burrowers was not fully realized until the Ordovician period. Suchanek (1983) and S. L. Williams and her collaborators (1985) documented the disruptive effects of burrowing on sediment-rooted plants in the Virgin Islands. There is debate about whether food or safety was the deciding enticement for animals to evolve the habit of burrowing in sediments. For reasons discussed below, I favor the safety argument.

23. Thayer (1979, 1983) documented the trends toward increased per capita disturbance and burial depths by animals burrowing in sediment. Interpretation of Late Neoproterozoic trail-makers is based on the work of A. G. Collins et al. (2000); see also Runnegar (1982), who in many ways anticipated the later results. The record of sediment disturbance by animals in the Cambrian is ably summarized by Droser and Li (2001). Droser and colleagues (2002a,b) have recently added the important observation that most Early Cambrian burrows are permanent, and that burrowing for food is a largely post-Cambrian phenomenon.

24. Absence of Cambrian deposit-feeders: Butterfield (2002), Droser et al. (2002a,b). Tiny bivalves as particle-feeders: Reid et al. (1992). Waller's (1998) exceedingly careful and thorough analysis of bivalve phylogeny indicates that deposit-feeding members of the Protobranchia, one of the earliest clades to diverge within the Bivalvia, are derived relative to particle-feeders and suspension-feeders.

25. This account is based on Thayer's (1979, 1983) work and on my subsequent review (Vermeij 1987, chapter 5). The best evidence for bacterial culturing by callianassids comes from the careful work of Ziebis et al. (1996) in the Adriatic Sea. The Early Paleozoic history of sediment disturbance has been documented by Droser and Bottjer (1989) and Droser et al. (1994, 1999). Sediment disturbance by gray whales and walrus in the Bering Sea: Oliver et al. (1983a,b). Pemberton and colleagues (1984) documented lobster burrows 15 cm in diameter and 2 m deep from the Cardium Formation (Late Cretaceous) of Alberta. In an earlier paper, Pemberton and colleagues (1976) described the deepest marine burrows so far known, from the Canso Strait in Nova Scotia, made by a 6 cm long mud shrimp, *Axius serratus*.

26. S. L. Williams (1984) was the first to document the ability of *Caulerpa* and other siphonaceous green algae to acquire ammonium from the sand in which these algae are rooted; see also Williams and Fischer (1985) and Williams et al. (1985). Molecular phylogenetic analyses by J. L. Olsen et al. (1994) and R. L. Chapman et al. (1998) imply that the siphonaceous green algae date only from the Permian, but the siphonaceous order Dasycladales, comprising calcified seaweeds, are known with certainty back to the Middle Ordovician (Riding 2001). Two Cambrian genera may belong to this group as well, but Riding finds membership of these fossils in Dasycladales unconvincing.

27. The history of marine seed plants was reviewed by Vermeij and R. Dudley (2000); see also Vermeij (1987) for the role of seagrasses in expanding and increasing marine primary production. For an account of mangrove evolution, see Ricklefs and Latham (1993); for seagrasses, see den Hartog (1970).

28. Weathering by lichens and bryophytes: T. A. Jackson (1973), Jackson and Keller (1970), Moulton and Berner (1998). Earliest Ordovician burrows on land: R. B. McNaughton et al. (2002).

29. Data on Devonian rooting depth are from Retallack (1997) and Elick et al. (1998); those for tropical trees are from Leigh (1999).

30. Earliest glomalean fungi: Redecker et al. (2000). Increased chemical weathering as vascular plants evolved: Algeo and Scheckler (1998), Moulton and Berner (1998), Lenton (2001). Diversification and spread of flowering plants: Crane (1987), Crane and Lidgard (1989), Crane et al. (1995), Wing and Boucher (1998). Origin of seeds during latest Devonian: Gillespie et al. (1983). Increase in productivity in Middle Paleozoic and again during the later Mesozoic: Vermeij (1987), Bambach (1999). Importance of mycorrhizae in enabling plants to succeed on land: L. Simon et al. (1993). Interpretations from molecular genetic sequences have led Heckman and colleagues (2001) to the view that root-associated fungi and land plants both date back to the Late Cambrian, about 510 Ma. This is possible, but earlier members of these groups probably

bore little resemblance to their descendants and may not have had the intimate economic relationship they often do today.

31. History of freshwater sediment disturbance and of burrowing: M. F. Miller (1984), Miller et al. (2002).

32. Oldest multicellular algae: Butterfield (2000). Early diversification of algae: Xiao et al. (1998a,b), Yuan et al. (1999). Details of the evolution and consequences of herbivory are given by Vermeij (1977b) and Vermeij and Lindberg (2000). Bellwood (2003) has confirmed the late (Early Eocene) origin of herbivory in fishes in his comparisons of jaw function among well-preserved fish assemblages of Late Triassic, Late Jurassic, Early Eocene, and modern reefs. Streelman and colleagues (2002) provide details of the evolution of parrotfishes (family Scaridae), which comprise a derived group of wrasses.

33. Early history of herbivory by arthropods: Labandeira (1998a). Beck and Labandeira (1998) document leaf consumption by arthropods in the Early Permian of Texas. Labandeira (1998b) has sketched the subsequent Mesozoic history of herbivory by insects. The earliest gall, together with herbivory by early insects in general, was described by Labandeira and Phillips (2002).

34. For the fossil record of mouthpart architecture of herbivorous insects, see Labandeira and Phillips (2002) and especially Labandeira (1997, 1998a). The multiple origins of herbivory among insect clades were carefully documented by Mitter et al. (1988), who showed further that the diversity of living species in these clades is almost always far higher than that in sister clades of predatory or detritus-feeding insects.

35. Reisz and Sues (2000) provide an excellent review of the fifteen major tetrapod clades in which herbivory evolved during the interval from the latest Carboniferous to the Triassic. This account is largely consistent with, but more detailed than, Bakker's (1980) earlier rendition. Rybczynski and Reisz (2001) showed that early anomodonts chewed their plant food. Whether this ability was present also in advanced anomodonts (dicynodonts) of the Late Permian and Triassic is not clear.

36. Nondinosaurian Mesozoic herbivores were reviewed by Reisz and Sues (2000). My discussion of herbivorous dinosaurs is taken from reviews by Benton (1997), Upchurch and Barrett (2000), Weishampel and Jianu (2000), and Sereno (1999). The aberrant mid-Cretaceous herbivorous sphenodontian *Ankylosphenodon*, from central Mexico, may have chewed its food and have had semiaquatic habits (Renoso 2000). X.-C. Wu and colleagues (1995) described a Cretaceous herbivorous crocodile from China.

37. Janis (2000) has given a particularly fine account of the evolution of herbivorous mammals during the Paleocene and Eocene. Herbivory in later mammals was reviewed by MacFadden (2000). The habit obviously evolved many times among mammals. It is known among sloths, including giant Pliocene and Pleistocene groundsloths of South and North America; as well as in perissodactyl and artiodactyl ungulates, primates (from the Late Eocene onward), rodents, hares, elephants, and even some Carnivora (bears and pandas, for example). For herbivory in birds, see Morse (1975), E. S. Morton (1978), and R. Dudley and Vermeij (1992); for lizards, see Ostrom (1963); for turtles, see Shaffer et al. (1997).

38. Growth of tree lycopsids of the Paleozoic: DiMichele and Phillips (1985), Bateman (1994), Labandeira and Phillips (2002). Growth forms of palms: Corner (1964), Tomlinson (1973).

39. So much attention has been lavished by researchers on chemical deterrents to herbivory that architectural features of plants against consumption have received little comment. Corner (1964) drew attention to the basal growth of grasses as an antigrazing mechanism, made more effective by the presence of silica in the blades. Gandolfo and colleagues (1998) described the earliest known liliopsidan monocot (90 Ma), but Bremer's (2002) molecular work puts the date of origin of Poales, and thus potentially of the basal growth habit, at 115 Ma (Aptian stage). The earliest spikelet referable to the grass family Poaceae has a geological age of 55 Ma, the Paleocene-Eocene boundary (Crepet and Feldman 1991), but grass pollen is recorded from the Cretaceous (Bremer 2002).

40. Bakker (1978) and Krassilov (1981) suggested that weedy angiosperms evolved as low-browsing ornithischian dinosaurs diversified during the Early Cretaceous. Tiffney (1997) reviewed Mesozoic vegetations in relation to herbivores, and supports their speculation. He did not, however, discuss the absence of the grass habit. The ability to chew characterizes most mammals as well as several major groups of dinosaurs (Upchurch and Barrett 2000). The early anomodont *Suminia getshanovi* from the Late Permian of Russia likewise chewed its food (Rybczynski and Reisz 2001).

41. Mann (1973) first drew attention to the basal growth habit of kelps. J. A. Estes and Steinberg (1988) surmised that kelps evolved no earlier than the Late Miocene, but Domning (1989) and I (Vermeij 1992, 2001a) suggested an earlier date of origin (Late Oligocene) because of the close parallels between the geographic history of these large brown algae and that of molluscs with an abundant fossil record. Yoon and colleagues (2001) published a phylogeny, based mainly on molecular sequences, that largely corroborates these geographic and temporal inferences of kelp evolution. Domning (1978) gives a detailed account of the anatomy, habits, and likely foods of Steller's seacow, which until its extinction in the mid-eighteenth century must have been a major consumer of seaweeds in the North Pacific. Kelp-eating seacows are evidently derived from earlier warm-water sirenians whose diet was composed largely of seagrasses (Domning 2001a). For an account of desmostylians see Domning et al. (1986). The oldest sirenian (*Pezosiren portelli*) from the Early Eocene of Jamaica was recently described by Domning (2001b).

42. Steneck (1983, 1985) documented the adaptations of corallines to herbivores and briefly sketched the history of this red-algal group. Along with certain leathery seaweeds, including kelps, corallines are among the most herbivore-resistant seaweeds, as documented in Steneck and Watling's (1982) classification of seaweed architecture. Steneck and Watling (1982) surveyed marine herbivores according to which types of algae were susceptible to grazing. Almost all herbivores can consume filamentous and sheetlike seaweeds, but only a few are capable of dealing with leathery, articulated, and encrusting forms.

43. Roth-Nebelsick and colleagues (2001) comprehensively reviewed the types, functions, and history of venation patterns in leaves. Howland (1962) was among the first to investigate the distribution of waters to all parts of the leaf through a system of branching veins. The hypothesis that the open branching of the veins in early land plants arises from a developmental limitation was explored by Boyce and Knoll (2002), who further noted that network patterns can evolve when leaf growth occurs away from, as well as at, the leaf margins. My hypothesis that herbivory is largely responsible for the evolutionary success of network venation is complementary to, and therefore not inconsistent with, Boyce and Knoll's developmental explanation.

44. The absence of large decomposers in Carboniferous forests and the presence of highly decay-resistant compounds in the forest plants of that time were pointed out by J. M. Robinson (1990). Oribatid and cockroach consumption of dead internal tissues of Carboniferous plants was very carefully documented by Labandeira et al. (1997) and Labandeira and Phillips (1996, 2002).

45. It is, of course, difficult to estimate the rate of primary production in ancient vegetations, but plant architecture and the nature of ancient soils provide qualitative indications, which led Beerling and colleagues (1998) to suggest that Carboniferous forests were characterized by low productivity. For further important discussions of vegetational changes see Labandeira and Phillips (2002) for the Late Paleozoic, and Tiffney (1997) and Wing and Boucher (1998) for the Mesozoic and Cenozoic.

46. The account of the history of predation given here is based on my longer reviews (Vermeij 1987, 2002b). Bengtson and Zhao (1992) described the complete and incomplete holes in *Cloudina*. Similar forms of predation were described from Early Cambrian fossils by Conway Morris and Bengtson (1994). Following speculations by earlier authors, I argued for the primary role of predators in the evolution of skeletons (Vermeij 1989b).

47. Vacelet and Boury-Esnault (1995) described predatory sponges from the deep Mediterranean. Although predatory habits of this kind are probably evolutionarily derived in sponges, as they are also in deep-sea bivalves, they demonstrate the potential antiquity of predation as a mode of feeding in the most basal phylum of animals. The ancient and basal position of predators in the evolutionary tree of animals and of each of the major clades of animals was noted by Vermeij and Lindberg (2000).

48. Further details of the nature and history of predation on shell-bearing animals are given by Vermeij (1977b, 1987, 1993, 2002b).

49. Although naticid moon snails are known chiefly as drilling predators, many routinely suffocate prey in the very large foot (Vermeij 1993). Large-footed naticids such as *Polinices, Mammilla,* and *Sinum* originated during Paleocene time, and appear to be derived relative to smaller-footed types. The same is probably true for volutids and olivids, but further functional and phylogenetic research will be needed to verify this suspicion.

50. For an account of seastars feeding on clams by insertion of the stomach, see Blake (1990) and Blake and Guensburg (1994).

51 For reviews of drilling predation, see Kabat (1990) and Vermeij (2002b). Despite reports to the contrary, unquestioned naticids and muricids cannot be verified before Campanian time (Merle and Pacaud 2001; Kase and Ishikawa 2003). Deline and colleagues (2003) describe the earliest known instance of edge-drilling in an Early Carboniferous brachiopod. Harper (2003) notes the small size of Early Paleozoic drill holes, and emphasizes the continuity of drilling as a means of predation over the entire Phanerozoic eon. Great advances in our understanding of drilling predation are being made by Greg Herbert and Gregory Dietl.

52. For details on anomalocarid feeding, see Whittington and Briggs (1985), Babcock and Robison (1989), Chen et al. (1994), and D. Collins (1996). Trilobite feeding and the structure of the feeding organs of trilobites have been ably reviewed by Fortey and Owens (1999).

53. For reviews of cephalopod history, see Vermeij (1987), Engeser (1990), and J. A. Chamberlain (1991). The classification and interpretation of cephalopod jaws (aptychi and anaptychi) have been fraught with controversy. Dagys and colleagues (1989) interpret the calcified aptychi of Mesozoic ammonoids as cutting devices, but N. Morton and Nixon (1987) interpret them as shovels. A Silurian structure (*Aptychopsis*) has been interpreted as having a solely protective function similar to that of the gastropod operculum (Stridsberg 1984). If this is correct, a feeding-related, crushing function of Devonian to Cretaceous anaptychi would be derived from a passive protective role.

54. Chen and Huang (2002) described *Eognathacantha ercainella*, a fossil they interpret as the earliest chaetognath, from the Early Cambrian (520 Ma) Maotanshan Formation of Kunming, China. Sansom and colleagues (1996) inferred the presence of acanthodians and *Chondrichthyes* from fossils in the Harding Sandstone of Colorado. For the feeding habits and apparatus of Silurian and Devonian fishes, see Gross (1967), Denison (1978, 1979), and Moy-Thomas and Miles (1971).

55. The importance of protrusible upper and lower jaws in expanding the feeding repertoire of bony fishes and sharks was recognized by Schaeffer and Rosen (1961). Excellent accounts of the mechanisms and history of predation are available for bony fishes (Lauder 1982; Lauder and Liem 1983; Wainwright and Bellwood 2002), sharks and rays (Moss 1977, 1981; Thies and Reif 1985; Summers 2000), diapsids including dinosaurs (Bakker 1980; Morales 1997; P. J. Currie 1997; Farlow and Holtz 2002; Van Valkenburgh and Molnar 2002). Details of feeding mechanisms and inferred bite forces are available for *Tyrannnosaurus* (Erickson et al. 1996), *Allosaurus* (Rayfield et al. 2001), and *Sarcosuchus* (Sereno et al. 2001). Farlow and Holtz (2002) and Van Valkenburgh and Molnar (2002) argue that the per capita food requirements of predatory dinosaurs were low compared to those of mammals.

56. More detailed accounts are available in Vermeij (1977b, 1987, 2002b). Tintori (1998) published a detailed account of the early evolution of crushing among Triassic fish. Unfortunately, most studies of fossil fish and crustaceans emphasize classification and phylogeny over

the evolution of feeding, locomotion, and defense. Exceptions are Thies and Reif's (1985) study, which showed that shell crushing in the modern (Early Triassic to Recent) clade of chondrichthyans arose in several lineages during the Toarcian stage of the Early Jurassic (notably in batoid skates and rays and in heterodontid sharks), and in Bellwood's (1996) analysis of fish from the Early Eocene of Monte Bolca, Italy. Excellent accounts are available for the history and structure of shell-crushing jaws and dentition of the most powerful predators: eagle rays (Summers 2000), puffers and related triggerfishes and other plectognaths (Tyler 1980), and carpiliid crabs (Guinot 1968). Recent discoveries by Tyler and Sorbini (1996) have pushed the first appearance of plectognath fishes back to the Cenomanian stage of the Middle Cretaceous, but shell-crushing specialists in this group of bony fish are known only back to the Early Eocene. Oji and colleagues (2003) show that broken shells, which result from predation by fish and crustaceans rather than from damage through collisions with boulders in wave-swept environments, become prominent in the Eocene and especially the Neogene (the last 23 my of Earth history).

57. Detailed accounts of gastropod architecture through time are in Vermeij (1987). Trends in Paleozoic armor: Signor and Brett (1984). Brachiopod architecture and functional interpretation of spines and pleats as antipredatory: R. R. Alexander (1990), Leighton (2001). Bivalve trends: Vermeij (1987), Harper and Skelton (1993). Cephalopods: Ward (1986), W. B. Saunders et al. (1999); for the interpretation of complex septal junctions with the shell wall as antipredatory, see Vermeij (1987) and Daniel et al. (1997). Repeated evolution of the crab habit: Morrison et al. (2002). Earliest true crab: Förster (1979). Evolution of autotomizing crinoids: Oji (2001).

58. Details in Vermeij (1987, 2002b). Patterns in gastropod shell repair through time: Vermeij et al. (1981). The idea that frequencies of repair did not rise in the Cretaceous until the Campanian stage, about 80 Ma, is supported by low frequencies in Turonian Chinese gastropods (Pan 1991) and pre-Campanian inoceramid bivalves in the western United States (Ozanne and Harries 2002). Incidences of arm regeneration in crinoids: Oji (2001).

59. J. A. Chamberlain (1991) gives an excellent summary of the history and mechanics of locomotion in cephalopods. D. K. Jacobs (1992) and Jacobs and Landman (1993) emphasize the importance of increasing power as compared to decreasing drag in enhancing cephalopod swimming performance. Packard (1972) and Chamberlain noted the general equity of performance between cephalopods and fish despite the much greater efficiency of fish locomotion, which involves displacement of external water, than that of cephalopods, which expel smaller volumes of internal water.

60. Increased locomotor performance of burrowing snails through time: Vermeij 1987, 1993; for burrowing bivalves, see Vermeij 1987; Checa and Jiménez-Jiménez 2003. Oblique, burrowing-enhancing sculpture evolved in one or two Cretaceous bivalves, but it became common only in the Eocene, when lucinids and later several other groups evolved this type of sculpture (Checa 2002; Checa and Jiménez-Jiménez 2003) Crinoid locomotion: Meyer and Macurda 1977. Motile bryozoans: P. L. Cook and Chimonides 1983; F. K. McKinney and Jaklin 1993.

61. Mechanics and history of marine vertebrate locomotion: P. W. Webb (1984), Braun and Reif (1985), Cowen (1996), Motani (2002).

62. For general discussions of the history of locomotion in terrestrial vertebrates, see Bakker (1971, 1980, 1983). Berman and colleagues (2000) described *Eudibamus* as the earliest bipedal, erect-postured reptile. Locomotor trends in predatory dinosaurs: P. J. Currie (1997); in mammals: Janis and Wilhelm (1993), Van Valkenburgh (1999). Blob (2001) showed that limb posture in Paleozoic therapsid reptiles varied within individuals from sprawling to erect, suggesting that the multiple evolution of erect limb posture in archosauromorph reptiles (J. M. Parrish 1986) could similarly have begun as a facultative change in orientation perhaps attained during brief, rapid bursts of running (see also Carrano 2000).

63. The best account of the evolution of flight is insects is that of R. Dudley (2000b). Early gliding vertebrates include *Coelurosauravus* from the Late Permian (Frey et al. 1997), the Late Triassic peltopleuriform fish *Thoracopterus* (Tintori and Sassi 1992), Late Triassic gliding lizards

in the Chinese kuehneosaurid genus *Fulengia* (reviewed in R. Estes 1983), and an Eocene eomyid mammal (Storch et al. 1996).

64. Increased biological control of the calcium cycle: Knoll et al. (1993). Details of the Archean to Proterozoic transition in the mode of precipitation of calcium carbonate, and the role of microbial mats in trapping and cementing crystals of this mineral, are given by Grotzinger and Kasting (1993), Sumner and Grotzinger (1996), and Grotzinger and Knoll (1999). For the transition between Mesoproterozoic and Neoproterozoic microbial mats, see Knoll and Sergeev (1995). Grotzinger and colleagues (2000) and Wood and colleagues (2002) described latest Neoproterozoic reef structures in Namibia.

65. For the history of rock destruction by organisms, see Vermeij (1987, chapter 5). Ekdale and Bromley (2001) have given further details of the Early Paleozoic history of deep borings, including the description of an Arenigian (Early Ordovician) boring of the type known as *Gastrochaenolites*.

66. History of silica: Maliva et al. (1990), Maldonado et al. (1999). Racki and Cordey (2000) pointed out that the skeletons of sponges and radiolarians thinned as Cenozoic diatoms became more abundant, and that in turn diatoms have thinned as more of the dissolved silica is retained on land by grasses.

67. Raup and Crick (1981) carried out a highly detailed analysis of evolution in *Kosmoceras*. The work on the finches *Geospiza fortis* and *G. scandens* by Peter and Barbara Grant (2002) is part of a larger long-term study of evolution of Darwin's finches in the Galápagos (see also P. R. Grant 1986).

Chapter 11
The Future of Growth and Power

1. Arndt (1978:143); Daly (1996:167); see also Douthwaite (1992).

2. See Paepke (1993). The two quotes are on pages 138 and xxvi, respectively.

3. Henry MacGillavry, a Dutch oil-company geologist, was the first clearly to describe the pattern that Eldredge and Gould (1972) later called punctuated equilibria. Although McGillavry's (1968) paper is in English and contains a wealth of evidence, it was published in an obscure Dutch journal. Once Eldredge (1971) and Eldredge and Gould (1972) published their papers and gave the phenomenon a name, a fierce debate over the reality and generality of the pattern ensued, with many population geneticists denying the phenomenon altogether, and many paleontologists claiming that the fossil record often conceals slow gradual evolution in favor of episodic change. Although it is true that fossils record some intervals and some environments far better than others, the idea that the phenotypes of species are stable for long periods of time strikes me as difficult to deny. Eldredge and Gould's original claim that evolution takes place only when one lineage splits from another may not hold up, because the fossil record chronicles many cases where the characteristics change over short periods without evidence of lineage splitting (see, for example, Hallam 1978).

4. See Anbar and Knoll (2002). Tappan (1970) and Tappan and Loeblich (1973) had earlier suggested that Late Carboniferous and Permian seas were relatively starved of nutrients, because many nutrients on land were being buried in equatorial coal swamps. Bambach (1977) described the pattern of community diversity in level-bottom marine assemblages. T. J. Palmer uncovered a similar pattern in his study of Paleozoic and Mesozoic species on marine hardgrounds (Palmer 1982).

5. The equilibrial view of biotas and clades found its inspiration in MacArthur and Wilson's (1967) theory of island biogeography, in which the number of species on an island is determined by a balance between immigration and extinction. Surprisingly, the theory did not consider species formation on the island as another potential source of species. Stanley (1979) and Sepkoski

(1979, 1981, 1984) drew an analogy between the growth and subsequent stability of populations on the one hand, and the growth and stability of clades and biotas on the other. A nonequilibrial view of species diversity was advocated by Valentine (1969) and T. D. Walker and Valentine (1984), as well as by the parasitologist Klaus Rohde (1991, 1999). Abbott (1974) even questioned the applicability of equilibrial perspectives to the diversity of plants and insects on islands. My own sense is that stability and equilibria are concepts that apply more to large, inclusive, and powerful economic units—species of warm-blooded animals, and perhaps major clades—but even here the plateau may rise as the productivity in the system of which they are a part increases (see Vermeij 1991a; Sax et al. 2002).

6. Estimates of human population growth and of per capita energy use are from J. E. Cohen (1995). For the role of wood and forests in civilization, see Perlin (1989).

7. For an excellent treatment of soils and their sustainable use, see Hillel (1991).

8. The land-limitation hypothesis is thoroughly discussed by Pomeranz (2000). This author points out that Europe's military power, expressed through conquest and the forced labor of conquered peoples, was ultimately responsible for the economic colonization and subjugation of the New World and much of the rest of the tropics. Evidence of that military power, however, surfaced long before European technological and economic supremacy emerged with the Industrial Revolution in the early nineteenth century. The land-limitation hypothesis is therefore only part of the answer to the question of why Europe led the human species to unparalleled rapid growth and technological innovation. The quote is from Hawken (1993:80).

9. The calculations on which Wackernagel and colleagues (2002) base their conclusion that the use of land and ocean area has been unsustainable since the 1980s depend on many assumptions and on rough estimates of such quantities as energy foregone by increased carbon dioxide, area of land occupied by roads and buildings, area and yield of fishing and of unwanted bycatch, and so on; but the overall trend Wackernagel and colleagues identify is unmistakable: humanity is reaching or has reached the point where nature's ecosystems cannot replenish resources fast enough.

10. For the relative roles of humans and volcanoes in liberating carbon dioxide into the atmosphere, see Sigurdsson (1990). The quote is from Daly (1996:31).

11. For a penetrating and fascinating analysis of human alcoholism and its evolutionary roots, see R. Dudley (2000a). Nesse and Williams (1994) have likewise made the case that many substances we make are outside the evolutionary experience of living things, including us, and that they therefore pose a potential threat.

12. See T. P. Hughes (1994) for a general account of reef degradation in Jamaica and elsewhere. J.B.C. Jackson and colleagues (2001) have made the general case that overfishing has eliminated most top consumers in marine ecosystems, with dramatic consequences for the systems affected. For excellent accounts of the effects of Hurricane Alan, see J. W. Porter et al. (1981) and Knowlton et al. (1981).

13. For an excellent account of military and weapons development, see McNeill (1982). Hanson (2001) provides a highly readable and well-informed account of the development of the Western style of fighting by the Greeks. Colinvaux (1980) argues that war, and thus the development of weapons for waging it, is often chosen by the wealthiest members of society and by the wealthiest nations as a means of maintaining or gaining control.

14. The most compelling treatment of this problem I have read is Greider's (1997) book, *One World, Ready or Not*. Low demand is also at the root of modern-day economic recessions; for a good discussion of this problem, see Krugman (1999).

15. The social nature of the business cycle is compellingly laid out by Ormerod (1998). Chancellor's (1999) excellent book on speculation, and Greider's (1997) book on globalization, further document and elaborate the instability and potentially disastrous consequences of speculation.

16. See Bambach (2002) and Van Valkenburgh (1999). Leigh and Vermeij (2002) discuss the occurrence of monopolies in nature.

17. Arndt (1978:134).

18. Rapport and Turner (1977:342).

19. The importance of technological innovation for economic growth and for overcoming problems has been appreciated by many economists and was explicitly set out in a mathematical model by Romer (1986). Mokyr's (1990) book, *The Lever of Riches*, is an eloquent and detailed historical account of the relation between innovation and growth. See also Hodgson (1999), Ayres (1994), Landes (1998), and J. Jacobs (2000).

20. See J. L. Simon (1977) and Kremer (1993).

21. Landes (1998) has been especially sharp in his condemnation of theocracies. He notes that those theocracies in which women's participation in society is strongly discouraged suffer from the additional problem that the pool of potential talent is automatically halved. Hanson (2001) has pointed out that resentment and distrust by soldiers fighting for an authoritarian or theocratic regime often translate into long-term defeat. On the other hand, the conviction that Christianity is superior to all other systems of belief surely played a large role in motivating Spaniards to fight during their conquest of much of the New World. D. S. Wilson (2002) argues that religions are social adaptations, a view with which I agree. I have more to say about religion below.

22. The two quotes are from D. S. Wilson (2002), pages 42 and 229, respectively.

23. Tawney (1937:105).

24. Tawney (1937:111). This quote comes across to me as an adaptive interpretation of Calvinism and, by extension, of other religious doctrines to the societies in which they originate, spread, and persist.

25. The historical background of the Scottish Enlightenment is eloquently explored by Herman (2001). The quote is from D. S. Wilson (2002:101).

26. See Colinvaux's (1980) book, *The Fates of Nations*, for the argument that it is wealthy members of society who most often incite war and revolution. Greider (1997) has emphasized the widening of the gap between rich and poor, seen at the level of individuals as well as at the level of national economies, in the rapidly globalizing capitalist world. Stiglitz (2002) likewise notes the general tendency for greater inequality to accompany economic growth, but points to East Asia as an exception. Like Maynard Keynes and others before him, Stiglitz argues that the free market is poor at preventing unemployment and many other social ills, and that government action is necessary to reign in and augment the free market.

27. See Daly and Cobb (1994) for per capita incomes, and Landes (1998) for comparisons among nations.

28. Tawney (1937:166).

29. The pastoral and agricultural origins of the major religions were emphasized by Tawney (1937), Critchfield (1981), and Landes (1998). My hypothesis refers specifically to the major monotheistic religions, which are evolutionarily derived belief systems. D. S. Wilson (2002) and I agree that religion is an adaptation, but Wilson did not compare the effectiveness of different religions as means to stabilize and motivate groups. The study of religion would clearly benefit from a comparative evolutionary approach.

30. D. S. Wilson (2002:228).

31. Critchfield (1981:283).

32. See Putnam (1992:180).

LITERATURE CITED

Abbott, I. 1974. Numbers of plant, insect and land bird species on nineteen remote islands in the southern hemisphere. Biological Journal of the Linnean Society 6: 143–152.

Addicott, W. O. 1970. Miocene gastropods and biostratigraphy of the Kern River area, California. U.S. Geological Survey Professional Paper 642: 1–174.

Ager, D. V. 1961. The epifauna of a Devonian spiriferid. Journal of the Geological Society of London 117: 1–10.

Agrawal, A. A. 2001. Phenotypic plasticity in the interactions and evolution of species. Science 294: 321–326.

Alexander, R. M. 1977. Swimming. Pages 222–248 in R. M. Alexander and G. Goldspink (eds.), Mechanics and energetics of animal locomotion. London: Chapman and Hall.

Alexander, R. R., 1990. Mechanical strength of shells of selected extant articulate brachiopods: Implications for Paleozoic morphologic trends. Historical Biology 3: 169–188.

Alexander, R. R., and G. P. Dietl. 2001. Shell repair frequencies in New Jersey bivalves: A Recent baseline for tests of escalation with Tertiary, mid-Atlantic congeners. Palaios 16: 354–371.

Algaze, G. 2001. Initial social complexity in southwestern Asia: The Mesopotamian advantage. Current Anthropology 42: 199–233.

Algeo, T. J., and S. E. Scheckler. 1998. Terrestrial-marine teleconnections in the Devonian: Links between the evolution of land plants, weathering processes, and marine anoxic events. Philosophical Transactions of the Royal Society of London B 353: 113–130.

Allmon, W. D. 1992a. A causal analysis of stages in allopatric speciation. Oxford Surveys in Evolutionary Biology 8: 219–257.

———. 1992b. Role of temperature and nutrients in extinctions of turritelline gastropods: Cenozoic of the northwestern Atlantic and northeastern Pacific. Palaeogeography, Palaeoclimatology, Palaeoecology 92: 41–54.

Allmon, W. D., and R. M. Ross. 2001. Nutrients and evolution in the marine realm. Pages 105–148 in W. D. Allmon and D. J. Bottjer (eds.), Evolutionary paleoecology: The ecological context of macroevolution. New York: Columbia University Press.

Allmon, W. D., S. D. Emslie, D. S. Jones, and G. S. Morgan. 1996. Late Neogene oceanographic change along Florida's west coast: Evidence and mechanisms. Journal of Geology 104: 143–162.

Altmann, S. A. 1987. The impact of locomotor energetics on mammalian foraging. Journal of Zoology, London, 21: 215–225.

Alvarez, F., and P. D. Taylor. 1987. Epizoan ecology and interactions in the Devonian of Spain. Palaeogeography, Palaeoclimatology, Palaeoecology 61: 16–31.

Alvarez, W., and R. A. Muller. 1984. Evidence from crater ages for periodic impacts on the Earth. Nature 308: 718–720.

Amano, K., M. Ukita, and S. Sato. 1996. Taxonomy and distribution of the subfamily Ancistrolepidinae (Gastropoda: Buccinidae) from the Plio-Pleistocene of Japan. Transactions and Proceedings of the Palaeontological Society of Japan n.s. 182: 467–477.

Ambrose, S. H. 2001. Paleolithic technology and human evolution. Science 291: 1748–1753.

Amthor, J. E., J. P. Grotzinger, S. Schröder, S. A. Bowring, J. Ramezani, M. W. Martin, and A. Matter. 2003. Extinction of *Cloudina* and *Namacalathus* at the Precambrian-Cambrian boundary in Oman. Geology 31: 431–434.

Anbar, A. D., and A. H. Knoll. 2002. Proterozoic ocean chemistry and evolution: A bioinorganic bridge? Science 297: 1137–1142.

Anderson, E. 1984. Who's who in the Pleistocene: A mammalian bestiary. Pages 40–89 in P. S. Martin and R. H. Klein (eds.), Quaternary extinctions: A prehistoric revolution. Tucson: University of Arizona Press.

Anderson, L. C. 2001. Temporal and geographic size trends in Neogene Corbulidae (Bivalvia) of tropical America: Using environmental sensitivity to decipher causes of morphologic trends. Palaeogeography, Palaeoclimatology, Palaeoecology 166:101–120.

Anderson, R. P., and C. O. Handley, Jr. 2002. Dwarfism in insular sloths: Biogeography, selection, and evolutionary rate. Evolution 56: 1045–1058.

Ansell, A. D. 1967. Leaping and other movements in some cardiid bivalves. Animal Behaviour 15: 421–426.

———. 1969. Defensive adaptations to predation in the Mollusca. Proceedings of the Symposium on Mollusca 2: 487–512.

Ansell, A. D., and A. Trevallion. 1969. Behavioural adaptations of intertidal molluscs from a tropical sandy beach. Journal of Experimental Marine Biology and Ecology 4: 9–35.

Appleton, R. D., and A. R. Palmer. 1988. Water-borne stimuli released by predatory crabs and damaged prey induce more predator-resistant shells in a marine gastropod. Proceedings of the National Academy of Sciences of the USA 85: 4387–4391.

Arnaud, P. M. 1974. Contribution à la bionomie marine benthique des régions antarctiques et subantarctiques. Téthys 6: 567–653.

Arndt, H. W. 1978. The rise and fall of economic growth: A study in contemporary thought. Melbourne: Cheshire: 161 pp.

Arnqvist, G. 1998. Comparative evidence for the evolution of genitalia by sexual selection. Nature 393: 784–786.

Arnqvist, G., and L. Rowe. 2002. Antagonistic coevolution between the sexes in a group of insects. Nature 415: 787–789.

Aronson, R. B. 1987. Predation on fossil and Recent ophiuroids. Paleobiology 13: 187–192.

———. 1988. Palatability of five Caribbean ophiuroids. Bulletin of Marine Science 43: 93–97.

———. 1991. Escalating predation on crinoids in the Devonian: Negative community-level evidence. Lethaia 24: 123–128.

Arthur, M. A., W. E. Dean, and S. O. Schlanger. 1985. Variations in the global carbon cycle during the Cretaceous related to climate, volcanism, and changes in atmospheric CO_2. Geophysical Monographs 32: 504–529.

Arthur, M. A., J. C. Zachos, and D. S. Jones. 1987. Primary productivity and the Cretaceous/Tertiary boundary event in the oceans. Cretaceous Research 8: 43–54.

Arthur, W. B. 1989. Competing technologies, increasing returns, and lock-in by historical events. Economic Journal 99: 116–131.

———. 1999. Complexity and the economy. Science 284: 107–109.

Atkinson, I. A. E. 1985. The spread of commensal species of *Rattus* to oceanic islands and their effects on island avifaunas. Pages 35–81 in P. J. Moors (ed.), Conservation of island birds: Case studies for the management of threatened island species. Cambridge: International Council for Bird Preservation.

Atsatt, P. R. 1991. Fungi and the origin of land plants. Pages 301–315 in L. Margulis and R. Gester (eds.), Symbiosis as a source of evolutionary innovation: Speciation and morphogenesis. Cambridge: MIT Press.

Ayres, R. U. 1994. Information, entropy, and progress. New York: American Institute of Physics: 301 pp.

Babcock, L. E., and R. A. Robison. 1989. Preferences of Palaeozoic predators. Nature 337: 695–696.

Bains, S., R. M. Corfield, and R. D. Norris. 1999. Mechanisms of climate warming at the end of the Paleocene. Science 285: 724–727.

Bains, S., R. D. Norris, R. M. Corfield, and K. L. Faul. 2000. Termination of global warmth at the Paleocene/Eocene boundary through productivity feedback. Nature 407: 171–174.

Bakker, R. T. 1971. Dinosaur physiology and the origin of mammals. Evolution 25: 636–658.

———. 1973. Anatomical and ecological evidence of endothermy in dinosaurs. Nature 238: 81–85.

———. 1977. Tetrapod mass extinctions—a model of the regulation of speciation rates and immigration by cycles of topographic diversity. Pages 439–468 in A. Hallam (ed.), Patterns of evolution as illustrated by the fossil record. Amsterdam: Elsevier.

———. 1978. Dinosaur feeding behaviour and the origin of flowering plants. Nature 274: 661–663.

———. 1980. Dinosaur heresy—dinosaur renaissance: Why we need endothermic dinosaurs for a comprehensive theory of energetic evolution. Pages 351–462 in R.D.K. Thomas and E. C. Olson (eds.), A cold look at the warm-blooded dinosaurs. AAAS Selected Symposium 28. Boulder: Westview Press.

———. 1983. The deer flees, the wolf pursues: Incongruencies in predator-prey coevolution. Pages 350–382 in D. J. Futuyma and M. Slatkin (eds.), Coevolution. Sunderland: Sinauer.

Bakus, G. J. 1974. Toxicity in holothurians: A geographical pattern. Biotropica 6: 229–236.

Bakus, G. J., and G. Green. 1974. Toxicity in sponges and holothurians. Science 185: 951–953.

Bambach, R. K. 1977. Species richness in marine benthic habitats through the Phanerozoic. Paleobiology 3: 152–167.

———. 1993. Seafood through time: Changes in biomass, energetics, and productivity in the marine ecosystem. Paleobiology 19: 372–397.

———. 1999. Energetics in the global marine fauna: A connection between terrestrial diversification and change in the marine biosphere. Géobios 32: 131–144.

———. 2002. Supporting predators: Changes in the global ecosystem inferred from changes in predator diversity. Pages 319–351 in M. Kowalewski and P. H. Kelley (eds.), The fossil record of predation. Paleontological Society Papers 8.

Bambach, R. K., A. H. Knoll, and J. J. Sepkoski, Jr. 2002. Anatomical and ecological constraints on Phanerozoic animal diversity in the marine realm. Proceedings of the National Academy of Sciences of the USA 99: 6854–6859.

Barker Jorgensen, B. 2001. Space for hydrogen. Nature 412: 286–289.

Barnes, D. K., and M. H. Dick. 2000. Overgrowth competition between clades: Implications for interpretation of the fossil record and overgrowth indices. Biological Bulletin 199: 89–94.

Basu, A. R., M. I. Petaev, R. J. Poreda, S. B. Jacobsen, and L. Becker. 2003. Chondritic meteorite fragments associated with the Permian-Triassic boundary in Antarctica. Science 302: 1388–1392.

Bateman, R. M. 1994. Evolutionary-developmental change in the growth architecture of fossil rhizomorphic lycopsids: Scenarios constructed on cladistic foundations. Biological Reviews 69: 527–597.

Beamish, F.W.H. 1978. Swimming capacity. Pages 101–187 in W. S. Hoar and D. J. Randall (eds.), Fish physiology, Vol. 7 (Locomotion). New York: Academic Press.

Beard, C. 1998. East of Eden: Asia as an important center of taxonomic origination in mammalian evolution. Bulletin of Carnegie Museum of Natural History 34: 5–39.

Beck, A. L., and C. C. Labandeira. 1998. Early Permian insect folivory on a gigantopterid-dominated riparian flora from north-central Texas. Palaeogeography, Palaeoclimatology, Palaeoecology 142: 139–173.

Becker, G. S. 1976. The economic approach to human behavior. Chicago: University of Chicago Press: 314 pp.

Becker, L., R. J. Poreda, A. G. Hunt, T. E. Bunch, and R. Rampino. 2001. Impact event at the Permian-Triassic boundary: Evidence from extraterrestrial noble gases in fullerenes. Science 291: 1530–1531.

Beerling, D. J., and R. A. Berner. 2000. Impact of a Permo-Carboniferous high O_2 event on the terrestrial carbon cycle. Proceedings of the National Academy of Sciences of the USA 97: 12428–12434.

Beerling, D. J., F. I. Woodward, M. R. Lomas, M. A. Wills, W. P. Quick, and P. J. Valdes. 1998. The influence of Carboniferous palaeoatmospheres on plant function: An experimental and modeling assessment. Philosophical Transactions of the Royal Society of London B 353: 131–140.

Beerling, D. J., B. H. Lomax, D. L. Royer, G. R. Upchurch Jr., and L. R. Kump. 2002. An atmospheric pCO_2 reconstruction across the Cretaceous-Tertiary boundary from leaf megafossils. Proceedings of the National Academy of Sciences of the USA 99: 6836–6840.

Behrens Yamada, S., S. A. Navarrete, and C. Needham. 1998. Predation induced changes in behavior and growth rate in three populations of the intertidal snail, *Littorina sitkana* (Philippi). Journal of Experimental Marine Biology and Ecology 220: 213–226.

Bellwood, D. R. 1996. The Eocene fishes of Monte Bolca: The earliest coral reef fish assemblage. Coral Reefs 15: 11–19.

———. 2003. Origins and escalation of herbivory in fishes: A functional perspective. Paleobiology 29: 71–83.

Bengtson, S., and Y. Zhao. 1992. Predatorial borings in Late Precambrian mineralized exoskeletons. Science 257: 367–369.

Benson, W. W., K. S. Brown, Jr., and L. E. Gilbert. 1975. Coevolution of plants and herbivores: Passion flower butterflies. Evolution 29: 659–680.

Benton, M. J. 1985. Mass extinction among non-marine tetrapods. Nature 316: 811–814.

———. 1989. Mass extinction among tetrapods and the quality of the fossil record. Philosophical Transactions of the Royal Society of London B 325: 369–386.

———. 1997. Origin and early evolution of dinosaurs. Pages 204–215 in J. O. Farlow and M. K. Brett-Surman (eds.), The complete dinosaur. Bloomington: Indiana University Press.

Bercovici, D., and J. Mahoney. 1994. Double flood basalts and plume head separation at the 660-kilometer discontinuity. Science 266: 1367–1369.

Berger, J., and J. Caster. 1979. Convergent evolution between phyla: A test case of mimicry between caddisfly larvae (*Helicopsyche borealis*) and aquatic snails (*Physa integra*). Evolution 33: 511–513.

Berger, W. H., V. Smetacek, and G. Wefer (eds.). 1989. Productivity of the ocean: Past and present. Chichester: Wiley.

Berggren, W. A., and D. R. Prothero. 1992. Eocene-Oligocene climatic and biotic evolution: An overview. Pages 1–28 in D. R. Prothero and W. A. Berggren (eds.), Eocene-Oligocene climatic and biotic evolution. Princeton: Princeton University Press.

Berman, D. S., R. R. Reisz, D. Scott, A. C. Henrici, S. S. Sumida, and T. Martens. 2000. Early Permian bipedal reptile. Science 290: 969–972.

Bernays, E. A. 1986. Diet-induced head allometry among foliage-chewing insects and its importance for graminivores. Science 231: 495–497.

Bernays, E. A., and M. Graham. 1988. On the evolution of host specificity in phytophagous arthropods. Ecology 69: 886–892.

Berner, R. A. 1994. 3GEOCARB II: A revised model of atmospheric CO_2 over Phanerozoic time. American Journal of Science 294: 56–91.

———. 2002. Examination of hypotheses for the Permo-Triassic boundary extinction by global carbon cycle modeling. Proceedings of the National Academy of Sciences of the USA 99: 4172–4177.

———. 2003. The long-term carbon cycle, fossil fuels and atmospheric composition. Nature 426: 323–326.

Berner, R. A., and R. Raiswell. 1983. Burial of organic carbon and pyrite sulfur in sediments over Phanerozoic time: A new theory. Geochimica et Cosmochimica Acta 47: 855–862.

Berner, R. A., S. T. Petsch, J. A. Lake, D. J. Beerling, B. N. Popp, R. S. Lane, E. A. Laws, M. B. Westley, N. Cassar, F. I. Woodward, and W. P. Quick. 2000. Isotope fractionation and atmospheric oxygen: Implications for Phanerozoic O_2 evolution. Science 287: 1630–1633.

Berner, R. A., D. J. Beerling, R. Dudley, J. M. Robinson, and R. A. Wildman, Jr. 2003. Phanerozoic atmospheric oxygen. Annual Reviews of Earth and Planetary Science 31: 105–134.

Berry, W. B. N., and P. Wilde. 1978. Progressive ventilation of the oceans—an explanation for the distribution of the Lower Paleozoic black shales. American Journal of Science 278: 257–275.

Bertness, M. D. 1981a. Pattern and plasticity in tropical hermit crab growth and reproduction. American Naturalist 117: 754–773.

———. 1981b. Conflicting advantages in resource utilization: The hermit crab housing dilemma. American Naturalist 118: 432–437.

———. 1981c. Predation, physical stress, and the organization of a tropical rocky intertidal hermit crab community. Ecology 62: 411–425.

———. 1981d. Competitive dynamics of a tropical hermit crab assemblage. Ecology 62: 751–761.

———. 1982. Shell utilization, predation pressure, and thermal stress in Panamanian hermit crabs: An interoceanic comparison. Journal of Experimental Marine Biology and Ecology 64: 159–187.

———. 1985. Fiddler crab regulation of *Spartina alterniflora* production on a New England salt marsh. Ecology 66: 1042–1055.

Bertness, M. D., and R. Callaway. 1994. Positive interactions in communities. Trends in Ecology and Evolution 9: 191–193.

Bertness, M. D., and C. Cunningham. 1981. Crab shell-crushing predation and gastropod architectural defense. Journal of Experimental Marine Biology and Ecology 50: 213–230.

Bertness, M. D., and G. H. Leonard. 1997. The role of positive interactions in communities: Lessons from intertidal habitats. Ecology 78: 1976–1989.

Bertness, M. D., S. D. Garrity, and S. C. Levings. 1981. Predation pressure and gastropod foraging: A tropical-temperate comparison. Evolution 35: 995–1007.

Bird, K. I., and J. A. Cali. 1998. A million-year record of fire in sub-Saharan Africa. Nature 394: 767–769.

Birkeland, C. 1974. Interactions between a sea pen and seven of its predators. Ecological Monographs 44: 211–232.

———. 1977. The importance of rate of biomass accumulation in early successional stages of benthic communities to the survival of coral recruits. Proceedings of the Third International Coral Reef Symposium 1, Biology: 15–21.

———. 1982. Terrestrial runoff as a cause of outbreaks of *Acanthaster planci* (Echinodermata: Asteroidea). Marine Biology 69: 175–185.

———. 1989. Geographic comparisons of coral-reef community processes. Proceedings of the Sixth International Coral Reef Symposium 1: 211–220.

———. 1996. Why some species are especially influential on coral reef communities and others are not. Galaxea 13: 77–84.

Bjerrum, C. J., and D. E. Canfield. 2002. Ocean productivity before about 1.9 Gyr ago limited by phosphorus adsorption onto iron oxides. Nature 417: 159–162.

Blackstone, N. W. 1995. A units-of-evolution perspective on the endosymbiont theory of the origin of the mitochondrion. Evolution 49: 785–796.

———. 1998. Individuality in early eukaryotes and the consequences of metazoan development. Pages 23–43 in W.E.G. Müller (ed.), Progress in molecular and subcellular biology, Vol. 19. Berlin: Springer.

Blake, D. B. 1990. Adaptive zones of the class Asteroidea (Echinodermata). Bulletin of Marine Science 46: 701–718.

Blake, D. B., and T. E. Guensburg. 1994. Predation by the Ordovician asteroid *Promopalaeaster* on a pelecypod. Lethaia 27: 235–239.

Blob, R. W. 2001. Evolution of hindlimb posture in nonmammalian therapsids: Biomechanical tests of paleontological hypotheses. Paleobiology 27: 14–38.

Block, B. A., J. R. Finnerty, A.F.R. Stewart, and J. Kidd. 1993. Evolution of endothermy in fish: Mapping physiological traits on a molecular phylogeny. Science 260: 210–214.

Bloom, A. J., F. S. Chapin III, and H. A. Mooney. 1985. Resource limitation in plants—an economic analogy. Annual Reviews of Ecology and Systematics 16: 363–392.

Blundon, J. A. 1988. Morphology and muscle stress of chelae of temperate and tropical stone crabs *Menippe mercenaria*. Journal of Zoology, London, 215: 663–673.

Bobrick, B. 2001. Wide as the waters: The story of the English Bible and the revolution it inspired. New York: Simon and Schuster.

Bock, W. J. 1970. Microevolutionary sequences as a fundamental concept in macroevolutionary models. Evolution 24: 704–722.

Botsford, L. W., J. C. Castilla, and C. H. Peterson. 1997. The management of fisheries and marine ecosystems. Science 277: 509–515.

Bowen, L., and D. Van Vuren. 1997. Insular endemic plants lack defenses against herbivores. Conservation Biology 11: 1249–1254.

Bowler, J. M., H. Johnston, J. M. Olley, J. R. Prescott, R. G. Roberts, W. Shawcross, and N. A. Spooner. 2003. New ages for human occupation and climatic change at Lake Mungo, Australia. Nature 421: 837–840.

Bowring, S. A., D. H. Erwin, Y. G. Jin, M. W. Martin, K. Davidek, and W. Wang. 1998. U/Pb zircon geochronology and tempo of the end-Permian mass extinction. Science 280: 1039–1045.

Boyce, C. K., and A. H. Knoll. 2002. Evolution of developmental potential and the multiple independent origins of leaves in Paleozoic vascular plants. Paleobiology 28: 70–100.

Brady, S. G. 2003. Evolution of the army ant syndrome: The origin and long-term evolutionary stasis of a complex of behavioral and reproductive adaptations. Proceedings of the National Academy of Sciences of the USA 100: 6575–6579.

Bralower, T. J., D. J. Thomas, J. C. Zachos, M. M. Hirschmann, U. Röhl, H. Sigurdsson, E. Thomas, and D. L. Whitney. 1997. High-resolution records of the Late Paleocene thermal maximum and circum-Caribbean volcanism: Is there a causal link? Geology 25: 963–966.

Bramble, D. M. 1974. Emydid shell kinesis: Biomechanics and evolution. Copeia no. 3: 707–727.

Bramble, D. M., J. H. Hutchison, and J. M. Legler. 1984. Kinosternid shell kinesis: Structure, function and evolution. Copeia no. 2: 456–475.

Branch, G. M. 1979. Aggression by limpets against invertebrate predators. Animal Behaviour 27: 408–410.

———. 1981. The biology of limpets: Physical factors, energy flow, and ecological interactions. Oceanography and Marine Biology Annual Review 19: 235–380.

Brasier, M. D. 1995. Fossil indicators of nutrient levels. 1: Eutrophication and climate change. Pages 113–132 in D.W.J. Bosence and P. A. Allison (eds.), Marine palaeoenvironmental analysis from fossils. Geological Society Special Publication 83.

Brasier, M. D., and J. F. Lindsay. 1998. A billion years of environmental stability and the emergence of eukaryotes: New data from northern Australia. Geology 26: 555–558.

———. 2001. Did supercontinental amalgamation trigger the "Cambrian explosion"? Pages 69–89 in A. Yu. Zhuravlev and R. Riding (eds.), The ecology of the Cambrian radiation. New York: Columbia University Press.

Brasier, M. D., O. Green, and G. Shields. 1997. Ediacarian sponge spicule clusters from southwestern Mongolia and the origins of the Cambrian fauna. Geology 25: 303–306.

Braun, J., and W.-E. Reif. 1985. A survey of aquatic locomotion in fishes and tetrapods. Neues Jahrbuch für Geologie und Paläontologie Abhandlungen 169: 307–332.

Bremer, K. 2002. Gondwanan evolution of the grass alliance of families (Poales). Evolution 56: 1374–1387.

Brett, M. T., and C. R. Goldman. 1997. Consumer versus resource control in freshwater pelagic food webs. Science 275: 384–386.

Briggs, J. C. 1967a. Dispersal of tropical marine shore animals: Coriolis parameters or competition? Nature 216: 350.

———. 1967b. Relationships of the tropical shelf regions. Studies in Tropical Oceanography 5: 569–578.

Brocks, J. J., G. A. Logan, R. Buick, and R. E. Summons. 1999. Archean molecular fossils and the early rise of eukaryotes. Science 285: 1033–1036.

Brodie, E. D., III, and E. D. Brodie, Jr. 1991. Evolutionary response of predators to dangerous prey: Reduction of toxicity of newts and resistance of garter snakes in island populations. Evolution 45: 221–224.

Brongersma-Sanders, M. 1957. Mass mortality in the sea. Geological Society of America Memoir 67 (1): 941–1010.

Brooks, D. R., J. Collier, B. A. Maurer, J. D. H. Smith, and E. O. Wiley. 1989. Entropy and information in evolving biological systems. Biology and Philosophy 4: 407–432.

Brooks, W. R. 1988. The influence of the location and abundance of the sea anemone *Calliactis tricolor* (Le Sueur) in protecting hermit crabs from *Octopus* predators. Journal of Experimental Marine Biology and Ecology 116: 15–21.

Brower, L. P. 1958. Bird predation and foodplant specificity in closely related procryptic insects. American Naturalist 92: 183–187.

Brown, J. H. 1995. Macroecology. Chicago: University of Chicago Press: 269 pp.

Brown, K. M., and J. F. Quinn. 1988. The effect of wave action on growth in three species of intertidal gastropods. Oecologia (Berlin) 75: 420–425.

Brown, S. C., S. R. Cassuto, and R. W. Loos. 1979. Biomechanics of chelipeds in some decapod crustaceans. Journal of Zoology, London 188: 153–169.

Bryan, J. R., B. D. Carter, R. H. Fluegeman, Jr., D. K. Krumm, and T. A. Stemann. 1997. The Salt Mountain Limestone of Alabama. Tulane Studies in Geology and Paleontology 30: 1–60.

Bull, J. J. 1995. Virulence. Evolution 48: 1423–1430.

Bundle, M. W., H. Hoppeler, R. Vock, J. M. Tester, and P. G. Weyand. 1999. High metabolic rates in running birds. Nature 397: 31–32.

Bunt, J. S. 1975. Primary productivity of marine ecosystems. Pages 169–183 in H. Lieth and R. H. Whittaker (eds.), Primary productivity of the biosphere. New York: Springer.

Burness, G. P., J. Diamond, and T. Flannery. 2001. Dinosaurs, dragons, and dwarfs: The evolution of maximum body size. Proceedings of the National Academy of Sciences of the USA 98: 14518–14523.

Burney, D. A. 1999. Rates, patterns, and processes of landscape transformation and extinction in Madagascar. Pages 145–164 in R.D.E. MacPhee (ed.), Extinctions in near time: Causes, contexts, and consequences. New York: Kluwer/Plenum.

Buss, L. W. 1981. Group living, competition, and the evolution of cooperation in a sessile invertebrate. Science 213: 1012–1014.

———. 1987. The evolution of individuality. Princeton: Princeton University Press: 201 pp.

Buss, L. W., and A. Seilacher. 1994. The phylum Vendobionta: A sister group of the Eumetazoa? Paleobiology 20: 1–4.

Bustamante, R. H., G. M. Branch, and S. Eekhout. 1995a. Maintenance of an exceptional intertidal grazer biomass in South Africa: Subsidy by subtidal kelps. Ecology 76: 2314–2329.

Bustamante, R. H., G. M. Branch, S. Eekhout, B. Robertson, P. Zoutendyk, M. Schleyer, A. Dye, N. Hanekom, D. Keats, M. Jurd, and C. McQuaid. 1995b. Gradients of intertidal primary productivity around the coast of South Africa and their relationships with consumer biomass. Oecologia 102: 189–201.

Butterfield, N. J. 1994. Burgess Shale-type fossils from a Lower Cambrian shallow-shelf sequence in northwestern Canada. Nature 369: 477–479.

———. 1997. Plankton ecology in the Proterozoic-Phanerozoic transition. Paleobiology 23: 247–262.

Butterfield, N. J. 2000. *Bangiomorpha pubescens* n. gen., n. sp.: Implications for the evolution of sex, multicellularity, and the Mesoproterozoic-Neoproterozoic radiation of eukaryotes. Paleobiology 26: 386–404.

———. 2001. Ecology and evolution of Cambrian plankton. Pages 200–216 in A. Yu. Zhuravlev and R. Riding (eds.), The ecology of the Cambrian radiation. New York: Columbia University Press.

———. 2002. *Leanchoilia* guts and the interpretation of three-dimensional structures in Burgess Shale-type fossils. Paleobiology 28: 155–171.

Campbell, I. H., G. K. Czamanske, V. A. Fedorenko, R. I. Hill, and V. Stepanov. 1992. Synchronism of the Siberian Traps and the Permian-Triassic boundary. Science 258: 1760–1763.

Canfield, D. E. 1998. A new model for Proterozoic ocean chemistry. Nature 396: 450–453.

Canfield, D. E., and A. Teske. 1996. Late Proterozoic rise in atmospheric oxygen concentration inferred from phylogenetic and sulfur-isotope studies. Nature 382: 127–132.

Caplan, M. L., R. M. Bustin, and K. A. Grimm. 1996. Demise of a Devonian-Carboniferous carbonate ramp by eutrophication. Geology 24: 715–718.

Cardinale, B. J., M. A. Palmer, and S. L. Collins. 2002. Species diversity enhances ecosystem functioning through interspecific facilitation. Nature 415: 426–429.

Carlquist, S. 1966. The biota of long-distance dispersal. II Loss of dispersibility in Pacific Compositae. Evolution 20: 30–48.

Carlton, J. T., G. J. Vermeij, D. R. Lindberg, D. A. Carlton, and E. C. Dudley. 1991. The first historical extinction of a marine invertebrate in an ocean basin: The demise of the eelgrass limpet *Lottia alveus*. Biological Bulletin 180: 72–80.

Carlton, J. T., J. B. Geller, M. L. Reaka-Kudla, and E. A. Norse. 1999. Historical extinctions in the sea. Annual Reviews in Ecology and Systematics 30: 515–538.

Carpenter, G.D.H. 1942. The relative frequency of beak-marks on butterflies of different edibility to birds. Proceedings of the Zoological Society of London (A) 111: 223–231.

Carrano, M. T. 2000. Homoplasy and the evolution of dinosaur locomotion. Paleobiology 26: 489–512.

Carrier, D. R. 1987. The evolution of locomotor stamina in tetrapods: Circumventing a mechanical constraint. Paleobiology 13: 326–341.

Carroll, S. B. 2001. Chance and necessity: the evolution of morphological complexity and diversity. Nature 409: 1102–1109.

Caswell, H., F. Reed, S. N. Stephenson, and P. A. Werner. 1973. Photosynthetic pathways and selective herbivory: A hypothesis. American Naturalist 107: 465–480.

Catling, D. C., K. J. Zahnle, and C. P. McKay. 2001. Biogenic methane, hydrogen escape, and the irreversible oxidation of early Earth. Science 293: 839–843.

Cebrián, J. 1999. Patterns in the fate of production in plant communities. American Naturalist 154: 449–468.

Chadwick, O. A., L. A. Derry, P. M. Vitousek, B. J. Huebert, and L. O. Hedin. 1999. Changing sources of nutrients during four million years of ecosystem development. Nature 397: 491–497.

Chaffee, C., and D. R. Lindberg. 1986. Larval ecology of Early Cambrian molluscs: The implications of small body size. Bulletin of Marine Science 39: 536–549.

Chai, P., and R. Dudley. 1995. Limits to vertebrate locomotor energetics suggested by hummingbirds hovering in heliox. Nature 377: 722–725.

Chai, P., and R. B. Srygley. 1990. Predation and the flight, morphology, and temperature of neotropical rain-forest butterflies. American Naturalist 135: 748–765.

Chaisson, E. J. 2001. Cosmic evolution: The rise of complexity in nature. Cambridge: Harvard University Press: 274 pp.

Chamberlain, C. K. 1975. Recent lebenspuren in nonmarine aquatic environments. Pages 431–458 in R. W. Frey (ed.), The study of trace fossils. New York: Springer.

Chamberlain, J. A., Jr. 1991. Cephalopod locomotor design and evolution: The constraints of jet propulsion. Pages 57–98 in J. M. V. Rayner and R. J. Wootton (eds.), Biomechanics in evolution. Cambridge: Cambridge University Press.

Chancellor, E. 1999. Devil take the hindmost: A history of financial speculation. New York: Plume: 387 pp.

Chapelle, F. H., K. O'Neill, P. M. Bradley, B. A. Methé, S. A. Ciufo, L. L. Knobel, and D. R. Lovley. 2002. A hydrogen-based subsurface microbial community dominated by methanogens. Nature 415: 312–315.

Chapin, F. S., III, B. H. Walker, R. J. Hobbs, D. U. Hooper, J. H. Lawton, O. E. Sala, and D. Tilman. 1997. Biotic control over the functioning of ecosystems. Science 277: 500–504.

Chapman, R. L., M. A. Buchheim, C. F. Delwiche, T. Friedl, V.A.R. Huss, K. G. Karol, L. A. Lewis, J. Manhart, R. M. McCourt, J. L. Olsen, and D. A. Waters. 1998. Molecular systematics of the green algae. Pages 508–540 in D. E. Soltis, P. S. Soltis, and J. J. Doyle (eds.), Molecular systematics of plants II. Boston: Kluwer.

Chapman, T., and L. Partridge. 1996. Sexual conflict as fuel for evolution. Nature 381: 189–190.

Charnov, E. 1997. Trade-off-invariant rules for evolutionarily stable life histories. Nature 387: 393–394.

Chase, J. M., and M. A. Leibold. 2002. Spatial scale dictates the productivity-diversity relationship. Nature 416: 427–430.

Checa, A. G. 2002. Fabricational morphology of oblique ribs in bivalves. Journal of Morphology 254: 195–209.

Checa, A. G., and A. P. Jiménez-Jiménez. 2003. Evolutionary morphology of oblique ribs of bivalves. Palaeontology 46: 709–724.

Chen, J.-Y., and D.-Y. Huang. 2002. A possible Lower Cambrian chaetognath (arrow worm). Science 298: 187.

Chen, J.-Y., L. Ramsköld, and G.-Q. Zhou. 1994. Evidence for monophyly and arthropod affinity of Cambrian giant predators. Science 264: 1304–1308.

Cisne, J. L. 1974. Evolution of the world fauna of aquatic free-living arthropods. Evolution 28: 337–366.

Clack, J. A. 2002. An early tetrapod from "Romer's gap." Nature 418: 72–76.

Claeys, P., J.-G. Casier, and S. V. Margolis. 1992. Microtectites and mass extinctions: Evidence for a Late Devonian asteroid impact. Science 257: 1102–1104.

Clark, M. A., N. A. Moran, P. Baumann, and J. J. Wernegreen. 2000. Cospeciation between bacterial endosymbionts (*Buchnera*) and a recent radiation of aphids (*Uroleucon*) and pitfalls of testing for phylogenetic congruence. Evolution 54: 517–525.

Clark, N., and C. Juma. 1987. Long-run economics: An evolutionary approach to economic growth. London: Pinter: 206 pp.

Clarke, A. 1980. A reappraisal of the concept of metabolic cold adaptation in polar marine invertebrates. Biological Journal of the Linnean Society 14: 77–92.

———. 1983. Life in cold water: The physiological ecology of polar marine ectotherms. Oceanography and Marine Biology Annual Review 21: 341–453.

———. 1993. Temperature and extinction in the sea: A physiologist's view. Paleobiology 19: 499–518.

Clarke, A., and N. Johnson. 1999. Scaling of metabolic rate with body mass and temperature in teleost fish. Journal of Animal Ecology 68: 893–905.

Clarke, B., J. Murray, and M. S. Johnson. 1984. The extinction of endemic species by a program of biological control. Pacific Science 38: 97–104.

Cleal, C. J., R. M. James, and E. L. Zodrow. 1999. Variation in stomatal density in the Late Carboniferous gymnosperm frond *Neuropteris ovata*. Palaios 14: 180–185.

Clifton, K. E. 1997. Mass spawning by green algae on coral reefs. Science 275: 116–118.

Coates, A. G., and J.B.C. Jackson. 1987. Clonal growth, algal symbiosis, and reef formation by corals. Paleobiology 13: 361–378.

Coddington, J. A. 1988. Cladistic tests of adaptational hypotheses. Cladistics 4: 3–22.

Codispoti, L. A. 1997. The limits to growth. Nature 387: 237–238.

Coffin, M. F., and O. Eldholm. 1993. Scratching the surface: Estimating dimensions of large igneous provinces. Geology 21: 515–519.

———. 1994. Large igneous provinces: Crustal structure, dimensions, and external consequences. Reviews of Geophysics 32: 1–36.

Cohen, A. S., and A. L. Coe. 2002. New geochemical evidence for the onset of volcanism in the Central Atlantic Magmatic Province and environmental change at the Triassic-Jurassic boundary. Geology 30: 267–270.

Cohen, J. E. 1995. How many people can the Earth support? New York: Norton: 532 pp.

Colbath, G. K. 1986. Abrupt terminal Ordovician extinction in phytoplankton associations, southern Appalachians. Geology 14: 943–946.

Coley, P. D., and J. A. Barone. 1996. Herbivory and plant defenses in tropical forests. Annual Reviews of Ecology and Systematics 27: 305–335.

Coley, P. D., J. P. Bryant, and F. S. Chapin III. 1985. Resource availability and plant antiherbivore defense. Science 230: 895–899.

Colinvaux, P. 1980. The fates of nations: A biological theory of history. New York: Simon and Schuster: 383 pp.

Collin, R., and C. M. Janis. 1997. Morphological constraints on tetrapod feeding mechanisms: Why were there no suspension-feeding marine reptiles? Pages 451–466 in J. M. Callaway and E. L. Nicholls (eds.), Ancient marine reptiles. San Diego: Academic Press.

Collins, A. G., J. H. Lipps, and J. W. Valentine. 2000. Modern mucociliary creeping trails and the body plans of Neoproterozoic trace-makers. Paleobiology 26: 47–55.

Collins, D. 1996. The "evolution" of *Anomalocaris* and its classification in the arthropod class Dinocarida (nov.) and order Radiodonta (nov.). Journal of Paleontology 70: 280–293.

Condie, K. C. 1998. Episodic continental growth and supercontinents: A mantle avalanche connection. Earth and Planetary Science Letters 163: 97–108.

Connell, J. H., and R. O. Slatyer. 1977. Mechanisms of succession in natural communities and their role in community stability and organization. American Naturalist 111: 1119–1144.

Conner, J., and T. Eisner. 1983. Capture of bombardier beetles by ant lion larvae. Psyche 90: 175–178.

Conway Morris, S., and S. Bengtson. 1994. Cambrian predators: Possible evidence for boreholes. Journal of Paleontology 68: 1–23.

Cook, C.D.K. 1999. The number and kinds of embryo-bearing plants which have become aquatic: A survey. Perspectives in Plant Ecology, Evolution and Systematics 2: 79–102.

Cook, P. L., and P. J. Chimonides. 1983. A short history of the lunulite Bryozoa. Bulletin of Marine Science 33: 566–581.

Cooper, S. M., and N. Owen-Smith. 1986. Effects of plant spinescence on large mammalian herbivores. Oecologia (Berlin) 68: 446–455.

Cope, J. C. W. 1996. The early evolution of the Bivalvia. Pages 361–370 in J. D. Taylor (ed.), Origin and evolutionary radiation of the Mollusca. Oxford: Oxford University Press.

Corner, E.J.H. 1964. The life of plants. New York: World Publishing Company.

Cornette, J. L., B. S. Lieberman, and R. H. Goldstein. 2002. Documenting a significant relationship between macroevolutionary origination and Phaneroozic CO_2 levels. Proceedings of the National Academy of Sciences of the USA 99: 7832–7835.

Corruccini, R. S., and R. M. Beecher. 1982. Occlusal variation related to soft diet in a nonhuman primate. Science 218: 74–76.

Côté, I. M. 1995. Effects of predatory crab effluent on byssus production in mussels. Journal of Experimental Marine Biology and Ecology 188: 233–241.

Courtillot, V., J. Besse, D. Vandamme, R. Montigny, J. J. Jaeger, and H. Cappetta. 1986. Deccan flood-basalts at the Cretaceous/Tertiary boundary? Earth and Planetary Science Letters 80: 361–374.

Cowen, R. 1973. Respiration in metazoan evolution. Evolution 27: 696–701.

———. 1981. Crinoid arms and banana plantations: An economic harvesting analogy. Paleobiology 7: 332–343.

———. 1983. Algal symbiosis and its recognition in the fossil record. Pages 431–478 in M.J.S. Tevesz and P. L. McCall (eds.), Biotic interactions in Recent and fossil benthic communities. New York: Plenum.

———. 1996. Locomotion and respiration in aquatic air-breathing vertebrates. Pages 337–352 in D. Jablonski, D. H. Erwin, and J. H. Lipps (eds.), Evolutionary paleobiology: In honor of James W. Valentine. Chicago: University of Chicago Press.

Cox, P. A. 1983. Extinction of the Hawaiian avifauna resulted in a change of pollinators for the ieie, *Freycinetia arborea*. Oikos 41: 195–199.

Crane, P. R. 1987. Vegetational consequences of the angiosperm diversification. Pages 108–144 in E. M. Friis, W. G. Chaloner, and P. R. Crane (eds.), The origins of angiosperms and their biological consequences. Cambridge: Cambridge University Press.

Crane, P. R., and S. Lidgard. 1989. Angiosperm diversification and paleolatitudinal gradients in Cretaceous floristic diversity. Science 246: 675–678.

Crane, P. R., E. M. Friis, and K. R. Pedersen. 1995. The origin and early diversification of angiosperms. Nature 374: 27–33.

Crepet, W. L., and G. D. Feldman. 1991. The earliest remains of grasses in the fossil record. American Journal of Botany 78: 1010–1014.

Cressler, W. L. 2001. Evidence of earliest known wildfires. Palaios 16: 171–174.

Cristoffer, C., and C. A. Peres. 2003. Elephants versus butterflies: The ecological role of herbivores in the evolutionary history of two tropical worlds. Journal of Biogeography 30: 1357–1380.

Critchfield, R. 1981. Villages. New York: Doubleday: 366 pp.

Cronin, T. M., and H. J. Dowsett. 1996. Biotic and oceanographic response to the Pliocene closing of the Central American isthmus. Pages 76–104 in J.B.C. Jackson, A. F. Budd, and A. G. Coates (eds.), Evolution and environment in tropical America. Chicago: University of Chicago Press.

Cronk, Q.C.B. 1980. Extinction and survival in the endemic vascular flora of Ascension Island. Biological Conservation 17: 207–219.

Crouch, E. M., K. Heilmann-Clausen, H. Brinkhuis, H.E.G. Morgans, K. M. Rogers, H. Egger, and B. Schmitz. 2001. Global dinoflagellate event associated with the Late Paleocene thermal maximum. Geology 29: 315–318.

Cullen, J. J. 1999. Iron, nitrogen and phosphorus in the ocean. Nature 402: 372.

Currie, C. R., B. Wong, A. E. Stuart, S. R. Schultz, S. A. Rehner, U. G. Mueller, G.-H. Sung, J. W. Spatafora, and N. A. Straus. 2003. Ancient tripartite coevolution in the attine ant-microbe symbiosis. Science 299: 386–388.

Currie, P. J. 1997. Theropods. Pages 217–233 in J. O. Farlow and K. Brett-Surman (eds.), The complete dinosaur. Bloomington: Indiana University Press.

Curry, G. B., A. D. Ansell, M. James, and L. Peck. 1989. Physiological constraints on living and fossil brachiopods. Transactions of the Royal Society of Edinburgh: Earth Sciences 80: 255–262.

Dagys, A. S., U. Lehmann, K. Bandel, K. Tanabe, and W. Weitschat. 1989. The jaw apparati of ectocochleate cephalopods. Paläontologische Zeitschrift 63: 41–53.

Dalsgaard, T., D. E. Canfield, J. Petersen, B. Thamdrup, and J. Acuña-González. 2003. N_2 production by the anammox reaction in the anoxic water column of Golfo Dulce, Costa Rica. Nature 422: 606–608.

Daly, H. E. 1996. Beyond growth: The economics of sustainable development. Boston: Beacon: 254 pp.

Daly, H. E., and J. B. Cobb, Jr. 1994. For the common good: Redirecting the economy toward community, the environment, and a sustainable future (second edition). Boston: Beacon: 534 pp.

Damuth, J. 1993. Cope's rule: The island rule and the scaling of mammalian population density. Nature 365: 748–750.

Daniel, T. L., B. S. Helmuth, W. B. Saunders, and P. D. Ward. 1997. Septal complexity in ammonoid cephalopods increased mechanical risk and limited depth. Paleobiology 23: 470–481.

Darveau, C.-A., R. K. Suarez, R. D. Andrews, and P. W. Hochachka. 2002. Allometric cascade as a unifying principle of body mass effects on metabolism. Nature 417: 166–170.

Darwin, C. 1859. The origin of species by natural selection or the preservation of favored races in the struggle for life. New York: Appleton: 501 pp.

Davidson, D. W. 1997. The role of resource imbalances in the evolutionary ecology of tropical arborial ants. Biological Journal of the Linnean Society 61: 153–181.

Davidson, D. W., and D. McKey. 1993. The evolutionary ecology of symbiotic ant-plant relationships. Journal of Hymenoptera Research 2: 13–83.

Davis, R. A., R.H.B. Fraaye, and C. H. Holland. 2001. Trilobites within nautiloid cephalopods. Lethaia 34: 37–45.

Dawkins, R. 1976. The selfish gene. Oxford: Oxford University Press: 215 pp.

———. 1995. River out of Eden: A Darwinian view of life. New York: Basic: 172 pp.

———. 1996. Climbing Mount Improbable. New York: Norton: 340 pp.

Dawkins, R., and J. R. Krebs. 1979. Arms races between and within species. Proceedings of the Royal Society of London B 205: 489–511.

Day, J. J., D. B. Norman, P. Upchurch, and H. P. Powell. 2002. Dinosaur locomotion from a new trackway. Nature 415: 494–495.

Day, R. W., A. Barkai, and P. A. Wickens. 1991. Trapping of three drilling whelks by two species of mussels. Journal of Experimental Marine Biology and Ecology 149: 109–122.

Dayton, P. K. 1971. Competition, disturbance, and community organization: The provision and subsequent utilization of space in a rocky intertidal community. Ecological Monographs 41: 351–389.

———. 1990. Polar benthos. Pages 631–685 in W. O. Smith Jr. (ed.), Polar oceanography, Part B: Chemistry, biology, and geology. San Diego: Academic Press.

Dayton, P. K., R. J. Rosenthal, L. C. Mahen, and T. Antezana. 1977. Population structure and foraging biology of the predaceous Chilean asteroid *Meyenaster gelatinosus* and the escape biology of its prey. Marine Biology 39: 361–370.

DeAngelis, D. L. 1992. Dynamics of nutrient cycling and food webs. London: Chapman and Hall: 270 pp.

Debrenne, F., and J. Reitner. 2001. Sponges, cnidarians, and ctenophores. Pages 301–325 in A. Yu. Zhuravlev and R. Riding (eds.), The ecology of the Cambrian radiation. New York: Columbia University Press.

De Groot, D. S. 1927. The California clapper rail: Its nesting habits, enemies and habitat. Condor 29: 259–270.

Deline, B., T. Baumiller, P. Kaplan, M. Kowalewski, and P. Hoffmeister. 2003. Edge-drilling on the brachiopod *Perditocardinia dubia* from the Mississippian of Missouri (USA). Palaeogeography, Palaeoclimatology, Palaeoecology 201: 211–219.

DeLucia, E. H., J. G. Hamilton, S. L. Naidu, R. B. Thomas, J. A. Andrews, A. Finzi, M. Lavine, R. Matamala, J. E. Mohan, G. R. Hendrey, and W. H. Schlesinger. 1999. Net primary production of a forest ecosystem with experimental CO_2 enrichment. Science 284: 1177–1179.

deMenocal, P. B. 2001. Cultural responses to climate change during the Late Holocene. Science 292: 667–673.

Denison, R. 1978. Placodermi. Handbook of Paleoichthyology, Vol. 2. Stuttgart: Gustav Fischer: 128 pp.

———. 1979. Acanthodii. Handbook of Paleoichthyology, Vol. 5. Stuttgart: Gustav Fischer: 62 pp.

Dennett, D. C. 1995. Darwin's dangerous idea: Evolution and the meaning of life. New York: Simon and Schuster: 587 pp.
Denny, M. W. 1988. Biology and the mechanics of the wave-swept environment. Princeton: Princeton University Press: 329 pp.
———. 1993. Air and water: The biology and physics of life's media. Princeton: Princeton University Press: 341 p.
———. 1999. The mechanical limits to size in wave-swept organisms. Journal of Experimental Biology 202: 3463–3466.
Denny, M. W., T. L. Daniel, and M.A.R. Koehl. 1985. Mechanical limits to size in wave-swept organisms. Ecological Monographs 55: 69–102.
Des Marais, D. J. 1994. Tectonic control of the crustal organic carbon reservoir during the Precambrian. Chemical Geology 114: 303–314.
Des Marais, D. J., H. Strauss, R. E. Summons, and J. M. Hayes. 1992. Carbon isotope evidence for the stepwise oxidation of the Proterozoic environment. Nature 359: 605–609.
D'Hondt, S., and M. A. Arthur. 1996. Late Cretaceous oceans and the cool tropic paradox. Science 271: 1838–1841.
D'Hondt, S., P. Donaghay, J. C. Zachos, D. Luttenberg, and M. Lindinger. 1998. Organic carbon fluxes and ecological recovery from the Cretaceous-Tertiary mass extinction. Science 282: 276–279.
Diamond, J. M. 1973. Distributional ecology of New Guinea birds. Science 179: 759–769.
———. 1974. Colonization of exploded volcanic islands by birds: The supertramp strategy. Science 184: 803–806.
———. 1984. Historic extinctions: A Rosetta Stone for understanding prehistoric extinctions. Pages 824–862 in P. S. Martin and R. G. Klein (eds.), Quaternary extinctions: A prehistoric revolution. Tucson: University of Arizona Press.
———. 1990. Biological effects of ghosts. Nature 345: 769–770.
———. 1992. Rubbish birds are poisonous. Nature 360: 19–20.
———. 1996. Competition for brain space. Nature 382: 756–757.
———. 1997. Guns, germs, and steel: The fate of human societies. New York: Norton: 480 pp.
———. 1998a. Peeling the Chinese onion. Nature 391: 433–434.
———. 1998b. Ants, crops, and history. Science 281: 1974–1975.
Diamond, J. M., and C. R. Veitsch. 1981. Extinctions and introductions in the New Zealand avifauna: Cause and effect? Science 211:499–501.
Dickens, G. R., J. R. O'Neil, D. K. Rea, and R. N. Owen. 1995. Dissociation of oceanic methane hydrate as a cause of the carbon isotope excursion at the end of the Paleocene. Paleoceanography 10: 965–971.
Dickens, G. R., M. M. Castillo, and J. C. G. Walker. 1997. A blast of gas in the latest Paleocene: Simulating first-order effects of massive dissociation of oceanic methane hydrate. Geology 25: 259–262.
Diekmann, O. E., R.P.M. Rak, L. Tonk, W. T. Stam, and J. L. Olsen. 2002. No habitat correlation of zooxanthellae in the coral genus *Madracis* on a Curaçao reef. Marine Ecology Progress Series 227: 221–232.
Dietl, G. P. 2003. Interaction strength between a predator and dangerous prey: *Sinistrofulgur* predation on *Mercenaria*. Journal of Experimental Marine Biology and Ecology 289: 287–301.
DiMichele, W. A., and T. L. Phillips. 1985. Arborescent lycopod reproduction and paleoecology in a coal swamp environment of Late Middle Pennsylvanian age (Herrin Creek, Illinois, U.S.A.). Review of Palaeobotany and Palynology 44: 1–26.
Dingus, L. 1984. Effects of stratigraphic completeness on interpretations of extinction rates across the Cretaceous-Tertiary boundary. Paleobiology 10: 420–438.
Dingus, L., and P. M. Sadler. 1982. The effects of stratigraphic completeness on estimates of evolutionary rate. Systematic Zoology 31: 400–412.

Dockery, D. T., III. 1998. Molluscan faunas across the Paleocene/Eocene series boundary in the North American Gulf Coastal Plain. Pages 296–322 in M.-P. Aubry, S. Lucas, and W. A. Berggren (eds.), Late Paleocene–early Eocene climatic and biotic events in the marine and terrestrial records. New York: Columbia University Press.

Dockery, D. T., III, and P. Lozouet. 2003. Molluscan faunas across the Eocene/Oligocene boundary in the North American Gulf coastal plain, with comparisons to those of the Eocene and Oligocene of France. Pages 303–340 in D. R. Prothero, L. C. Ivany, and E. A. Nesbitt (eds.), From greenhouse to icehouse: The marine Eocene-Oligocene transition. New York: Columbia University Press.

Domning, D. P. 1978. Sirenian evolution in the North Pacific Ocean. University of California Publications in Geological Sciences 118: 1–176.

———. 1989. Kelp evolution: A comment. Paleobiology 15:53–56.

———. 2001a. Sirenians, seagrasses, and Cenozoic ecological change in the Caribbean. Palaeogeography, Palaeoclimatology, Palaeoecology 166: 27–50.

———. 2001b. The earliest known fully quadrupedal sirenian. Nature 413: 625–627.

Domning, D. P., C. E. Ray, and M. C. McKenna. 1986. Two new Oligocene desmostylians and a discussion of tethytherian systematics. Smithsonian Contributions to Paleobiology 59: 1–56.

Donn, T. E., and S. F. Els. 1990. Burrowing times of *Donax serra* from the south and west coasts of South Africa. Veliger 33: 355–358.

Douglas, A. E. 1994. Symbiotic interactions. Oxford: Oxford University Press: 148 pp.

Douthwaite, R. 1992. The growth illusion. Bideford: Green Books: 367 pp.

Downing, A. L., and M. A. Leibold. 2002. Ecosystem consequences of species richness and composition in pond food webs. Nature 416: 837–844.

Dransfield, J., J. R. Flenley, S. M. King, D. D. Harkness, and S. Rapu. 1984. A recently extinct palm from Easter Island. Nature 312: 750–752.

Droser, M. L., and D. J. Bottjer. 1989. Ordovician increase in extent and depth of bioturbation: Implications for understanding Early Paleozoic space utilization. Geology 17: 850–852.

Droser, M. L., and X. Li. 2001. The Cambrian radiation and the diversification of sedimentary fabrics. Pages 137–169 in A. Yu. Zhuravlev and R. Riding (eds.), The ecology of the Cambrian radiation. New York: Columbia University Press.

Droser, M. L., and P. M. Sheehan. 1997. Paleoecology of the Ordovician radiation; resolution of large scale patterns with individual clade histories, paleogeography and environments. Géobios M.S. 20: 221–229.

Droser, M. L., N. C. Hughes, and P. A. Jell. 1994. Infaunal communities and tiering in Early Paleozoic nearshore clastic environments: Trace-fossil evidence from the Cambro-Ordovician of New South Wales. Lethaia 27: 273–283.

Droser, M. L., J. G. Gehling, and S. Jensen. 1999. When the worm turned: Concordance of Early Cambrian ichnofabric and trace fossil record in siliciclastic rocks of South Australia. Geology 27: 625–628.

Droser, M. L., S. Jensen, J. G. Gehling, P. M. Myrow, and G. M. Narbonne. 2002a. Lowermost Cambrian ichnofabrics from the Chapel Island Formation, Newfoundland: Implications for Cambrian substrates. Palaios 17: 3–15.

Droser, M. L., S. Jensen, and J. G. Gehling. 2002b. Trace fossils and substrates of the terminal Proterozoic-Cambrian transition: Implications for the record of early bilaterians and sediment mixing. Proceedings of the National Academy of Sciences of the USA 99: 12572–12576.

Duarte, C. M. 1995. Submerged aquatic vegetation in relation to different nutrient regimes. Ophelia 41: 87–112.

Duarte, C. M., and S. Agustí. 1998. The CO_2 balance of unproductive aquatic ecosystems. Science 281: 234–236.

Dudley, E. C., and G. J. Vermeij. 1989. Shell form and burrowing performance in gastropods from Pacific Panama, with comments on regional differences in functional specialization. Veliger 32: 284–287.

Dudley, R. 1995. Extraordinary flight performance of orchid bees (Apidae: Euglossini) hovering in heliox (80% He/20% O_2). Journal of Experimental Biology 198: 1065–1070.

———. 1998. Atmospheric oxygen, giant Paleozoic insects and the evolution of aerial locomotor performance. Journal of Experimental Biology 201: 1043–1050.

———. 2000a. Evolutionary origins of human alcoholism in primate frugivory. Quarterly Review of Biology 75: 3–15.

———. 2002b. The biomechanics of insect flight: Form, function, evolution. Princeton: Princeton University Press: 476 pp.

Dudley, R., and P. DeVries. 1990. Tropical rain forest structure and the geographical distribution of gliding vertebrates. Biotropica 22: 432–434.

Dudley, R., and G. J. Vermeij. 1992. Do the power requirements of flapping flight constrain folivory in flying animals? Functional Ecology 6: 101–104.

Dugdale, R. C., and F. P. Wilkerson. 1998. Silicate regulation of new production in the equatorial Pacific upwelling. Nature 391: 270–273.

Duggins, D. O., C. A. Simenstad, and J. A. Estes. 1989. Magnification of secondary production by kelp detritus in coastal marine ecosystems. Science 245: 170–173.

Dumbacher, J. P. B. M. Beehler, T. F. Spande, H. M. Garraffo, and J. W. Daly. 1992. Homobatrachotoxin in the genus *Pitohui*: Chemical defense in birds? Science 258: 799–801.

Dunn, D. F., D. M. Devaney, and B. Roth. 1981. *Stylobates*: A shell-forming sea anemone (Coelenterata: Anthozoa, Actiniidae). Pacific Science 34: 379–388.

Eaton, R. L. 1974. The cheetah: The biology, ecology, and behavior of an endangered species. New York: Van Nostrand Reinhold: 178 pp.

Edmunds, M. 1974. Defense in animals: A survey of anti-predator defenses. Harlow: Longman: 357 pp.

Eigen, M., with R. Winkler-Oswatitsch. 1992. Steps toward life: A perspective on evolution. (Translated by P. Woolley.) Oxford: Oxford University Press: 173 pp.

Ekdale, A. A., and R. G. Bromley. 2001. Bioerosional innovation for living in carbonate hardgrounds in the Early Ordovician of Sweden. Lethaia 34: 1–12.

Eldredge, N. 1971. The allopatric model and phylogeny in Paleozoic invertebrates. Evolution 25: 156–167.

Eldredge, N., and S. J. Gould. 1972. Punctuated equilibria: An alternative to phyletic gradualism. Pages 82–115 in T.J.M. Schopf (ed.), Models in paleobiology. San Francisco: Freeman, Cooper, and Company.

Elick, J. M., S. G. Driese, and C. I. Mora. 1998. Very large plant and root traces from the Early to Middle Devonian: Implications for early terrestrial ecosystems and atmospheric $p(CO_2)$. Geology 26: 143–146.

Elliott, J. P., I. M. Cowan, and C. S. Holling. 1977. Prey capture by the African lion. Canadian Journal of Zoology 55: 1811–1828.

Ellwood, B. B., S. L. Benoist, L. El Hassani, C. Wheeler, and R. E. Crick. 2003. Impact ejecta layer from the mid-Devonian: Possible connection to global mass extinctions. Science 300: 1735–1738.

El-Nakhal, H., and K. Bandel. 1991. Geographical distribution of the small gastropod genus *Scaliola*. Micropaleontology 38: 423–424.

Elner, R. W., and A. Campbell. 1981. Force, function and mechanical advantage in the chelae of the American lobster *Homarus americanus* (Decapoda: Crustacea). Journal of Zoology, London, 193: 269–286.

Emmons, L. H., and A. H. Gentry. 1983. Tropical forest structure and the distribution of gliding and prehensile-tailed vertebrates. American Naturalist 121: 513–524.

Endler, J. A. 1986. Defense against predators. Pages 109–134 in M. E. Feder and G. V. Lauder (eds.), Predator-prey relationships: Perspectives and approaches from the study of lower vertebrates. Chicago: University of Chicago Press.

Engeser, T. S. 1990. Major events in cephalopod evolution. Pages 119–138 in P. D. Taylor and G. P. Larwood (eds.), Major evolutionary radiations. Oxford: Clarendon.

Erickson, G. M., S. D. Van Kirk, G. Su, M. E. Levenston, W. E. Caler, and D. R. Carter. 1996. Bite-force estimation of *Tyrannosaurus rex* from tooth-marked bones. Nature 382: 706–708.

Erwin, D. H. 1999. The origin of body plans. American Zoologist 39: 617–629.

Erwin, D. H., and T. A. Vogel. 1992. Testing for causal relationships between large pyroclastic volcanic eruptions and mass extinctions. Geophysical Research Letters 19: 893–896.

Estes, J. A., and J. F. Palmisano. 1974. Sea otters: Their role in structuring nearshore communities. Science 185: 1058–1060.

Estes, J. A., and P. D. Steinberg. 1988. Predation, herbivory, and kelp evolution. Paleobiology 14: 19–36.

Estes, J. A., D. O. Duggins, and G. B. Rathbun. 1989. The ecology of extinctions in kelp forest communities. Conservation Biology 3: 252–264.

Estes, J. A., M. T. Tinker, T. M. Williams, and D. F. Doak. 1998. Killer whale predation on sea otters linking oceanic and nearshore ecosystems. Science 282: 473–476.

Estes, R. 1983. *Sauria terrestria*, Amphisbaenia. Encyclopedia of Palaeoherpetology 10–11. Stuttgart: Gustav Fischer: 249 pp.

Estes, R. D., and J. Goddard. 1967. Prey selection and hunting behavior of the African wild dog. Journal of Wildlife Management 31: 52–70.

Eston, V. R. de, M. E. Migatto, E. C. de Oliveira Filho, S. de Almeida Rodrigeus, and J. C. de Freitas. 1986. Vertical distribution of benthic marine organisms on rocky coasts of the Fernando de Noronha archipelago (Brazil). Boletim do Instituto Oceanografico de São Paulo 34: 37–53.

Fagan, B. 2000. The Little Ice Age: How climate made history, 1300–1850. New York: Basic Books.

Falkowski, P. G. 1997. Evolution of the nitrogen cycle and its influence on the biological sequestration of CO_2 in the ocean. Nature 387: 272–275.

Farlow, J. O. 1981. Estimates of dinosaur speeds from a new trackway site in Texas. Nature 294: 747–748.

Farlow, J. O., and T. R. Holtz. 2002. The fossil record of predation in dinosaurs. Pages 251–265 in M. Kowalewski and P. H. Kelley (eds.), The fossil record of predation. Paleontological Society Papers 8.

Farmer, C. G. 1999. Evolution of the vertebrate cardio-pulmonary system. Annual Reviews of Physiology 61: 573–592.

Farquhar, G. D., and M. L. Roderick. 2003. Pinatubo, diffuse light, and the carbon cycle. Science 299: 1997–1998.

Farrell, B. D., D. E. Dussourd, and C. Mitter. 1991. Escalation of plant defense: Do latex and resin canals spur plant diversification? American Naturalist 138: 881–900.

Farrell, B. D., A. S. Sequeira, B. C. O'Meara, B. B. Normark, J. H. Chung, and B. H. Jordal. 2001. The evolution of agriculture in beetles (Curculionidae: Scolytinae and Platypodinae). Evolution 55: 2011–2027.

Fehr, E., and U. Fischbacher. 2003. The nature of human altruism. Nature 425: 785–791.

Feifarek, B. P. 1987. Spines and epibionts as antipredator defenses in the thorny oyster *Spondylus americanus* Hermann. Journal of Experimental Marine Biology and Ecology 105: 39–56.

Feinstein, N., and S. D. Cairns. 1998. Learning from the collector: A survey of azooxanthellate corals affixed by *Xenophora* (Gastropoda: Xenophoridae), with an analysis and discussion of attachment patterns. Nautilus 112: 73–83.

Fensome, R. A., R. A. MacRae, J. M. Moldowan, F.J.R. Taylor, and G. L. Williams. 1996. The Early Mesozoic radiation of dinoflagellates. Paleobiology 22: 329–338.

Field, C. B. 2001. Global change: Sharing the garden. Science 294: 2490–2491.

Field, C. B., M. J. Behrensfeld, J. T. Randerson, and P. Falkowski. 1998. Primary production of the biosphere: Integrating terrestrial and oceanic components. Science 281: 237–240.

Filippelli, G. M. 1997. Intensification of the Asian monsoon and a chemical weathering event in the Late Miocene-Early Pliocene: Implications for Late Neogene climate change. Geology 25: 27–30.

Filippelli, G. M., and M. L. Delaney. 1994. The oceanic phosphorus cycle and continental weathering during the Neogene. Paleoceanography 9: 643–652.

Finlay, B. J. 2002. Global dispersal of free-living microbial eukaryote species. Science 296: 1061–1063.

Finlay, B. J., and K. J. Clarke. 1999. Ubiquitous dispersal of microbial species. Nature 400: 828.

Fischer, A. G. 1984. Biological innovations and the sedimentary record. Pages 145–157 in H. D. Holland and A. F. Trendall (eds.), Patterns of change in Earth evolution. Berlin: Springer.

Fischer, D. H. 1996. The great wave: Price revolutions and the rhythm of history. New York: Oxford University Press: 536 pp.

Fisher, A.G.B. 1935. The clash of progress and security. New York: Augustus M. Kelley (reprinted 1966): 234 pp.

Fisher, R. A. 1958. The genetical theory of natural selection (second edition). New York: Dover: 291 pp.

Fishlin, D. A., and D. W. Phillips. 1980. Chemical camouflaging and behavioral defenses against a predatory seastar by three species of gastropods from the surfgrass *Phyllospadix* community. Biological Bulletin 158: 34–48.

Flannery, T. F., and R. G. Roberts. 1999. Late Quaternary extinctions in Australia. Pages 239–255 in R.D.E. MacPhee (ed.), Extinctions in near time: Causes, contexts, and consequences. New York: Kluwer/Plenum.

Föllmi, K. B. 1995. 160 m.y. record of marine sedimentary phosphorus burial: Coupling of climate and continental weathering under greenhouse and icehouse conditions. Geology 23: 859–862.

Förster, R. 1979. *Eocarcinus praecursor* Withers (Decapoda, Brachyura) from the Lower Pliensbachian of Yorkshire and the early crabs. Neues Jahrbuch für Geologie und Paläontologie Monatshefte 1: 15–27.

Fortey, R. A., and R. M. Owens. 1999. Feeding habits in trilobites. Palaeontology 42: 429–465.

Fox, L. R. 1981. Defense and dynamics in plant-herbivore systems. American Zoologist 21: 853–864.

Fox, L. R., and P. A. Morrow. 1981. Specialization: Species property or local phenomenon? Science 211: 887–893.

Fraaye, R., and M. Jäger. 1995. Decapods in ammonite shells: Examples of inquilinism from the Jurassic of England and Germany. Palaeontology 38: 63–75.

Francisco-Ortega, J., R. K. Jansen, and A. Santos-Guerra. 1996. Chloroplast DNA evidence of colonization, adaptive radiation, and hybridization in the evolution of the Macaronesian flora. Proceedings of the National Academy of Sciences of the USA 93: 4085–4090.

Frank, S. A. 1996. The design of natural and artificial adaptive systems. Pages 451–505 in M. R. Rose and G. V. Lauder (eds.), Adaptation. San Diego: Academic Press.

Freed, A. N. 1980. Prey selection and feeding behavior of the green treefrog (*Hyla cinerea*). Ecology 61: 461–465.

Frey, E., H.-D. Sues, and W. Munk. 1997. Gliding mechanism in the Late Permian reptile *Coelurasauravus*. Science 275: 1450–1452.

Friis, E. M., K. R. Pedersen, and P. R. Crane. 2001. Fossil evidence of water lilies (Nymphaeales) in the Early Cretaceous. Nature 410: 357–360.

Fritz, R. S., and D. H. Morse. 1985. Reproductive success and foraging of the crab spider *Misumena vatia*. Oecologia (Berlin) 65: 194–200.

Froelich, P. N., M. L. Bender, N. A. Luedtke, G. R. Heath, and T. DeVries. 1982. The marine phosphorus cycle. American Journal of Science 282: 474–511.

Futuyma, D. J. 1983. Evolutionary interactions among herbivorous insects and plants. Pages 207–231 in D. J. Futuyma and M. Slatkin (eds.), Coevolution. Sunderland: Sinauer.

Galbraith, J. K. 1973. Economics and the public purpose. Boston: Houghton Mifflin: 334 pp.
———. 1983. The anatomy of power. Boston: Houghton Mifflin: 189 pp.
———. 1984. The affluent society (fourth edition). Boston: Houghton Mifflin: 277 pp.
Gale, A. S. 2000. The Cretaceous world. Pages 4–19 in S. J. Culver and P. F. Rawson (eds.), Biotic response to global change: The last 145 million years. Cambridge: Cambridge University Press.
Gandolfo, M. A., K. C. Nixon, W. L. Crepet, D. W. Stevenson, and E. M. Friis. 1998. Oldest known fossils of monocotyledons. Nature 394: 532–533.
Gans, C., R. Dudley, N. M. Aguilar, and J. B. Graham. 1999. Late Paleozoic atmosphere and biotic evolution. Historical Biology 13: 199–219.
Garland, T., Jr. 1983a. The relation between maximal running speed and body mass in terrestrial mammals. Journal of Zoology, London 199: 157–170.
———. 1983b. Scaling the ecological cost of transport to body mass in terrestrial mammals. American Naturalist 121: 571–587.
Garrity, S. D. 1984. Some adaptations of gastropods to physical stress on a tropical rocky shore. Ecology 65: 559–574.
Garrity, S. D., and S. C. Levings. 1981. A predator-prey interaction between two physically and biologically constrained tropical rocky shore gastropods: Direct, indirect and community effects. Ecological Monographs 51: 267–286.
Gatesy, S. M., and K. P. Dial. 1996. Locomotor modules and the evolution of avian flight. Evolution 50: 331–340.
Gattuso, J.-P., and R. W. Buddemeier. 2000. Calcification and CO_2. Nature 407: 311–313.
Gehling, J. G., and J. K. Rigby. 1996. Long expected sponges from the Neoproterozoic Ediacara fauna of South Australia. Journal of Paleontology 70: 185–195.
Geider, R. J., and H. L. MacIntyre. 2002. Physiology and biochemistry of photosynthesis and algal carbon acquisition. Pages 44–77 in P. J. le B. Williams, D. N. Thomas, and C. S. Reynolds (eds.), Phytoplankton productivity: Carbon assimilation in marine and freshwater ecosystems. Oxford: Blackwell Science.
Geller, J. B. 1982. Chemically mediated avoidance response of a gastropod, *Tegula funebralis* (A. Adams), to a predatory crab, *Cancer antennarius* (Stimpson). Journal of Experimental Marine Biology and Ecology 65: 19–27.
Gersonde, R., F. T. Kyte, U. Bleil, B. Diekmann, J. A. Flores, K. Gohl, G. Grahl, R. Hagen, G. Kuhn, F. J. Sierro, D. Völker, A. Abelmann, and J. H. Bostwick. 1997. Geological record and reconstruction of the Late Pliocene impact of the Eltanin asteroid in the Southern Ocean. Nature 390: 357–363.
Ghiselin, M. T. 1974. The economy of nature and the evolution of sex. Berkeley: University of California Press: 346 pp.
Gilbert, L. E. 1971. Butterfly-plant coevolution: Has *Passiflora adenopoda* won the selectional race with heliconiine butterflies? Science 172: 585–586.
———. 1983. The role of insect-plant coevolution in the organization of ecosystems. Pages 399–415 in V. Labyrie (ed.), Comportement des insectes et milieu trophique. Paris: CNRS.
Gillespie, W. H., G. W. Rothwell, and S. E. Scheckler. 1983. The earliest seeds. Nature 293: 462–464.
Gillooly, J. F., J. H. Brown, G. B. West, V. M. Savage, and E. L. Charnov. 2001. Effects of size and temperature on metabolic rate. Science 293: 2248–2251.
Gillooly, J. F., E. L. Charnov, G. B. West, V. M. Savage, and J. H. Brown. 2002. Effects of size and temperature on developmental time. Nature 417: 70–73.
Gingerich, P. D. 1983. Rates of evolution: Effects of time and temporal scaling. Science 222: 159–161.
Givnish, T. J. 1978. On the adaptive significance of compound leaves with particular reference to tropical trees. Pages 351–380 in P. B. Tomlinson and R. H. Zimmermann (eds.), Tropical trees as living systems. Cambridge: Cambridge University Press.

Gladenkov, A. Yu., A. E. Oleinik, L. Marincovich, Jr., and K. B. Barinov. 2002. A refined age for the earliest opening of Bering Strait. Palaeogeography, Palaeoclimatology, Palaeoecology 183: 321–328.

Glikson, A. Y. 1999. Oceanic mega-impacts and crustal evolution. Geology 27: 387–390.

Glynn, P. W. 1983. Increased survivorship in corals harboring crustacean symbionts. Marine Biology Letters 4: 105–111.

Glynn, P. W. (ed.) 1990. Global ecological consequences of the 1982–83 El Niño-Southern Oscillation. Amsterdam: Elsevier.

Glynn, P. W., H. H. Stewart, and J. E. McCosker. 1972. Pacific coral reefs of Panama: Structure, distribution and predators. Geologische Rundschau 61: 481–519.

Godfrey, H.C.J. 1997. Making life simpler. Nature 387: 351–352.

Goldman, J. C., J. J. McCarthy, and D. G. Peavey. 1979. Growth rate influence on the chemical composition of phytoplankton in oceanic waters. Nature 279: 210–215.

Gonor, J. J. 1965. Predator-prey reactions between two marine prosobranch gastropods. Veliger 7: 228–232.

Goslow, G. E., Jr. 1971. The attack and strike of some North American raptors. Auk 88: 815–827.

Goss, R. J. 1965. Mammalian regeneration and its phylogenetic relationships. Pages 33–38 in V. Kiortsis and H. A. B. Trampusch (eds.), Regeneration in animals and related problems. Amsterdam: North-Holland.

Gould, S. J. 1977. Ontogeny and phylogeny. Cambridge: Belknap Press of Harvard University: 501 pp.

———. 1985. The paradox of the first tier: An agenda for paleobiology. Paleobiology 11: 2–12.

———. 1988. Trends as changes in variance: A new slant on progress and directionality in evolution. Journal of Paleontology 62: 319–329.

———. 1989. Wonderful life: The Burgess Shale and the nature of history. New York: Norton: 347 pp.

———. 1996. Full house: The spread of excellence from Plato to Darwin. New York: Harmony: 244 pp.

———. 1999. Rocks of ages: Science and religion in the fullness of life. New York: Ballantine: 241 pp.

———. 2000. Beyond competition. Paleobiology 26: 1–6.

Graham, J. B., R. Dudley, N. L. Aguilar, and C. Gans. 1995. Implications of the Late Palaeozoic oxygen pulse for physiology and evolution. Nature 375: 117–120.

Graham, L. E., M. E. Cook, and J. S. Busse. 2000. The origin of plants: Body plan changes contributing to a major evolutionary radiation. Proceedings of the National Academy of Sciences of the USA 97: 4535–4540.

Grant, P. R. 1986. Ecology and evolution of Darwin's finches. Princeton: Princeton University Press: 458 pp.

Grant, P. R., and I. Abbott. 1980. Interspecific competition, island biogeography and null hypothesis. Evolution 34: 332–341.

Grant, P. R., and B. R. Grant. 2002. Unpredictable evolution in a 30-year study of Darwin's finches. Science 296: 707–711.

Graus, R. R. 1974. Latitudinal trends in the shell characteristics of marine gastropods. Lethaia 7: 303–314.

Gray, K., M. R. Walter, and C. R. Calver. 2003. Neoproterozoic biotic diversification: Snowball Earth or aftermath of the Acraman impact? Geology 31: 459–462.

Greene, H. W. 1997. Snakes: The evolution of mystery in nature. Berkeley: University of California Press: 351 pp.

Greider, W. 1997. One world: Ready or not: The manic logic of global capitalism. New York: Simon and Schuster: 528 pp.

Griffis, K., and D. J. Chapman. 1988. Survival of phytoplankton under prolonged darkness: Implications for the Cretaceous-Tertiary boundary darkness hypothesis. Palaeogeography, Palaeoclimatology, Palaeoecology 67: 305–314.

Griffiths, D. 1980. Foraging costs and relative prey size. American Naturalist 116: 743–752.

Grime, J. P. 1977. Evidence for the existence of three primary strategies in plants and its relevance to ecological and evolutionary theory. American Naturalist 111: 1169–1194.

Grosberg, R. K., and R. R. Strathmann. 1998. One cell, two cell, red cell, blue cell: The persistence of a unicellular stage in multi-cellular life histories. Trends in Ecology and Evolution 13: 112–116.

Gross, W. 1967. Über das Gebisz der Acanthodier und Placodermen. Journal of the Linnaean Society, Zoology, 47: 121–130.

Grotzinger, J. P., and J. F. Kasting. 1993. New constraints on Precambrian ocean composition. Journal of Geology 101: 235–243.

Grotzinger, J. P., and A. H. Knoll. 1999. Stromatolites in Precambrian carbonates: Evolutionary milestones or environmental dipsticks? Annual Reviews of Earth and Planetary Sciences 27: 313–358.

Grotzinger, J. P., W. A. Watters, and A. H. Knoll. 2000. Calcified metazoans in thrombolite-stromatolite reefs of the terminal Proterozoic Nama Group, Namibia. Paleobiology 26: 334–359.

Grünebaum, H., G. Bergman, D. P. Abbott, and J. C. Ogden. 1978. Intraspecific agonistic behavior in the rock-boring sea urchin *Echinometra lucunter* (L.) (Echinodermata: Echinoidea). Bulletin of Marine Science 28: 181–188.

Gu, L., D. D. Baldocchi, S. C. Wofsy, J. W. Munger, J. J. Michalsky, S. P. Urbanski, and T. A. Boden. 2003. Response of a deciduous forest to the Mount Pinatubo eruption: Enhanced photosynthesis. Science 299: 2035–2038.

Guerrero, R., C. Pedros-Alió, I. Esteve, J. Mas, D. Chase, and L. Margulis. 1986. Predatory prokaryotes: Predation and primary consumption evolved in bacteria. Proceedings of the National Academy of Sciences of the USA 83: 2138–2142.

Guinot, D. 1968. Recherches préliminaires sur les groupements naturels chez les crustacés brachyures. VI Les Carpilinae. Bulletin du Museum national d'Histoire naturelle (Zoologie) (ser. 2) 40: 320–334.

Guo, Z. T., W. F. Ruddiman, Q. Z. Hao, H. B. Wu, Y. S. Qiao, B. Y. Yuan, and T. S. Liu. 2002. Onset of Asian desertification by 22 Myr ago inferred from loess deposits in China. Nature 416: 159–163.

Hadlock, R. P. 1980. Alarm response of the intertidal snail *Littorina littorea* (L.) to predation by the crab *Carcinus maenas* (L.). Biological Bulletin 159: 269–279.

Hafner, M. S., P. D. Sudman, F. X. Villablanca, T. A. Spradling, J. W. Demastes, and S. A. Nadler. 1994. Disparate rates of molecular evolution in cospeciating hosts and parasites. Science 265: 1087–1090.

Hairston, N. G., F. E. Smith, and L. Slobodkin. 1960. Community structure, population control, and competition. American Naturalist 94: 421–425.

Hall, C., P. Tharakan, J. Hallock, C. Cleveland, and M. Jefferson. 2003. Hydrocarbons and the evolution of human culture. Nature 426: 318–322.

Hallam, A. 1978. How rare is phyletic gradualism and what is its evolutionary significance? Evidence from Jurassic bivalves. Paleobiology 4: 16–25.

———. 1992. Phanerozoic sea-level changes. New York: Columbia University Press: 266 pp.

———. 1994. An outline of Phanerozoic biogeography. Oxford: Oxford University Press: 246 pp.

Hallam, A., and P. B. Wignall. 1997. Mass extinction and its aftermath. Oxford: Oxford University Press: 320 pp.

Hallock, W., and W. Schlager. 1986. Nutrient excess and the demise of coral reefs and carbonate platforms. Palaios 1: 389–398.

Hammond, K. A., and J. Diamond. 1997. Maximal sustained energy budgets in humans and animals. Nature 386: 457–462.

Handler, P. 1989. The effect of volcanic aerosols on global climate. Journal of Volcanology and Geothermal Research 37: 233–249.

Hansen, T. A., P. H. Kelley, V. D. Melland, and S. E. Graham. 1999. Effect of climate-related mass extinctions on escalation in molluscs. Geology 27: 1139–1142.

Hanson, V. D. 2001. Carnage and culture: Landmark battles in the rise of Western culture. New York: Doubleday: 492 pp.

Haq, B. U., J. Hardenbol, and P. R. Vail. 1987. Chronology of fluctuating sea levels since the Triassic. Science 235: 1156–1167.

Harper, E. M. 2003. Assessing the importance of drilling predation over the Paleozoic and Mesozoic. Palaeogeography, Palaeoclimatology, Palaeoecology, 201:185–198.

Harper, E. M., and P. W. Skelton. 1993. The Mesozoic marine revolution and epifaunal bivalves. Scripta Geologica, Special Issue 2: 127–153.

Harrison, I. J., and M.L.J. Stiassny. 1999. The quiet crisis: A preliminary listing of the freshwater fishes of the world that are extinct or "missing in action." Pages 271–331 in R.D.E. MacPhee (ed.), Extinctions in near time: Causes, contexts, and consequences. New York: Kluwer/Plenum.

Harrison, K. G. 2000. Role of increased marine silica input on paleo-pCO_2 levels. Paleoceanography 15: 292–298.

Harrold, C. 1982. Escape responses and prey availability in a kelp-forest predator-prey system. American Naturalist 119: 132–135.

Hartog, C. den. 1970. The sea-grasses of the world. Amsterdam: North-Holland: 275 pp.

Haszprunar, G. 1992. The first molluscs—small animals. Bollettino Zoologico 59: 1–16.

Haszprunar, G., L. von Salvini-Plawen, and R. M. Rieger. 1995. Larval planktotrophy—a primitive trait in the Bilateria? Acta Zoologica (Stockholm) 76: 141–154.

Hawken, P. 1993. The ecology of commerce: a declaration of sustainability. New York: HarperBusiness: 250 pp.

Hay, M. E. 1981. The functional morphology of turf-forming seaweeds: Persistence in stressful marine habitats. Ecology 62: 739–750.

Hay, M. E., and W. Fenical. 1988. Marine plant-herbivore interactions: The ecology of chemical defense. Annual Reviews of Ecology and Systematics 19: 111–145.

Hay, M. E., and P. D. Steinberg. 1992. The chemical ecology of plant-herbivore interactions in marine versus terrestrial communities. Pages 371–413 in G. A. Rosenthal and M. R. Berenbaum (eds.), Herbivores: Their interactions with secondary plant metabolites, second edition, Vol. II: Ecological and evolutionary processes. San Diego: Academic Press.

Hay, M. E., J. R. Pawlik, J. E. Duffy, and W. Fenical. 1989. Seaweed-herbivore-predator interactions: Host-plant specialization reduces predation on small herbivores. Oecologia (Berlin) 81: 418–427.

Hay, M. E., J. E. Duffy, V. J. Paul, P. E. Renaud, and W. Fenical. 1990a. Specialist herbivores reduce their susceptibility to predation by feeding on the chemically defended seaweed *Avrainvillea longicaulis*. Limnology and Oceanography 35: 1734–1743.

Hay, M. E., J. E. Duffy, and W. Fenical. 1990b. Host-plant specialization decreases predation on a marine amphiopod: an herbivore in plant's clothing. Ecology 71: 733–743.

Hay, W. W., M. J. Rosol, J. L. Sloan II, and D. E. Jory. 1988. Plate tectonic control of global patterns of detrital and carbonate sedimentation. Pages 1–34 in L. J. Doyle and H. H. Roberts (eds.), Carbonate-clastic transitions. Developments in Sedimentology 42. Amsterdam: Elsevier.

Hayes, J. M. 2002. A low-down on oxygen. Nature 417: 127–128.

Hayes, J. P., and T. Garland, Jr. 1995. The evolution of endothermy: Testing the aerobic capacity model. Evolution 49: 836–847.

Heaney, L. R. 1978. Island area and body size of insular mammals: Evidence from the tricolored squirrel (*Callosciurus prevosti*) of southeast Asia. Evolution 32: 29–44.

Heaney, L. R. 1986. Biogeography of mammals in SE Asia: Estimates of rates of colonization, extinction and speciation. Biological Journal of the Linnean Society 28: 127–165.

———. 2000. Dynamic disequilibrium: A long-term, large-scale perspective on the equilibrium model of island biogeography. Global Ecology and Biogeography 9: 59–74.

Heatwole, H., and R. Levins. 1972. Biogeography of the Puerto Rican Bank: Flotsam transport of terrestrial animals. Ecology 53: 112–117.

Heckman, D. S., D. M. Geiser, B. R. Eidell, R. L. Stauffer, N. L. Kardos, and S. B. Hedges. 2001. Molecular evidence for the early colonization of land by fungi and plants. Science 293: 1129–1133.

Hector, A., B. Schmid, M. Beierkuhnlein, M. C. Caldeira, M. Diemer, P. G. Dimitrakopoulos, J. A. Finn, H. Freitas, P. S. Giller, J. Good, R. Harris, P. Högberg, K. Huss-Danell, I. Joshi, A. Jumpponen, C. Kömer, P. W. Leadley, M. Loreau, A. Minns, C.P.H. Mulder, G. O'Donovan, S. J. Otway, J. S. Pereira, A. Prinz, D. J. Read, M. Scheder-Lorenzen, E. D. Schulze, A.-S.D. Siamantziouras, E. M. Spehn, A. C. Terry, A. Y. Troumbis, F. I. Woodward, S. Yachi, and J. H. Lawton. 1999. Plant diversity and productivity experiments in European grasslands. Science 286: 1123–1127.

Heinrich, B. 1993. The hot-blooded insects: Strategies and mechanisms of thermoregulation. Cambridge: Harvard University Press: 601 pp.

Heinrich, B., and G. A. Bartholomew. 1979. Roles of endothermy and size in inter- and intraspecific competition for elephant dung in an African dung beetle, *Scarabaeus laevistriatus*. Physiological Zoology 52: 484–496.

Heinrich, B., and S. L. Collins. 1983. Caterpillar leaf damage, and the game of hide-and-seek with birds. Ecology 64: 592–602.

Heinrich, B., and S. P. Mommsen. 1985. Flight of winter moths near 0° C. Science 228: 177–179.

Hennemann, W. W. III. 1983. Relationship among body mass, metabolic rate and the intrinsic rate of natural increase in mammals. Oecologia (Berlin) 56: 104–108.

Herman, A. 2001. How the Scots invented the modern world. New York: Three Rivers.

Herre, E. A. 1999. Laws governing species interactions? Encouragement and caution from figs and their associates. Pages 209–237 in L. Keller (ed.), Levels of selection in evolution. Princeton: Princeton University Press.

Herre, E. A., N. Knowlton, U. G. Mueller, and S. A. Rehner. 1999. The evolution of mutualisms: Exploring the paths between conflict and cooperation. Trends in Ecology and Evolution 14: 49–53.

Hertz, P. E., R. B. Huey, and T. Garland, Jr. 1988. Time budgets, thermoregulation, and maximal locomotor performance: Are reptiles olympians or Boy Scouts? American Zoologist 28: 927–938.

Hesselbo, S. P., D. R. Gröcke, H. C. Jenkyns, C. J. Bjerrum, P. Farrimond, H. S. Morgansbell, and O. R. Green. 2000. Massive dissociation of gas hydrate during a Jurassic oceanic anoxic event. Nature 406: 392–395.

Hesselbo, S. P., S. A. Robinson, F. Surlyk, and S. Piasecki. 2002. Terrestrial and marine extinction at the Triassic-Jurassic boundary synchronized with major carbon-isotope perturbation: A link to initiation of massive volcanism? Geology 30: 251–254.

Hibberd, J. M., and W. P. Qu. 2002. Characteristics of C_4 photosynthesis in stems and petioles of C_3 flowering plants. Nature 415: 451–454.

Hibbett, D. S., and M. Binder. 2001. Evolution of marine mushrooms. Biological Bulletin 201: 319–322.

Hickman, C. S. 2003. Evidence for abrupt Eocene-Oligocene molluscan faunal change in the Pacific Northwest. Pages 71–87 in D. R. Prothero, L. C. Ivany, and E. A. Nesbitt (eds.), From greenhouse to icehouse: The marine Eocene-Oligocene transition. New York: Columbia University Press.

Highsmith, R. C. 1980. Geographic patterns of coral bioerosion: A productivity hypothesis. Journal of Experimental Marine Biology and Ecology 46: 177–196.

Hillebrand, H., B. Worm, and H. K. Lotze. 2000. Marine microbenthic community structure regulated by nitrogen loading and grazing pressure. Marine Ecology Progress Series 204: 27–38.

Hillel, D. 1991. Out of the Earth: Civilization and the life of the soil. Berkeley: University of California Press: 321 pp.

Hirtle, R.W.M., and K. H. Mann. 1978. Distance chemoreception and vision in the selection of prey by American lobster (*Homarus americanus*). Journal of the Fisheries Research Board of Canada 35: 1006–1008.

Hochachka, P. W., and G. N. Somero. 1973. Strategies of biochemical adaptation. Philadelphia: W. B. Saunders: 358 pp.

Hodgson, G. M. 1993. Economics and evolution: Bringing life back into economics. Cambridge: Polity: 381 pp.

———. 1999. Economics and utopia: Why the learning economy is not the end of history. London: Routledge: 337 pp.

———. 2001. How economics forgot history: The problem of historical specificity in social science. London: Routledge: 422 pp.

Hodych, J. P., and G. R. Dunning. 1992. Did the Manicouagan impact trigger end-of-Triassic mass extinction? Geology 20: 51–54.

Hoehler, T. M., B. M. Bebout, and D. J. Des Marais. 2001. The role of microbial mats in the production of reduced gases on the early Earth. Nature 412: 324–327.

Hoernle, K., P. van den Bagaard, R. Werner, B. Lissinna, F. Hauff, G. Alvarado, and D. Garbe-Schönberg. 2002. Missing history (16–71 Ma) of the Galápagos hotspot: Implications for the tectonic and biological evolution of the Americas. Geology 30: 795–798.

Hoffman, A. 1989. Arguments on evolution: A paleontologist's perspective. New York: Oxford University Press: 274 pp.

Hoffman, P. F., A. J. Kaufman, G. P. Halverson, and D. P. Schrag. 1998. A Neoproterozoic Snowball Earth. Science 281: 1342–1346.

Hoffmann, C., V. Courtillot, F. Géraud, P. Rochette, G. Yirgu, E. Ketefo, and R. Pik. 1997. Timing of the Ethiopian flood basalt event and implications for plume birth and global change. Nature 389: 838–841.

Holland, B., and W. R. Rice. 1998. Chase-away sexual selection: Antagonistic seduction versus resistance. Evolution 52: 1–7.

Holland, H. D. 1984. The chemical evolution of the atmosphere and oceans. Princeton: Princeton University Press: 582 pp.

Hölldobler, B., and E. O. Wilson. 1990. The ants. Cambridge: Belknap Press of Harvard University: 732 pp.

Holling, C. S. 1966. The functional response of invertebrate predators to prey density. Memoirs of the Entomological Society of Canada 47: 3–86.

Holt, R. D. 1996. Demographic constraints in evolution: Towards unifying the evolutionary theories of senescence and niche conservatism. Evolutionary Ecology 10: 1–11.

Holt, R. D., and R. Gomulkiewicz. 1997. How does immigration influence local adaptation? A reexamination of a familiar paradigm. American Naturalist 149: 563–572.

Holt, R. D., and M. E. Hochberg. 1997. When is biological control evolutionarily stable (or is it?)? Ecology 78: 1673–1683.

Holt, R. D., and J. H. Lawton. 1994. The ecological consequences of shared natural enemies. Annual Reviews of Ecology and Systematics 25: 495–520.

Honegger, R. E. 1981. List of amphibians and reptiles either known or thought to have become extinct since 1600. Biological Conservation 19: 141–158.

Hooper, D. U., and P. M. Vitousek. 1997. The effects of plant composition and diversity on ecosystem processes. Science 277: 1302–1305.

Hooper, P. R. 1990. The timing of crustal extension and the eruption of continental flood basalts. Nature 345: 246–249.

Horn, H. S. 1971. The adaptive geometry of trees. Princeton: Princeton University Press: 144 pp.

Howland, H. C. 1962. Structural, hydraulic, and "economic" aspects of leaf venation and shape. Pages 183–191 in E. E. Bernard and M. R. Kare (eds.), Biological prototypes and synthetic systems, Vol. 1. New York: Plenum.

Hsü, K. J. 1986. Environmental changes in times of biotic crisis. Pages 297–312 in D. M. Raup and D. Jablonski (eds.), Patterns and processes in the history of life. Berlin: Springer.

Hsü, K. J., Q. He, J. A. McKenzie, H. Weissert, K. Perch-Nielsen, H. Oberhänsli, K. Kelts, J. LaBrecque, L. Tauxe, U. Krähenbühl, S. F. Percival, Jr., R. Wright, A. M. Karpoff, N. Petersen, P. Tucker, R. Z. Poore, A. M. Gombos, K. Pisciotto, M. F. Carman Jr., and E. Schreiber. 1982. Mass mortality and its environmental and evolutionary consequences. Science 216: 249–256.

Hsü, K. J., H. Oberhänsli, J. Y. Gao, S. Shu, C. Haihong, and U. Krähenbühl. 1985. "Strangelove ocean" before the Cambrian explosion. Nature 316: 809–811.

Huey, R. B., E. R. Pianka, and L. J. Vitt. 2001. How often do lizards "run on empty"? Ecology 82: 1–7.

Huff, W. D., S. M. Bergström, and D. R. Kolata. 1992. Giant Ordovician volcanic ash fall in North America and Europe: Biological, tectonomagmatic, and event-stratigraphic significance. Geology 20: 875–878.

Hughes, R. N. 1985. Predatory behaviour of *Natica unifasciata* feeding intertidally on gastropods. Journal of Molluscan Studies 51: 331–335.

Hughes, T. P. 1994. Catastrophes, phase shifts, and large-scale degradation of a Caribbean coral reef. Science 265: 1547–1551.

Hurst, J. 1974. Selective epizoan encrustation of some Silurian brachiopods from Gotland. Palaeontology 16: 423–429.

Huston, M. A. 1993. Biological diversity, soils, and economics. Science 262: 1676–1680.

Hutchinson, G. E. 1960. Evolutionary euryhalinity. American Journal of Science 258: 98–103.

Hutchinson, J. R., and M. Garcia. 2002. *Tyrannosaurus* was not a fast runner. Nature 415: 1018–1021.

Il'ina, L. G. 1987. Evidence of boring in shells of brackish-water gastropods. Paleontological Journal 21: 23–30.

Ivanov, B. A., and H. J. Melosh. 2003. Impacts do not initiate volcanic eruptions: Eruptions close to the crater. Geology 31: 369:872.

Ivany, L. C., E. A. Nesbitt, and D. R. Prothero. 2003. The marine Eocene-Oligocene transition: a synthesis. Pages 522–534 in D. R. Prothero, L. C. Ivany, and E. A. Nesbitt (eds.), From greenhouse to icehouse: The marine Eocene-Oligocene transition. New York: Columbia University Press.

Jablonski, D. 1986. Larval ecology and macroevolution in marine invertebrates. Bulletin of Marine Science 39: 565–587.

———. 1993. The tropics as a source of evolutionary novelty through geological time. Nature 364: 142–144.

Jablonski, D., S. Lidgard, and P. D. Taylor. 1997. Comparative ecology of bryozoan radiations: origin of novelties in cyclostomes and cheilostomes. Palaios 12: 505–523.

Jackson, J.B.C. 1977a. Competition on marine hard substrata: The adaptive significance of solitary and colonial strategies. American Naturalist 111: 743–767.

———. 1977b. Habitat area, colonization, and development of epibenthic community structure. Proceedings of the llth European Marine Biological Symposium, Galway, Ireland, October, 1976: 349–358.

———. 1979. Morphological strategies of sessile animals. Pages 499–555 in G. Larwood and B. R. Rosen (eds.), Biology and systematics of colonial organisms. London: Academic Press.

———. 1983. Biological determinants of present and past sessile animal distributions. Pages 39–120 in M.J.S. Tevesz and P. L. McCall (eds.), Biotic interactions in Recent and fossil benthic communities. New York: Plenum.

———. 1997. Reefs since Columbus. Coral Reefs 16 Supplement: S23-S32.

Jackson, J.B.C., and F. K. McKinney. 1990. Ecological processes and progressive macroevolution of marine clonal benthos. Pages 173–209 in R. M. Ross and W. D. Allmon (eds.), Causes of evolution: A paleontological perspective. Chicago: University of Chicago Press.

Jackson, J.B.C., M. X. Kirby, W. H. Berger, K. A. Bjorndal, L. W. Botsford, B. J. Bourque, R. H. Bradbury, R. Cooke, J. Erlandson, J. A. Estes, T. P. Hughes, S. Kidwell, C. B. Lang, H. S. Lenihan, J. M. Pandolfi, C. H. Peterson, R. S. Steneck, M. J. Tegner, and R. R. Warner. 2001. Historical overfishing and the recent collapse of coastal ecosystems. Science 293: 629–638.

Jackson, T. A. 1973. "Humic" matter in the bitumen of ancient sediments: Variations through geologic time. A new approach to the study of pre-Paleozoic (Precambrian) life. Geology 1: 163–166.

Jackson, T. A., and W. D. Keller. 1970. A comparative study of the role of lichens and "inorganic" processes in the chemical weathering of Recent Hawaiian lava flows. American Journal of Science 269: 446–466.

Jacobs, D. K. 1992. Shape, drag, and power in ammonoid swimming. Paleobiology 18: 203–220.

Jacobs, D. K., and N. H. Landman. 1993. *Nautilus*—a poor model for the function and behavior of ammonoids? Lethaia 26: 101–111.

Jacobs, D. K., C. G. Wray, C. J. Wedeen, R. Kostriken, R. DeSalle, J. L. Staton, R. D. Gates, and D. R. Lindberg. 2000. Molluscan *Engrailed* expression, serial organization, and shell evolution. Evolution and Development 2: 340–347.

Jacobs, J. 2000. The nature of economies. New York: Random House: 190 pp.

James, H. F., and D. A. Burney. 1997. The diet and ecology of Hawaii's extinct flightless waterfowl: Evidence from coprolites. Biological Journal of the Linnean Society 62:279–297.

Janis, C. 1982. Evolution of horns in ungulates: Ecology and paleoecology. Biological Reviews 57: 261–317.

———. 2000. Patterns in the evolution of herbivory in large terrestrial mammals: The Paleogene of North America. Pages 168–222 in H.-D. Sues (ed.), Evolution of herbivory in terrestrial vertebrates: Perspectives from the fossil record. Cambridge: Cambridge University Press.

Janis, C. M., and P. B. Wilhelm. 1993. Were there mammalian pursuit predators in the Tertiary? Journal of Mammalian Evolution 1: 103–125.

Janzen, D. H. 1974. Tropical blackwater rivers, animals, and mast fruiting in the Dipterocarpaceae. Biotropica 6: 69–103.

———. 1976. The depression of reptile biomass by large herbivores. American Naturalist 110: 371–400.

Janzen, D. H., and P. S. Martin. 1982. Neotropical anachronisms: The fruits the gomphotheres ate. Science 215: 19–27.

Jaramillo, C. A. 2002. Response of tropical vegetation to Paleogene warming. Paleobiology 28: 222–243.

Järvi, T., B. Sillén-Tullberg, and C. Wiklund. 1981. The cost of being aposematic. An experimental study of predation on larvae of *Papilio machaon* by the great tit *Parus major*. Oikos 36: 267–272.

Javaux, E. J., A. J. Knoll, and M. R. Walter. 2001. Morphological and ecological complexity in early eukaryotic ecosystems. Nature 412: 66–69.

Jeanne, R. L. 1979. A latitudinal gradient in rates of ant predation. Ecology 60: 1211–1224.

Jeffery, C. H. 2001. Heart urchins at the Cretaceous/Tertiary boundary: A tale of two clades. Paleobiology 27: 140–158.

Jeffries, M. J., and J. H. Lawton. 1984. Enemy free space and the structure of ecological communities. Biological Journal of the Linnean Society 23: 269–286.

Jerison, J. H. 1973. Evolution of the brain and intelligence. New York: Academic Press: 482 pp.

Jin, Y. G., Y. Wang, W. Wang, Q. H. Shang, C. Q. Cao, and D. H. Erwin. 2000. Pattern of marine mass extinction near the Permian-Triassic boundary in South China. Science 289: 432–436.

Joachimski, M. J., and W. Buggisch. 1993. Anoxic events in the Late Frasnian—causes of the Frasnian-Famennian faunal crisis? Geology 21: 675–678.

Johnson, K. R., and B. Ellis. 2002. A tropical rainforest in Colorado 1.4 million years after the Cretaceous-Tertiary boundary. Science 296: 2379–2383.

Johnson, K. S., F. P. Chavez, and G. E. Friederich. 1999. Continental-shelf sediment as a source of iron for coastal phytoplankton. Nature 398: 697–700.

Johnson, N. K., J. A. Marten, and C. J. Ralph. 1989. Genetic evidence for the origin and relationships of Hawaiian honeycreepers (Aves: Fringillidae). Condor 91: 379–396.

Johnson, S. 2001. Emergence: The connected lives of ants, brains, cities, and software. New York: Scribner.

Jokiel, P. L. 1984. Long distance dispersal of reef corals by rafting. Coral Reefs 3:113–116.

———. 1989. Rafting of reef corals and other organisms at Kwajalein Atoll. Marine Biology 101: 483–493.

Jones, M. L. 1969. Boring of shell by *Caobangia* in freshwater snails of southeast Asia. American Zoologist 9: 829–835.

Kabat, A. R. 1990. Predatory ecology of naticid gastropods with a review of shell boring predation. Malacologia 32: 155–193.

Kadko, D., J. Baross, and J. Alt. 1995. The magnitude and global implications of hydrothermal flux. Geophysical Monographs 91: 446–466.

Kaplan, H. S., and A. J. Robson. 2002. The emergence of humans: The coevolution of intelligence and longevity with intergenerational transfers. Proceedings of the National Academy of Sciences of the USA 99: 10221–10226.

Karlstrom, K. E., S. A. Bowring, C. M. Dehler, A. H. Knoll, S. M. Parker, D. J. Des Marais, A. B. Weil, Z. D. Sharp, J. W. Geissman, M. B. Elrick, J. M. Timmons, L. J. Crossey, and K. L. Davidek. 2000. Chuar Group of the Grand Canyon: Record of breakup of Rodinia, associated change in the global carbon cycle, and ecosystem expansion by 740 Ma. Geology 28: 619–622.

Karr, J. R. 1982a. Avian extinction on Barro Colorado Island, Panama: A reassessment. American Naturalist 119: 220–239.

———. 1982b. Population variability and extinction in the avifauna of a tropical land bridge island. Ecology 63: 1975–1978.

Kase, T., and M. Ishikawa. 2003. Mystery of naticid predation history solved: Evidence from a "living fossil" species. Geology 31: 403–406.

Kasting, J. F., and J. L. Siefert. 2002. Life and the evolution of Earth's atmosphere. Science 296: 1066–1068.

Kaufman, L. 1977. The three spot damselfish: Effects on benthic biota of Caribbean coral reefs. Proceedings of the Third International Coral Reef Symposium, 1: 559–564.

Kauffman, S. A. 1993. The origins of order: Self-organization and selection in evolution. New York: Oxford University Press: 709 pp.

———. 2000. Investigations. Oxford: Oxford University Press: 287 pp.

Kay, R. F., C. Ross, and B. A. Williams. 1997. Anthropoid origins. Science 275: 797–804.

Kelley, P. H., and T. A. Hansen. 1996. Naticid gastropod prey selectivity through time and the hypothesis of escalation. Palaios 11: 437–445.

Kelt, D. A., and D. H. Van Vuren. 2001. The ecology and macroecology of mammalian home range area. American Naturalist 157: 637–645.

Kemp, A.E.S., R. B. Pearce, I. Koizumi, J. Pike, and S. J. Rance. 1999. The role of mat-forming diatoms in the formation of Mediterranean sapropels. Nature 398: 57–61.

Kemper, E. 1982. Apt und Alb—Beginn einer neuen Zeit. Geologisches Jahrbuch (A) 65: 681–693.

Kennett, J. P., and L. D. Stott. 1991. Abrupt deep-sea warming, palaeoceanographic changes and benthic extinctions at the end of the Palaeocene. Nature 353: 225–229.

Kent, B. W. 1983. Patterns of coexistence in busyconine whelks. Journal of Experimental Marine Biology and Ecology 66: 257–283.

Kerfoot, W. C., D. L. Kellogg, Jr., and J. R. Strickler. 1980. Visual observations of live zooplankters: Evasion, escape, and chemical defenses. Pages 10–27 in W. C. Kerfoot (ed.), Evolution and ecology of zooplankton communities. Hanover: University Press of New England.

Kidwell, S. M., and P. J. Brenchley. 1994. Patterns in bioclastic accumulation through the Phanerozoic: Changes in input or in destruction? Geology 22: 1139–1143.

———. 1996. Evolution of the fossil record: Thickness trends in skeletal accumulations and their implications. Pages 299–336 in D. Jablonski, D. H. Erwin, and J. H. Lipps (eds.), Evolutionary paleobiology: In honor of James W. Valentine. Chicago: University of Chicago Press.

Kilar, J. A., and R. M. Lou. 1986. The subtleties of camouflage and dietary preference of the decorator crab, *Microphrys bicornutus* Latreille (Decapoda: Brachyura). Journal of Experimental Marine Biology and Ecology 101: 143–160.

Kiltie, R. A. 1981. The function of interlocking canines in rain forest peccaries (Tayassuidae). Journal of Mammology 62: 459–469.

———. 1982. Bite force as a basis for niche differentiation between rain forest peccaries (*Tayassu tajacu* and *T. pecari*). Biotropica 14: 188–195.

Kim, S.-C., D. J. Crawford, and J. Francisco-Ortega. 1996. A common origin for woody *Sonchus* and five related genera in the Macaronesian islands: Molecular evidence for extensive radiation. Proceedings of the National Academy of Sciences of the USA 93: 7743–7748.

Kimura, H., and Y. Watanabe. 2001. Oceanic anoxia at the Precambrian-Cambrian boundary. Geology 29: 995–998.

Kimura, M. 1983. The neutral theory of molecular evolution. Cambridge: Cambridge University Press: 367 pp.

King, D. A. 1990. The adaptive significance of tree height. American Naturalist 135: 809–828.

Kirby, M. X. 2001. Differences in growth rate and environment between Tertiary and Quaternary *Crassostrea* oysters. Paleobiology 27:84–103.

Kirchner, J. W. 2002. Evolutionary speed limits inferred from the fossil record. Nature 415: 65–68.

Kirschner, M., and J. Gerhart. 1998. Evolvability. Proceedings of the National Academy of Sciences of the USA 95: 8420–8427.

Kitchell, J. A., D. L. Clark, and A. M. Gombos. 1986. Biological selectivity of extinction: A link between background and mass extinction. Palaios 1: 504–511.

Kleypas, J. A., R. W. Buddemeier, D. Archer, J.-P. Gattuso, C. Langdon, and B. N. Opdyke. 1999. Geochemical consequences of increased atmospheric carbon dioxide on coral reefs. Science 284: 118–120.

Klironomos, J. N., and M. M. Hart. 2001. Animal nitrogen swap for plant carbon. Nature 410: 651–652.

Knoll, A. H. 1984. Patterns of extinction in the fossil record of vascular plants. Pages 21–68 in M. H. Nitecki (ed.), Extinctions. Chicago: University of Chicago Press.

———. 1992. The early evolution of eukaryotes: A geological perspective. Science 256: 622–627.

———. 1994. Proterozoic and Early Cambrian protists: Evidence for accelerating evolutionary tempo. Proceedings of the National Academy of Sciences of the USA 91: 6743–6750.

———. 1996. Breathing room for early animals. Nature 382: 111–112.

———. 1999. A new molecular window on early life. Science 285: 1025–1026.

Knoll, A. H., and R. K. Bambach. 2000. Directionality in the history of life: diffusion from the left wall or repeated scaling of the right? Paleobiology 26 (supplement to No. 4): 1–14.

Knoll, A. H., and D. E. Canfield. 1998. Isotopic inferences on early ecosystems. Paleontological Society Papers 4: 212–243.

Knoll, A. H., and V. N. Sergeev. 1995. Taphonomic and evolutionary changes across the Mesoproterozoic-Neoproterozoic transition. Neues Jahrbuch für Geologie und Paläontologie Abhandlungen 195: 289–302.

Knoll, A. H., K. J. Niklas, and B. H. Tiffney. 1979. Phanerozoic land plant diversity in North America. Science 206: 1400–1402.

Knoll, A. H., I. J. Fairchild, and K. Swett. 1993. Calcified microbes in Neoproterozoic carbonates: implications for our understanding of the Proterozoic/Cambrian transition. Palaios 8: 512–525.

Knoll, A. H., R. K. Bambach, D. E. Canfield, and J. P. Grotzinger. 1996. Comparative Earth history and Late Permian mass extinction. Science 273: 452–457.

Knowlton, N., J. C. Lang, M. C. Rooney, and P. Clifford. 1981. Evidence for delayed mortality in hurricane-damaged Jamaican staghorn corals. Nature 294: 251–252.

Knutson, R. M. 1974. Heat production and temperature in eastern skunk cabbage. Science 186: 746–747.

Koehl, M.A.R., and R. S. Alberte. 1988. Flow, flapping, and photosynthesis of *Nereocystis luetkeana*: A functional comparison of undulate and flat blade morphologies. Marine Biology 99: 535–544.

Kohn, A. J. 1978a. Ecological shift and release in an isolated population: *Conus miliaris* at Easter Island. Ecological Monographs 48: 323–336.

———. 1978b. Gastropods as predators and prey at Easter Island. Pacific Science 32: 35–37.

Kolber, Z. S., F. G. Plumley, A. S. Lang, J. T. Beatty, R. E. Blankenship, C. L. VanDover, C. Vetriani, M. Koblizek, C. Rathgeber, and P. G. Falkowski. 2001. Contribution of aerobic photoheterotrophic bacteria to the carbon cycle in the ocean. Science 292: 2492–2495.

Kornfield, I., D. C. Smith, P. S. Gagnon, and J. N. Taylor. 1982. The cichlid fish of Cuatro Cienegas, Mexico: Direct evidence for conspecificity among distinct trophic morphs. Evolution 36: 658–664.

Kowalke, T. 2001. Protoconch morphology, ontogenetical development and ecology of three species of the genus *Potamides* Brongniart, 1810, and a discussion of the evolutionary history of the Potamididae (Caenogastropoda: Cerithimorpha). Freiberger Forschungshefte 492: 27–42.

Krassilov, V. A. 1981. Changes of Mesozoic vegetation and the extinction of dinosaurs. Palaeogeography, Palaeoclimatology, Palaeoecology 34: 207–224.

Kremer, M. 1993. Population growth and technological change: One million B.C. to 1990. Quarterly Journal of Economics 108: 681–716.

Kropp, R. K. 1983. Responses of five holothurian species to attacks by a predatory gastropod, *Tonna perdix*. Pacific Science 36: 445–452.

Krugman, P. 1996. The self-organizing economy. Cambridge, MA: Blackwell: 122 pp.

———. 1999. The return of depression economics. New York: Norton: 176 pp.

Kruuk, H. 1972. The spotted hyena: A study of predation and social behavior. Chicago: University of Chicago Press: 335 pp.

Kump, L. R., and F. T. McKenzie. 1996. Regulation of atmospheric O_2: Feedback in the microbial feedbag. Science 271: 459–460.

Kump, L. R., M. A. Arthur, M. E. Patzkowsky, M. T. Gibbs, D. S. Pinkus, and P. M. Sheehan. 1999. A weathering hypothesis for glaciation at high atmospheric pCO_2 during the Late Ordovician. Palaeogeography, Palaeoclimatology, Palaeoecology 152: 173–187.

Kuypers, M.M.M., A. O. Sliekers, G. Lavik, M. Schmid, B. Barker Jorgensen, J. G. Kuenens, J. S. Sinninghe-Damsté, M. Strous, and M.S.M. Jetten. 2003. Anaerobic ammonium oxidation by anammox bacteria in the Black Sea. Nature 422: 607–611.

Labandeira, C. C. 1997. Insect mouthparts: Ascertaining the paleobiology of insect feeding strategies. Annual Review of Ecology and Systematics 28: 153–193.

———. 1998a. Early history of arthropod and vascular plant associations. Annual Reviews of Earth and Planetary Sciences 26: 329–377.

———. 1998b. The role of insects in Late Jurassic to middle Cretaceous ecosystems. New Mexico Museum of Natural History and Science Bulletin 14: 195–224.

Labandeira, C. C., and T. L. Phillips. 1996. Insect fluid feeding on Upper Pennsylvanian tree ferns (Palaeodictyoptera: Marattiales) and the early history of the piercing-and-sucking functional group. Annals of the Entomological Society of America 89: 157–183.

———. 2002. Stem borings and petiole galls from Pennsylvanian tree ferns of Illinois, USA: Implications for the origin of the borer and galler functional-feeding-groups and holometabolous insects. Palaeontographica A 254: 1–84.

Labandeira, C. C., and J. J. Sepkoski, Jr. 1993. Insect diversity in the fossil record. Science 261: 310–315.

Labandeira, T. L. Phillips, and R. A. Norton. 1997. Oribatid mites and the decomposition of plant tissues in Paleozoic coal-swamp forests. Palaios 12: 319–353.

Labandeira, C. C., K. R. Johnson, and P. Wilf. 2002. Impact of the terminal Cretaceous event on plant-insect associations. Proceedings of the National Academy of Sciences of the USA 99: 2061–2066.

LaBarbera, M. 1977. Brachiopod orientation to water movement 1. Theory, laboratory behavior, and field orientations. Paleobiology 3: 270–287.

———. 1984. Feeding currents and particle capture mechanisms in suspension feeding animals. American Zoologist 24: 71–84.

———. 1990. Principles of design of fluid transport systems in zoology. Science 249: 992–1000.

Lamb, H. H. 1982. Climate, history and the modern world. London: Methuen: 387 pp.

Lande, R. 1978. Evolutionary mechanisms of limb loss in tetrapods. Evolution 32: 73–92.

Landes, D. S. 1998. The wealth and poverty of nations: Why some are so rich and some so poor. Norton: New York: 650 pp.

Landis, G. P., J. K. Rigby, R. E. Sloan, R. Hengst, and L. W. Snee. 1996. Pele hypothesis: ancient atmospheres and geologic-geochemical controls on evolution, survival, and extinction. Pages 519–556 in N. McLeod and G. Keller (eds.), Cretaceous-Tertiary mass extinctions: Biotic and environmental changes. London: Norton.

Langley, W. 1981. The effect of prey defenses on the attack behavior of the southern grasshopper mouse (*Onychomys torridus*). Zeitschrift für Tierpsychologie 56: 115–127.

Larson, A., and J. B. Losos. 1996. Phylogenetic systematics of adaptation. Pages 187–220 in M. R. Rose and G. V. Lauder (eds.), Adaptation. San Diego: Academic Press.

Larson, R. L. 1991a. Latest pulse of Earth: Evidence for a mid-Cretaceous superplume. Geology 19: 547–550.

Larson, R. L. 1991b. Geological consequences of superplumes. Geology 19: 963–966.

Latimer, J. C., and G. M. Filippelli. 2001. Terrigenous input of paleoproductivity in the southern ocean. Paleoceanography 16: 627–643.

Lauder, G. V. 1982. Patterns of evolution in the feeding mechanism of actinopterygian fishes. American Zoologist 22: 275–285.

Lauder, G. V., and K. F. Liem. 1983. The evolution and interrelationships of the actinopterygian fishes. Bulletin of the Museum of Comparative Zoology 150: 95–197.

Laverack, M. S. 1988. The diversity of chemoreceptors. Pages 287–312 in J. Atema, R. R. Fay, A. N. Popper, and N. Tavolga (eds.), Sensory biology of aquatic animals. New York: Springer.

Lawton, P. 1989. Predatory interaction between the brachyuran crab *Cancer pagurus* and decapod crustacean prey. Marine Ecology Progress Series 52: 169–179.

Leal, J. H. 1991. Marine prosobranch gastropods from oceanic islands off Brazil: Species composition and biogeography. Oegstgeest: Universal Book Services/Dr. W. Backhuys: 418 pp.

Leckie, R. M., T. J. Bralower, and R. Cashman. 2002. Oceanic anoxic events and plankton evolution: Biotic response to tectonic forcing during the mid-Cretaceous. Paleoceanography 17: 13-1–13-29.

Leigh, E. G., Jr. 1994. Do insect pests promote mutualism among tropical trees? Journal of Ecology 82: 677–680.

Leigh, E. G., Jr. 1999. Tropical forest ecology: A view from Barro Colorado Island. New York: Oxford University Press: 245 pp.

Leigh, E. G., Jr., and T. E. Rowell. 1995. The evolution of mutualism and other forms of harmony at various levels of biological organization. Ecologie 26: 131–152.

Leigh, E. G., Jr., and G. J. Vermeij. 2002. Does natural selection organize ecosystems for the maintenance of high productivity and diversity? Philosophical Transactions of the Royal Society of London B 357: 709–718.

Leigh, E. G., Jr., R. T. Paine, J. F. Quinn, and T. H. Suchanek. 1987. Wave energy and intertidal productivity. Proceedings of the National Academy of Sciences of the USA 84: 1314–1318.

Leigh, E. G., Jr., S. J. Wright, E. A. Herre, and F. E. Putz. 1993. The decline of tree diversity on newly isolated tropical islands: A test of a null hypothesis and some implications. Evolutionary Ecology 7: 76–102.

Leighton, L. R. 2001. New example of Devonian predatory boreholes and the influence of brachiopod spines on predator success. Palaeogeography, Palaeoclimatology, Palaeoecology 165: 53–69.

Lenski, R. E., C. Ofria, R. T. Pennock, and C. Adami. 2003. The evolutionary origin of complex features. Nature 423: 139–144.

Lenton, T. M. 1998. Gaia and natural selection. Nature 394: 439–447.

———. 2001. The role of land plants, phosphorus weathering and regulation of atmospheric oxygen. Global Change Biology 7: 613–629.

Leonard, G. H., J. M. Levine, P. R. Schmidt, and M. D. Bertness. 1998. Flow-driven variation in intertidal community structure in a Maine estuary. Ecology 79: 1395–1411.

Leonard, G. H., M. D. Bertness, and P. O. Yund. 1999. Crab predation, waterborne cues, and inducible defenses in the blue mussel, *Mytilus edulis*. Ecology 80: 1–14.

Les, D. H., and M. A. Cleland. 1997. Phylogenetic studies in Alismatidae, II: Evolution of marine angiosperms (seagrasses) and hydrophily. Systematic Botany 22: 443–463.

Lescinsky, H. L. 1996. Early brachiopod associates: Epibionts on Middle Ordovician brachiopods. Pages 169–173 in P. Copper and J. Jin (eds.), Brachiopods. Rotterdam: Balkema.

———. 1997. Epibiont communities: Recruitment and competition on North American Carboniferous brachiopods. Journal of Paleontology 71: 34–53.

Lescinsky, H. L., E. Edinger, and M. J. Risk. 2002. Mollusc shell encrustation and bioerosion rates in a modern epeiric sea: Taphonomy experiments in the Java Sea, Indonesia. Palaios 17: 171–191.

Lessios, H. A., M. J. Garrido, and B. D. Kessing. 2001. Demographic history of *Diadema antillarum*, a keystone herbivore on Caribbean reefs. Proceedings of the Royal Society of London B 268: 2347–2353.

Levin, D. A. 1975. Pest pressure and recombination systems in plants. American Naturalist 109: 437–451.

———. 1976. Alkaloid-bearing plants: An ecogeographic perspective. American Naturalist 110: 261–284.

Levin, D. A., and B. M. York. 1978. The toxicity of plant alkaloids: An ecogeographic perspective. Biochemical Systematics and Ecology 6: 61–76.

Levin, S. A. 1999. Fragile dominion: Complexity and the commons. Reading: Perseus: 250 pp.

Levins, R. 1968. Evolution in changing environments: Some theoretical explorations. Princeton: Princeton University Press: 119 pp.

Levinton, J. S., and M. L. Judge. 1993. The relationship of closing force to body size for the major claw of *Uca pugnax* (Decapoda: Ocypodidae). Functional Ecology 7: 339–345.

Levinton, J. S., M. L. Judge, and J. P. Kurdziel. 1995. Functional differences between the major and minor claws of fiddler crabs (*Uca*, family Ocypodidae, order Decapoda, subphylum Crustacea): A result of selection or developmental constraint? Journal of Experimental Marine Biology and Ecology 193: 147–160.

Levitan, D. 1992. Community structure in times past: Human fishing pressure on alga-urchin interactions. Ecology 73: 1597–1605.

Lewis, D. H. 1991. Mutualistic symbioses in the origin and evolution of land plants. Pages 288–300 in L. Margulis and R. Gester (eds.), Symbiosis as a source of evolutionary innovation: speciation and morphogenesis. Cambridge: MIT Press.

Lewis, W. A. 1955. The theory of economic growth. Homewood: Richard D. Erwin: 453 pp.

Lewontin, R. C. 1965. Selection for colonizing ability. Pages 77–91 in H. G. Baker and L. G. Stebbins (eds.), The genetics of colonizing species. New York: Academic Press.

Lichtenegger, H. C., T. Schöberl, H. Bartl, H. Waite, and G. D. Stucky. 2002. High abrasion resistance with sparse mineralization: Copper biomineral in worm jaws. Science 298: 389–392.

Lidgard, S. 1990. Growth in encrusting cheilostome bryozoans: II. Circum-Atlantic distribution patterns. Paleobiology 16: 304–321.

Lidgard, S., F. K. McKinney, and P. D. Taylor. 1993. Competition, clade replacement, and a history of cyclostome and cheilostome bryozoan diversity. Paleobiology 19: 352–371.

Lieth, H. 1975. Primary production of the major vegetation units of the world. Pages 203–215 in H. Lieth and R. H. Whittaker (eds.), Primary productivity of the biosphere. New York: Springer.

Lilburn, T. G., K. S. Kim, N. E. Ostrom, K. R. Byzek, J. R. Leadbetter, and J. A. Breznak. 2001. Nitrogen fixation by symbiotic and free-living spirochetes. Science 292: 2495–2498.

Lill, J. T., R. J. Marquis, and R. E. Ricklefs. 2002. Host plants influence parasitism of forest caterpillars. Nature 417: 170–173.

Lindberg, D. R., and W. F. Ponder. 2001. The influence of classification on the evolutionary interpretation of structure—a re-evaluation of the evolution of the pallial cavity of gastropod molluscs. Organisms, Diversity, and Evolution 1: 273–299.

Lindstedt, S. L., J. Hokanson, D. J. Wells, S. D. Swain, H. Hoppeler, and V. Navarro. 1991. Running energetics in the pronghorn antelope. Nature 353: 748–750.

Lister, A. M. 1989. Rapid dwarfing of red deer on Jersey in the last interglacial. Nature 342: 539–542.

Littler, M. M., and D. S. Littler. 1980. The evolution of thallus form and survival strategies in benthic marine macroalgae: Field and laboratory tests of a functional form model. American Naturalist 116: 25–44.

Livezey, B. C. 1992. Flightlessness in the Galápagos cormorant (*Compsohalieus* (*Nannopterum*) *harrisi*): Heterochrony, gigantism and specialization. Zoological Journal of the Linnean Society 205: 155–224.

———. 1993. An ecomorphological review of the dodo (*Raphus cucullatus*) and solitaire (*Pezophaps solitaria*), flightless Columbiformes of the Mascarene Islands. Journal of Zoology, London, 230: 247–292.

Logan, G. A., J. M. Hayes, G. B. Hieshima, and R. E. Summons. 1995. Terminal Proterozoic reorganization of biogeochemical cycles. Nature 376: 51–56.

Looy, C. V., R. J. Twitchett, D. L. Dilcher, J. H. A. Koningenburg-Van Cittert, and H. Visscher. 2001. Life in the end-Permian dead zone. Proceedings of the National Academy of Sciences of the USA 98: 7879–7883.

Loreau, M., S. Naeem, P. Inchausti, J. Bengtsson, J. P. Grime, A. Hector, D. U. Hooper, M. A. Huston, D. Raffaelli, and D. A. Wardle. 2001. Biodiversity and ecosystem functioning: Current knowledge and future challenges. Science 294: 804–808.

Losos, J. B., and D. Schluter. 2000. Analysis of an evolutionary species-area relationship. Nature 408: 847–850.

Losos, J. B., J. J. Irschick, and T. W. Schoener. 1994. Adaptation and constraint in the evolution of specialization of Bahamian *Anolis* lizards. Evolution 48: 1786–1798.

Lowenstam, H. A. 1981. Minerals formed by organisms. Science 211: 1126–1131.

Lozouet, P., and J. Le Renard. 1998. Les Coralliophilidae, Gastropoda de l'Oligocène et du Miocène inférieur d'Aquitaine (sud-ouest de la France): Systématique et coraux hôtes. Géobios 31: 171–184.

Lubchenco, J. 1978. Plant species diversity in a marine intertidal community: Importance of herbivore food preference for algal competitive abilities. American Naturalist 112: 23–39.

———. 1983. *Littorina* and *Fucus*: Effects of herbivores, substratum heterogeneity, and plant escapes during succession. Ecology 64: 1116–1123.

Lynch, M. 1999. The age and relationships of the major animal phyla. Evolution 53: 319–325.

MacArthur, R. H. 1965. Patterns of species diversity. Biological Reviews 40: 510–533.

———. 1972. Geographical ecology: Patterns in the distribution of species. New York: Harper and Row: 269 pp.

MacArthur, R. H., and E. O. Wilson. 1967. The theory of island biogeography. Princeton: Princeton University Press: 203 pp.

MacFadden, B. J. 2000. The evolution of the grazing guild in Cenozoic New World terrestrial mammals. Pages 223–244 in H.-D. Sues (ed.), Evolution of herbivorous terrestrial vertebrates: Perspectives from the fossil record. Cambridge: Cambridge University Press.

MacGillavry, H. J. 1968. Modes of evolution mainly among marine invertebrates: An observational approach. Bijdragen tot de Dierkunde 38: 69–74.

MacLeod, N. 2003. The causes of Phanerozoic extinctions. Pages 253–227 in L. Rothschild and A. Lister (eds.), Evolution on planet earth: The impact of physical environment. Amsterdam: Academic Press.

Magaritz, M. 1989. ^{13}C minima follow extinction events: A clue to faunal radiation. Geology 17: 337–340.

Maldonado, M., M. C. Carmona, M. J. Uriz, and A. Cruzado. 1999. Decline in Mesozoic reef-building sponges explained by silicon limitation. Nature 401: 785–788.

Maliva, R. G., A. H. Knoll, and R. Siever. 1990. Secular change in chert distribution: A reflection of evolving biological participation in the silica cycle. Palaios 4: 519–532.

Mancinelli, R. O. 2003. What good is nitrogen: An evolutionary perspective. Pages 25–34 in L. J. Rothschild and A. M. Lister (eds.), Evolution on planet Earth: The impact of the physical environment. London: Academic Press.

Mann, K. H. 1973. Seaweeds: Their productivity and strategy of growth. Science 182: 975–981.

Mann, K. H., A.R.O. Chapman, and J. A. Gagné. 1980. Productivity of seaweeds: The potential and the reality. Pages 363–380 in P. G. Falkowski (ed.), Primary productivity in the sea. New York: Plenum.

Marden, J. H., and P. Chai. 1991. Aerial predation and butterfly design: How palatability, mimicry, and the need for evasive flight constrain mass allocation. American Naturalist 138: 15–36.

Margalef, R. 1991. Networks in ecology. Pages 41–54 in M. Higashi and T. P. Burns (eds.), Theoretical studies of ecosystems: The network perspective. Cambridge: Cambridge University Press.

Margolin, A. S. 1964a. The mantle response of *Diodora aspera*. Animal Behaviour 12: 187–194.

———. 1964b. A running response of *Acmaea* to seastars. Ecology 45: 191–193.

———. 1975. Responses to sea stars by three naticid gastropods. Ophelia 14: 85–92.

Margulis, L. 1981. Symbiosis in cell evolution: Life and its environment on the early Earth. San Francisco: Freeman: 419 pp.

———. 1991. Symbiogenesis and symbionticism. Pages 1–14 in L. Margulis and R. Gester (eds.), Symbiosis as a source of evolutionary innovation: Speciation and morphogenesis. Cambridge: MIT Press.

Marincovich, L., Jr. 2000. Central American paleogeography controlled Pliocene Arctic Ocean molluscan migrations. Geology 28: 551–554.

Marincovich, L., Jr., and A. Yu. Gladenkov. 1999. Evidence for an early opening of the Bering Strait. Nature 397: 149–151.

Mariscal, R. N. 1971. Experimental studies on the protection of anemone fishes from sea anemones. Pages 283–315 in J. C. Eng (ed.), Aspects of the biology of symbiosis. Baltimore: University Park Press.

Markel, R. P. 1971. Temperature relations in two species of tropical West American littorines. Ecology 52: 1126–1130.

Marshall, A. T. 1996. Calcification in hermatypic and ahermatypic corals. Science 271: 637–639.

Marshall, L. G. 1981. The Great American interchange—an invasion induced crisis for South American mammals. Pages 133–229 in M. H. Nitecki (ed.), Biotic crises in ecological and evolutionary time. New York: Academic Press.

Martin, J. H., and S. E. Fitzwater. 1988. Iron deficiency limits phytoplankton growth in the northeast Pacific subarctic. Nature 331: 341–343.

Martin, P. S. 1984. Prehistoric overkill: The global model. Pages 354–403 in P. S. Martin and R. G. Klein (eds.), Quaternary extinctions: A prehistoric revolution. Tucson: University of Arizona Press.

Martin, P. S., and D. W. Steadman. 1999. Prehistoric extinctions on islands and continents. Pages 17–55 in R.D.E. MacPhee (ed.), Extinctions in near time: Causes, contexts, and consequences. New York: Kluwer/Plenum.

Martin, R. E. 1995. Cyclic secular variation in microfossil biomineralization: Clues to the biogeochemical evolution of Phanerozoic oceans. Global and Planetary Change 11: 2–23.

———. 1998. Catastrophic fluctuations in nutrient levels as an agent of mass extinction: Upward scaling of ecological processes? Pages 405–429 in M. L. McKinney and J. A. Drake (eds.), Biodiversity dynamics: Turnover of populations, taxa, and communities. New York: Columbia University Press.

Martin, R. E., G. J. Vermeij, D. Dorritie, K. Caldeira, M. R. Rampino, A. H. Knoll, R. K. Bambach, D. Canfield, J. P. Grotzinger, P. B. Wignall, and R. J. Twitchett. 1996. Late Permian extinctions. Science 274: 1549–1552.

Martin, W., and M. Müller. 1998. The hydrogen hypothesis for the first eukaryote. Nature 392: 37–41.

Martin, W., and M. J. Russell. 2003. On the origins of cells: A hypothesis for the evolutionary transitions from abiotic geochemistry to chemoautotrophic prokaryotes, and from prokaryotes to nucleated cells. Philosophical Transactions of the Royal Society of London B 358: 59–85.

Marzoli, A., P. R. Renne, E. M. Piccirillo, M. Ernesto, G. Bellieni, and A. de Min. 1999. Extensive 200-million-year-old continental flood basalts of the Central Atlantic Magmatic Province. Science 284: 616–618.

Max, M. D., W. P. Dillon, C. Nishimura, and B. G. Hurdle. 1999. Sea-floor methane blow-out and global firestorm at the K-T boundary. Geo-Marine Letters 18: 285–291.

Maynard Smith, J. 1989. The causes of extinction. Philosophical Transactions of the Royal Society of London B 325: 241–252.

———. 1991. A Darwinian view of symbiosis. Pages 26–39 in L. Margulis and R. Gester (eds.), Symbiosis as a source of evolutionary innovation: Speciation and morphogenesis. Cambridge: MIT Press.

Maynard Smith, J., and E. Szathmáry. 1995. The major transitions in evolution. Oxford: W. H. Freeman, Spektrum: 346 pp.

———. 1999. The origins of life—from the birth of life to the origin of language. Oxford: Oxford University Press: 180 pp.

Mazancourt, C. de, M. Loreau, and U. Dieckmann. 2001. Can the evolution of plant defense lead to plant-herbivore mutualism? American Naturalist 158: 109–123.

McCall, P. L., and M.J.S. Tevesz. 1982. The effects of benthos on physical properties of freshwater sediments. Pages 105–176 in P. L. McCall and M.J.S. Tevesz (eds.), Animal-sediment relations: The biogenic alteration of sediments. New York: Plenum.

McConnaughey, T. 1991. Calcification in *Chara corallina*: CO_2 hydroxylation generates protons for bicarbonate assimilation. Limnology and Oceanography 36: 619–628.

McConnaughey, T. A. 1994. Calcification, photosynthesis, and global carbon cycles. Bulletin de l'Institut d'Oceanographie Monaco, Numéro Special 13: 137–161.

McConnaughey, T. A., and R. F. Falk. 1991. Calcium-proton exchange during algal calcification. Biological Bulletin 180: 185–195.

McElwain, J. C. 1998. Do fossil plants signal palaeoatmospheric CO_2 concentration in the geologic past? Proceedings of the Royal Society of London B 353: 83–96.

McElwain, J. C., D. J. Beerling, and F. I. Woodward. 1999. Fossil plants and global warming at the Triassic-Jurassic boundary. Science 285: 1386–1390.

McFall-Ngai, M. J. 1990. Crypsis in the pelagic environment. American Zoologist 30: 175–188.

McGhee, G. R. 1988. Late Devonian extinction event: Evidence for abrupt ecosystem collapse. Paleobiology 14: 250–257.

McGowan, J. A., and P. A. Walker. 1993. Pelagic diversity patterns. Pages 203–214 in R. E. Ricklefs and D. Schluter (eds.), Species diversity in ecological communities: Historical and geographical perspectives. Chicago: University of Chicago Press.

McGrady-Steed, J., P. M. Harris, and P. J. Morin. 1997. Biodiversity regulates ecosystem predictability. Nature 390: 162–165.

McIlroy, D., and G. A. Logan. 1999. The impact of bioturbation on infaunal ecology and evolution during the Proterozoic-Cambrian transition. Palaios 14: 58–72.

McKey, D. 1979. The distribution of secondary compounds within plants. Pages 55–133 in G. A. Rosenthal and D. H. Janzen (eds.), Herbivores: Their interaction with secondary plant metabolites. New York: Academic Press.

McKillup, S. C. 1982. The selective advantage of avoidance of the predatory whelk *Lepsiella venosa* Lamarck by *Littorina unifasciata* Philippi. Journal of Experimental Marine Biology and Ecology 63: 59–66.

McKinney, F. K. 1984. Feeding currents of gymnolaemate bryozoans: Better organization with higher colonial integration. Bulletin of Marine Science 34: 315–319.

———. 1992. Competitive interactions between related clades: Evolutionary interactions of overgrowth interactions between encrusting cyclostome and cheilostome Bryozoa. Marine Biology 114: 645–652.

———. 1993. A faster-paced world?: Contrasts in biovolume and life-process rates in cyclostome (class Stenolaemata) and cheilostome (class Gymnolaemata) bryozoans. Paleobiology 19: 335–351.

———. 1995. One hundred million years of competitive interactions between bryozoan clades: Asymmetrical but not escalating. Biological Journal of the Linnean Society 56: 465–481.

McKinney, F. K., and A. Jaklin. 1993. Living populations of free-lying bryozoans: Implications for post-Paleozoic decline of the growth habit. Lethaia 26: 171–179.

McKinney, F. K., S. Lidgard, J. J. Sepkoski Jr., and P. D. Taylor. 1998. Decoupled temporal patterns of evolution and ecology in two post-Paleozoic clades. Science 281: 807–809.

McLaughlin, P. A., and T. Murray. 1990. *Clibanarius fonticola*, new species (Anomura: Paguridea: Diogenidae), from a fresh-water pool on Espiritu Santo, Vanuatu. Journal of Crustacean Biology 10: 695–702.

McLean, D. M. 1985. Mantle degassing induced dead ocean in the Cretaceous-Tertiary transition. Geophysical Monographs 32: 493–503.

McLean, R. B. 1974. Direct shell acquisition by hermit crabs from gastropods. Experientia 30: 206–208.

———. 1983. Gastropod shells: A dynamic resource that helps shape benthic community structure. Journal of Experimental Marine Biology and Ecology 69: 151–174.

McLennan, S. M. 1993. Weathering and global denudation. Journal of Geology 101: 295–303.

McNab, B. K. 1980. Food habits, energetics, and the population biology of mammals. American Naturalist 116: 106–124.

———. 1994a. Energy conservation and the evolution of flightlessness in birds. American Naturalist 144: 628–642.

———. 1994b. Resource use and the survival of land and freshwater vertebrates on oceanic islands. American Naturalist 144: 643–660.

McNaughton, R. B., J. M. Cole, R. Dalrymple, S. J. Braddy, D.E.G. Briggs, and S. D. Lukie. 2002. First steps on land: Arthropod trackways in Cambrian-Ordovician eolian sandstone, southeastern Ontario, Canada. Geology 30: 391–394.

McNaughton, S. J. 1984. Grazing lawns: Animals in herds, plant form, and coevolution. American Naturalist 124: 863–866.

McNaughton, S. J., F. F. Banyikwa, and M. M. McNaughton. 1997. Promotion of the cycle of diet-enhancing nutrients by African grazers. Science 278: 1798–1800.

McNeill, W. H. 1982. The pursuit of power: Technology, armed force, and society since A.D. 1000. Chicago: University of Chicago Press: 405 pp.

McShea, D. W. 1994. Mechanisms of large-scale evolutionary trends. Evolution 48: 1747–1763.

———. 1996. Metazoan complexity and evolution: Is there a trend? Evolution 50: 477–492.

———. 1998. Possible largest-scale trends in organismal evolution: Eight "live" hypotheses. Annual Reviews of Ecology and Systematics 29: 293–318.

———. 2002. A complexity drain on cells in the evolution of multicellularity. Evolution 56: 441–452.

Mech, L. D. 1966. The wolves of Isle Royale. Washington: Government Printing Office: 210 pp.

Melillo, J. M., M. D. McGuire, D. W. Kicklighter, B. Moore III, C. J. Vorosmarty, and A. L. Schloss. 1993. Global climate change and terrestrial net primary production. Nature 363: 234–240.

Merle, D., and J.-M. Pacaud. 2001. The first record of *Poirieria subcristata* (d'Orbigny, 1850) (Muricidae: Muricinae) in the early Cuisian of the Paris Basin (Celles-sur-Aisne, Aizy Formation), with comments on the sculptural evolution of some Palaeocene and Eocene *Poirieria* and *Paziella*. Tertiary Research 21: 19–27.

Meyer, D. L. 1985. Evolutionary implications of predation on Recent comatulid crinoids from the Great Barrier Reef. Paleobiology 11: 154–164.

Meyer, D. L., and W. I. Ausich. 1983. Biotic interactions among Recent and among fossil crinoids. Pages 377–427 in M.J.S. Tevesz and P. L. McCall (eds.), Biotic interactions in Recent and fossil benthic communities. New York: Plenum.

Meyer, D. L., and D. B. Macurda. 1977. Adaptive radiation of the comatulid crinoids. Paleobiology 3: 74–82.

Michod, R. E. 1997. Evolution of the individual. American Naturalist 150: S5–S21.

Miller, G. F. 2000. The mating mind: How sexual choice shaped the evolution of human nature. New York: Doubleday: 503 pp.

Miller, M. F. 1984. Distribution of biogenic structures in Paleozoic nonmarine and marine-margin sequences: An actualistic model. Journal of Paleontology 58: 550–570.

Miller, M. F., T. McDowell, S. E. Smail, Y. Shyr, and N. R. Kemp. 2002. Hardly used habitats: Dearth and distribution of burrowing in Paleozoic and Mesozoic stream and lake deposits. Geology 30: 527–530.

Miller, S. L. 1974. Adaptive design of locomotion and foot form in prosobranch gastropods. Journal of Experimental Marine Biology and Ecology 14: 99–156.

Mitchell, J. G. 2002. The energetics and scaling of search strategies in bacteria. American Naturalist 160: 727–740.

Mittelbach, G. G., C. F. Steiner, S. M. Scheiner, K. L. Gross, H. L. Reynolds, R. B. Waide, M. R. Willig, S. I. Dodson, and L. Gough. 2001. What is the observed relationship between species richness and productivity? Ecology 82: 2381–2396.

Mitter, C., B. Farrell, and B. Wiegmann. 1988. The phylogenetic study of adaptive zones: Has phytophagy promoted insect diversification? American Naturalist 132: 107–128.

Mokyr, J. 1990. The lever of riches: Technological creativity and economic progress. New York: Oxford University Press: 349 pp.

———. 1991. Evolutionary biology, technological change and economic history. Bulletin of Economic Research 43: 127–149.

Molbo, D., C. Machado, J. G. Sevenster, L. Keller, and E. A. Herre. 2003. Cryptic species of fig-pollinating wasps: Implications for the evolution of fig-wasp mutualism, sex allocation, and precision of adaptation. Proceedings of the National Academy of Sciences of the USA 100: 5867–5872.

Monod, J. 1971. Chance and necessity: An essay on the natural philosophy of modern biology. (Translated by A. Wainhouse.) New York: Knopf: 199 pp.

Moores, E. M. 1993. Neoproterozoic crustal thinning, emergence of continents, and origin of the Phanerozoic ecosystem: a model. Geology 21: 5–8.

———. 2002. Pre-l Ga (pre-Rodinian) ophiolites: Their tectonic and environmental implications. Geological Society of America Bulletin 114: 89–95.

Morales, M. 1997. Major groups of non-dinosaurian vertebrates of the Mesozoic era. Pages 605–626 in J. O. Farlow and M. K. Brett-Surman (eds.), The complete dinosaur. Bloomington: Indiana University Press.

Moran, N. A., and A. Telange. 1998. Bacteriocyte-associated symbionts of insects: A variety of insect groups harbor ancient prokaryotic endosymbionts. BioScience 48: 295–304.

Morgan, S. G. 1995. Life and death in the plankton: Larval mortality and adaptation. Pages 279–321 in L. McEdward (ed.), Ecology of invertebrate larvae. Boca Raton: CRC Press.

Morowitz, H. J. 1968. Energy flow in biology: Biological organization as a problem in thermal physics. New York: Academic Press: 179 pp.

Morrison, C. L., A. W. Harvey, S. Lavery, K. Tieu, Y. Huang, and C. W. Cunningham. 2002. Mitochondrial gene rearrangements confirm the parallel evolution of the crab-like form. Proceedings of the Royal Society of London B 269: 345–350.

Morse, D. H. 1975. Ecological aspects of adaptive radiation in birds. Biological Reviews 50: 167–214.

Morton, B., and J. C. Britton. 1993. The ecology, diet and foraging strategy of *Thais orbita* (Gastropoda: Muricidae) on a rocky shore of Rottnest Island, Western Australia. Pages 539–563 in F. E. Wells, D. I. Walker, H. Kirkman, and R. Lethbridge (eds.), The marine flora and fauna of Rottnest Island, Western Australia: Proceedings of the Fifth International Marine Biological Workshop, held at Rottnest Island in January 1991. Perth: Western Australian Museum.

Morton, E. S. 1978. Avian arborial folivores: Why not? Pages 123–130 in G. G. Montgomery (ed.), The ecology of arborial folivores. Washington: Smithsonian Press.

Morton, N., and M. Nixon. 1987. Size and function of ammonite aptychi in comparison with buccal masses of modern cephalopods. Lethaia 20: 231–238.

Moss, S. A. 1977. Feeding mechanisms in sharks. American Zoologist 17: 355–364.

———. 1981. Shark feeding mechanisms. Oceanus 24: 23–29.

Motani, R. 2002. Swimming speed estimation of extinct marine reptiles: Energetic approach revisited. Paleobiology 28: 251–262.

Moulton, K. L., and R. A. Berner. 1998. Quantification of the effect of plants on weathering: Studies in Iceland. Geology 26: 895–898.

Moynihan, M. 1971. Successes and failures of tropical mammals and birds. American Naturalist 105: 371–383.

Moy-Thomas, J. A., and R. S. Miles. 1971. Palaeozoic fishes. London: Chapman and Hall.

Mueller, U. G., and N. Gerado. 2002. Fungus-farming insects: Multiple origins and diverse evolutionary histories. Proceedings of the National Academy of the USA 99: 15247–15249.

Mueller, U. G., S. H. Rehner, and T. R. Schultz. 1998. The evolution of agriculture in ants. Science 281: 2034–2038.

Muirhead, A., P. A. Tyler, and M. A. Thurston. 1986. Reproductive biology and growth of the genus *Epizoanthus* (Zoantharia) from the north-east Atlantic. Journal of the Marine Biological Association of the United Kingdom 66: 131–143.

Mukhopadhyay, S., K. A. Farley, and A. Montanari. 2001. A short duration of the Cretaceous-Tertiary boundary event: Evidence from extra-terrestrial helium-3. Science 291: 1952–1955.

Mulcahy, D. L. 1979. The rise of the angiosperms: A genecological factor. Science 206: 1023–1024.

Nadkarni, N. M. 1981. Canopy roots: Convergent evolution in rainforest nutrient cycles. Science 214: 1023–1024.

Naeem, S., and S. Li. 1997. Biodiversity enhances ecosystem reliability. Nature 390: 507–509.

Naeem, S., L. J. Thompson, S. P. Lawler, J. H. Lawton, and R. M. Woodfin. 1994. Declining biodiversity can alter the performance of ecosystems. Nature 368: 734–737.

Nagy, L. M., and T. A. Williams. 2001. Comparative limb development as a tool for understanding the evolutionary diversification of limbs in arthropods: Challenging the modularity paradigm. Pages 455–488 in G. P. Wagner (ed.), The character concept in evolutionary biology. San Diego: Academic Press.

Nardi, F., G. Spinsanti, J. L. Boore, A. Carapelli, R. Dallai, and F. Frati. 2003. Hexapod origins: Monophyletic or paraphyletic? Science 299: 1587–1589.

Nentwig, W. 1983. The prey of web-building spiders compared with feeding experiments (Araneae: Araneidae, Linyphiidae, Pholcidae, Agelenidae). Oecologia (Berlin) 56: 132–139.

Nesse, R. M., and G. C. Williams. 1994. Why we get sick: The new science of Darwinian medicine. New York: Random House: 291 pp.

Newman, S. A., and G. B. Müller. 2001. Epigenetic mechanisms for character origination. Pages 559–579 in G. P. Wagner (ed.), The character concept in evolutionary biology. San Diego: Academic Press.

Nielsen, C. 1975. Observations on *Buccinum undatum* L. attacking bivalves and on prey responses, with a short review on attack methods of other prosobranchs. Ophelia 13: 87–108.

Nisbet, E. G., and N. H. Sleep. 2001. The habitat and nature of early life. Nature 409: 1083–1091.

Norris, K. S., and B. Mohl. 1983. Can odontocetes debilitate prey with sound? American Naturalist 122: 85–104.

Norris, R. D. 1996. Symbiosis as an evolutionary innovation in the radiation of Paleocene planktic Foraminifera. Paleobiology 22: 461–480.

Norris, R. D., and U. Röhl. 1999. Carbon cycling and chronology of climate warming during the Palaeocene/Eocene transition. Nature 401: 775–78.

Norris, R. D., and P. A. Wilson. 1998. Low-latitude sea-surface temperatures for the mid-Cretaceous and the evolution of planktic Foraminifera. Geology 26: 823–826.

Norris, R. D., J. Firth, J. S. Blusztajn, and G. Ravizza. 2000a. Mass failure of the North Atlantic margin in the Cretaceous-Paleogene bolide impact. Geology 28: 1119–1122.

Norris, R. D., R. M. Corfield, and K. Hayes-Baker. 2000b. Mountains and Eocene climate. Pages 161–196 in B. T. Huber, K. G. MacLeod, and S. L. Wing (eds.), Warming climates in Earth history. Cambridge: Cambridge University Press.

Novotny, V., Y. Basset, S. E. Miller, G. D. Weiblen, B. Bremer, L. Cizek, and P. Drozd. 2002. Low host specificity of herbivorous insects in a tropical forest. Nature 416: 841–844.

Nowak, M. A., M. A. Boerlijst, J. Cooke, and J. Maynard Smith. 1997. Evolution of genetic redundancy. Nature 388: 167–171.

Nowicki, S., and T. Eisner. 1983. Predatory capture of bombardier beetles by a tabanid fly larva. Psyche 90: 119–122.

Oberbeck, V. R., J. R. Marshall, and H. Aggrawal. 1993. Impacts, tillites, and the breakup of Gondwanaland. Journal of Geology 101: 1–19.

Odum, E. P. 1969. Strategy of ecosystem development. Science 164: 162–170.

Odum, H. T. 1983. Systems ecology: An introduction. New York: Wiley: 644 pp.

Ofek, H. 2001. Second nature: Economic origins of human evolution. Cambridge: Cambridge University Press: 254 pp.

Officer, C. B., and C. L. Drake. 1985. Terminal Cretaceous environmental events. Science 227: 1161–1167.

Officer, C. B., A. Hallam, C. L. Drake, and J. D. Devine. 1987. Late Cretaceous and paroxysmal Cretaceous/Tertiary extinctions. Nature 326: 143–149.

Oji, T. 1996. Is predation intensity reduced with increasing depth? Evidence from the west Atlantic stalked crinoid *Endoxocrinus parrae* (Gervais) and implications for the Mesozoic marine revolution. Paleobiology 22: 339–351.

———. 2001. Fossil record of echinoderm regeneration with special regard to crinoids. Microscopy Research and Technique 55: 397–402.

Oji, T., and T. Okamoto. 1994. Arm autotomy and arm branching pattern as anti-predatory adaptations in stalked and stalkless crinoids. Paleobiology 20: 27–39.

Oji, T., C. Ogaya, and T. Sato. 2003. Increase of shell-crushing predation recorded in fossil shell fragmentation. Paleobiology 29: 520–526.

Olive, C. W. 1980. Foraging specializations in orb-weaving spiders. Ecology 60: 1133–1144.

Oliver, J. S., P. N. Slattery, E. F. O'Connor, and L. F. Lowry. 1983a. Walrus, *Odobenus rosmarus*, feeding in the Bering Sea: A benthic perspective. Fisheries Bulletin 81: 501–512.

Oliver, J. S., P. N. Slattery, M. A. Silberstein, and E. F. O'Connor. 1983b. A comparison of gray whale, *Eschrichtius robustus*, feeding in the Bering Sea and Baja California. Fisheries Bulletin 81: 513–522.

Olivera, B. M. 2002. *Conus* venom peptides: Reflections from the biology of clades and species. Annual Review of Ecology and Systematics 33: 25–47.

Olsen, J. L., W. T. Stam, S. Berger, and D. Menzel. 1994. 18S rDNA and evolution in the Dasycladales (Chlorophyta): Modern living fossils. Journal of Phycology 30: 729–744.

Olsen, P. E. 1999. Giant lava flows, mass extinctions, and mantle fluxes. Science 284: 604–605.

Olsen, P. E., D. V. Kent, H.-D. Sues, C. Koeberl, H. Huber, A. Montanari, E. A. Rainforth, S. J. Fowell, M. J. Szajna, and B. W. Hartline. 2002. Ascent of dinosaurs linked to an iridium anomaly at the Triassic-Jurassic boundary. Science 296: 1305–1307.

Olson, M. 1982. The rise and decline of nations: Economic growth, stagflation, and social rigidities. New Haven: Yale University Press: 273 pp.

Olson, S. L. 1985. The fossil record of birds. Pages 79–252 in D. S. Farmer, R. King, and K. C. Parkes (eds.), Avian biology, Vol. 8. New York: Academic Press.

Olson, S. L., and H. F. James. 1991. Descriptions of thirty-two new species of bird from the Hawaiian Islands: part I. Non-Passeriformes. Ornithological Monographs 45: 1–88.

Onuf, C. P., J. M. Teal, and I. Valiela. 1977. Interactions of nutrients, plant growth and herbivory in a mangrove ecosystem. Ecology 58: 514–526.

Orians, G. H., and D. H. Janzen. 1974. Why are embryos so tasty? American Naturalist 108: 581–592.

Orians, G. H., and O. T. Solbrig. 1977. A cost-income model of leaves and roots with special reference to arid and semiarid areas. American Naturalist 111: 677–690.

Ormerod, P. 1998. Butterfly economics: A new general theory of social and economic behavior. New York: Pantheon: 217 pp.

Osborne, C. P., and D. J. Beerling. 2002. Sensitivity of tree growth to a high CO_2 environment: Consequences for interpreting the characteristics of fossil woods from ancient "greenhouse" worlds. Palaeogeography, Palaeoclimatology, Palaeoecology 182: 15–29.

Osman, R. W., and J. A. Haugsness. 1981. Mutualism among sessile invertebrates: a mediator of competition and predation. Science 211: 846–848.

Ostrom, J. H. 1963. Further comments on herbivorous lizards. Evolution 17: 368–369.

Otto, C., and B. S. Svensson. 1980. The significance of case material selection for the survival of caddis larvae. Journal of Animal Ecology 49: 855–865.

Ozanne, C. R., and P. J. Harries. 2002. Role of predation and parasitism in the extinction of the inoceramid bivalves: An evaluation. Lethaia 35: 1–19.

Packard, A. 1972. Cephalopods and fish: The limits of convergence. Biological Reviews 47: 241–307.

Paepke, C. O. 1993. The evolution of progress: The end of economic growth and the beginning of human transformation. New York: Random House: 383 pp.

Pagani, M., M. A. Arthur, and K. H. Freeman. 1999a. Miocene evolution of carbon dioxide. Paleoceanography 14: 273–292.

Pagani, M., K. H. Freeman, and M. A. Arthur. 1999b. Late Miocene atmospheric CO_2 concentrations and the expansion of C_4 grasses. Science 285: 876–879.

Paige, K. N., and T. G. Whitham. 1987. Overcompensation in response to mammalian herbivory: The advantage of being eaten. American Naturalist 129: 407–416.

Paine, R. T. 1966. Food web complexity and species diversity. American Naturalist 100: 65–75.

———. 1969. A note on trophic complexity and community stability. American Naturalist 103: 91–93.

———. 1974. Intertidal community structure: Experimental studies on the relationship between a dominant competitor and its principal predator. Oecologia (Berlin) 15: 93–120.

———. 1976. Biological observations on a subtidal *Mytilus californianus* bed. Veliger 19: 125–130.

———. 1980. Food webs: Linkage, interaction strength and community infrastructure. Journal of Animal Ecology 49: 667–685.

———. 2002. Trophic control of production in a rocky intertidal community. Science 296: 736–739.

Pálfy, J., and P. L. Smith. 2000. Synchrony between Early Jurassic extinction, oceanic anoxic event, and the Karoo-Ferrar flood basalt volcanism. Geology 28: 747–750.

Pálfy, J., A. Demény, J. Haas, M. Hetényi, M. J. Orchard, and I. Vető. 2001. Carbon isotope anomaly and other geochemical changes at the Triassic-Jurassic boundary from a marine section in Hungary. Geology 29: 1047–1050.

Palmer, A. R. 1979. Fish predation and the evolution of gastropod shell sculpture: Experimental and geographic evidence. Evolution 33: 697–713.

———. 1981. Do carbonate skeletons limit the rate of body growth? Nature 292: 150–152.

———. 1982. Predation and parallel evolution: Recurrent parietal plate reduction in balanomorph barnacles. Paleobiology 8: 31–44.

———. 1990. Effect of crab effluent and scent of damaged conspecifics on feeding, growth, and shell morphology of the Atlantic dogwhelk *Nucella lapillus* (L.). Hydrobiologia 293: 155–182.

———. 1992. Calcification in marine molluscs: How costly is it? Proceedings of the National Academy of Sciences of the USA 89: 1379–1382.

Palmer, T. J. 1982. Cambrian to Cretaceous changes in hardground communities. Lethaia 15: 309–323.

Pan, H.-Z. 1991. Lower Turonian gastropod ecology and biotic interaction in *Helicaulax* community from western Tarim Basin, southern Xinjiang, China. Paleoecology of China 1: 266–280.

Parrish, J. M. 1986. Locomotor adaptations in the hindlimb and the pelvis of the Thecodontia. Hunteria 1 (2): 1–35.

Parrish, J. T. 1987. Paleo-upwelling and the distribution of organic-rich rocks. Pages 199–205 in J. Brooks and A. J. Fleet (eds.), Marine petroleum source rocks. Geological Society Special Publication 26.

Pearson, P. N., and M. R. Palmer. 1999. Middle Eocene seawater pH and atmospheric carbon dioxide concentrations. Science 284: 1824–1826.

———. 2000. Atmospheric carbon dioxide concentrations over the past 60 million years. Nature 406: 695–699.

Peck, L. S., and L. Z. Conway. 2000. The myth of metabolic cold adaptation: oxygen consumption in stenothermal Antarctic bivalves. Pages 441–450 in E. M. Harper, J. D. Taylor, and J. A. Crame (eds.), The evolutionary biology of the Bivalvia. Geological Society Special Publication 177.

Peizhen, Z., P. Molnar, and W. R. Downs. 2001. Increased sedimentation rates and grain sizes 2–4 Myr ago due to the influence of climate change on erosion rates. Nature 410: 891–897.

Pelechaty, S. M. 1996. Stratigraphic evidence for the Siberia-Laurentia connection and Early Cambrian rifting. Geology 24: 719–722.

Pellmyr, O., J. N. Thompson, J. M. Brown, and R. G. Harrison. 1996. Evolution of pollination and mutualism in the yucca moth lineage. American Naturalist 148: 827–847.

Pemberton, S. G., M. J. Risk, and D. E. Buckley. 1976. Supershrimp: Deep bioturbation in the Strait of Canso, Nova Scotia. Science 192: 790–791.

Pemberton, S. G., R. W. Frey, and R. G. Walker. 1984. Probable lobster burrows in the Cardium Formation (Upper Cretaceous) of southern Alberta, Canada, and comments on modern burrowing decapods. Journal of Paleontology 58: 1422–1435.

Pennings, S. C., E. L. Siska, and M. D. Bertness. 2001. Latitudinal differences in plant palatability in Atlantic coast salt marshes. Ecology 82: 1344–1359.

Perlin, J. 1989. A forest journey: The role of wood in the development of civilization. Cambridge: Harvard University Press: 445 pp.

Persson, A., L.-A. Hansson, C. Brönmark, P. Lundberg, L. B. Petersson, L. Greenberg, P. A. Nilsson, P. Nyström, P. Romare, and L. Tranvik. 2001. Effects of enrichment on simple aquatic food webs. American Naturalist 157: 654–669.

Peterson, C. H., and M. Quammen. 1982. Siphon nipping: Its importance to small fishes and its impact on growth of the bivalve *Protothaca staminea* (Conrad). Journal of Experimental Marine Biology and Ecology 63: 249–268.

Petit, J. R., J. Jouzel, D. Raynaud, N. I. Barkov, J.-M. Barnola, I. Basila, M. Bender, J. Chappellaz, M. Davis, G. Delaygue, M. Delmotte, M. Kotlyakov, M. Legrand, V. Y. Lipenkov, C. Lorius, L. Pépin, C. Ritz, E. Saltzman, and M. Stievenard. 1999. Climate and atmospheric history of the past 420,000 years from the Vostok Ice Core, Antarctica. Nature 399: 429–436.

Petuch, E. J. 1993. Patterns of diversity and extinction in Transmarian muricacean, buccinacean, and conacean gastropods. Nautilus 106: 155–173.

Pfeiffer, T., S. Schuster, and S. Bonhoeffer. 2001. Cooperation and competition in the evolution of ATP-producing pathways. Science 292: 504–506.

Pierrehumbert, R. T. 2002. The hydrologic cycle in deep-time climate problems. Nature 419: 191–198.

Piersma, T., A. Koolhaas, and A. Dekinga. 1993. Interactions between stomach structure and diet choice in shorebirds. Auk 110: 552–564.

Pigliucci, M. 2001. Characters and environments. Pages 363–388 in G. P. Wagner (ed.), The character concept in evolutionary biology. San Diego: Academic Press.

Poag, C. W., E. Mankinen, and R. D. Norris. 2003. Eocene impacts: Geologic record, correlation, and paleoenvironmental consequences. Pages 495–510 in D. R. Prothero, L. C. Ivany, and E. A. Nesbitt (eds.), From greenhouse to icehouse: The marine Eocene-Oligocene transition. New York: Columbia University Press.

Podolsky, R. D. 1994. Temperature and water viscosity: Physiological versus mechanical effects on suspension feeding. Science 265: 100–103.

Polis, G. A., and S. D. Hurd. 1996. Linking marine and terrestrial food webs: Allochthonous input from the ocean supports secondary productivity on small islands and coastal land communities. American Naturalist 147: 396–423.

Polis, G. A., W. B. Anderson, and R. D. Holt. 1997. Toward an integration of landscape and food web ecologies: The dynamics of spatially subsidized food webs. Annual Reviews of Ecology and Systematics 28: 289–316.

Polyak, V. J., and Y. Asmerom. 2001. Late Holocene climate and cultural changes in the southwestern United States. Science 294: 148–151.

Pomeranz, K. 2000. The great divergence: China, Europe, and the making of the modern world economy. Princeton: Princeton University Press: 382 pp.

Pope, K. O. 2002. Impact dust not the cause of the Cretaceous-Tertiary mass extinction. Geology 30: 99–102.

Popper, K. 1964. The poverty of historicism (fourth edition). New York: Harper and Row: 166 pp.

Porter, J. W., J. D. Woodley, G. J. Smith, J. E. Neigel, J. F. Battey, and D. G. Dallmeyer. 1981. Population trends among Jamaican reef corals. Nature 294: 249–250.

Porter, K. G. 1973. Selective grazing and differential digestion of algae by zooplankton. Nature 244: 179–180.

———. 1976. Enhancement of algal growth and productivity by grazing zooplankton. Science 192: 1332–1334.

Porter, S. M., and A. H. Knoll. 2000. Testate amoebae in the Neoproterozoic era: Evidence from vase-shaped microfossils in the Chuar Group, Grand Canyon. Paleobiology 26: 360–385.

Pough, F. H. 1980. The advantages of ectothermy for tetrapods. American Naturalist 115: 92–112.

Power, M. E. 1984. Depth distributions of armored catfish: Predator-induced resource avoidance? Ecology 65: 523–528.

Preston, S. J., I. C. Revie, J. F. Orr, and D. Roberts. 1996. A comparison of the strengths of gastropod shells with forces generated by potential crab predators. Journal of Zoology, London 238: 181–193.

Price, P. W. 1980. Evolutionary biology of parasites. Princeton: Princeton University Press: 237 pp.

Prothero, D. R. 1985. North American mammalian diversity and Eocene-Oligocene extinctions. Paleobiology 11: 389–405.

Putnam, H. 1992. Renewing philosophy. Cambridge: Harvard University Press: 234 pp.

Queiroz, A. de. 1999. Do image-forming eyes promote evolutionary diversification? Evolution 53: 1654–1664.

Racki, G. 1998. Frasnian-Famennian biotic crisis: Undervalued tectonic control. Palaeogeography, Palaeoclimatology, Palaeoecology 141: 177–198.

———. 1999. Silica-secreting biota and mass extinctions: Survival patterns and processes. Palaeogeography, Palaeoclimatology, Palaeoecology 154: 107–132.

Racki, G., and F. Cordey. 2000. Radiolarian paleoecology and radiolarites: Is the present the key to the past? Earth-Science Reviews 83–120.

Radwin, G. E., and H. W. Wells. 1968. Comparative radular morphology and feeding habits of muricid gastropods from the Gulf of Mexico. Bulletin of Marine Science 18: 72–85.

Raff, R. A. 1996. The shape of life: Genes, development, and the evolution of animal form. Chicago: University of Chicago Press: 520 pp.

Rampino, M. R., and K. Caldeira. 1993. Major episodes of geologic change: Correlations, time structure and possible causes. Earth and Planetary Science Letters 110: 215–227.

Rampino, M. R., and S. Self. 1992. Volcanic winter and accelerated glaciation following the Toba super-eruption. Nature 359: 50–52.

Rampino, M. R., and R. B. Stothers. 1988. Flood basalt volcanism during the past 250 million years. Science 241: 663–668.

Rampino, M. R., A. Prokoph, and A. Adler. 2000. Tempo of the end-Permian event: High-resolution cyclostratigraphy at the Permian-Triassic boundary. Geology 28: 643–646.

Randall, J. E. 1964. Contribution to the biology of the queen conch, *Strombus gigas*. Bulletin of Marine Science of the Gulf and Caribbean 14: 246–295.

Rapport, D. J., and J. E. Turner. 1977. Economic models in ecology. Science 195: 367–373.

Raup, D. M. 1992. Large-body impact and extinction in the Phanerozoic. Paleobiology 18: 80–88.

Raup, D. M., and R. E. Crick. 1981. Evolution of single characters in the Jurassic ammonite *Kosmoceras*. Paleobiology 7: 200–215.

Raup, D. M., and S. J. Gould. 1974. Stochastic simulation and evolution of morphology—towards a nomothetic paleontology. Systematic Zoology 23: 305–322.

Raup, D. M., and A. Seilacher. 1969. Fossil foraging behavior: Computer simulation. Science 166: 994–995.

Raup, D. M., and J. J. Sepkoski, Jr. 1982. Mass extinctions in the marine fossil record. Science 215: 1501–1503.

Raup, D. M., S. J. Gould, T. J. M. Schopf, and D. S. Simberloff. 1973. Stochastic models of phylogeny and the evolution of diversity. Journal of Geology 81: 525–542.

Raven, J. A. 1998. Extrapolating feedback processes from the present to the past. Philosophical Transactions of the Royal Society of London B 353: 19–28.

———. 2002. Evolutionary options. Nature 415: 375–377.

Ravizza, J., and Peucker-Ehrenbrink, B. 2003. Chemostratigraphic evidence of Deccan volcanism from the marine osmium isotope record. Science 302: 1392–1395.

Rayfield, E. J., D. B. Norman, C. C. Horner, J. R. Horner, P. M. Smith, J. J. Thomason, and P. Upchurch. 2001. Cranial design and function in a large theropod dinosaur. Nature 409: 1033–1037.

Raymo, M. E., and W. F. Ruddiman. 1992. Tectonic forcing of Late Cenozoic climate. Nature 359: 117–122.

Raymo, M. E., B. Grant, M. Horowitz, and G. H. Rau. 1996. Mid-Pliocene warmth: Stronger greenhouse and stronger conveyor. Marine Micropaleontology 27: 313–326.

Rea, D. K. 1994. The paleoclimatic record provided by eolian deposition in the deep sea: The geologic history of wind. Reviews of Geophysics 32: 159–195.

Rea, D. K., J. C. Zachos, R. M. Owen, and P. D. Gingerich. 1990. Global change at the Paleocene-Eocene boundary: Climatic and evolutionary consequences of tectonic events. Palaeogeography, Palaeoclimatology, Palaeoecology 79: 117–128.

Redecker, D., R. Kodner, and L. E. Graham. 2000. Glomalian fungi from the Ordovician. Science 289: 1920–1921.

Rees, P. M. 2002. Land-plant diversity and the end-Permian mass extinction. Geology 30: 827–830.

Reeve, H. K., and P. W. Sherman. 1993. Adaptation and the goals of evolutionary research. Quarterly Review of Biology 68: 2–32.

———. 2001. Optimality and phylogeny. A critique of current thought. Pages 64–113 in S. H. Orzack and E. Sober (eds.), Adaptationism and optimality. Cambridge: Cambridge University Press.

Reich, P. B., M. B. Walters, and D. S. Ellsworth. 1992. Leaf life-span in relation to leaf, plant, and stand characteristics among diverse ecosystems. Ecological Monographs 62: 365–392.

Reichow, M. K., A. D. Saunders, R. V. White, M. S. Pringle, A. F. Al'mukhamedov, A. I. Medvedev, and N. P. Kirda. 2002. $^{40}Ar/^{39}Ar$ dates from the West Siberian Basin: Siberian flood basalt province doubled. Science 296: 1846–1849.

Reid, R.G.B., R. F. McMahon, Ó Foighil, D., and R. Finnigan. 1992. Anterior inhalant currents and pedal feeding in bivalves. Veliger 35: 93–104.

Reinfelder, J. R., A.M.L. Kraepiel, and F.M.M. Morel. 2000. Unicellular C_4 photosynthesis in a marine diatom. Nature 407: 996–999.

Reisz, R. R., and H.-D. Sues. 2000. Herbivory in Late Paleozoic and Triassic terrestrial vertebrates. Pages 9–41 in H.-D. Sues (ed.), Evolution of herbivory in terrestrial vertebrates: Perspectives from the fossil record. Cambridge: Cambridge University Press.

Renne, P. R., M. Ernesto, I. G. Pacca, R. S. Coe, J. M. Glen, M. Prévot, and M. Perrin. 1992. The age of Paraná flood volcanism, rifting of Gondwanaland, and the Jurassic-Cretaceous boundary. Science 259: 975–979.

Renoso, V. H. 2000. An unusual aquatic sphenodontian (Reptilia: Diapsida) from the Tlayua Formation (Albian), central Mexico. Journal of Paleontology 74: 133–148.

Retallack, G. J. 1995. Permian-Triassic life crisis on land. Science 267: 77–80.

———. 1997. Early forest soils and their role in Devonian global change. Science 276: 583–585.

Retallack, G. J., J. J. Veevers, and R. Morante. 1996. Global coal gap between Permian-Triassic extinction and Middle Triassic recovery of peat-forming plants. Geological Society of America Bulletin 108: 195–207.

Rhoades, D. F. 1985. Offensive-defensive interactions between herbivores and plants: Their relevance in herbivore population dynamics and ecological theory. American Naturalist 125: 205–238.

Rhoads, D. C. 1970. Mass properties, stability, and ecology of marine muds related to burrowing activity. Pages 391–406 in T. P. Crimes and J. C. Harper (eds.), Trace fossils. Geological Journal, Special Issue 3.

Rhoads, D. C., and D. K. Young. 1970. The influence of deposit-feeding organisms on sediment stability and community trophic structure. Journal of Marine Research 28: 150–178.

Rhodes, M. C., and C. W. Thayer. 1991. Effects of turbidity on suspension feeding: Are brachiopods better than bivalves? Pages 191–196 in D. I. MacKinnon, D. E. Lee, and J. D. Campbell (eds.), Brachiopods through time. Rotterdam: Balkema.

Rhodes, M. C., and R. J. Thompson. 1993. Comparative physiology of suspension-feeding in living brachiopods and bivalves: Evolutionary implications. Paleobiology 19: 322–334.

Rice, W. R. 1996. Sexually antagonistic male adaptation triggered by experimental arrest of female evolution. Nature 381: 232–234.

Richards, R. P. 1972. Autecology of Richmondian brachiopods (Late Ordovician of Indiana and Ohio). Journal of Paleontology 46: 386–405.

———. 1974. Ecology of the Cornulitidae. Journal of Paleontology 48: 514–523.

Richerson, P. J., R. Boyd, and R. L. Bettinger. 2001. Was agriculture impossible during the Pleistocene but mandatory during the Holocene? A climate change hypothesis. American Antiquity 66: 387–411.

Ricklefs, R. E., and R. E. Latham. 1993. Global patterns of diversity in mangrove floras. Pages 215–229 in R. E. Ricklefs and D. Schluter (eds.), Species diversity in ecological communities: Historical and geographical perspectives. Chicago: University of Chicago Press.

Riding, R. 2001. Calcified algae and bacteria. Pages 445–473 in A. Yu. Zhuravlev and R. Riding (eds.), The ecology of the Cambrian radiation. New York: Columbia University Press.

Riebessel, U. 2000. Carbon fix for a diatom. Nature 407: 959–960.

Riebessel, U., D. A. Wolf-Gladrow, and V. Smetacek. 1993. Carbon dioxide limitation of marine phytoplankton growth rates. Nature 361: 249–251.

Riebessel, U., I. Zondervan, B. Rost, P. D. Tortell, R. E. Zeebe, and F.M.M. Morel. 2000. Reduced calcification of marine plankton in response to increased atmospheric CO_2. Nature 407: 364–367.

Riedl, R. 1978. Order in living organisms. (Translated by R. P. S. Jeffries.) New York: Wiley: 313 pp.

Riggs, S. R. 1984. Paleoceanographic model of Neogene phosphorite deposition, U.S. Atlantic continental margin. Science 223: 123–131.

Rittschof, D., C. M. Kratt, and A. S. Clare. 1990. Gastropod predation sites: The role of predator and prey in chemical attraction of the hermit crab *Clibanarius vittatus*. Journal of the Marine Biological Association of the United Kingdom 70: 583–596.

Robbins, R. K. 1981. The "false head" hypothesis: Predation and wing pattern variation of lycaenid butterflies. American Naturalist 118: 779–785.

Roberts, R. G., T. F. Flannery, L. K. Ayliffe, H. Yoshida, J. M. Olley, G. J. Prideaux, G. M. Laslett, A. Baynes, M. A. Smith, R. Jones, and B. L. Smith. 2001. New ages for the last Australian megafauna: Continent-wide extinction about 46,000 years ago. Science 292: 1888–1892.

Robinson, J. M. 1990. The burial of organic carbon as affected by the evolution of land plants. Historical Biology 3: 189–201.

Robinson, M. H. 1969. Defenses against visually hunting predators. Evolutionary Biology 3: 225–250.

Robinson, M. H., L. G. Abele, and B. Robinson. 1970. Attack autotomy: A defense against predators. Science 169: 300–301.

Roces, F., and J.R.B. Lighton. 1995. Larger bites of leaf-cutting ants. Nature 373: 392–393.

Rohde, K. 1991. Intra- and interspecific interactions in low density populations in resource-rich habitats. Oikos 60: 91–104.

———. 1999. Latitudinal gradients in species diversity and Rapoport's Rule revisited: A review of recent work and what can parasites teach us about the causes of the gradients? Ecography 22: 593–613.

Rojstaczer, S., S. M. Sterling, and N. J. Moore. 2001. Human appropriation of photosynthesis production. Science 294: 2549–2552.

Romer, P. M. 1986. Increasing returns and long-run growth. Journal of Political Economy 94: 1002–1037.

Roopnarine, P. D. 1996. Systematics, biogeography and extinction of chionine bivalves (Bivalvia: Veneridae) in tropical America: Early Oligocene-Recent. Malacologia 38: 103–142.

Rosen, B. R. 2000. Algal symbiosis, and the collapse and recovery of reef communities: Lazarus corals across the K-T boundary. Pages 164–180 in S. J. Culver and P. F. Rawson (eds.), Biotic response to global change: The last 145 million years. Cambridge: Cambridge University Press.

Rosenzweig, M. L. 1995. Species diversity in space and time. Cambridge: Cambridge University Press: 436 pp.

Rosenzweig, M. L., and S. Abramsky. 1993. How are diversity and productivity related? Pages 52–65 in R. E. Ricklefs and D. Schluter (eds.), Species diversity in ecological communities: Historical and geographical perspectives. Chicago: University of Chicago Press.

Ross, D. M. 1971. Protection of hermit crabs (*Dardanus* spp.) from *Octopus* by commensal sea anemones (*Calliactis* spp.). Nature 230: 401–402.

———. 1974. Evolutionary aspects of associations between crabs and sea anemones. Pages 111–125 in W. B. Vernberg (ed.), Symbiosis in the sea. Columbia: University of South Carolina Press.

Ross, H. H. 1964. Evolution of caddisworm cases and nets. American Zoologist 4: 209–220.

Roth-Nebelsick, A., D. Uhl, V. Mosbrugger, and H. Kerp. 2001. Evolution and function of leaf venation architecture: A review. Annals of Botany 87: 553–566.

Rothschild, L. J., and R. L. Mancinelli. 2001. Life in extreme environments. Nature 409: 1092–1011.

Rowan, R., and D. A. Powers. 1991. A molecular genetic classification of zooxanthellae and the evolution of animal-algal symbioses. Science 251: 1348–1351.

Rowan, R., N. Knowlton, A. Baker, and J. Jara. 1997. Landscape ecology of algal symbionts increases variation in episodes of coral bleaching. Nature 388: 265–269.

Rowe, L., and G. Arnqvist. 2002. Sexually antagonistic coevolution in a mating system: Combining experimental and comparative approaches to address evolutionary processes. Evolution 56: 754–767.

Ruben, J. A. 1995. The evolution of endothermy in mammals and birds: From physiology to fossils. Annual Reviews of Physiology 57: 69–95.

Rubenstein, D. I., and M.A.R. Koehl. 1977. The mechanisms of filter feeding: Some theoretical considerations. American Naturalist 111: 981–994.

Ruiz-Trillo, I., J. Paps, M. Loukota, C. Ribera, K. Jondelius, J. Baguñà, and M. Riutort. 2002. A phylogenetic analysis of myosin heavy chain type II sequences corroborates that Acoela and Nemertodermatida are basal bilaterians. Proceedings of the National Academy of Sciences of the USA 99: 11246–11251.

Runnegar, B. 1982. The Cambrian explosion: Animals or fossils? Journal of the Geological Society of Australia 29: 395–411.

Russ, G. R. 1987. Is rate of removal of algae by grazers reduced inside territories of tropical damselfishes? Journal of Experimental Marine Biology and Ecology 110: 1–17.

Rybczynski, N., and R. R. Reisz. 2001. Earliest evidence for efficient oral processing in a terrestrial herbivore. Nature 411: 684–687.

Rypstra, A. L. 1984. A relative measure of predation on web-spiders in temperate and tropical forests. Oikos 43: 129–132.

Ryskin, G. 2003. Methane-driven oceanic eruptions and mass extinctions. Geology 31: 741–744.

Sachs, J. P., and D. J. Repeta. 1999. Oligotrophy and nitrogen fixation during eastern Mediterranean sapropel events. Science 286: 2485–2488.

Sadler, P. M. 1981. Sediment accumulation rates and the completeness of stratigraphic sections. Journal of Geology 89: 569–584.

Sagan, C. 1997. Billions and billions: Thoughts on life and death at the brink of the millennium. New York: Random House: 241 pp.

Sager, W. W., and H. C. Han. 1993. Rapid formation of the Shatsky Rise Oceanic Plateau inferred from its magnetic anomaly. Nature 364: 610–613.

Saito, T., T. Yamanoi, and K. Kaiho. 1986. End-Cretaceous devastation of terrestrial flora in the boreal Far East. Nature 323: 253–255.

Saltzman, M. R. 2003. Late Paleozoic ice age: Oceanic gateway or pCO_2? Geology 31: 151–154.

Sandoval, J., L. O'Dogherty, and J. Guex. 2001. Evolutionary rates of Jurassic ammonites in relation to sea-level fluctuations. Palaios 16: 311–335.

Sansom, I. J., M. M. Smith, and M. P. Smith. 1996. Scales of thelodont and shark-like fishes from the Ordovician of Colorado. Nature 379: 628–630.

Saunders, P. T. 1999. Darwinism and economic theory. Pages 259–278 in P. Koslowski (ed.), Sociobiology and bioeconomics: The theory of evolution in biological and economic theory. Berlin: Springer.

Saunders, W. B., D. M. Work, and S. V. Nikolaeva. 1999. Evolution of complexity in Paleozoic ammonoid sutures. Science 286: 760–763.

Savazzi, E., and R. A. Reyment. 1989. Subaerial hunting behaviour in *Natica gualtieriana* (naticid gastropod). Palaeogeography, Palaeoclimatology, Palaeoecology 74: 355–364.

Sax, D. F., S. D. Gaines, and J. H. Brown. 2002. Species invasions exceed extinctions on islands worldwide: A comparative study of plants and birds. American Naturalist 160: 766–783.

Schaeffer, B., and D. E. Rosen. 1961. Major adaptive levels in the evolution of the actinopterygian feeding mechanism. American Zoologist 1: 187–204.

Schäfer, W. 1954. Form und Funktion der brachyuren Scheere. Abhandlungen der Senckenbergischen Naturforschenden Gesellschaft 489: 1–66.

Schaller, G. B. 1972. The Serengeti lion: A study of predator-prey relations. Chicago: University of Chicago Press: 480 pp.

Scheffer, M., S. Carpenter, J. A. Foley, C. Folke, and B. Walker. 2001. Catastrophic shifts in ecosystems. Nature 413: 591–596.

Schindel, D. E. 1982a. The gaps in the fossil record. Nature 297: 282–284.

———. 1982b. Resolution analysis: A new approach to the gaps in the fossil record. Paleobiology 8: 340–353.

Schmitt, R. J., C. W. Osenberg, and M. G. Bercovitch. 1983. Mechanisms and consequences of shell fouling in the kelp snail, *Norrisia norrisi* (Sowerby) (Trochidae): Indirect effects of *Octopus* drilling. Journal of Experimental Marine Biology and Ecology 99: 267–281.

Schneider, D. 1982. Escape response of an infaunal clam *Ensis directus* Conrad 1843, to a predator snail, *Polinices duplicatus* Say 1822. Veliger 24: 371–372.

Schoener, T. W. 1969. Optimal size and specialization in constant and fluctuating environments: An energy-time approach. Brookhaven Symposium in Biology 22: 103–114.

———. 1974a. Resource partitioning in ecological communities. Science 185: 27–37.

Schoener, T. W. 1974b. The compression hypothesis of temporal resource partitioning. Proceedings of the National Academy of Sciences of the USA 71: 4169–4172.

———. 1983. Field experiments on interspecific competition. American Naturalist 122: 240–285.

Schoener, T. W., and A. Schoener. 1983. The time to extinction of a colonizing propagule of lizards increases with island area. Nature 302: 332–334.

Schoener, T. W., and D. A. Spiller. 1995. Effect of predators and area on invasion: An experiment with island spiders. Science 267: 1811–1813.

Schoener, T. W., and C. A. Toft. 1983. Spider populations: Extraordinarily high densities on islands without top predators. Science 219: 1353–1355.

Schoener, T. W., D. A. Spiller, and J. B. Losos. 2001. Predators increase the risk of catastrophic extinction of prey populations. Nature 412: 183–186.

Schopf, T. J. M. 1980. Paleoceanography. Cambridge: Harvard University Press: 341 pp.

Schultz, P. H., M. Zarate, W. Hames, C. Camilión, and J. King. 1998. A 3.3-Ma impact in Argentina and possible consequences. Science 282: 2061–2063.

Schwartz, M. W., and J. D. Hoeksema. 1998. Specialization and resource trade: biological markets as a model of mutualism. Ecology 79: 1029–1038.

Schwartzman, D. 1999. Life, temperature, and the Earth: The self-organizing biosphere. New York: Columbia University Press: 241 pp.

Seebacher, F. 2003. Dinosaur body temperatures: The occurrence of endothermy and ectothermy. Paleobiology 29: 105–122.

Seed, R., and R. N. Hughes. 1995. Criteria and prey size-selection in molluscivorous crabs with contrasting claw morphologies. Journal of Experimental Marine Biology and Ecology 193: 177–195.

Seger, J., and J. W. Stubblefield. 1996. Optimization and adaptation. Pages 93–123 in M. R. Rose and G. V. Lauder (eds.), Adaptation. San Diego: Academic Press.

Seilacher, A. 1989. Vendozoa: Organismic construction in the Proterozoic biosphere. Lethaia 22: 229–239.

Sellwood, B. W., G. D. Price, and P. J. Valdes. 1994. Cooler estimates of Cretaceous temperatures. Nature 370: 453–455.

Sepkoski, J. J., Jr. 1979. A kinetic model of Phanerozoic taxonomic diversity. II. Early Phanerozoic families and multiple equilibria. Paleobiology 5: 222–251.

——— 1981. A factor analytic description of the Phanerozoic marine fossil record. Paleobiology 7: 36–53.

——— 1984. A kinetic model of Phanerozoic taxonomic diversity. III. Post-Paleozoic families and mass extinctions. Paleobiology 10: 246–267.

——— 1990. The taxonomic structure of periodic extinction. Geological Society of America Special Paper 247: 33–44.

——— 1993. Ten years in the library: New data confirm paleontological patterns. Paleobiology 19: 43–51.

Sepkoski, J. J., Jr., F. K. McKinney, and S. Lidgard. 2000. Competitive displacement among post-Paleozoic cyclostome and cheilostome bryozoans. Paleobiology 26: 7–18.

Sereno, P. C. 1999. The evolution of dinosaurs. Science 284: 2137–2146.

Sereno, P. C., H. C. E. Larsson, C. A. Sidor, and B. Gado. 2001. The giant crocodyliform *Sarcosuchus* from the Cretaceous of Africa. Science 294: 1516–1519.

Seymour, R. S., and P. Schultze-Motel. 1996. Thermoregulating lotus flowers. Nature 383: 305.

Seymour, R. S., C. R. White, and M. Gibernau. 2003. Heat reward for insect pollinators. Nature 426: 243–244.

Shaffer, H. B., P. Meylan, and M. L. McKnight. 1997. Tests of turtle phylogeny: Molecular, morphological, and paleontological approaches. Systematic Biology 46: 235–268.

Shear, W. A. 1991. The early development of terrestrial ecosystems. Nature 351: 283–289.

Sheehan, P. M. 2001. The Late Ordovician mass extinction. Annual Reviews of Earth and Planetary Sciences 29: 331–364.

Sheehan, P. M., and T. A. Hansen. 1986. Detritus feeding as a buffer to extinction at the end of the Cretaceous. Geology 14: 868–870.

Sheehan, P. M., P. J. Coorough, and D. E. Fastovsky. 1996. Biotic selectivity during the K/T and Late Ordovician extinction events. Geological Society of America Special Paper 307: 477–489.

Sheets, H. D., and C. E. Mitchell. 2001. Uncorrelated change produces the apparent dependence of evolutionary rate on interval. Paleobiology 27: 429–444.

Shen, Y. A., R. Buick, and D. E. Canfield. 2001. Isotopic evidence for microbial sulfate reduction in the early Archean sea. Nature 410:77–81.

Sheridan, R. E. 1997. Pulsation tectonics as a control on the dispersal and assembly of supercontinents. Journal of Geodynamics 23: 173–196.

Shields, W., and K. L. Bildstein. 1979. Birds versus bats: Behavioral interactions at a localized food source. Ecology 60: 468–474.

Shoji, S., M. Nanzyo, and R. Dahlgren. 1993. Volcanic ash soils: Genesis, properties and utilization. Amsterdam: Elsevier: 288 pp.

Shubin, N. H., and H.-D. Sues. 1991. Biogeography of Early Mesozoic continental tetrapods: Patterns and implications. Paleobiology 17: 214–230.

Shubin, N. H., C. Sabin, and S. Carroll. 1997. Fossils, genes and the evolution of animal limbs. Nature 388: 639–648.

Sidor, C. A. 2001. Simplification as a trend in tetrapod cranial evolution. Evolution 55: 1419–1442.

Sigman, D. M., and E. A. Boyle. 2000. Glacial/interglacial variations in atmospheric carbon dioxide. Nature 407: 859–869.

Signor, P. W., III, and C. E. Brett. 1984. The mid-Paleozoic precursor to the Mesozoic marine revolution. Paleobiology 10: 229–245.

Signor, P. W., III, and G. J. Vermeij. 1994. The plankton and the benthos: Origins and early history of an evolving relationship. Paleobiology 20: 297–319.

Sigurdsson, H. 1990. Assessment of the atmospheric impact of volcanic eruptions. Geological Society of America Special Paper 247: 99–110.

Silliman, B. R., and M. D. Bertness. 2002. A trophic cascade regulates salt marsh primary production. Proceedings of the National Academy of Sciences of the USA 99: 10500–10505.

Simberloff, D. S., and E. O. Wilson. 1969. Experimental zoogeography of islands: The colonization of empty islands. Ecology 50: 278–295.

Simenstad, C. A., J. A. Estes, and K. W. Kenyon. 1978. Aleuts, sea otters, and alternate stable state communities. Science 200: 403–411.

Simkiss, K. 1989. Biomineralisation in the context of geological time. Transactions of the Royal Society of Edinburgh (Earth Sciences) 80: 193–199.

Simon, J. L. 1977. The economics of population growth. Princeton: Princeton University Press: 555 pp.

Simon, L., J. Bousquet, R. C. Lévesque, and M. Lalonde. 1993. Origin and diversification of endomycorrhizal fungi and coincidence with vascular land plants. Nature 363: 67–69.

Singer, M. C., C. D. Thomas, and C. Parmesan. 1993. Rapid human-induced evolution of insect-host associations. Nature 366: 681–683.

Slawson, W. D. 1981. The new inflation: The collapse of free markets. Princeton: Princeton University Press: 424 pp.

Smetacek, V. 1998. Diatoms and the silicate factor. Nature 391: 224–225.

———. 2001. A watery arms race. Nature 411: 745.

———. 2002. The ocean's veil. Nature 419: 565.

Smith, A. 1776. An inquiry into the nature and causes of the wealth of nations (1971 edition, with introduction by E. Cannon and notes by M. Lerner). New York: Random House: 976 pp.

Smith, A. B., and C. H. Jeffery. 1998. Selectivity of extinction among sea urchins at the end of the Cretaceous period. Nature 392: 69–71.

Smith, A. R. 1972. Comparison of fern and flowering plant distributions with some evolutionary interpretations for ferns. Biotropica 4: 4–9.

Smith, B. N. 1976. Evolution of C_4 photosynthesis in response to changes in carbon and oxygen concentrations in the atmosphere through time. BioSystems 8: 24–32.

Smith, L. D. 1995. Effects of limb autotomy and tethering on juvenile blue crab survival from cannibalism. Marine Ecology Progress Series 116: 65–74.

Smith, L. D., and J. A. Jennings. 2000. Induced defensive responses by the bivalve *Mytilus edulis* to predators with different attack modes. Marine Biology 136: 461–469.

Smith, L. D., and A. R. Palmer. 1994. Effects of manipulated diet on size and performance of brachyuran crab claws. Science 264: 710–712.

Smith, R.M.H., and P. D. Ward. 2001. Pattern of vertebrate extinctions across an event bed at the Permian-Triassic boundary in the Karoo Basin of South Africa. Geology 29: 1147–1150.

Smith, S. V. 1981. Marine macrophytes as a global carbon sink. Science 211: 838–840.

Smits, J. D., F. Witte, and F. G. van Veen. 1996. Functional changes in the anatomy of the pharyngeal jaw apparatus of *Astatoreochromis alluaudi* (Pisces: Cichlidae), and their effects on adjacent structures. Biological Journal of the Linnean Society 59: 389–409.

Sober, E. 1984. The nature of selection: Evolutionary theory in philosophical focus. Cambridge: MIT Press: 838 pp.

Sober, E., and D. S. Wilson. 1998. Unto others: The evolution and psychology of unselfish behavior. Cambridge: Harvard University Press: 394 pp.

Soden, B. J., R. T. Wetherald, G. L. Stenchikov, and A. Robock. 2002. Global cooling after the eruption of Mount Pinatubo: a test of climate feedback by water vapor. Science 296: 727–730.

Sousa, W. P., S. C. Schroeter, and S. D. Gaines. 1981. Latitudinal variation in intertidal algal community structure: the influence of grazing and vegetative propagation. Oecologia (Berlin) 48: 297–307.

Sparks, D. K., R. D. Hoare, and R. D. Kesling. 1980. Epizoans on the brachiopod *Paraspirifer bownockeri* (Stewart) from the Middle Devonian of Ohio. Papers on Paleontology, University of Michigan Museum of Paleontology 23: 1–105.

Spray, J. G., S. P. Kelley, and D. B. Rowley. 1998. Evidence for a Late Triassic multiple impact event on Earth. Nature 392: 171–173.

Srygley, R. B., and P. Chai. 1990. Predation and the elevation of thoracic temperatures in brightly colored Neotropical butterflies. American Naturalist 135: 766–787.

Stachowicz, J. J., and M. E. Hay. 2000. Geographic variation in camouflage specialization by a decorator crab. American Naturalist 156: 59–71.

Stallard, R. F. 1988. Weathering and erosion in the humid tropics. Pages 225–246 in A. Lerman and M. Meybeck (eds.), Physical and chemical weathering in geochemical cycles. Dordrecht: Kluwer.

———. 1992. Tectonic processes, continental freeboard, and the rate-controlling step for continental denudation. Pages 93–121 in S. S. Butcher, R. J. Charlson, G. H. Orians, and G. V. Wolfe (eds.), Global biogeochemical cycles. London: Academic Press.

Stanley, S. M. 1970. Relation of shell form to life habits of the Bivalvia (Mollusca). Geological Society of America Memoir 125: 1–296.

———. 1973. An explanation for Cope's Rule. Evolution 28: 1–26.

———. 1979. Macroevolution: Pattern and process. San Francisco: Freeman: 332 pp.

———. 1986. Anatomy of a regional mass extinction: Plio-Pleistocene decimation of the western Atlantic bivalve fauna. Palaios 1: 17–36.

Stasek, C. R. 1967. Autotomy in the Mollusca. Occasional Papers of the California Academy of Sciences 61: 1–44.

Steadman, D. W. 1995. Prehistoric extinctions of Pacific island birds: Biodiversity meets zooarcheology. Science 267: 1123–1131.

Steadman, D. W., and P. S. Martin. 1984. Extinction of birds in the Late Pleistocene of North America. Pages 466–477 in P. S. Martin and R. G. Klein (eds.), Quaternary extinctions: A prehistoric revolution. Tucson: University of Arizona Press.

Steele-Petrovič, M. 1979. The physiological differences between articulate brachiopods and filter-feeding bivalves as a factor in the evolution of marine level-bottom communities. Palaeontology 22: 101–134.

Stehli, F. G., and S. D. Webb (eds.) 1985. The Great American biotic interchange. New York: Plenum.

Steneck, R. S. 1983. Escalating herbivory and resulting adaptive trends in calcareous algal crusts. Paleobiology 9: 44–61.

———. 1985. Adaptations of crustose coralline algae to herbivory: Patterns in space and time. Pages 352–366 in D. F. Toomey and M. H. Nitecki (eds.), Paleoalgology: Contemporary research and applications. Berlin: Springer.

Steneck, R. S., and M. N. Dethier. 1994. A functional group approach to the structure of algal-dominated communities. Oikos 69: 476–498.

Steneck, R. S., and L. Watling. 1982. Feeding capabilities and limitations of herbivorous molluscs: A functional group approach. Marine Biology 68: 299–315.

Sterrer, W. 1992. Prometheus and Proteus: The creative, unpredictable individual in evolution. Evolution and Cognition 1: 101–129.

———. 1993. Human economics: A non-human perspective. Ecological Economics 7: 183–202.

———. 2002. On the origin of sex as vaccination. Journal of Theoretical Biology 216: 387–396.

Steward, V. B., K. G. Smith, and F. M. Stephen. 1988. Red-winged blackbird predation on periodical cicadas (Cicadidae: Magicicada spp.): Bird behavior and cicada responses. Oecologia (Berlin) 76: 348–352.

Stiglitz, J. E. 2002. Globalization and its discontents. New York: Norton.

Stimson, J. 1973. The role of the territory in the ecology of the intertidal limpet *Lottia gigantea* (Gray). Ecology 54: 1020–1030.

Storch, G., B. Engesser, and M. Wuttke. 1996. Oldest fossil record of gliding in rodents. Nature 379: 439–441.

Storey, M., J. J. Mahoney, A. D. Saunders, R. A. Duncan, S. P. Kelley, and M. F. Coffin. 1995. Timing of hot spot-related volcanism and the breakup of Madagascar and India. Science 267: 852–855.

Strathmann, R. R. 1978a. The evolution and loss of feeding larval stages of marine invertebrates. Evolution 32: 894–906.

———. 1978b. Progressive vacating of adaptive types during the Phanerozoic. Evolution 32: 907–914.

———. 1987. Larval feeding. Pages 465–550 in A. C. Giese, J. S. Pearse, and B. Pearse (eds.), Reproduction of marine invertebrates, Vol. IX, General aspects, seeking unity in diversity. Palo Alto: Blackwell/Boxwood.

———. 1993. Hypotheses on the origins of marine larvae. Annual Reviews of Ecology and Systematics 24: 89–117.

Streelman, J. T., M. Alfaro, M. W. Westneat, D. R. Bellwood, and S. A. Karl. 2002. Evolutionary history of the parrotfishes: Biogeography, ecomorphology, and comparative diversity. Evolution 56: 961–971.

Stridsberg, S. 1984. Aptychopsid plates—jaw elements or protective operculum? Lethaia 17: 93–98.

Strong, D. R., Jr., E. D. McCoy, and J. R. Rey. 1977. Time and the number of herbivore species: The pests of sugar cane. Ecology 58: 167–175.

Strong, D. R., L. A. Liszka, and D. S. Simberloff. 1979. Tests of community-wide character displacement against null hypothesis. Evolution 33: 897–913.

Suchanek, T. 1983. Control of seagrass communities and sediment distribution by *Callianassa* (Crustacea, Thalassinidea) bioturbation. Journal of Marine Research 41: 281–298.

Summers, A. P. 2000. Stiffening the stingray skeleton—an investigation of durophagy in myliobatid stingrays (Chondrichthyes, Batoidea, Myliobatidae). Journal of Morphology 243: 113–126.

Summons, R. E., L. L. Jahnke, J. M. Hope, and G. A. Logan. 1999. 2-methylhopanoids as biomarkers of cyanobacterial oxygenic photosynthesis. Nature 409: 554–557.

Sumner, D. Y., and J. P. Grotzinger. 1996. Were kinetics of Archean calcium carbonate precipitation related to oxygen concentration? Geology 24: 119–122.

Sun, G., Q. Ji, D. L. Dilcher, S. Zheng, K. C. Nixon, and X. Wan. 2002. Archaeofructaceae, a new basal angiosperm family. Science 296: 899–904.

Sutherland, J. P. 1974. Multiple stable points in natural communities. American Naturalist 108: 859–873.

Swain, T. 1978. Plant-animal coevolution: A synthetic view of the Paleozoic and Mesozoic. Pages 3–19 in J. B. Harborne (ed.), Biochemical aspects of plant and animal coevolution. London: Academic Press.

Takeda, S. 1998. Influence of iron availability on nutrient consumption ratio of diatoms in oceanic waters. Nature 393: 774–777.

Tanner, W., and H. Beevers. 2001. Transpiration, a prerquisite for long-distance transport of minerals in plants? Proceedings of the National Academy of Sciences of the USA 98: 9443–9447.

Tappan, H. 1970. Phytoplankton abundance and Late Paleozoic extinctions: A reply. Palaeogeography, Palaeoclimatology, Palaeoecology 8: 49–66.

Tappan, H., and A. R. Loeblich, Jr. 1973. Evolution of the oceanic plankton. Earth Science Reviews 9: 207–240.

Tarduno, J. A., W. V. Sliter, L. Kroenke, M. Leckie, H. Mayer, J. J. Mahoney, R. Musgrave, M. Storey, and E. L. Winterer. 1991. Rapid formation of Ontong Java Plateau by Aptian mantle plume volcanism. Science 254: 399–403.

Tarduno, J. A., D. B. Brinkman, P. R. Renne, R. D. Cottrell, H. Scher, and P. Castillo. 1998. Evidence for extreme climatic warmth from Late Cretaceous Arctic vertebrates. Science 282: 241–244.

Tardy, Y., R. N'kounkou, and J.-L. Probst. 1989. The global water cycle and continental erosion during Phanerozoic time (570 my). American Journal of Science 289: 455–483.

Targett, N. M., L. D. Coen, A. A. Boettcher, and C. E. Tanner. 1992. Biogeographic comparisons of marine algal polyphenolics: Evidence against a latitudinal trend. Oecologia (Berlin) 89: 464–470.

Tawney, R. H. 1937. Religion and the rise of capitalism. New York: Harcourt, Brace: 337 pp.

Taylor, G. M. 2000. Maximum force production: Why are crabs so strong? Proceedings of the Royal Society of London B 267: 1475–1480.

———. 2001. The evolution of armament strength: Evidence for a constraint on the biting performance of claws of durophagous decapods. Evolution 55: 550–560.

Taylor, J. D. 1980. Diets of sublittoral predatory gastropods of Hong Kong. Pages 907–920 in B. S. Morton and C. K. Tseng (eds.), Proceedings of the First International Marine Biological Workshop: The marine flora and fauna of Hong Kong and southern China, Hong Kong, 1980. Hong Kong: University of Hong Kong Press.

———. 1984. A partial food web involving predatory gastropods on a Pacific fringing reef. Journal of Experimental Marine Biology and Ecology 74: 273–290.

———. 1986. Diets of sand-living predatory gastropods at Piti Bay, Guam. Asian Marine Biology 3: 47–58.

———. 1989. Diet of coral-reef Mitridae (Gastropoda) from Guam; with a review of other species of the family. Journal of Natural History 23: 261–278.

Taylor, J. D., E. A. Glover, and C. J. R. Braithwaite. 1999. Bivalves with "concrete overcoats": *Granocorium* and *Samarangia*. Acta Zoologica (Stockholm) 80: 285–300.

Taylor, M. A. 1986. Stunning whales and deaf squids. Nature 323: 298–299.

Taylor, P. D. 1994. Evolutionary paleoecology of symbioses between bryozoans and hermit crabs. Historical Biology 9: 157–205.

Temple, S. A. 1977. Plant-animal mutualism: Coevolution with dodo leads to near extinction of plant. Science 197: 885–886.

Terborgh, J. 1985. The vertical component of plant species diversity in temperate and tropical forests. American Naturalist 126: 760–776.

Terborgh, J., L. Lopez, P. Nuñez V., M. Rao, G. Shahabuddin, G. Orihuela, M. Riveros, R. Ascanio, G. H. Adler, T. D. Lambert, and L. Balbas. 2001. Ecological melt-down in predator-free forest fragments. Science 294: 1923–1926.

Thayer, C. W. 1979. Biological bulldozers and the evolution of marine benthic communities. Science 203: 458–461.

———. 1983. Sediment-mediated biological disturbance and the evolution of marine benthos. Pages 479–625 in M.J.S. Tevesz and P. L. McCall (eds.), Biotic interactions in Recent and fossil benthic communities. New York: Plenum.

———. 1985. Brachiopods versus mussels: Competition, predation, and palatability. Science 228: 1527–1528.

———. 1986. Are brachiopods better than bivalves? Mechanisms of turbidity tolerance and their interaction with feeding in articulates. Paleobiology 12: 161–174.

Théry, M., and J. Kasas. 2002. Predator and prey views of spider camouflage. Nature 415: 133.

Thierstein, H. R. 1982. Terminal Cretaceous planktonic extinctions: A critical assessment. Geological Society of America Special Paper 190: 385–399.

Thies, D., and W.-E. Reif. 1985. Phylogeny and evolutionary ecology of Mesozoic Neoselachii. Neues Jahrbuch für Geologie und Paläontologie Abhandlungen 169: 333–361.

Thompson, D. W. 1942. On growth and form. London: Cambridge University Press: 1116 pp.

Thompson, J. N. 1994. The coevolutionary process. Chicago: University of Chicago Press: 376 pp.

———. 1998. The population biology of coevolution. Researches on Population Ecology Review 40: 159–166.

———. 1999a. The evolution of species interactions. Science 284: 2116–2118.

———. 1999b. Coevolution and escalation: Are ongoing coevolutionary meanderings important? American Naturalist 153: S92-S93.

Thompson, J. N., and B. M. Cunningham. 2002. Geographic structure and dynamics of coevolutionary selection. Nature 417: 735–738.

Tiffney, B. H. 1981. Diversity and major events in the evolution of land plants. Pages 193–230 in K. J. Niklas (ed.), Paleobotany, paleoecology, and evolution, Vol. 2. New York: Praeger.

———. 1997. Land plants as food and habitat in the age of dinosaurs. Pages 352–370 in J. O. Farlow and M. K. Brett-Surman (eds.), The complete dinosaur. Bloomington: Indiana University Press.

Tiffney, B. H., and K. J. Niklas. 1985. Clonal growth in land plants: a paleobotanical perspective. Pages 35–66 in J.B.C. Jackson, L. W. Buss, and R. E. Cook (eds.), The population biology and evolution of clonal organisms. New Haven: Yale University Press.

Tilman, D., and J. A. Downing. 1994. Biodiversity and stability in grasslands. Nature 367: 363–365.

Tilman, D., D. Wedin, and J. Knops. 1996. Productivity and sustainability influenced by biodiversity in grassland ecosystems. Nature 379: 718–720.

Tilman, D., J. Knops, D. Wedin, P. Reich, M. Ritchie, and E. Siemann. 1997. The influence of functional diversity and composition on ecosystem processes. Science 277: 1300–1302.

Tilman, D., P. B. Reich, J. Knops, D. Wedin, T. Mielke, and C. Lehman. 2001. Diversity and productivity in a long-term grassland experiment. Science 294: 843–845.

Tintori, A. 1998. Fish biodiversity in the marine Norian (Late Triassic) of northern Italy: The first neopterygian radiation. Italian Journal of Zoology 65, supplement: 193–199.

Tintori, A., and D. Sassi. 1992. *Thoracopterus* Bronn (Osteichthyes: Actinopterygii): A gliding fish from the Upper Triassic of Europe. Journal of Vertebrate Paleontology 12: 265–283.

Todd, J. A., J.B.C. Jackson, K. G. Johnson, H. M. Fortunato, A. Heisz, M. Alvarez, and P. Jung. 2002. The ecology of extinction: Molluscan feeding and faunal turnover in the Caribbean Neogene. Proceedings of the Royal Society of London B 269: 571–577.

Togeweiler, J. V. 1999. An ultimate limiting nutrient. Nature 400: 511–512.

Tomlinson, P. B. 1973. The monocotyledons; their evolution and comparative biology. III. Branching in monocotyledons. Quarterly Review of Biology 48: 458–466.

Torsvik, V., L. Øvreås, and S. F. Thingstad. 2002. Prokaryotic diversity—magnitude, dynamics, and controlling factors. Science 296: 1064–1066.

Towe, K. M. 1970. Oxygen-collagen priority and the early metazoan fossil record. Proceedings of the National Academy of Sciences of the USA 65: 781–788.

Tréguer, P., and P. Pondaven. 2000. Silica control of carbon dioxide. Nature 406: 358–359.

Tréguer, P., D. M. Nelson, A. J. van Bennekom, D. D. DeMaster, A. Leynaert, and B. Quéguiner. 1995. The silica balance in the world ocean: A reestimate. Science 268: 375–379.

Trueman, E. R., and A. C. Brown. 1989. The effect of shell shape on the burrowing performance of species of *Bullia* (Gastropoda: Nassariidae). Journal of Molluscan Studies 55: 129–131.

Trussell, G. C. 1996. Phenotypic plasticity in an intertidal snail: The role of a common crab predator. Evolution 50: 448–454.

———. 1997. Phenotypic plasticity in the foot size of an intertidal snail. Ecology 78: 1033–1048.

———. 2000. Phenotypic clines, plasticity, and morphological trade-offs in an intertidal snail. Evolution 54: 151–166.

Trussell, G. C., and L. D. Smith. 2000. Induced defenses in response to an invading crab predator: An explanation of historical and geographic phenotypic change. Proceedings of the National Academy of Sciences of the USA 97: 2123–2127.

Tschudy, R. H., C. L. Pillmore, C. J. Orth, J. S. Gilmore, and J. D. Knight. 1984. Disruption of the terrestrial plant ecosystem at the Cretaceous-Tertiary boundary, Western Interior. Science 225: 1030–1032.

Tyler, J. C. 1980. Osteology, phylogeny, and higher classification of the fishes of the order Plectognathi (Tetraodontiformes). NOAA Technical Report, NMFS Circular 434: 1–422.

Tyler, J. C., and L. Sorbini. 1996. New superfamily and three new families of tetraodontiform fishes from the Upper Cretaceous: The earliest and morphologically most primitive plectognaths. Smithsonian Contributions to Paleobiology 82: 1–59.

Tyrrell, T. 1999. The relative influences of nitrogen and phosphorus on oceanic primary production. Nature 400: 525–531.

Upchurch, P., and P. M. Barrett. 2000. The evolution of sauropod feeding mechanisms. Pages 79–122 in H.-D. Sues (ed.), Evolution of herbivory in terrestrial vertebrates: Perspectives from the fossil record. Cambridge: Cambridge University Press.

Vacelet, J., and N. Boury-Esnault. 1995. Carnivorous sponges. Nature 373: 333–335.

Vale, F. K., and M. A. Rex. 1988. Repaired shell damage in deep-sea prosobranch gastropods from the western North Atlantic. Malacologia 28: 65–79.

Valentine, J. W. 1969. Patterns of taxonomic and ecological structure of the shelf benthos during Phanerozoic time. Palaeontology 12: 684–709.

———. 1986. Permian-Triassic extinction event and invertebrate developmental modes. Bulletin of Marine Science 39: 607–615.

Valentine, J. W., D. Jablonski, and D. H. Erwin. 1999. Fossils, molecules, and embryos: New perspective on the Cambrian explosion. Development 126: 851–859.

Van Alstyne, K. L., and V. J. Paul. 1990. The biogeography of polyphenolic compounds in marine macroalgae: Temperate brown algal defenses deter feeding by tropical herbivorous fishes. Oecologia (Berlin) 84: 158–163.
Vannier, J., and J.-Y. Chen. 2000. The Early Cambrian colonization of pelagic niches exemplified by *Isoxys* (Arthropoda). Lethaia 33: 295–311.
Van Valen, L. 1971. Group selection and the evolution of dispersal. Evolution 25: 591–598.
———. 1973. A new evolutionary law. Evolutionary Theory 1: 1–18.
———. 1976. Energy and evolution. Evolutionary Theory 1: 179–220.
Van Valkenburgh, B. 1999. Major patterns in the history of carnivorous mammals. Annual Reviews of Earth and Planetary Sciences 27: 463–493.
Van Valkenburgh, B., and F. Hertel. 1993. Tough times at La Brea: Tooth breakage in large carnivores of the Late Pleistocene. Science 261: 456–459.
Van Valkenburgh, B., and I. Jenkins. 2002. Evolutionary patterns in the history of Permo-Triassic and Cenozoic synapsid predators. Pages 267–288 in M. Kowalewski and P. H. Kelley (eds.), The fossil record of predation. Paleontological Society Papers 8.
Van Valkenburgh, B., and R. E. Molnar. 2002. Dinosaurian and mammalian predators compared. Paleobiology 28: 527–543.
Veevers, J. J. 1989. Middle/Late Triassic (230 ± 5 Ma) singularity in the stratigraphic and magmatic history of the Pangean heat anomaly. Geology 17: 784–787.
———. 1990. Tectonic-climatic supercycle in the billion-year plate-tectonic eon: Permian Pangean icehouse alternates with Cretaceous dispersed-continents greenhouse. Sedimentary Geology 68: 1–16.
Vermeij, G. J. 1971. Gastropod evolution and morphological diversity in relation to shell geometry. Journal of Zoology, London 163: 215–223.
———. 1973a. Morphological patterns in high intertidal gastropods: Adaptive strategies and their limitations. Marine Biology 20: 319–346.
———. 1973b. Adaptation, versatility, and evolution. Systematic Zoology 22: 466–477.
———. 1974. Marine faunal dominance and molluscan shell form. Evolution 28: 656–664.
———. 1977a. Patterns in crab claw size: The geography of crushing. Systematic Zoology 26: 138–151.
———. 1977b. The Mesozoic marine revolution: Evidence from snails, predators and grazers. Paleobiology 3: 245–258.
———. 1978. Biogeography and adaptation: Patterns of marine life. Cambridge: Harvard University Press: 332 pp.
———. 1982a. Unsuccessful predation and evolution. American Naturalist 120: 701–720.
———. 1982b. Gastropod shell form, repair, and breakage in relation to predation by the crab *Calappa*. Malacologia 23: 1–12.
———. 1983. Intimate associations and coevolution in the sea. Pages 311–327 in D. J. Futuyma and M. Slatkin (eds.), Coevolution. Sunderland: Sinauer.
———. 1987. Evolution and escalation: An ecological history of life. Princeton: Princeton University Press: 527 pp.
———. 1989a. Interoceanic differences in adaptation: Effects of history and productivity. Marine Ecology Progress Series 57: 293–305.
———. 1989b (1990). The origin of skeletons. Palaios 5: 585–589.
———. 1990a. An ecological crisis in an evolutionary context: El Niño in the eastern Pacific. Pages 505–517 in P. W. Glynn (ed.), Global ecological consequences of the 1982–83 El Niño–Southern Oscillation. Amsterdam: Elsevier.
———. 1990b. Tropical Pacific pelecypods and productivity. Bulletin of Marine Science 47: 62–67.
———. 1991a. Anatomy of an invasion: The Trans-Arctic interchange. Paleobiology 17: 281–307.
———. 1991b. When biotas meet: Understanding biotic interchange. Science 253: 1099–1104.

Vermeij, G. J.. 1992. Time of origin and biogeographical history of specialized relationships between northern marine plants and herbivorous molluscs. Evolution 46: 657–664.

———. 1993. A natural history of shells. Princeton: Princeton University Press: 207 pp.

———. 1994. The evolutionary interaction among species: Selection, escalation, and coevolution. Annual Reviews of Ecology and Systematics 25: 219–236.

———. 1995. Economics, volcanoes, and Phanerozoic revolutions. Paleobiology 21: 125–152.

———. 1996a. Adaptations of clades: Resistance and response. Pages 363–380 in M. R. Rose and G. V. Lauder (eds.), Adaptation. San Diego: Academic Press.

———. 1996b. Animal origins. Science 274: 525–526.

———. 1998a. *Sabia* on shells: A specialized Pacific-type commensalism in the Caribbean Neogene. Journal of Paleontology 72: 465–472.

———. 1998b. Fossils and the social future of science. Science 281: 1444–1445.

———. 1999. Inequality and the directionality of history. American Naturalist 153: 243–253.

———. 2001a. Community assembly in the sea: Geologic history of the living shore biota. Pages 39–60 in M. D. Bertness, S. D. Gaines, and M. E. Bay (eds.), Marine community ecology. Sunderland: Sinauer.

———. 2001b. Innovation and evolution at the edge: Origins and fates of gastropods with a labral tooth. Biological Journal of the Linnean Society 72: 461–508.

———. 2002a. Characters in context: Molluscan shells and the forces that mold them. Paleobiology 28: 41–54.

———. 2002b. Evolution in the consumer age: Predators and the history of life. Pages 375–393 in M. Kowalewski and P. H. Kelley (eds.), The fossil record of predation. Paleontological Society Papers 8.

———. 2002c. The geography of evolutionary opportunity: Hypothesis and two cases in gastropods. Integrative and Comparative Biology 42: 935–940.

———. 2003. Temperature, tectonics, and evolution. Pages 209–232 in L.J. Rothschild and A. M. Lister (eds.), Evolution on planet Earth: The impact of the physical environment. London: Academic Press.

———. 2004. Island life: A view from the sea. L.R. Heaney and M.V. Lomolino (eds.), Frontiers of biogeography: New directions in the geography of nature. Sunderland: Sinauer.

Vermeij, G. J., and A. P. Covich. 1978. Coevolution of freshwater gastropods and their predators. American Naturalist 112: 833–843.

Vermeij, G. J., and J. D. Currey. 1980. Geographical variation in the strength of thaidid snail shells. Biological Bulletin 158: 383–389.

Vermeij, G. J., and E. C. Dudley. 1985. Distributions of adaptations: A comparison between functional shell morphology of freshwater and marine pelecypods. Pages 461–478 in E. R. Trueman (ed.), Biology of the Mollusca, Vol. 10, Evolution. London: Academic Press.

Vermeij, G. J., and R. Dudley. 2000. Why are there so few transitions between aquatic and terrestrial ecosystems? Biological Journal of the Linnean Society 70: 541–554.

Vermeij, G. J., and L. Leighton. 2003. Does global diversity mean anything? Paleobiology 29: 3–7.

Vermeij, G. J., and D. R. Lindberg. 2000. Delayed herbivory and the assembly of marine benthic ecosystems. Paleobiology 26: 419–430.

Vermeij, G. J., and E. J. Petuch. 1986. Differential extinction in tropical American molluscs: Endemism, architecture, and the Panama land bridge. Malacologia 27: 29–41.

Vermeij, G. J., and G. Rosenberg. 1993. Giving and receiving: The tropical Atlantic as donor and recipient region for invading species. American Malacological Bulletin 10: 181–194.

Vermeij, G. J., and E. Zipser. 1986. Burrowing performance of some tropical Pacific gastropods. Veliger 29: 200–206.

Vermeij, G. J., D. E. Schindel, and E. Zipser. 1981. Predation through geological time: Evidence from gastropod shell repair. Science 214: 1024–1026.

Vermeij, G. J., R. B. Lowell, L. J. Walters, and J. A. Marks. 1987. Good hosts and their guests: Relations between trochid gastropods and the epizoic limpet *Crepidula adunca*. Nautilus 101: 69–74.

Vermeij, G. J., E. C. Dudley, and E. Zipser. 1989. Successful and unsuccessful drilling predation in Recent pelecypods. Veliger 32: 266–273.

Versluis, M., B. Schmitz, A. von der Heydt, and D. Lohse. 2000. How snapping shrimps snap: Through cavitating bubbles. Science 289: 2114–2117.

Villareal, T. A., C. Pilskaln, M. Brzezinski, F. Lipschultz, M. Dennett, and G. B. Gardner. 1999. Upward transport of oceanic nitrate by migrating diatom mats. Nature 397: 423–425.

Vitousek, P. M., H. A. Mooney, J. Lubchenco, and J. Melillo. 1997a. Human domination of Earth's ecosystems. Science 277: 494–499.

Vitousek, P. M., C. L. D'Antonio, L. L. Loope, M. Rejmánek, and R. Westbrooks. 1997b. Introduced species: A significant component of human-caused global change. New Zealand Journal of Ecology 21: 1–16.

Vogel, S. 1970. Convective air cooling at low air speeds and the shapes of broad leaves. Journal of Experimental Botany 21: 91–101.

Voznesenskaya, E. V., V. R. Franceschi, O. Kiirats, H. Freitag, and G. E. Edwards. 2001. Kranz anatomy is not essential for terrestrial C_4 plant photosynthesis. Nature 414: 543–546.

Vrba, E. S. 1983. Macroevolutionary trends: New perspectives on the roles of adaptation and incidental effect. Science 221: 387–389.

Waal, F. B. M. de. 1996. Good natured: The origins of right and wrong in humans and other animals. Cambridge: Harvard University Press: 296 pp.

Wackernagel, M., N. B. Schulz, D. Deumling, A. Callejas Linares, M. Jenkins, V. Kapos, C. Monfreda, J. Loh, N. Myers, R. Norgaard, and J. Randers. 2002. Tracking the ecological overshoot of the human economy. Proceedings of the National Academy of Sciences of the USA 99: 9266–9271.

Waddington, C. H. 1962. New patterns in genetics and development. New York: Columbia University Press: 271 pp.

Waggoner, B. 1999. Biogeographic analyses of the Ediacara biota: A conflict with paleotectonic reconstructions. Paleobiology 25: 440–458.

Wainwright, P. C., and D. R. Bellwood. 2002. Ecomorphology of feeding in coral reef fishes. Pages 33–55 in P. F. Sale (ed.), Coral reef fishes. Dynamics and diversity in a complex ecosystem. San Diego: Academic Press.

Waldrop, M. M. 1992. Complexity: The emerging science at the edge of order and chaos. New York: Simon and Schuster.

Walker, S. E., K. Parsons-Hubbard, E. Powell, and C. E. Brett. 2002. Predation on experimentally deployed molluscan shells from shelf to slope depths in a tropical carbonate environment. Palaios 17: 147–170.

Walker, T. D., and J. W. Valentine. 1984. Equilibrium models and evolutionary species diversity and the number of empty niches. American Naturalist 124: 887–899.

Wallace, C. C. 1985. Reproduction, recruitment and fragmentation in nine sympatric species of the coral genus *Acropora*. Marine Biology 88: 217–233.

Waller, T. R. 1998. Origin of the molluscan class Bivalvia and a phylogeny of major groups. Pages 1–45 in P. A. Johnston and J. W. Haggart (eds.), Bivalves: An eon of evolution—paleobiological studies honoring Norman D. Newell. Calgary: University of Calgary Press.

Walsh, J. J. 1988. On the nature of continental shelves. San Diego: Academic Press: 520 pp.

Wang, K. 1992. Glassy microspherules (microtectites) from an Upper Devonian limestone. Science 256: 1547–1550.

Wang, K., C. J. Orth, M. Attrep, Jr., B.D.E. Chatterton, H. Hou, and H.H.J. Geldsetzer. 1991. Geological evidence for a catastrophic biotic event at the Frasnian/Famennian boundary in South China. Geology 19: 776–779.

Wang, K., B.D.E. Chatterton, M. Attrep, Jr., and C. J. Orth. 1993. Late Ordovician mass extinction in the Selwyn Basin, northwestern Canada: Geochemical, sedimentological, and paleontological evidence. Canadian Journal of Earth Sciences 30: 1879–1880.

Ward, P. 1986. Cretaceous ammonite shell shapes. Malacologia 27:3–28.

Ward, P. D., D. R. Montgomery, and R. Smith. 2000. Altered river morphology in South Africa related to the Permian-Triassic extinction. Science 289: 1740–1743.

Ward, P. D., J. W. Haggart, E. S. Carter, D. Wilbur, H. W. Tipper, and T. Evans. 2001. Sudden productivity collapse associated with the Triassic-Jurassic boundary mass extinction. Science 292: 1147–1151.

Warner, G. F., and A. R. Jones. 1976. Leverage and muscle type in crab chelae (Crustacea: Brachyura). Journal of Zoology, London 180: 57–68.

Warren, J. H. 1985. Climbing as an avoidance behavior in the salt marsh periwinkle, *Littorina irrorata* (Say). Journal of Experimental Marine Biology and Ecology 89: 11–28.

Watanabe, Y., J. E. J. Martini, and H. Ohmoto. 2000. Geochemical evidence for terrestrial ecosystems 2.6 billion years ago. Nature 408: 574–578.

Watson, A. J. 1997. Volcanic iron, CO_2, and climate. Nature 385: 587–588.

Wayne, T. A. 1987. Responses of a mussel to shell-boring snails: Defensive behavior in *Mytilus edulis?* Veliger 30: 138–147.

Webb, P. W. 1984. Body form, locomotion and foraging in aquatic vertebrates. American Zoologist 24: 107–120.

Webb, S. D. 1991. Ecogeography and the Great American interchange. Paleobiology 17: 266–280.

Weber, B. H., D. J. Depew, C. Dyke, S. N. Salthe, E. D. Schneider, R. E. Ulanowicz, and J. S. Wicken. 1989. Evolution in thermodynamic perspective: An ecological approach. Biology and Philosophy 4: 373–405.

Weishampel, D. B., and C.-M. Jianu. 2000. Plant-eaters and ghost lineages: Dinosaurian herbivory revisited. Pages 123–143 in H.-D. Sues (ed.), Evolution of herbivory in terrestrial vertebrates: Perspectives from the fossil record. Cambridge: Cambridge University Press.

Weiss, H., and R. S. Bradley. 2001. What drives societal collapse? Science 291: 609–610.

Weldon, P. J. 1981. Biting as a defense in gastropods of the genus *Busycon* (Prosobranchia: Melongenidae). Veliger 23: 357–360.

Wells, M. J., and R. K. O'Dor. 1991. Jet propulsion and the evolution of the cephalopods. Bulletin of Marine Science 49: 419–432.

Wells, M. L., G. K. Vallis, and E. A. Silva. 1999. Tectonic processes in Papua New Guinea and past productivity in the eastern equatorial Pacific Ocean. Nature 398: 601–604.

Werner, R., K. Hoernle, P. van den Bogaard, C. Ranero, R. von Huene, and D. Korich. 1999. Drowned 14-m.y-old Galápagos archipelago off the coast of Costa Rica: Implications for tectonic and evolutionary models. Geology 27: 499–502.

Wesselingh, F. P., G. C. Cadée, and W. Renema. 1999. Flying high: On the airborne dispersal of aquatic organisms as illustrated by the distribution histories of the gastropod genera *Tryonia* and *Planorbarius*. Geologie en Mijnbouw 78: 165–174.

West, G. B., J. H. Brown, and B. J. Enquist. 1997. A general model for the origin of allometric scaling laws in biology. Science 276: 122–126.

———. 2001. A general model for ontogenetic growth. Nature 413: 628–631.

West, L. 1988. Prey selection by the tropical snail *Thais melones*: A study of individual variation. Ecology 69: 1839–1854.

West-Eberhard, M. J. 1983. Sexual selection, social competition, and speciation. Quarterly Review of Biology 58: 155–183.

Weyl, P. K. 1968. The role of the oceans in climatic change: A theory of the ice ages. Meteorological Monographs 8: 37–62.

White, L. A. 1943. Energy and the evolution of culture. American Anthropologist 45: 335–356.

———. 1959. The evolution of culture. New York: McGraw-Hill: 378 pp.

White, T.C.R. 1978. The importance of a relative shortage of food in animal ecology. Oecologia (Berlin) 33: 71–86.

Whittington, H. B., and D.E.G. Briggs. 1985. The largest Cambrian animal, *Anomalocaris*, Burgess Shale, British Columbia. Philosophical Transactions of the Royal Society of London B 309: 569–609.

Wiens, J. J., and J. L. Slingluff. 2001. How lizards turn into snakes: A phylogenetic analysis of body-form evolution in anguid lizards. Evolution 55: 2303–2318.

Wignall, P. B. 2001. Large igneous provinces and mass extinctions. Earth Science Reviews 53: 1–33.

Wignall, P. B., and A. Hallam. 1992. Anoxia as a cause of the Permian/Triassic mass extinction: Facies evidence from northern Italy and the western United States. Palaeogeography, Palaeoclimatology, Palaeoecology 93: 21–46.

Wignall, P. B., and R. J. Twitchett. 1996. Oceanic anoxia and the end Permian mass extinction. Science 272: 1155–1158.

Wiklund, C., and T. Jårvi. 1982. Survival of distasteful insects after being attacked by naive birds: A reappraisal of the theory of aposematic coloration evolving through individual selection. Evolution 36: 998–1002.

Wiklund, C., and B. Sillén-Tullberg. 1985. Why distasteful butterflies have aposematic larvae and adults, but cryptic pupae: Evidence from predation experiments on the monarch and the European swallowtail. Evolution 39: 1155–1158.

Wilber, T. P., and W. F. Herrnkind. 1984. Predaceous gastropods regulate new-shell supply to saltmarsh hermit crabs. Marine Biology 79: 145–150.

Wilde, P., and W. B. N. Berry. 1984. Destabilization of the oceanic density structure and its significance to marine "extinction" events. Palaeogeography, Palaeoclimatology, Palaeoecology 48: 143–162.

Wiley, E. O., and D. R. Brooks. 1982. Victims of history—a nonequilibrium approach to evolution. Systematic Zoology 31: 1–24.

Wilf, P., and C. C. Labandeira. 1999. Response of plant-insect associations to Paleocene-Eocene warming. Science 284: 2153–2156.

Wilf, P., C. C. Labandeira, K. R. Johnson, P. D. Coley, and A. D. Cutter. 2001. Insect herbivory, plant defense, and Early Cenozoic climate change. Proceedings of the National Academy of Sciences of the USA 98: 6221–6226.

Wilf, P., K. R. Johnson, and B. T. Huber. 2003. Correlated terrestrial and marine evidence for global climate changes before mass extinction at the Cretaceous-Paleocene boundary. Proceedings of the National Academy of the USA 100: 599–604.

Wilken, G. B., and C. C. Appleton. 1991. Avoidance responses of some indigenous and exotic freshwater pulmonate snails to leech predation in South Africa. South African Journal of Science 26: 6–10.

Williams, G. C. 1992. Natural selection: Domains, levels, and challenges. New York: Oxford University Press: 208 pp.

Williams, S. L. 1984. Uptake of sediment ammonium and translocation in a marine green macroalga *Caulerpa cupressoides*. Limnology and Oceanography 29: 374–379.

Williams, S. L., and T. R. Fischer. 1985. Kinetics of nitrogen-15 labeled ammonium uptake by *Caulerpa cupressoides* (Chlorophyta). Journal of Phycology 21: 287–296.

Williams, S. L., V. A. Breda, T. W. Anderson, and B. B. Nyden. 1985. Growth and sediment disturbances of *Caulerpa* spp. (Chlorophyta) in a submarine canyon. Marine Ecology Progress Series 21: 275–281.

Wilson, D. S. 1997. Biological communities as functionally organized units. Ecology 78: 2018–2024.

———. 2002. Darwin's cathedral: Evolution, religion, and the nature of society. Chicago: University of Chicago Press: 268 pp.

Wilson, E. O. 1961. The nature of the taxon cycle in the Melanesian ant fauna. American Naturalist 95: 179–193.
———. 1971. The insect societies. Cambridge: Belknap Press of Harvard University: 548 pp.
———. 1998. Consilience: The unity of knowledge. New York: Knopf: 332 pp.
Wilson, P. A., and R. D. Norris. 2001. Warm tropical ocean surface and global anoxia during the mid-Cretaceous period. Nature 412:425–429.
Wing, S. L. 1998. Late Paleocene–Early Eocene floral and climatic change in the Bighorn Basin, Wyoming. Pages 380–400 in M.-P. Aubry, S. Lucas, and W. A. Berggren (eds.), Late Paleocene–Early Eocene climatic and biotic events in the marine and terrestrial records. New York: Columbia University Press.
Wing, S. L., and L. Boucher. 1998. Ecological aspects of the Cretaceous flowering plant radiation. Annual Reviews of Earth and Planetary Sciences 26: 379–421.
Wing, S. L., and G. J. Harrington. 2001. Global response to rapid warming in the earliest Eocene and implications for concurrent faunal change. Paleobiology 27: 539–563.
Winston, J. E. 1986. Victims of avicularia. Marine Ecology 7: 193–199.
Witmer, L. M., and K. D. Rose. 1991. Biomechanics of the jaw apparatus of the giant Eocene bird *Diatryma*: Implications for diet and mode of life. Paleobiology 17: 95–120.
Wonders, J.B.W. 1977. The role of benthic algae in the shallow reef of Curaçao (Netherlands Antilles) III. The significance of grazing. Aquatic Botany 3: 357–390.
Wood, R. 1993. Nutrients, predation and the history of reef-building. Palaios 8: 526–543.
Wood, R. A., J. P. Grotzinger, and J.A.D. Dickson. 2002. Proterozoic modular biomineralized metazoan from the Nama Group, Namibia. Science 296: 2383–2386.
Woodin, S. A. 1977. Algal "gardening" behavior by nereid polychaetes: Effects on soft-bottom community structure. Marine Biology 44: 39–42.
Woodring, W. P. 1966. The Panama land bridge as a sea barrier. American Philosophical Society Proceedings 110: 425–433.
Woods, A. D., D. J. Bottjer, M. Mutti, and J. Morrison. 1999. Lower Triassic large sea-floor carbonate cements: Their origin and a mechanism for the prolonged biotic recovery from the end-Permian mass extinction. Geology 27: 645–648.
Worm, B. 2000. Consumer versus resource control in rocky shore foodwebs: Baltic Sea and northwest Atlantic Ocean. Berichte aus dem Institut für Meereskunde an der Christian-Albrechts-Universität Kiel no. 316: 1–147.
Worm, B., H. K. Lotze, and U. Sommer. 2000. Coastal food web structure, carbon storage, and nitrogen retention regulated by consumer pressure and nutrient loading. Limnology and Oceanography 45: 339–349.
Worm, B., H. K. Lotze, H. Hillebrand, and U. Sommer. 2002. Consumer versus resource control of species diversity and ecosystem functioning. Nature 417: 842–851.
Worthy, T. H., and R. N. Holdaway. 2002. The lost world of the moa: Prehistoric life of New Zealand. Bloomington: Indiana University Press: 719pp.
Wrangham, R. W. 2001. Out of the *Pan*, into the fire: How our ancestors' evolution depended on what they ate. Pages 120–143 in F.B.M. de Waal (ed.), Tree of origin: What primate behavior can tell us about human social evolution. Cambridge: Harvard University Press.
Wrangham, R. W., J. Jones Holland, G. Baden, D. Pilbeam, and N. Conklin-Brittain. 1999. The raw and the stolen: Cooking and the ecology of human origins. Current Anthropology 40: 566–594.
Wray, G. A., J. S. Levinton, and L. H. Shapiro. 1996. Molecular evidence for deep Precambrian divergences among metazoan phyla. Science 274: 568–573.
Wright, R. 2000. Nonzero: The logic of human destiny. New York: Pantheon: 435 pp.
Wright, S. J. 1979. Competition between insectivorous lizards and birds in central Panama. American Zoologist 19: 1145–1156.
Wu, J., W. Sunda, E. A. Boyle, and D. M. Karl. 2000. Phosphate depletion in the western North Atlantic Ocean. Science 289: 759–762.

Wu, X.-C., H.-D. Sues, and A. Sun. 1995. A plant-eating crocodyliform reptile from the Cretaceous of China. Nature 376: 678–680.

Xiao, S., Y. Zhang, and A. H. Knoll. 1998a. Three-dimensional preservation of algae and animal embryos in a Neoproterozoic phosphorite. Nature 391: 553–558.

Xiao, S., A. H. Knoll, and X. Yuan. 1998b. Morphological reconstruction of *Miaohephyton bifurcatum*, a possible brown alga from the Neoproterozoic Doushantuo Formation, South China. Journal of Paleontology 72: 1072–1086.

Xiong, J., W. M. Fischer, K. Inoue, M. Nakahara, and C. E. 2000. Molecular evidence for the early evolution of photosynthesis. Science 289: 1724–1730.

Yoon, H. S., J. Y. Le, S. M. Boo, and D. Bhattacharya. 2001. Phylogeny of Alariaceae, Laminariaceae, and Lessoniaceae (Phaeophyceae) based on plastid-encoded RuBisCo spacer and nuclear-encoded ITS sequence comparisons. Molecular Phylogenetics and Evolution 21: 231–243.

Yuan, X., J. Li, and R. Cao. 1999. A diverse metaphyte assemblage from the Neoproterozoic black shales of South China. Lethaia 32: 143–155.

Yuan, X., S. Xiao, R. L. Parsons, C. Zhou, Z. Chen, and J. Hu. 2002. Tiering sponges in an Early Cambrian Lagerstätte: disparity between nonbilaterian and bilaterian epifaunal tierers at the Neoproterozoic-Cambrian transition. Geology 30: 363–366.

Zachos, J. C., K. C. Lohmann, J. C. G. Walker, and S. W. Wise. 1993. Abrupt climate change and transient climate during the Paleogene: A marine perspective. Journal of Geology 101: 191–213.

Zachos, J. C., M. Pagani, L. Sloan, E. Thomas, and K. Billups. 2001. Trends, rhythms, and aberrations in global climate 65 Ma to present. Science 292: 687–693.

Zazula, G. D., G. E. Froese, C. E. Schweger, R. W. Mathewes, A. B. Baudoin, A. M. Telka, C. R. Harington, and J. A. Westgate. 2003. Ice-age steppe vegetation in east Beringia. Nature 423: 603.

Zhu, R. X., K. A. Hoffman, R. Potts, C. L. Deng, Y. X. Pan, B. Guo, C. D. Shi, Z. T. Guo, B. Y. Yuan, Y. M. Hou, and W. W. Wang. 2001. Earliest presence of humans in northeast Asia. Nature 413: 413–417.

Zhuravlev, A. Yu., and R. A. Wood. 1996. Anoxia as the cause of the mid-Early Cambrian (Botomian) extinction event. Geology 24: 311–314.

Ziebis, W., S. Forster, M. Huettel, and B. B. Jorgensen. 1996. Complex burrows of the mud shrimp *Callianassa truncata* and their geochemical impact in the sea bed. Nature 382: 619–622.

Ziegler, A. C. 1971. A theory of the evolution of therian dental formulas and replacement patterns. Quarterly Review of Biology 46: 226–249.

Zielinski, G. A., P. A. Mayewski, L. D. Meeker, S. Whitlow, M. S. Twickler, M. Morrison, D. A. Meese, A. J. Gow, and R. V. Alley. 1994. Record of volcanism since 7000 B.C. from the GISP 2 Greenland ice core and implications for the volcano-climate system. Science 264: 948–952.

Zimov, P. A., V. I. Chuprynin, A. A. Oreshko, P. S. Chapin III, J. F. Reynolds, and M. C. Chapin. 1995. Steppe-tundra transition: a herbivore-driven biome shift at the end of the Pleistocene. American Naturalist 156: 765–794.

Zipser, E., and G. J. Vermeij. 1978. Crushing behavior of tropical and temperate crabs. Journal of Experimental Marine Biology and Ecology 32: 155–172.

INDEX